Handbook of Turbulence

Volume 1

Fundamentals and Applications

HANDBOOK OF TURBULENCE

Volume 1: Fundamentals and Applications
Edited by Walter Frost and Trevor H. Moulden

Volume 2: Modeling and Measurement
Edited by Walter Frost, Trevor H. Moulden, and Jurgen Bitte

Handbook of Turbulence

Volume 1

Fundamentals and Applications

Edited by

Walter Frost and Trevor H. Moulden

The University of Tennessee Space Institute at Tullahoma

Plenum Press · New York and London

Library of Congress Cataloging in Publication Data

Main entry under title:

Handbook of turbulence.

Originally presented as lectures in a short course program held at the University of Tennessee Space Institute.
Includes bibliographical references and index. Contents: v. 1. Fundamentals and applications.
1. Turbulence. I. Frost, Walter, 1935- II. Moulden, Trevor H. III. University of Tennessee (System). Space Institute.
TL574.T8H36 532'.0527 77-23781
ISBN 0-306-31004-X

A Division of Plenum Publishing Corporation
227 West 17th Street, New York, N.Y. 10011

Printed in the United States of America

Contributors

R. BETCHOV Professor, Department of Aerospace and Mechanical Engineering, University of Notre Dame, Notre Dame, Indiana 46556

J. BITTE Assistant Professor of Engineering Science and Mechanics, University of Tennessee Space Institute, Tullahoma, Tennessee 37388

W. C. CLIFF Space Scientist, NASA Marshall Space Flight Center, Huntsville, Alabama 35812

J. R. CONNELL Associate Professor of Physics, The University of Tennessee Space Institute, Tullahoma, Tennessee 37388

R. G. DEISSLER Staff Scientist, Physical Science Division, NASA Lewis Research Center, Cleveland, Ohio 44135

R. C. ELSTNER Principal, Wiss, Janney, Elstner, and Associates, Inc., Northbrook, Illinois 60062

G. H. FICHTL Chief, Environmental Dynamics Branch, NASA Marshall Space Flight Center, Huntsville, Alabama 35812

W. FROST Director of Atmospheric Science Division and Professor of Mechanical Engineering, The University of Tennessee Space Institute, Tullahoma, Tennessee 37388

P. T. HARSHA R & D Associates, P.O. Box 3580, Santa Monica, California 9043. *Present address*: Science Applications, Inc., Woodland Hills, California

A. H. GARNER Summer Visiting Student, The University of Tennessee Space Institute, Tullahoma, Tennessee 37388

W. S. LEWELLEN Senior Consultant, Aeronautical Research Associates of Princeton, Inc., Princeton, New Jersey 08540

T. H. MOULDEN Assistant Professor of Aerospace Engineering, The University of Tennessee Space Institute, Tullahoma, Tennessee 37388

S. A. ORSZAG Professor of Applied Mathematics, Department of Mathematics, Massachusetts Institute of Technology, Cambridge, Massachusetts 02139

M. PERLMUTTER Research Associate, The University of Tennessee Space Institute, Tullahoma, Tennessee 37388

V. A. SANDBORN Professor of Aerospace Engineering, Colorado State University, Fort Collins, Colorado 80521

H. TENNEKES Professor of Aerospace Engineering, Pennsylvania State University, University Park, Pennsylvania 16802. *Present address*: Royal Netherlands Meteorological Institute, de Bilt, the Netherlands

Preface

Turbulence takes place in practically all flow situations that occur naturally or in modern technological systems. Therefore, considerable effort is being expended in an attempt to understand this very complex physical phenomenon and to develop both empirical and mathematical models for its description. Such numerical and analytical computational schemes would allow the reliable prediction and design of turbulent flow processes to be carried out. The purpose of this book is to bring together, in a usable form, some of the fundamental concepts of turbulence along with turbulence models and experimental techniques. It is hoped that these have "general applicability" in current engineering design. The phrase "general applicability" is highlighted because the theory of turbulence is still so much in a formative stage that completely general analyses are not available now, nor will they be available in the immediate future.

The concepts and models described herein represent the state-of-the-art methods that are now being used to give answers to turbulent flow problems.

As in all turbulent flow analysis, the methods are a blend of analytical and empirical input, and the reader should be cognizant of the simplification and restrictions imposed upon the methods when applying them to physical situations different from those for which they have been developed.

The book does not attempt to discuss all on-going research in turbulence but only to present a unified description of several of the more useful fundamental concepts, turbulence models, and experimental techniques available. It is felt that the reader who studies the text will have a workable knowledge of turbulence and the necessary background to readily grasp the essentials of future research and also the current research that, through space limitation, has been excluded from this book.

The first four chapters of the book provide a development of the governing equations of fluid mechanics (this development, although complete, is not detailed since it is assumed that the reader has some formal training in graduate level fluid mechanics) and a detailed description of the statistical concepts associated with random turbulence processes. Chapter 5 describes some conflicts between the use of deterministic equations of fluid mechanics on one hand and of the probability theory of random processes on the other and, finally, hints at the possibility of quantum effects. Chapter 6 summarizes some concepts related to the transition from laminar to turbulent flow, while Chapters 7–10 describe turbulence models of varying degrees of complexity. Chapters 11 and 12 describe experimental techniques for use in the natural environment and in the laboratory, respectively. Chapter 13 details some of the recent advances in the measurement of turbulence with instruments employing acoustic and electromagnetic wave phenomena. Methods of analog and digital simulation of random turbulence signals having all the correct statistics of a given turbulence process are discussed in Chapter 14. Finally, Chapter 15 describes some day-to-day problems created by turbulence in building design as viewed by a practicing engineer.

The book is an outgrowth of a short course held at The University of Tennessee Space Institute and each chapter represents a lecture given by a leading authority on that particular subject. The editors are greatly indebted to these lecturers for their significant contributions. Also, Jules Bernard, manager of the short course program, has been most helpful in the organization of a unified series of lectures necessary for the development of the completed book. Finally the authors acknowledge the foresight of Dean Emeritus Dr. B. H. Goethert, who originated the short course series at The University of Tennessee Space Institute, and of Dean Charles Weaver, who continues to promote this effective method of disseminating information to the industrial and research communities.

Much filling, deleting, and retyping of the original lectures was necessary to provide continuity and uniformity of the text while at the same time avoiding duplication. Thus, considerable typing effort was expended by Rena Northcutt, Marie Henderson, June Jarrell, Judy Wright, and Fay Fuller. For this the editors are greatly indebted.

A word about the organization of the book is necessary before this preface is complete. Each chapter includes its own list of references. In addition, each chapter contains its own list of nomenclature; this is necessary because, despite efforts to make the nomenclature consistent, slight discrepancies occur owing to preferences of individual authors. Equation and figure numbers, on the other hand, are prefixed with chapter numbers to allow cross-referencing between chapters.

Walter Frost
Trevor H. Moulden

Contents

Chapter 11

Laboratory Instrumentation in Turbulence Measurements 315

V. A. Sandborn

Chapter 12

Techniques for Measuring Atmospheric Turbulence 369

J. R. Connell

Chapter 13

Optical and Acoustical Measuring Techniques 403

William C. Cliff

Chapter 14

Monte Carlo Turbulence Simulation 433

G. H. Fichtl, Morris Perlmutter, and Walter Frost

CHAPTER 1

The Complexity of Turbulent Fluid Motion

TREVOR H. MOULDEN, WALTER FROST, and ALBERT H. GARNER

1.1. Introduction

Turbulence, a state of fluid motion, will be left as an intuitive concept without formal definition. Here we will only discuss the characteristics of turbulent motion and put forward techniques for its description. Understanding and predicting turbulence is the subject of the present work.

The discussion will start with a treatment of the fundamental aspects of turbulent fluid motion and the need for its statistical description. This background will lead into a discussion of the calculation techniques currently under development for the prediction of turbulence and, finally, into a review of some experimental methods available to turbulence research. The individual treatments will be selective rather than attempting all-inclusive coverage of the subject. Even Monin and Yaglom,[1] after two thousand pages, cannot claim completeness.

Turbulent motion is chaotic. Yet the term "chaotic" must be used almost as a synonym for the word "turbulent." This *is* the substance of the motion. At the same time, the designation "random" is often applied as a description of turbulent motion, but it should be pointed out that the motion

TREVOR H. MOULDEN, WALTER FROST, and ALBERT H. GARNER • The University of Tennessee Space Institute, Tullahoma, Tennessee 37388

can never be totally random if its velocity components are to abide by the conservation laws. Thus, if one velocity component assumes a random behavior, then the remaining components must have a restricted range of fluctuation in conformity with the conservation equations. What is certain, from a practical point of view at least, is that the motion is not deterministic and must be treated within a statistical framework.

What characteristics of fluid motion should be used to designate the flow as a turbulent one? One basic requirement is for the velocity at a point in the flow to be time dependent. Although this requirement is a necessary one, it is not sufficient to describe turbulent motion. More specifically, the velocity fluctuations at a given point should exhibit that measure of chaos that is characteristic of turbulence. The velocity fluctuations should not correlate in any way with some imposed time dependence in the motion—such as would occur, for instance, in the laminar flow over a slowly oscillating

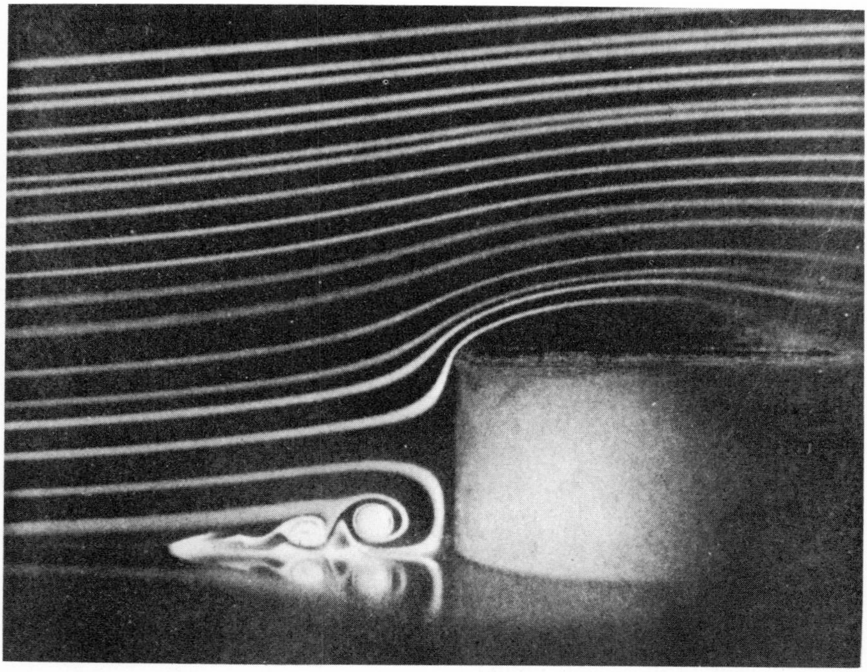

Fig. 1.1. Low-speed boundary-layer flow over a stubby cylinder mounted normal to the flow when the boundary-layer thickness and cylinder height are of the same order. The free-stream velocity is very low. Reproduced by permission of E. P. Sutton, Cambridge University Engineering Laboratory; further details are given in *Incompressible Aerodynamics* (B. Thwaites, ed.), Oxford University Press (1960).

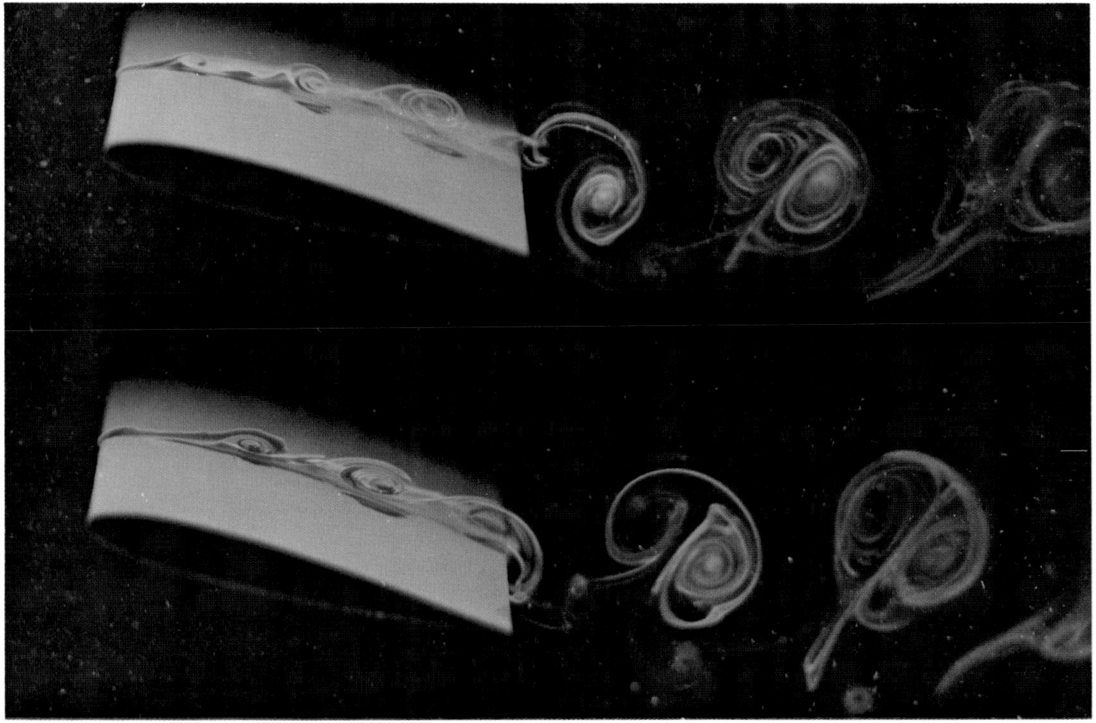

Fig. 1.2. Low-speed flow over an airfoil oscillating in pitch and the unsteady vortex separation so produced. Reproduced by permission of H. Werlé (ONERA, Paris). Further details given in Hydrodynamic flow visualization, in: *Annual Review of Fluid Mechanics*, Vol. 5, Annual Reviews Inc., Palo Alto, California (1973), pp. 361–382.

body. One obvious contradiction is evident here when the turbulent flow over an oscillating body is considered. Under these conditions there would be a correlation between the boundary conditions and the local flow velocity at the specific frequency associated with the boundary motion, but the broad-band spectrum of the turbulence would not show significant correlation with the applied temporal behavior. The velocity fluctuations of the turbulence appear to evolve within the motion itself owing to events whose history is lost, and these fluctuations are not, therefore, directly coupled to any external agency.

Some other features of fluid motion that have a more or less passing connection with the subject in hand can be brought out. Figure 1.1 shows the laminar shear flow past a cylindrical stub mounted normally to a solid surface. An interesting feature of this flow is the generation of a bound vortex system around the base of the body. Here the distributed vorticity in the approaching shear layer is remodeled as a discrete vortex pattern by the velocity gradients established near the body. The next example of fluid flow considered is the vortex disposition associated with the flow over a lifting wing as shown by Werlé's photographs in Figure 1.2.

The significance of these flows, apart from their intrinsic interest and beauty, lies in the fact that the fluid motion is far more complex than a cursory evaluation of the imposed boundary conditions would indicate. While an intuitive understanding of fluid flow could lead to the anticipation of a vortex structure like that presented in Figure 1.1, it is unlikely that a similar exercise in connection with the situation of Figure 1.2 would be rewarding. The streamlines in the body of the fluid have little sympathy with the imposed boundary conditions. The same remarks hold true for the flows shown in Figure 1.3. This latter figure shows one, two, and three circular cylinders immersed in a low-speed flow. It is evident from this figure that the vortex pattern is repeated over and over. It persists for a length scale that is significantly greater than the characteristic length of the body inducing the motion. While the regular vortex pattern is very persistent, it does not endure throughout the entire flow but finally degenerates into a more chaotic flow. The start of this latter process is shown in Figure 1.3b.

It may be legitimately asked whether these deliberations, which are primarily related to laminar flows, have any connection with the subject in hand—turbulence. The answer must be in the affirmative and without qualification. This is largely because vorticity dynamics hold a central position in turbulence theory. Not only is the fine structure of turbulence strongly controlled by the properties of vortex interactions, but there is also, and often, a large-scale coherent structure superimposed upon the motion. Such coherent structures are essentially vortical in nature, as indicated by Roshko's flow studies (Figure 1.4). Like their counterpart in laminar flow, these vortex structures also persist for a substantial length scale. Coherent flow structures in turbulence have been observed for a long time, but their study has not attracted attention until recently.[2] Thus Prandtl[3] demonstrated the presence of gigantic internal structures in the turbulent wall boundary layer, while Nikuradse[4] found large eddy structures in channel flows. Recent advances in conditional sampling techniques, applied to the analysis of turbulent flows,[5] provide the most suitable vehicle for the evaluation of coherent structures. To what extent they feature in calculation techniques is not certain.

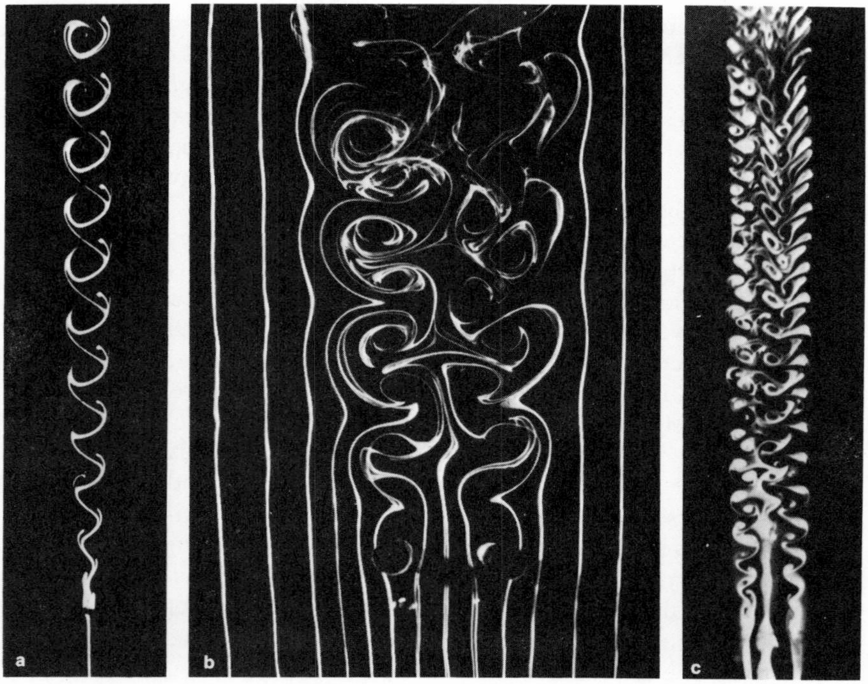

Fig. 1.3. Low-speed flow over various circular cylinder configurations. (a) Visualization of the wake behind a circular cylinder (low-speed water-tunnel study, Re ~ 100). From M. Gaster (Aerodynamics Division, National Physical Laboratory, Teddington, England), Vortex shedding from slender cones at low Reynolds numbers, *J. Fluid Mech.* **38**, 565–576 (1969). (b) Flow past two circular cylinders (wind-tunnel study, $Re \sim 2 \times 10^4$). From P. W. Bearman and A. J. Wadcock (Aeronautics Department, Imperial College, London), The interaction between a pair of circular cylinders normal to a stream, *J. Fluid Mech.* **61**, 499–511 (1973). (c) Interaction between three circular cylinders (wind-tunnel study, Re ~ 300). From M. M. Zdravkovich (University of Belgrade), Smoke observations of the wake of a group of three cylinders at low Reynolds numbers, *J. Fluid Mech.* **32**, 339–351 (1968). (All photographs reproduced by permission of Cambridge University Press.)

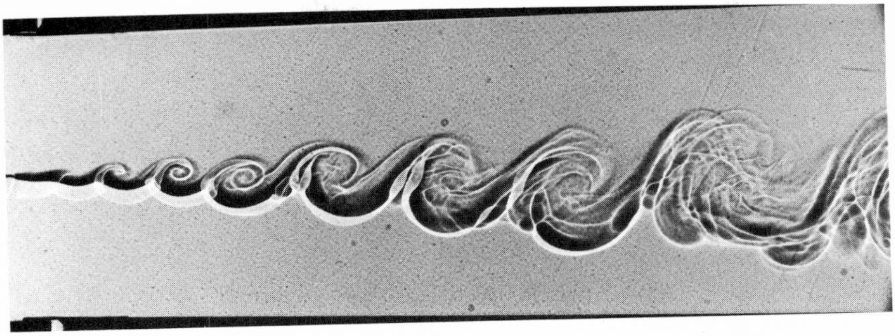

Fig. 1.4. Shear layer developing at the mixing interface between two dissimilar gases; photographs show the coherent structure within the shear layer between helium and nitrogen ($Re \sim 10^5$). From A. Roshko and G. Brown (California Institute of Technology), On density effects and large structures in turbulent mixing layers, *J. Fluid Mech.* **64**, 775–816 (1974) (reproduced by permission of Cambridge University Press).

Further comments about the nature of turbulent flow will be reserved until after the basic equations of motion have been presented. It is not the aim of the present work to provide a detailed derivation of these equations, since many excellent treatments are available in the literature.[6–8] Rather, our interest centers around developing an understanding of fluid motion and turbulent motion in particular. It is expedient, however, if some crucial points in the development of the equations of continuum fluid motion are outlined. In this way a complete background to some fundamental remarks about turbulent flow becomes available.

1.2. On Continuum Fluid Motion

Let $I_N \equiv \{n : n = \text{integer} \leq N\}$ define an index set of length N and let $R \equiv \{r : r \text{ real}\}$ denote the infinite collection of real numbers. The triple $\{r_i\}$, with $i \in I_3$, will specify a point in space whose coordinates are expressed as r_i with respect to a suitable origin. An infinite set of such triples will represent the entire Euclidean space $E(\equiv R \times R \times R)$ and it is supposed that fluid motion takes place in a subdomain, X, of E.

A fluid domain, D, will consist of a set of fluid elements that can be labeled consecutively in the form ζ_k; $k \in I_\infty$. In addition to this the subdomain D_l of D will be understood by its configuration, i.e., the subdomain X_l of X in physical space. It is further supposed that there exists a one-to-one mapping of the elements ζ_k of D_l onto the space X_l and that the fluid element may be represented by the corresponding coordinate triple $\{\bar{x}_i\}$ associated with some initial space point $\bar{x}_i$. Careful distinction should be made between an element of fluid and the location $\{x_i\} \in X$ occupied by the fluid element.

Axiom 1. The fluid is considered to be a continuum in the sense that adjacent fluid elements within D_l have a one-to-one correspondence with the coordinate triples $\{x_i\}$ in X_l, i.e.,

$$\{x_i\} = Z(\zeta_k)$$

$$\zeta_k = Z^{-1}(\{x_i\}) \qquad \blacksquare$$

Axiom 1 suggests that fluid elements are distributed continuously throughout the space domain X in accordance with the real numbers used to form the coordinates x_i. Thus, if $S_r(\bar{p}) \equiv \{\bar{x} : d(\bar{x}, \bar{p}) < r\}$ is a sphere of space in X, with $d(\bar{x}, \bar{p})$ a suitable metric on X, then $\lim_{r \to 0} S_r(\bar{p})$ is the vector point $\bar{p}$ in space occupied by the fluid element $\zeta_{\bar{p}}$. Such concepts of a continuum fluid are in strong contrast to the physically recognized molecular structure of the real fluid, and the theory should be approached within this context. Again, Axiom 1 implies that individual fluid elements always remain distinct throughout the entire motion. It may also be remarked that the set of fluid elements that constitute D_l is invariant with time. This latter condition is a fundamental requirement for the application of conservation principles to a body of fluid.

Further restrictions must be placed upon the specification of the fluid subdomains, D_l, namely,

$$\bigcup_{l=1,\infty} D_l = D$$

and $D_l \cap D_k = 0$ for every l, $k \in I_\infty$ but $l \neq k$.

In general, as the fluid motion progresses, a given fluid element ζ_k (at location $\{\bar{x}_i\}$) will map into a new space location $\{x_i\}$. Then

$$\{x_i\} = \chi(\{\bar{x}_i\}; t)$$

and, inversely,

$$\{\bar{x}_i\} = \chi^{-1}(\{x_i\}; t)$$

It is required that the mappings χ and χ^{-1} be sufficiently differentiable for the subsequent analysis to proceed.

Corollary 1. The transformation Jacobian $J(\xi_i; \bar{\xi}_i) \equiv \partial(\xi_i)/\partial(\bar{\xi}_i)$ between any two configurations, X_l and $\bar{X}_l$, of the subdomain D_l is nonzero. Since $(\partial(\zeta)/\partial(\bar{\zeta})) \times (\partial(\bar{\zeta})/\partial(\zeta)) = 1$, then the Jacobian is also bounded. $\blacksquare$

Fluid motion is now treated as a sequence of mappings of a given subdomain through space where time acts as the parameter. That is, the motion is specified as the mapping of the configuration of subdomain D_l at time t into its new configuration at time $t+\Delta t$.

Axiom 2. The physical attributes of the fluid are assumed to be such that all required derivatives of fluid properties exist. ∎

Let M_l denote the mass of fluid within the subdomain D_l; then Axiom 2 and the Radon–Nikodym theorem[9] assure the existence of a function $\rho(\xi_i)$ such that

$$M_l = \int_{D_l} \rho(\xi_i)\, d\xi_i; \quad i \in I_3$$

The function $\rho(\xi_i)$ will be referred to as the density function for the fluid and will be a local property of the fluid. Greater physical meaning can be given to the density function if appeal is made to the lemma concerning the mean value of integrals, when it is suggested that

$$M_l = \rho(\bar{\zeta})V_l$$

where

$$V_l \equiv \int_{D_l} d\xi_i$$

is the volume of the subdomain D_l and $\bar{\zeta}$ is some vector point within D_l. Now the density function is just the ratio of the mass of the fluid particle to its volume so that the concept of a continuum, as embodied in Axiom 1, demands that the limit

$$\rho(\bar{\zeta}) \equiv \lim_{D_l \to 0} \frac{M_l}{V_l} \qquad \text{for each } D_l \subset D$$

exists and is the value of the density function at the point $\bar{\zeta}$. It should be noted that this limit is a direct statement and does not involve a statistical representation of molecular motion.

The adoption of a continuum fluid theory removes any dilemma concerning the size of the subdomain D_l when specifying conditions at a point. It does not, however, remove the physical reality that involves the molecular motion of a real fluid when the length scale of the motion becomes very small. Of particular moment in the study of turbulent flows is the magnitude of the fine-eddy structure relative to the magnitude of the molecular mean free path. It is clear that the total characteristics of the real fluid flow must reside in the properties of the molecular motion and that some pertinent feature may be occluded by the adoption of a continuum model. There is, however, a fairly definite minimum eddy size found in turbulent flow. Under rather extreme pipe flow conditions (Re $\sim 10^6$ in a 10-mm-diameter pipe) the minimum eddy is of the order of 10^{-4} mm in diameter. The molecular mean free path is of the order of 10^{-7} mm in liquids and 10^{-5} mm in gas flows under the above conditions. These observations suggest that the mean free path will be smaller than the turbulent eddy size and do not directly condemn the continuum model.

A further remark can be made concerning the volume of the fluid element V_l. Let V_0 be the volume of D_l at some reference condition; then

$$V_0 = \int_{D_0} d\bar{\xi}_i = \int_{D_l} J(\bar{\xi}_i\,;\,\xi_i)\,d\xi_i$$

and the mean-value lemma shows that

$$V_0 = J(\bar{\xi}_i\,;\,\bar{\zeta})\,V_l$$

Now the transformation Jacobian is directly the ratio of volume elements as the fluid subdomain moves through space and the Jacobian is constant for an isochoric motion.

In the same spirit as that in which the continuum fluid model was adopted, it is assumed that fluid motion can be described within the framework of certain conservation axioms. That is, any uncertainty in the empirical assessment that mass, for example, is conserved, is removed by the adoption of an axiomatic statement concerning mass conservation. Now a body of theory for continuum fluid mechanics can be established on a logical basis and its fruits tested against any observation of real flow situations.

Axiom 3. The identifiable collection of fluid particles within the subdomain D_l are mass invariant with respect to time. ∎

This axiom concerning the conservation of mass asserts that

$$\frac{d}{dt}\int_{D_l} \rho(\xi_i, t)\,d\xi_i = 0 \tag{1.1}$$

where d/dt denotes the substansive derivative. It is the very substance of fluid motion that the subdomain D_l will deform as the motion progresses. For this reason, the derivative in Equation (1.1) can be more easily evaluated if the motion is mapped onto a reference state at $t = t_0$. This reference state will consist of the domain D_0 and will be delineated by the local coordinates ζ_i. Now the statement

$$\frac{d}{dt}\int_{D_0} \rho(\zeta_i, t_0) J(\xi_i(t)\,;\,\zeta_i)\,d\zeta_i = 0 \tag{1.1a}$$

where $J(\xi_i\,;\,\zeta_i) \equiv \partial(\xi_i)/\partial(\zeta_i)$ is the instantaneous transformation Jacobian between D_0 and D_l, is identical to that contained in Equation (1.1). The indicated differentiation then shows that

$$\int_{D_0}\left(\frac{d\rho}{dt}J+\rho\frac{dJ}{dt}\right)d\zeta_i=0$$

Appeal to Euler's identity[10]

$$\frac{dJ}{dt}=J\frac{\partial v_i}{\partial x_i}$$

where v_i are the velocity components of the fluid particle with respect to the space frame, shows that

$$\int_{D_0}\left(\frac{d\rho}{dt}+\rho\frac{\partial v_i}{\partial x_i}\right)J(\xi_i;\zeta_i)\,d\zeta_i\equiv\int_{D_l}\left(\frac{d\rho}{dt}+\rho\frac{\partial v_i}{\partial x_i}\right)d\xi_i=0$$

Now, provided that the functions under the integral sign satisfy the necessary continuity requirements (which is guaranteed by Axiom 2), the Dubois–Reymond lemma[11] gives—since the subdomain D_l was arbitrarily selected within D—the result

$$\frac{d\rho}{dt}+\rho\frac{\partial v_i}{\partial x_i}=0$$

and the continuity requirement in differential form is recovered. Expanding the substansive derivative yields the alternative form

$$\frac{\partial\rho}{\partial t}+\frac{\partial\rho v_i}{\partial x_i}=0 \tag{1.2}$$

A further reduction of the continuity equation is possible when the motion is isopycnic, in which case Equation (1.2) becomes

$$\frac{\partial v_i}{\partial x_i}=0 \tag{1.3}$$

Here the mass conservation axiom reduces to a purely kinematic statement about the motion without any reference to the fluid density. Equation (1.1a) further implies that the transformation Jacobian, J, is constant when the fluid density is constant so that isopycnic flow is also isochoric.

Progress toward the development of a momentum-conservation equation must depend upon further assumptions. The first major assumption is embodied in Euler's generalization of Newton's hypothesis for rigid-body motion: Force and rate of momentum change are directly proportional. This hypothesis can be applied to each fluid subdomain, D_l, provided that proper recognition of the surface force that each subdomain applies to its neighbor is incorporated. Specification of this surface force constitutes the second major assumption required.

Axiom 4. Two fluid subdomains that are in contact will both experience the same force field on their mutual interface (with due regard for the sign of the force). ∎

Axiom 5. The resultant total force, $\bar{F}_l$, on a fluid subdomain D_l produces a motion of that subdomain in the direction of the force such that

$$F_l \propto \frac{d}{dt}(M_l U_l)$$

where U_l is the velocity vector of the fluid subdomain in the direction of F_l. ∎

If f_i denotes a body force per unit mass and if τ_{ij} is the surface stress tensor, then the momentum-conservation axiom implies the equation

$$\frac{d}{dt}\int_{D_l} \rho v_j \, d\xi_i = \int_{D_l} \rho f_j \, d\xi_i + \int_{\partial D_l} \tau_{kj} n_k \, d\sigma \qquad (i, j, k \in I_3)$$

Here, n_k is the unit normal to the surface ∂D_l of subdomain D_l and $d\sigma$ is an element of that surface. By similar arguments to those used above, it can be shown that

$$\begin{aligned} \frac{d}{dt}\int_{D_l} \rho v_j \, d\xi_i &\equiv \int_{D_l} \left[\rho \frac{dv_j}{dt} + v_j\left(\frac{d\rho}{dt} + \rho \frac{\partial v_k}{\partial x_k}\right)\right] d\xi_i \\ &= \int_{D_l} \rho \frac{dv_j}{dt} \, d\xi_i \end{aligned}$$

when the continuity requirement is acknowledged. Provided that the stress tensor τ_{ik} is a sufficiently smooth function throughout the fluid domain it is possible to write

$$\int_{\partial D_l} \tau_{kj} \eta_k \, d\sigma = \int_{D_l} \frac{\partial \tau_{kj}}{\partial x_k} \, d\xi_i$$

in conformity with the divergence theorem. With the incorporation of this result, the momentum-conservation axiom is stated in the form of the volume integral:

$$\int_{D_l} \left(\rho \frac{dv_j}{dt} - \rho f_j - \frac{\partial \tau_{kj}}{\partial x_k}\right) d\xi_i = 0$$

Application of the Dubois–Reymond lemma then provides a result for the differential form of the momentum equation, namely;

$$\rho\frac{\partial v_j}{\partial t}+\rho v_i\frac{\partial v_j}{\partial x_i}=\rho f_j+\frac{\partial \tau_{kj}}{\partial x_k} \tag{1.4}$$

If P is the fluid static pressure and δ_{ij} denotes Kronecker's symbol, then, on the basis of Stokes' relation for the stress tensor, it is possible to state the following axiom.[12]

Axiom 6. The local stress tensor is linearly related to the local rate of strain. ∎

As an equation, Axiom 6 can be written

$$\tau_{ij}=-P\delta_{ij}+2\mu(V_{ij}-\tfrac{1}{3}V_{kk}\delta_{ij}) \tag{1.5}$$

where

$$V_{ij}\equiv\frac{1}{2}\left(\frac{\partial v_i}{\partial x_j}+\frac{\partial v_j}{\partial x_i}\right)$$

is the rate of deformation tensor and μ represents the fluid viscosity. It is recognized that Equation (1.5) is nothing more than an assumption and that its adequacy[13] for a complete description of fluid motion can be questioned. For the applications in hand, experience suggests that it should be an entirely adequate statement.

The mathematical formulation of the problem is complete upon specification of the boundary conditions to be imposed upon the motion. There is:

Axiom 7. For a solid body, B, immersed in an expanse of fluid there is no relative motion between the fluid and the surface ∂B of B. ∎

Development of the classical equations of fluid motion is completed save for deliberations on an energy equation, which need not be repeated here. These equations were derived without any reference to the nature of the motion, other than that it should conform to conservation principles and to a stress relation such as is represented by Equation (1.5). In view of this there are several questions that present themselves as being pertinent to the subject of turbulence. The classical equations are usually accepted as being applicable to turbulent flow without any further comment. Thus Panchev[14] adopts the Navier–Stokes equations axiomatically and justifies this on the basis of good agreement between theory and experiment. Leslie[15] makes some additional comments in support of the adoption of the Navier–Stokes equations, while other writers[16,17] do not appear to inquire if the equations have any applicability to turbulent motion before utilizing them in this context.

Suppose that the entire set of boundary conditions imposed upon the motion within the domain D are identically independent of time. Then, without the benefit of experimental evidence pointing to the contrary, it is natural to suppose (when restricted to isopycnic motion) that

$$\frac{\partial v_i}{\partial x_i}=0$$

$$\rho v_j\frac{\partial v_i}{\partial x_j}+\frac{\partial P}{\partial x_i}=\mu\frac{\partial^2 v_i}{\partial x_j \partial x_j} \tag{1.6}$$

is an adequate statement of the motion when considered as a mathematical boundary-value problem. That the real motion is found to depend upon time t—a parameter not appearing in the problem as posed—should be cause for careful deliberation. Thus, it is determined experimentally that if the Reynolds number is greater than some critical value, then the flow will degenerate into a turbulent one. This observation is somewhat related to the comments made in Section 1.1 with regard to the flow shown in Figures 1.1–1.4, namely, that the real flow is far more complex than the boundary conditions imposed upon the motion would lead one to expect. The mathematical boundary-value problem contained in Equation (1.6) should not be left without the remark that, in any real flow situation, it is never possible to make the boundary conditions identically constant in time. Also a real body surface can never be truly smooth at a molecular level. These facts may well be the reconciliation between the ideal and the practical; it is the stability of the motion that should be investigated as leading into the initiation of turbulence.

Another possible concern at this juncture is the uniqueness of the solution to Equation (1.6). In other words, is the disposition of fluid at any time instant uniquely determinable from that at the previous instant? Such questions are not resolved in the literature and will not be of great concern in the present work; but they should be borne in mind in turbulence research and in the application of semiempirical models. It is also of interest to speculate whether, even if a complete space–time velocity field were determinable, a turbulent flow can be calculated without some statistical input.

A preliminary requirement for the development of the conservation equations was that the fluid subdomain, D_l, should always consist of the same elements as the motion proceeds. This requirement is also fundamental to the application of the differential equations of motion since use of the Dubois–Reymond lemma in the restatement of the integral relation as a differential equation does not remove this requirement. Of course it can be argued that the subdomain D_l can be taken as small as required before application of the Dubois–Reymond lemma, but the relevant question for turbulent motion concerns the lower limit to the subdomain size. A lot needs to be understood about turbulent motion before any comment can be made in an attempt to resolve this question. Certainly, measured spectra of turbulent quantities still contain energy to frequencies beyond 50 kHz. Owing to the highly chaotic mixing associated with turbulent motion, this typical fluid subdomain can be no larger than the smallest characteristic eddy size of the motion. For the continuum model this presents no difficulty in accord with Axiom 1, but it is less evident in a real fluid that the molecular nature of the motion may not contribute directly to the viscous dissipation in the small eddy structure and hence may feature in the model equations. In order for the continuum fluid to be used it must be supposed that the length scale of the smallest identifiable eddy is still sufficiently large for the flow in this structure to be considered as a continuum and be endowed with the physical characteristics usually adopted.

No lower limit to the length scale is implied in the development of the equations of motion within the continuum framework since any comparison with molecular motion, via a Knudsen number, does not enter into the discussion. Since the equations were developed solely on the basis of the surface interactions between adjacent subdomains D_l and D_j no mention is made of the motion within these subdomains. In essence, this remark is related to the adequacy of the supposed shear stress relation and its applicability to any length scale in the motion. It is well documented (Lorenz[(18)]) that the characteristics of the small-eddy structure must be correctly represented in any calculation procedure since errors readily propagate from one eddy scale to the next. In particular, the eddy structure is highly three-dimensional even when the mean motion is two-dimensional. Further remarks will be made in later chapters.

As a final remark in this section, it can be noted that in general the Navier–Stokes equations give a very satisfactory level of agreement with experimental data. This is true of both laminar and turbulent motions. In this latter connection, the experimental work of Stewart[(19)] for isotropic turbulence gives a rewarding measure of agreement with the Kármán–Howarth theory.

1.3. Further Remarks on Turbulence

Section 1.1 discussed certain aspects of fluid flow and made some remarks about particular flows, but there is no clear understanding why turbulence is established in a fluid motion. There has recently been considerable increase in the understanding of how a laminar motion adjusts itself and takes on the characteristics of turbulence, and this subject will be reviewed in Chapter 6. Why the flow should degenerate into turbulence is still largely conjectural. The classical stability theory, based upon solutions of the Orr–Sommerfeld equations, is essentially linearized in the sense of being a small-disturbance theory. Most probably such a theory can give an adequate description of the initial phases of the breakdown of a laminar motion. It is quite clear, however, that the final stage of the transition phase is highly nonlinear and indeed exhibits many of the attributes of fully turbulent flow—at least within an intermittent framework.

Much has been written on hydrodynamic stability[20–22] and will not be repeated here. Before we leave the subject, however, two examples will be presented. Figure 1.5 shows, after Brennen, an interesting wave instability on the surface of a cavitation bubble; the waves travel at more or less the

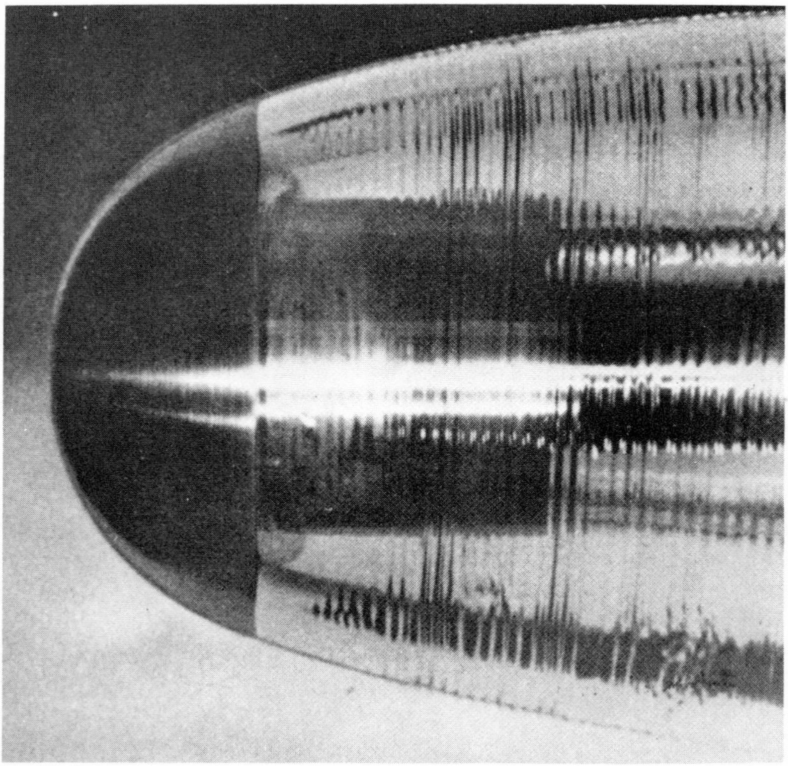

Fig. 1.5. Wave instability on the surface of a cavitation bubble behind a hemispherical head. From C. Brennen (California Institute of Technology), Cavity surface wave patterns and general appearance, *J. Fluid Mech.* **44**, 33–49 (1970) (reproduced by permission of Cambridge University Press).

fluid velocity but do not remain in planes normal to the motion. The process and transition from laminar to turbulent flow can be further demonstrated by the flow shown in Figure 1.6. This figure presents an interferometer trace of a jet flow that is progressing from laminar, through the transitional, and into the turbulent regime. Several dominant features of the flow require reiteration. Initially, the laminar motion progresses with only a very small diffusion rate and the contours of constant density remain essentially straight. Then, slowly at first, an oscillation of the jet structure takes place, develops into a convolution, and then finally degenerates into fully turbulent motion. The width of the jet increases by an order of magnitude during this "transition" process; the mixing rate becoming significantly greater in the turbulent flow.

Obstacles placed in the flow are a ready means of rendering a laminar motion turbulent. Figure 1.7 shows the flow over an array of rectangular blocks and the large turbulent wake created downstream of them. There is, in addition, evidence of a large-scale motion being established by the obstruction with a length scale of the same order as the characteristic length of the body. The full three-dimensional nature of the flow is not evident from this photograph; in particular, the trailing vortex system developed from the corners of the structure is not delineated. The type of flow shown in this figure is also of interest in connection with atmospheric motion around buildings and the type of force field that may be generated on the building. At the same time, it is clear what type of wind environment exists between adjacent buildings and suggests some consequences for habitation in this area.

The wall turbulent boundary layer holds a central position in fluid mechanics and contains two regions of especial interest. These are the wall region (see Figure 1.8), where the large vorticity production breaks down to feed momentum into the main body of the turbulent flow, and the intermittent edge structure in the superlayer, which delineates the outer bound to the turbulent motion. This latter characteristic is somewhat evident in Figure 1.6, where it is seen that, downstream of the laminar motion, the boundary between the jet structure and the quiescent fluid is highly convoluted. The situation is more dramatically presented in Figure 1.9. A turbulent shear layer does not possess a smooth, steady, interface with the external flow. Rather, there is a highly intermittent, but sharp, demarcation between the turbulent motion in the shear layer and the external fluid.

The turmoil at the outer edge of the turbulent wall shear layer is reflected in the structure of the flow very close to the wall. This does not necessarily imply a correlation between events occurring at the inner and outer edges of the flow. Indeed two-point velocity correlations[(23)] show a lack of strong correlation between such events.

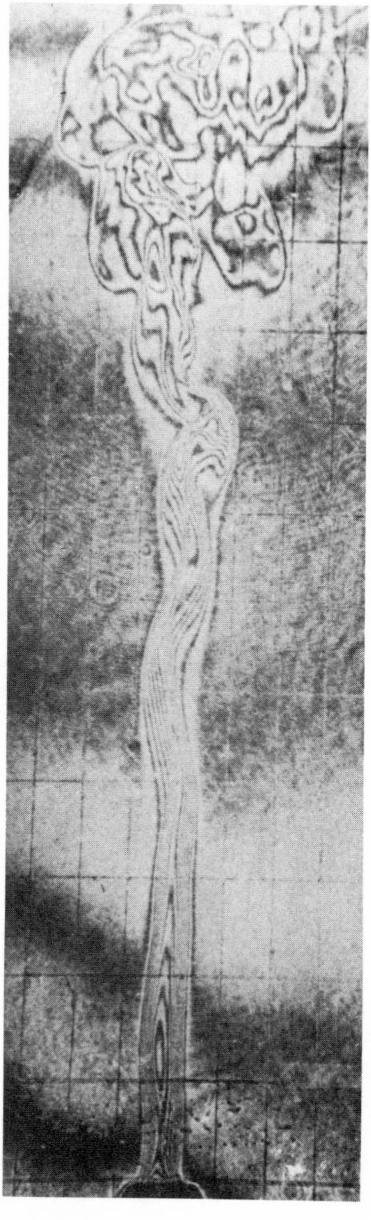

Fig. 1.6. Instability in a heated laminar jet (low Reynolds number, ~250). From J. C. Mollendorf (Western Electric Co.) and B. Gebhart (Cornell University), An experimental and numerical study of the viscous stability of a round laminar vertical jet with and without thermal buoyancy for symmetric and asymmetric disturbances, *J. Fluid Mech.* **61**, 367–399 (1973) (reproduced by permission of Cambridge University Press).

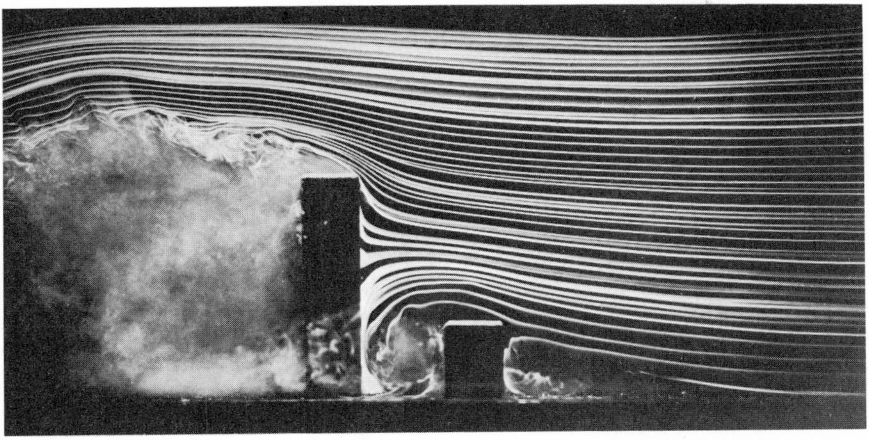

Fig. 1.7. Low-speed flow over a simulated building. Photograph by Ingelman-Sundberg (Eidgenössische Technische Hochschule, Zürich), from Thomann, H., Wind effects on buildings and structures, *Am. Sci.* **63**, 278–287 (1975) (reproduced by permission of *American Scientist*).

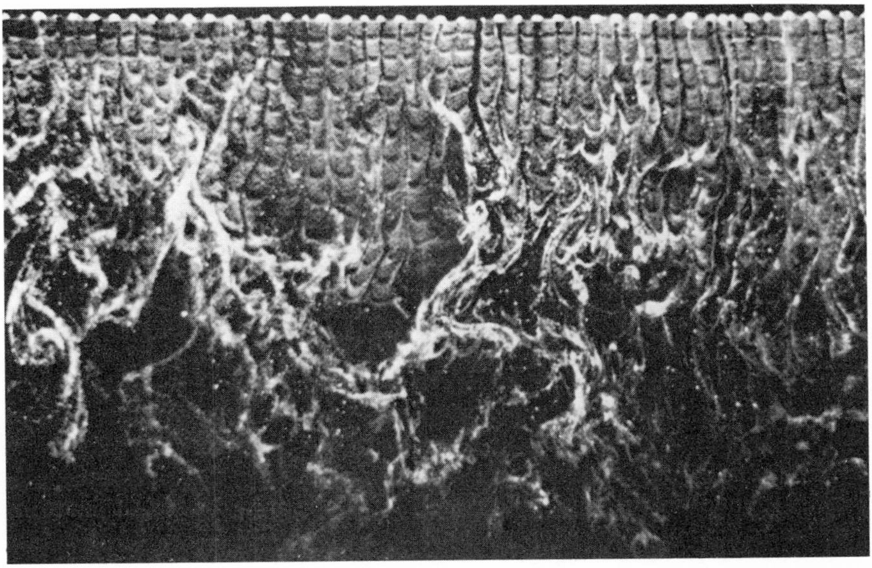

Fig. 1.8. Hydrogen bubble study of flow in boundary layer in water. Wire located at $y^+ = 82$ within the log layer. Zero-pressure gradient flow in low-speed water channel. From S. J. Kline, W. C. Reynolds, F. A. Schraub, and P. W. Runstadler (Stanford University), The structure of turbulent boundary layers, *J. Fluid Mech.* **30**, 741–773 (1967) (reproduced by permission of Cambridge University Press).

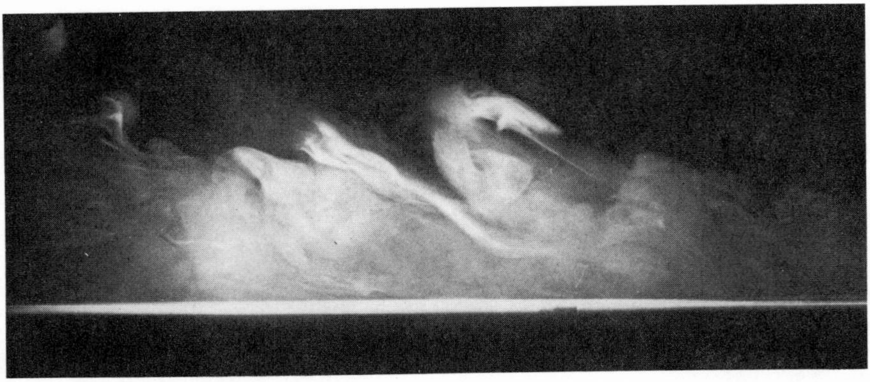

Fig. 1.9. A smoke-filled low-speed boundary layer. Flow from right to left. Shows the highly intermittent and convoluted outer edge of a low-speed turbulent shear layer. From P. Bradshaw (Aeronautics Department, Imperial College, London), The understanding and prediction of turbulent flow, *Aeronaut. J.* **76**, 403–418 (1972) (reproduced by permission of the Royal Aeronautical Society).

The intermittent flow that appears at the outer edge of a turbulent shear layer is associated with the large-scale structure within the body of the turbulent flow. This structure may be identified with the eddy cascade process assumed in a turbulent motion, or it may be associated with some coherent structure as shown by the flow in Figures 1.4 or 1.10.

In wall turbulence, where large quantities of vorticity are generated at the wall, it is of interest to study how this vorticity is worked into the structure of the turbulent layer, and Kline's hydrogen bubble studies give a nice visual idea of the process. Figure 1.8 shows a typical photograph with the wire located at $y^+ = 82$. Large streamwise filaments of vorticity are ejected from the lower regions into the outer wake regions of the layer. This should be a deterministic process in itself (although not readily calculated), but the location and the timing of such burst events is random in nature, and therefore the process must ultimately be described statistically.[24]

Confirmation that the turbulent structure does cause the propagation of disturbances into the quiescent external flow is shown in Figure 1.11, where a supersonic turbulent jet flow is presented. Turbulent jet noise is here to stay.

1.4. Looking Onward

These brief comments have shown the complexity of fluid motion and implied the need for the statistical description of turbulent flows. The

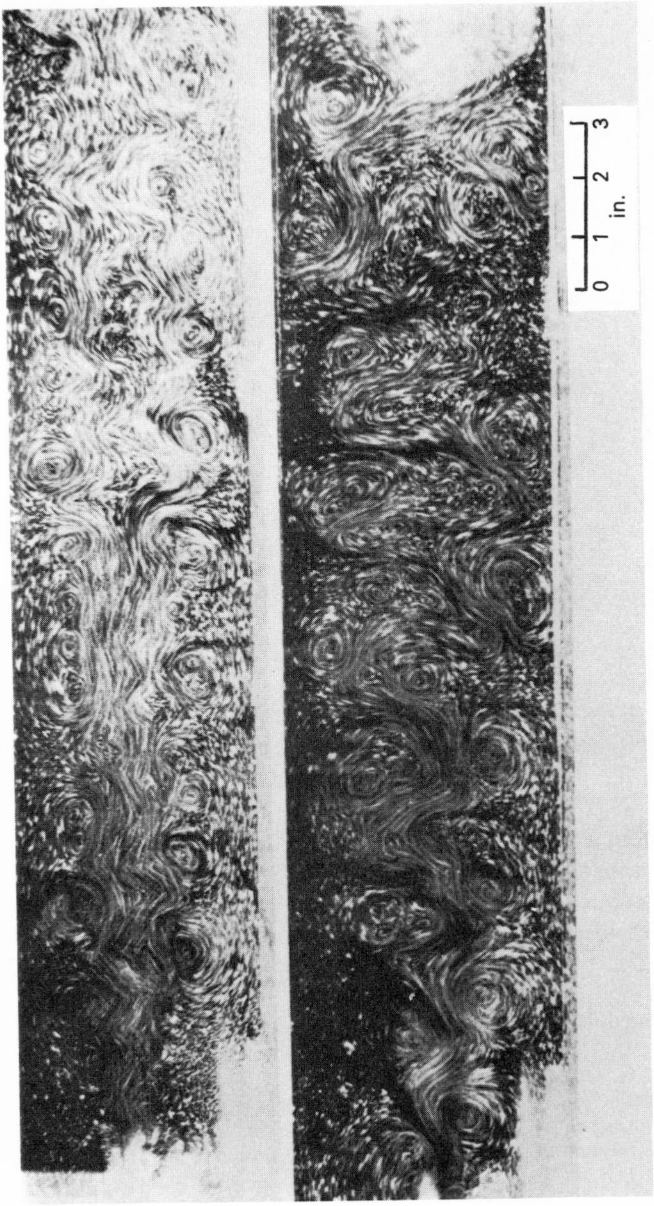

Fig. 1.10. Vortex street in a turbulent wake (Re ~ 10^4). From D. P. Papailiou and P. S. Lykoudis (Purdue University), Turbulent vortex streets and the entrainment mechanism of the turbulent wake, *J. Fluid Mech.* **62**, 11–31 (1974) (reproduced by permission of Cambridge University Press).

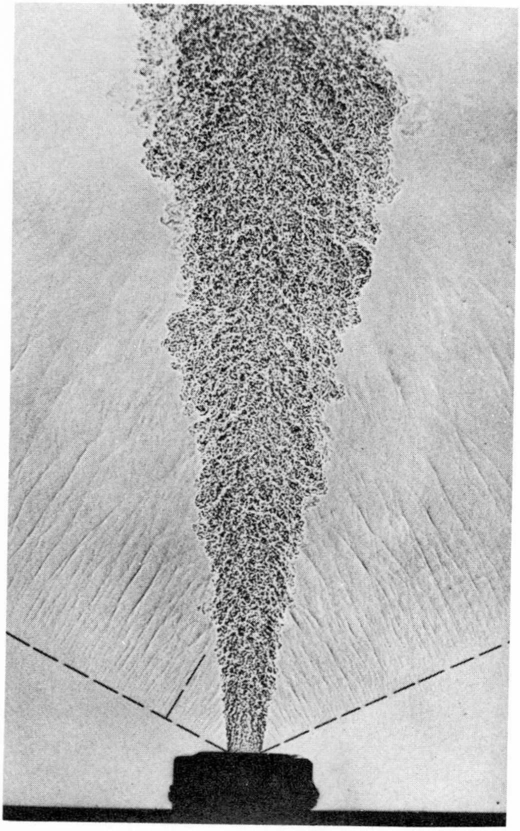

Fig. 1.11. Sound radiated from a supersonic helium jet with total pressure of 54.7 psia. From C. K. W. Tam (Florida State University), Directional acoustic radiation from a supersonic jet generated by shear layer instability, *J. Fluid Mech.* **46**, 757–768 (1971) (reproduced by permission of Cambridge University Press).

following chapters will take up both the basic theory of statistical methods in fluid dynamics and the modeling of the Navier–Stokes equations into viable, calculative techniques. Finally, measurement techniques adapted for turbulent flows will be discussed.

Notation

D	Fluid domain
E	Euclidean space
$F; f$	Force; force/unit mass
i, j, k, l	Indices
I	Index set
J	Transformation Jacobian
M	Mass of fluid element
n, N	Integer
P	Fluid static pressure
t	Time
v_i	Velocity components along directions x_i

V_{ij}	Rate of deformation tensor	μ	Fluid viscosity coefficient
V	Volume of fluid element	τ_{ij}	Stress tensor
X	Subset of E	∂D	Boundary of domain D
$x_i; \xi_i; \zeta_i$	Rectangular coordinates	$\cup$	Set union
δ_{ij}	Kronecker's symbol	$\cap$	Set intersection
ρ	Fluid density		

References

1. Monin, A. S., and Yaglom, A. M., *Statistical Fluid Mechanics: Mechanics of Turbulence*, (two volumes) MIT Press, Cambridge, Massachusetts (1971).
2. Davies, P. O. A. L., and Yule, A. J., Coherent structures in turbulence, *J. Fluid Mech.* **69**, 513–537 (1975).
3. Prandtl, L., Neuere Ergebnisse der Turbulenzforschung, *Z. Verein Dtsch. Ing.* **77**, 105–114 (1933).
4. Nikuradse, J., Kinematographische Aufnahme einer turbulenten Strömung, *Z. Angew. Math. Mech.* **9**, 495–496 (1929).
5. Van Atta, C. W., Sampling techniques in turbulence measurements, *Annual Review of Fluid Mechanics*, Vol. 6, (M. Van Dyke, W. G. Vincenti, and J. V. Wehausen, eds.), Annual Reviews Inc., Palo Alto, California (1974), pp. 75–91.
6. Serrin, J., Mathematical principles of classical fluid mechanics, *Handbuch der Physik*, Vol. 8/1 (S. Flügge, ed.), Springer-Verlag, Berlin (1959), pp. 125–263.
7. Truesdell, C., *The Elements of Continuum Mechanics*, Springer-Verlag, Berlin (1966).
8. Shinbrot, M., *Lectures on Fluid Mechanics*, Gordon and Breach, London (1973).
9. Kolmogorov, A. M., and Fomin, S. V., *Introductory Real Analysis*, Dover, New York (1975).
10. Jeffreys, H., and Jeffreys, B. S., *Methods of Mathematical Physics*, third edition, Cambridge University Press, Cambridge (1956).
11. Lin, C. C., and Segel, L. A., *Mathematics Applied to Deterministic Problems in the Natural Sciences*, Macmillan Publishing Company, New York (1974).
12. Aris, R., *Vectors, Tensors, and the Basic Equations of Fluid Mechanics*, Prentice-Hall Inc., Englewood Cliffs, New Jersey (1962).
13. Rosenhead, L., The second coefficient of viscosity: A brief review of fundamentals (and other papers on the same subject), *Proc. R. Soc.* (*London*), **A226**, 1–69 (1954).
14. Panchev, S., *Random Functions and Turbulence*, Pergamon Press, Oxford (1971).
15. Leslie, D. C., *Developments in the Theory of Turbulence*, Clarendon Press, Oxford (1973).
16. Batchelor, G. K., *The Theory of Homogeneous Turbulence*, Cambridge University Press, Cambridge (1970).
17. Townsend, A. A., *The Structure of Turbulent Shear Flow* (second edition), Cambridge University Press, Cambridge (1976).
18. Lorenz, E. N., Investigating the predictability of turbulent motion, in: *Statistical Models and Turbulence* (M. Rosenblatt and C. W. Van Atta, eds.), Springer-Verlag, Berlin (1972), pp. 195–204.
19. Stewart, R. W., Triple velocity correlations in isotropic turbulence, *Proc. Cambridge Phil. Soc.* **47**, 146–157 (1951).
20. Chandrasekhar, S., *Hydrodynamic and Hydromagnetic Stability*, Oxford University Press, Oxford (1961).
21. Lin, C. C., *The Theory of Hydrodynamic Stability*, Cambridge University Press, Cambridge (1966).

22. Stewart, J. T., Nonlinear stability theory, *Annual Review of Fluid Mechanics*, Vol. 3 (M. Van Dyke, W. G. Vincenti, and J. V. Wehauser, eds.), Annual Reviews, Inc., Palo Alto, California (1971), pp. 347–370.
23. Cliff, W. C., and Sandborn, V. A., Correlation between the outer flow and the turbulence production in a boundary layer, NASA Report No. TM X 64935 (1975).
24. Chorin, A. J., *Lectures on Turbulence Theory*, Publish or Perish Inc., Boston, Massachusetts (1975).

CHAPTER 2

An Introduction to Turbulence Phenomena

TREVOR H. MOULDEN

2.1. Introduction

Investigations into turbulence phenomena have produced an unresolved dichotomy.

> So as they eddied past on the whirling tide,
> I raised my voice: "O souls that wearily rove,
> Come to us, speak to us—if it be not denied"*

Turbulence, like any physical process that is not strictly deterministic, must be discussed within a statistical framework. Useful studies in turbulence will include parallel treatments of both the experimental and theoretical aspects of the subject. Experimental studies usually measure a temporal mean of the physical quantities under investigation. On the other hand, a theoretical presentation is more logically developed within the context of probability theory since then the difficulties contained in the specification of an averaging weight function are circumvented. The first statement of the dichotomy is resolved if the turbulence process is ergodic.[1] The deeper

* Dante Alighieri, *L'Inferno* Canto V, c. 1300, translation from Italian by Dorothy L. Sayers.

TREVOR H. MOULDEN • The University of Tennessee Space Institute, Tullahoma, Tennessee 37388 philosophical question of whether theory and experiment can ever be made to have meaningful comparisons will be left open.

It was intimated above that a logical theoretical development could be pursued and the results of such a study shown either to agree or to not agree with observed reality. However, neither the form of the required axioms nor the direction that the theoretical development should take are clear without some firm understanding of the physical process involved. Since the turbulence process is not fully known, any theoretical treatment of the subject must rely to a large extent upon conjecture. Indeed, the entire "practical turbulence theory" used in wall turbulence calculations depends upon empiricism in one form or another. The nature of these various empiricisms will be discussed later.

In simple terms we have that the logical theoretical treatment couched in probabilistic terms cannot proceed to completion without conceptual effort provided by experimental studies. It is encouraging that the physical understanding of turbulence is advancing rapidly with the introduction of more sophisticated experimental techniques. This is particularly needed in the case of nonhomogeneous turbulence.

A general property of fluid flow that is particularly pertinent to turbulence can easily be demonstrated. The following remarks are intended to be illustrative and not to model an actual flow situation. Let a viscous fluid of constant physical properties be in motion within a two-dimensional domain D. If ζ is the vorticity of the flow, then the equations of motion can be expressed as

$$\frac{\partial \zeta}{\partial t}+u\frac{\partial \zeta}{\partial x}+v\frac{\partial \zeta}{\partial y}\equiv\frac{D\zeta}{Dt}=\lambda\nabla^2\zeta \tag{2.1}$$

with

$$\zeta=-\nabla^2\psi$$

where ψ is the stream function and λ denotes the inverse Reynolds number in terms of a characteristic length and velocity of the problem.

If the boundary conditions are such that the upstream flow is irrotational and steady, then the decomposition

$$\psi=\psi_0+\psi_1, \qquad \zeta=\zeta_0+\zeta_1 \tag{2.2}$$

where $\nabla^2\psi_0\equiv 0$ and ψ_0 satisfies the far-field boundary conditions, shows that

$$\zeta_0=0, \qquad \zeta_1=-\nabla^2\psi_1 \tag{2.3}$$

and therefore the entire vorticity of the flow is associated with the function ψ_1.

Now Equation (2.1) becomes

$$\frac{D\zeta_1}{Dt} = \lambda \nabla^2 \zeta_1 \tag{2.4}$$

and indicates that both changes in vorticity, and any time dependence that may develop in the flow, are intimately related to the viscous property of the fluid; for, if $\lambda \equiv 0$, then $D\zeta_1/Dt \equiv 0$, and the upstream boundary condition demands that the vorticity vanish throughout the flow field.

This dependence of turbulent motion upon the fluid viscosity is true for the initial creation of turbulence adjacent to a solid surface. It is also true for the ultimate energy dissipation in the small-scale-eddy structure. The above two-dimensional model fails to admit much that is fundamental to turbulent motion. The whole phenomenon of vortex stretching is a pertinent example of this failure. This stretching mechanism is related to the cascade of eddy sizes that features in turbulent flows and determines the transfer of energy from the external driving mechanism to the fine-eddy structure where the viscous dissipation takes place. The total energy content of the large eddies remains approximately constant during this process (Gartshore[(2)]). Further discussion of vortex stretching is given in Chapter 3.

Nonhomogeneous wall turbulence suffers flow breakdown at the outer edge of the viscous sublayer. Here, the significant vorticity created on the wall develops into pronounced streamwise vorticity, which breaks away from the wall sublayer and injects momentum into the outer wake region. The studies by Kline *et al.*,[(3)] by Kim *et al.*,[(4)] and by Corino and Brodkey,[(5)] all indicate this structure in a visual sense. The hot-wire studies of Willmarth and Lu[(6)] amplify the understanding of the phenomenon, while Gupta *et al.*[(7)] discuss the spanwise structure of the wall flow. The earlier visual studies of Fernholz[(8)] should also be recognized.

Far from the wall it is found that the turbulent structure takes on a form similar to that associated with wake flow (Coles[(9)]) and adopts a strongly intermittent character. Fiedler and Head,[(10)] Kovasznay *et al.*[(11)] and Antonia,[(12)] among many others, have studied this intermittent structure in various ways. The large-scale intermittent structure in the flow is not restricted to the outer reaches of the turbulence region and is most readily studied with conditional sampling techniques (Laufer[(13)] and Davies and Yule[(14)]), a discussion of which is deferred until later.

Certain overall features of turbulence can be set down. Taylor suggested that the flow behaved as if it possessed an infinite number of degrees of freedom. While this may be a useful idea for introducing the need for a statistical description of the flow, it carries the connotation of a structureless meandering within the flow. The above remarks indicate that a turbulent flow (particularly wall turbulence) does, on the contrary, have a well-defined structure. Even without such a structure, the flow must conform to the laws of conservation. It is probably most useful to consider a turbulent field as being indeterminate in the sense that an entire set of initial conditions can never be specified. This has somewhat the same content as the problem of statistical mechanics, but now the physical situation is much different.

Turbulence can be characterized by its highly diffusive nature and by its large fluctuating vorticity. These processes are accompanied by energy dissipation in the small-eddy structure and signify that turbulence can only be maintained if a continuous supply of energy is available. The larger rates of diffusion and energy dissipation that are associated with a turbulent flow, as compared to a laminar one, are well documented.

It is perhaps not out of order to note (Tennekes and Lumley[(15)]) that turbulence is a feature of the flow and not a physical property of the fluid.

The energy used to maintain the turbulent motion can be taken from a buoyancy force or from a shear in the mean flow. External forces (magnetic, for example) can also influence the structure of the energy cascade if this imposed force is responsible for significant energy transfer.

2.2. On the Basic Equations of Motion

2.2.1. General Considerations

The development contained herein will revolve around the assumption of a continuum fluid flow. This assumption immediately restricts the range of Knudsen numbers for which the discussion is pertinent. In addition, it implies a lower limit to the length scales associated with the turbulent motion. The Kolmogorov inner scale (for instance) must be large compared to the molecular mean free path. This latter requirement is satisfied in most flow situations of interest.

The continuum flow assumption renders the consideration of molecular ensembles unnecessary and allows us to adopt the equations of motion established by Stokes and Navier. The corresponding relief of statistical complexity is welcome here, but one comment is in order concerning the implications of this transition from molecular to continuum considerations. The initial-value problem in kinetic theory admits a unique solution, whereas that for the Navier–Stokes equations may not (Shinbrot[(16)]). The approximations involved in the specification of a distribution function and the subsequent first-order integration of Boltzmann's equation may have changed the mathematical content of the problem.

To begin the equation development we delineate the four-dimensional space–time domain D with the rectangular coordinates (x_i, t) $(i = 1, 2, 3)$

and accept the suffix summation convention usually adopted when discussing Cartesian tensor quantities. The space domain Ω will be assumed to have the character of an Euclidian E^3 space. Then fluid motion may be discussed in terms of an operator Γ_t, which maps an element $\delta\Omega$ of E^3 into E^3. Suitable restrictions concerning the smoothness of the operator Γ_t, and its inverse, must be imposed (Meyer,[17] for instance). Since the differential equations describing the motion are derived under such assumptions, it is clear that the ensuing solutions to these equations must be likewise restricted. It is, of course, still necessary to discuss the existence and uniqueness of the solutions to these equations—see Ladyzhenskaya[18,19] and Shinbrot.[16]

The equations will be stated under certain restrictions. These specifically contain the idea of a fluid with constant physical properties and the supposed absence of an imposed force field on the motion. These restrictions are adopted from the view of the resulting simplification rather than from necessity. Indeed, there is concern that fluid flow (particularly of a gaseous fluid) will never be at a strictly constant fluid density. This fact could have consequences when the detailed structure of a small-eddy configuration with significant vorticity is under study. When density fluctuations are also present, the theory must discuss correlations between density and velocity components as dictated by the compressible flow equations. Again, the neglect of external forces does not imply that flows such as those encountered in magneto-fluid-mechanics are of no interest or concern.

Thus, while considerable simplification is rendered by the assumptions imposed, the results of the theoretical treatment must be evaluated in light of the possible consequences of these assumptions. If a dominant feature of an experimental situation has been assumed away in the theoretical development, then such a study is not likely to mirror the "real" flow. Specifically, a turbulent flow is *never* two-dimensional in nature even if the boundary conditions imposed upon the flow would intuitively lead to that expectation. Similarly, it may not be possible to discuss the energy dissipation in a fine-eddy structure without appeal to the molecular nature of the fluid, or at least to its nonconstant physical properties. Despite this, useful results are still obtained from the continuum model.

The way in which long-molecule polymer solutes may influence a fluid flow is another subject and will not be considered here.

2.2.2. Detailed Development

Without giving any details of the derivation, the equations of motion, subject to the assumptions specified above, have the following form[20]:

Continuity:

$$\frac{\partial v_i}{\partial x_i} = 0 \tag{2.5}$$

Momentum:

$$\rho\frac{\partial v_i}{\partial t} + \rho v_j\frac{\partial v_i}{\partial x_j} + \frac{\partial P}{\partial x_i} = \frac{\partial}{\partial x_j}\left(\mu\frac{\partial v_i}{\partial x_j}\right) \tag{2.6}$$

Energy:

$$\Phi + k\frac{\partial^2 T}{\partial x_i \partial x_i} = \rho\frac{\partial e}{\partial t} + \rho v_i\frac{\partial e}{\partial x_i} \tag{2.7}$$

In this latter equation e denotes the specific internal energy of the fluid. The dissipation function has been given the label Φ.

Then

$$\Phi = \bar{\sigma}_{ij}\frac{\partial u_i}{\partial x_j} \tag{2.8a}$$

where

$$\bar{\sigma}_{ij} = \mu\left(\frac{\partial v_i}{\partial x_j} + \frac{\partial v_j}{\partial x_i}\right) \tag{2.8b}$$

is the stress tensor component devoid of the pressure term.

The divergence of Equation (2.6) gives

$$\nabla^2 P = -\rho\frac{\partial^2 v_i v_j}{\partial x_i \partial x_j} \equiv -\rho\frac{\partial v_j}{\partial x_i}\frac{\partial v_i}{\partial x_j} \tag{2.9}$$

which is a Poisson equation for the pressure field in terms of gradients of the velocity field. It should be noted that Equation (2.9) makes no reference to the fluid viscosity. The viscosity does, however, enter into the Neumann boundary conditions imposed upon the solution to Equation (2.9).

The above equations contain Newton's suggestion that a fluid motion is subjected to a linear relation between stress and rate of strain. Without overemphasizing the point, it is clear that such a specification could be challenged in many situations of interest. This would pertain to situations where high shear rates occur (in small-scale eddies, for example) or where the flow is not of a uniphase nature. For now, we will accept the equations as they stand and discuss the consequences that ensue.

Some support for the acceptance of the Navier–Stokes equations as an adequate statement for turbulent fluid flow was presented by Stewart.[21] Stewart shows experimental verification of the von Kármán–Howarth dynamic equation for isotropic turbulence, which is a form of the Navier–Stokes equations.

For later applications it is useful to note how the use of Equation (2.5) yields an alternate form for Equation (2.6):

$$\rho\frac{\partial v_i}{\partial t}+\rho\frac{\partial v_i v_j}{\partial x_j}+\frac{\partial P}{\partial x_i}=\frac{\partial}{\partial x_j}\left(\mu\frac{\partial v_i}{\partial x_j}\right) \tag{2.6a}$$

Just as the divergence of Equation (2.6) produced an expression for the pressure field, so the curl of the same equation produces an equation for the variation of vorticity (Tennekes and Lumley[15]). Now

$$\rho\frac{\partial \omega_i}{\partial t}+\rho u_j\frac{\partial \omega_i}{\partial x_j}=\rho\omega_j\frac{\partial u_i}{\partial x_j}+\mu\frac{\partial^2 \omega_i}{\partial x_j\,\partial x_j} \tag{2.10}$$

Equations (2.5)–(2.7) and (2.9) constitute a boundary-value problem for the specified flow variables once appropriate boundary conditions are provided.

In many problems of practical interest, it can be assumed that the far field is completely steady. The continuum fluid flow model demands that the fluid be at rest relative to any solid surface placed within the flow. The turbulent fluctuations must also decay to zero on the solid surface. In this context it was shown experimentally by Laufer[22] that close to the wall $u_2/x_2 \to 0$, while u_1/x_2 and u_3/x_2 adopt constant values. The ratio u_1/U_1 also assumes constancy. The way in which the large quantity of vorticity generated on the surface breaks down into the turbulence structure (Kline *et al.*,[3] for example) is not a feature of the boundary conditions. It *is* a crucial component of wall turbulence.

Specifically, the boundary conditions state

$$\left.\begin{aligned} &\lim_{\mathbf{r}\to\infty} v_i \to U_\infty \\ &v_i = 0 \qquad \text{on the body} \end{aligned}\right\} \tag{2.11}$$

In the far field the vorticity will assume the value specified (usually zero) while on a solid surface the fluctuating vorticity is unknown and relates to the local skin friction.

The boundary conditions to be imposed upon the pressure field are obtained from the momentum equation (2.6) restricted to conditions on the wall, i.e.,

$$\left.\frac{\partial P}{\partial x_i}\right|_{\text{wall}}=\mu\left.\frac{\partial^2 v_i}{\partial x_j\,\partial x_j}\right|_{\text{wall}} \tag{2.12}$$

In general, the pressure on the wall will not be steady since the velocity gradient is associated with the turbulence structure. Willmarth and Lu[6] measured significant velocity fluctuations in the sublayer. Further studies of the wall pressure field have been made by Lilley and Hodgson,[23] Willmarth and Wooldridge,[24] and Wills.[25] In general terms, the space–time correlation contains a sharp peak at the origin and falls away rapidly with increased space–time separation, the spatial separation for zero correlation being of the order of one boundary-layer thickness (Willmarth[26]). At the same time, the pressure pulses move over the wall at about eight-tenths of the free-stream velocity.

2.3. Reynolds' Decomposition

The innovation of discussing turbulent fluid motion as a random variation about a mean value was first put forward by Reynolds.[27] The exact requirements for the specification of the mean-value operator are not obvious, however. A general expression for such a concept takes the form (Monin and Yaglom[28])

$$U = \langle u, f \rangle \equiv \int_D u(x_i - \xi_i, t - \tau) f(\xi_i, \tau)\, d\Xi_i \tag{2.13}$$

with a suitable normalization expressed by

$$\int_D f(\xi_i, \tau)\, d\Xi_i \equiv 1 \tag{2.14}$$

Ξ_i denotes the set (ξ_i, τ).

Any $f(\xi_i, \tau)$ for which the integral in Equation (2.13) has meaning (the sense of Lebesgue included) will generate the relations

$$\begin{aligned} &\text{(a)} \quad \langle u + v, f \rangle = \langle u, f \rangle + \langle v, f \rangle \\ &\text{(b)} \quad \langle au, f \rangle = a \langle u, f \rangle \\ &\text{(c)} \quad \langle a, f \rangle = a \\ &\text{(d)} \quad \left\langle \frac{\partial u}{\partial x_i}, f \right\rangle = \frac{\partial}{\partial x_i} \langle u, f \rangle \end{aligned} \tag{2.15}$$

The quantity a in these expressions is a constant.

Once the form of the weight function f has been specified, it is convenient to suppress this quantity from the notation; $\langle u \rangle$ then denotes the mean value. A straightforward temporal mean, denoted by an overbar, e.g.,

$$\bar{u}_i = \frac{1}{2T} \int_{-T}^{T} u_i(\tau)\, d\tau$$

predominates in experimental work. If this latter form of the averaging process is to have meaning, then the physical situation to which it is applied must possess certain characteristics. Simply, the period of the fluctuations being studied must be small compared to the period of the mean flow. In other words, there needs to be a large separation between the high- and low-frequency components of the flow (Monin and Yaglom[28]). This condition is not always satisfied in reality but is often a sufficiently good representation of the flow.

As an example of the implementation of the mean-value operation, let U_i denote the mean of v_i so that

$$U_i \equiv \langle v_i, f\rangle \tag{2.16}$$

for the selected weight f. Then, the fluctuation u_i is defined so that

$$v_i = U_i + u_i \tag{2.17}$$

Applying Equation (2.13) to this relation gives

$$\langle v_i, f\rangle = \langle U_i, f\rangle + \langle u_i, f\rangle \equiv U_i$$

Provided an f can be selected with the property

$$\langle U_i, f\rangle = U_i \tag{2.18}$$

it follows that $\langle u_i, f\rangle \equiv 0$.

Recognition of the continuity equation (2.5) shows that the mean value, U_i, satisfies the condition

$$\left\langle \frac{\partial v_i}{\partial x_i}, f\right\rangle = \frac{\partial}{\partial x_i}\langle v_i, f\rangle = \frac{\partial U_i}{\partial x_i} = 0 \tag{2.19}$$

and Equation (2.17) readily gives

$$\frac{\partial u_i}{\partial x_i} = 0$$

a result that is true for any f, provided the integral in Equation (2.13) exists, without the need for Equation (2.18).

The specification of the weight function f will depend upon the information to be extracted from the subject under study. In its simplest form, the choice

$$\begin{aligned} f &= \delta(\xi_i)/2T, \qquad |\tau| < T \\ f &= 0, \qquad\qquad\quad |\tau| > T \end{aligned} \tag{2.20}$$

gives the temporal mean. This form of weight function satisfies the requirements of Equation (2.18) but does not remove the practical problems discussed above for unsteady flows.

The interval T required before the mean value dictated by Equation (2.20) becomes stationary may be sufficiently large that interesting features of the flow are lost (Laufer[13]). Intermittency measurements are an obvious example in this respect. Suppose that a probe is placed near the outer edge of a shear layer, within a transition region, or in the sublayer of wall turbulence. For a steady mean flow, it is clear that the average given by Equation (2.20) is a function of T. Changes in T, while T is small, will produce significant fluctuations in the mean value, but a stationary result would develop as $T\to\infty$. The physical reason for this is that the flow is turbulent for a fraction of the total sample time only. A more rewarding study would select those portions of the sample record where turbulence was present and average this abridged sample. The ratio of the length of the abridged sample to the entire sample would be a direct measure of the intermittency.

In conditional sampling techniques only selected sections of a measured record are utilized in the determination of statistical quantities. Implementation of such techniques requires sophisticated data-reduction equipment. In these studies the form of the weight function is determined from the physical process being investigated. In principle, a triggered signal selects either $f=1$ or $f=0$ depending upon whether or not the local flow satisfies the predetermined condition. In their study of the intermittent outer reaches of wall turbulence, Kovasznay *et al.*[11] used a measure of the local vorticity as the trigger signal. When the local flow contains vorticity above the predetermined level, then the probe signal is admitted to the data processing. Otherwise it is rejected. Kaplan and Laufer[29] used a similar technique. Sufficient signal record is processed to allow the mean value to become stationary.

It is evident that the conditional sampling technique is artificial insofar as arbitrary limits are imposed when defining the trigger signal. Again it is the specification of the weight function that presents the difficulty. Despite this somewhat artificial specification of the conditional mean, it is clear that significantly greater understanding of the flow is possible with these techniques. Much interesting insight into the nature of vortex breakdown in the sublayer of wall turbulence is obtainable and its possible relation to the intermittency at the outer edge of the flow can be determined.

2.3.1. The Mean-Value Equations

Application of the mean-value operator to the momentum equation (2.6a) requires the condition of Equation (2.18). One can proceed to show this by selecting the nonlinear term

$$I\equiv\frac{\partial}{\partial x_j}(v_iv_j)=\frac{\partial}{\partial x_j}[(U_i+u_i)(U_j+u_j)]$$

whence

$$\langle I,f\rangle=\frac{\partial}{\partial x_j}\{\langle U_iU_j,f\rangle+\langle U_iu_j,f\rangle+\langle U_ju_i,f\rangle+\langle u_iu_j,f\rangle\}$$

In general it is clear that, provided Equation (2.18) is enforced and the domain D is sufficiently large for the velocity components U_i and u_i to be uncorrelated, then

$$\langle U_i u_j, f\rangle = 0 \tag{2.21}$$

which produces the result

$$\langle I\rangle = \frac{\partial}{\partial x_j}\{U_i U_j + \langle u_i u_j\rangle\}$$

The term f is now removed from the notation since the specific forms required by Equation (2.18) are invoked. The covariance $\tau_{ij} \equiv \langle u_i u_j\rangle$ will be zero if v_i and v_j are not correlated.

Further discussion of correlation techniques will be delayed until Section 2.4.

The momentum equation for the mean motion is

$$\rho\frac{\partial U_i}{\partial t} + \rho\frac{\partial U_i U_j}{\partial x_j} = \frac{\partial \sigma_{ij}}{\partial x_j} \tag{2.22}$$

where the total stress tensor σ_{ij} has the form

$$\sigma_{ij} = \mu\frac{\partial U_i}{\partial x_j} - \rho\tau_{ij} - \langle P\rangle\delta_{ij}$$

An equation similar to Equation (2.22) can be derived for the actual velocity fluctuations. Equation (2.6a) gives

$$\rho\frac{\partial U_i}{\partial t} + \rho\frac{\partial u_i}{\partial t} + \rho\frac{\partial}{\partial x_j}(U_i U_j + U_i u_j + U_j u_i + u_i u_j) + \frac{\partial\langle P\rangle}{\partial x_i} + \frac{\partial p}{\partial x_i}$$
$$= \frac{\partial}{\partial x_j}\mu\left(\frac{\partial U_i}{\partial x_j} + \frac{\partial u_i}{\partial x_j}\right)$$

or, when Equation (2.22) is acknowledged,

$$\rho\frac{\partial u_i}{\partial t} + \rho\frac{\partial}{\partial x_k}(U_i u_k + U_k u_i) + \frac{\partial p}{\partial x_i} = \frac{\partial}{\partial x_k}\left(\mu\frac{\partial u_i}{\partial x_k} - \rho u_i u_k + \rho\tau_{ij}\right) \tag{2.23}$$

A similar equation can be obtained for the mean vorticity Ω_i from Equation (2.10):

$$\rho\frac{\partial \Omega_i}{\partial t} + \rho U_i\frac{\partial \Omega_i}{\partial x_j} + \rho\left\langle u_j\frac{\partial \omega_i'}{\partial x_j}\right\rangle = \rho\left\langle \omega_j'\frac{\partial u_i}{\partial x_j}\right\rangle + \rho\Omega_j\frac{\partial U_i}{\partial x_j} + \mu\frac{\partial^2 \Omega_i}{\partial x_j\,\partial x_j} \tag{2.24}$$

where the fluctuating vorticity ω_i' is given as

$$\omega_i' = \omega_i - \Omega_i$$

Then it follows that

$$\rho\frac{\partial\omega_i'}{\partial t}+\rho U_j\frac{\partial\omega_i'}{\partial x_j}+\rho u_j\frac{\partial\Omega_i}{\partial x_j}+\rho\left\langle u_j\frac{\partial\omega_i'}{\partial x_j}\right\rangle=\rho\omega_j'\frac{\partial U_i}{\partial x_j}+\rho\Omega_j\frac{\partial U_i}{\partial x_j}+\rho\left\langle\omega_j'\frac{\partial u_i}{\partial x_j}\right\rangle+\mu\frac{\partial^2\omega_i'}{\partial x_j\,\partial x_j} \tag{2.24a}$$

is the equation that describes the vorticity fluctuations.

The equations for the velocity fluctuations, like those for the mean values, contain the covariance $\tau_{ij}=\langle u_iu_j\rangle$. The similar equations for vorticity involve correlations of the velocity components with velocity second derivatives, i.e., with second derivatives of the stress tensor τ_{ij}. An excellent discussion of the role of vorticity dynamics in turbulence is given by Tennekes and Lumley.[15]

The central difficulty in turbulence theory now emerges. The introduction of the Reynolds decomposition and the attendant averaging has created a new and unknown stress tensor τ_{ij}. While essentially arbitrary, the decomposition has some merit from a physical point of view since it highlights the nature of the turbulence transport process. Much of this book deals with methods of modeling the stress tensor τ_{ij}.

As will be shown later, moments of the momentum equation can be taken in order to generate equations for the covariance. These equations again involve triple products of the form $\langle u_iu_ju_k\rangle$ and introduce the closure problem of turbulence.

2.3.2. Some Comments

Equation (2.22) contains some interesting features. For a weight function with the assumed property $\langle U_i, f\rangle = U_i$, the mean-value equation differs from the original equation (2.6) by an additional term associated with the nonlinearity in the inertia term $\partial U_iU_j/\partial x_j$. In general, all covariance terms (with arbitrary function g_i) of the form $\langle u_ig_i\rangle$ arise from determining the mean value of a nonlinear term. For any other form of f than that specified by Equation (2.18), additional terms may also enter the mean-value equation. They would not be physically real in the sense that they contain a dependence upon the form of f.

The term $\partial(v_iv_j)/\partial x_j$ is the only nonlinear one in Equation (2.6) for a constant property fluid. Of course, if v_i and v_j were not correlated or were orthogonal, then the term $\langle u_iu_j\rangle$ would vanish. The fact that v_i and v_j do correlate is responsible for the large differences in the diffusion rates between laminar and turbulent flows. Experiment shows that (except very close to a wall in nonhomogeneous turbulence) τ_{ij} is orders of magnitude greater than the corresponding molecular shear stress.

The quantity $\partial\langle u_i u_j\rangle/\partial x_j$ has been combined with the viscous stress term in Equation (2.22). Doing this has more content than merely a mathematical convenience. The quantity $\langle u_i u_j\rangle$—referred to as the Reynolds stress tensor—expresses the transference of momentum as a result of the correlation between the fluctuations u_i and u_j. It is recognized that the fluid viscosity does not enter into the Reynolds stress term, even though the fluid must be viscous if turbulent motion is to be generated.

There is some analogy with the kinetic theory of gases wherein small "lumps" of fluid transport energy and momentum in a somewhat random fashion. Too close an analogy with kinetic theory, however, is dangerous. Fluid flow is *not* a system of noninteracting particles meandering in space. It is, for our purposes, a continuum wherein all fluid elements are governed by Equations (2.5)–(2.7).

Direct appeal to an analogy with kinetic theory concepts introduces such quantities as the mixing length and attempts to force upon the covariance $\langle u_i u_j\rangle$ a structure incompatible with the expressed governing equations.

2.3.3. On Pressure Fluctuations

So far little mention has been made of the fluctuating pressure field which is of interest in aerodynamic noise studies. Equation (2.9) stated that

$$\nabla^2 P = -\rho\frac{\partial^2 v_i v_j}{\partial x_i\,\partial x_j}$$

from which follows the equation

$$\nabla^2(\langle P\rangle + p) = -\rho\frac{\partial^2}{\partial x_i\,\partial x_j}(U_i U_j + U_i u_j + U_j u_i + u_i u_j)$$

When the condition of Equation (2.18) is imposed upon the mean-value operation, the result

$$\nabla^2\langle P\rangle = -\rho\frac{\partial^2}{\partial x_i\,\partial x_j}(U_i U_j + \langle u_i u_j\rangle) \tag{2.25}$$

follows. If this equation is then subtracted from the above, an expression is produced for the pressure fluctuations in the form

$$\nabla^2 p = -\rho\frac{\partial^2}{\partial x_i\,\partial x_j}(U_i u_j + U_j u_i + u_i u_j - \langle u_i u_j\rangle) \tag{2.26}$$

The wall boundary condition imposed upon p follows from Equation (2.11), namely,

$$\left.\frac{\partial p}{\partial x_i}\right|_{\text{wall}} = \mu\left.\frac{\partial^2 u_i}{\partial x_j\,\partial x_i}\right|_{\text{wall}} \tag{2.26a}$$

Equation (2.26) shows that the pressure fluctuations depend upon quadratic terms in the velocity components and upon the Reynolds stress tensor.

Once the mean velocity field and the Reynolds stress tensor are available, the mean pressure may be obtained from Equation (2.25) by applying Green's theorem:

$$\langle P\rangle = \Gamma(x_i) - \rho \int_D G(x_i;\xi_i)\frac{\partial^2}{\partial\xi_i\,\partial\xi_j}(U_iU_j + \langle u_iu_j\rangle)\,d\xi_i \tag{2.27}$$

In this result, the surface integral term involving the boundary conditions has been represented by the function $\Gamma(x_i)$; $G(x_i;\xi_i)$ denotes the Green's function for the Laplacian operator in domain D.

A similar procedure shows that

$$p = \bar{\Gamma}(x_i) - \rho \int_D G(x_i;\xi_i)\frac{\partial^2}{\partial\xi_i\,\partial\xi_j}(U_ju_i + U_iu_j + u_iu_j - \langle u_iu_j\rangle)\,d\xi_j \tag{2.28}$$

represents the solution for the pressure fluctuation field and shows that the pressure fluctuations at a given point depend upon an integral of the velocity fluctuations throughout the entire domain. Equation (2.28) can be used as the starting point for deriving pressure–velocity correlations. Adopting the notation

$$N(\xi_i) \equiv \frac{\partial^2}{\partial\xi_i\,\partial\xi_j}(U_iu_j + u_ju_i + u_iu_j - \langle u_iu_j\rangle)$$

it follows that

$$\langle u_ip\rangle = \langle u_i\bar{\Gamma}(x_i)\rangle - \rho\int_D G(x_i;\xi_i)\langle u_iN\rangle\,d\xi_i \tag{2.29}$$

for example. Near the wall it is expected (Willmarth and Wooldridge[24]) that the first term in this equation will be small.

It can be shown for isotropic homogeneous turbulence (Panchev[30]) that the pressure does not correlate with the velocity components. In nonhomogeneous turbulence (particularly near a wall), the pressure–velocity correlation is more significant. The wall pressure field is discussed in greater detail by Lilley and Hodgson.[23]

2.3.4. Passage to Statistical Theory

Questions concerning the form of the weight function f that occurs in Equation (2.13) have not been resolved. The only restriction so far imposed upon f is that it should satisfy the relation $\langle U_i, f\rangle \equiv U_i$. There is, however, no reason to suppose that, if such a function exists, it is unique. That at least one such function exists is easily shown, as the above temporal mean indicates. The time interval associated with the temporal mean is again arbitrary (as is the specification of a reference level in the conditional mean). There would be advantages in having one unique specification for the mean value.

Suppose that the physical process is such that an ergodic hypothesis (von Neumann[31] and Birkhoff[32]) is realistic. Now the definition of a mean value in Equation (2.13) (ergodicity is defined in Chapter 3) can be replaced by an equivalent expression in terms of probability theory. In this formulation the weight function f is replaced by a probability density function P such that (Panchev[30] or Monin and Yaglom[28])

$$U_i = \int u_i P \, du_i$$

with the normalization

$$\int P \, du_i \equiv 1$$

The conditions expressed by Equation (2.15) still hold. At the same time, it is evident that

$$\int U_i P \, du_i = U_i \int P \, du_i = U_i$$

and the development of Section 2.3.2 can go through without restrictions of the form contained in Equation (2.18). The theory is self-consistent and devoid of the arbitrary specification of a weight function.

The practical problem reduces to one of specifying the conditions under which the ergodic hypothesis is valid. This is necessary since experiments mostly measure the simple temporal mean and not the ensemble average.

2.4. Correlations and the Closure Condition

Equations (2.22) and (2.19) do not constitute a deterministic system owing to the presence of the Reynolds stress tensor τ_{ij}. The device whereby the velocity is decomposed as $v_i = U_i + u_i$ with $U_i = \langle v_i, f \rangle$ has rendered the original problem [Equations (2.5) and (2.6)] indeterminate. In the sense that sufficient initial data could not be specified, it is true that the original problem was also indeterminate. Now the difficulty resides in the fact that the stress tensor τ_{ij} occurs with no closed system of equations to describe its behavior.

Two principal techniques have been pursued for the establishment of closure conditions: empiricism and the development of exact equations.

While the latter approach is clearly to be preferred, it cannot be carried out in practice. At the present time, all closure conditions have depended upon empiricism or approximation.

In an experimental situation, any two or more functions that correlate could be made the subject for a covariance. This could be achieved by using either the spatial or temporal separation between two records as a parameter when forming a covariance or by utilizing two velocity components at a given point. Two-point velocity correlations, when combined with conditional sampling techniques, can be useful in defining the structure of a turbulent flow (Laufer[13]). In this way the spanwise variations in the wall turbulence field were demonstrated by Gupta *et al.*[7]

The present chapter is restricted to a discussion of one-point correlations only.

The formal development of an equation for the one-point correlation tensor τ_{ij} proceeds from a restatement of Equation (2.23) (Hinze[33]):

$$\rho\frac{\partial u_i}{\partial t}+\rho\frac{\partial}{\partial x_k}(U_i u_k+U_k u_i)+\frac{\partial p}{\partial x_i}=\frac{\partial}{\partial x_k}\left(\mu\frac{\partial u_i}{\partial x_k}-\rho u_i u_k+\rho\tau_{ik}\right) \tag{2.30}$$

along with a similar statement for the velocity component u_j:

$$\rho\frac{\partial u_j}{\partial t}+\rho\frac{\partial}{\partial x_k}(U_j u_k+U_k u_j)+\frac{\partial p}{\partial x_j}=\frac{\partial}{\partial x_k}\left(\mu\frac{\partial u_j}{\partial x_k}-\rho u_j u_k+\rho\tau_{jk}\right) \tag{2.31}$$

Equation (2.30) is first multiplied by u_j and Equation (2.31) by u_i and the results added. Then, with rearrangements of the form

$$\frac{\partial}{\partial x_k}(u_i u_j U_k)=U_k\frac{\partial}{\partial x_k}(u_i u_j)$$

$$\rho u_i\frac{\partial}{\partial x_k}(u_j u_k)+\rho u_j\frac{\partial}{\partial x_k}(u_i u_k)=\rho\frac{\partial}{\partial x_k}(u_i u_j u_k)$$

which are easily verified by differentiation and appeal to continuity, it follows that

$$\begin{aligned}
&\rho\frac{\partial}{\partial t}(u_i u_j)+\rho(u_j u_k)\frac{\partial U_i}{\partial x_k}+\rho(u_i u_k)\frac{\partial U_j}{\partial x_k}+\rho U_k\frac{\partial u_i u_j}{\partial x_k}+\frac{\partial}{\partial x_i}(u_j p)\\
&\quad+\frac{\partial}{\partial x_j}(u_i p)-p\left(\frac{\partial u_j}{\partial x_i}+\frac{\partial u_i}{\partial x_j}\right)\\
&\quad=-\rho\frac{\partial}{\partial x_k}(u_i u_j u_k)+\mu\frac{\partial^2}{\partial x_k\,\partial x_k}(u_i u_j)-2\mu\frac{\partial u_i}{\partial x_k}\frac{\partial u_j}{\partial x_k}\\
&\quad+\rho\left(u_j\frac{\partial}{\partial x_k}\langle u_i u_k\rangle+u_i\frac{\partial}{\partial x_k}\langle u_j u_k\rangle\right)
\end{aligned} \tag{2.32}$$

Applying the mean-value operator to this equation gives the final result:

$$\rho\frac{\partial}{\partial t}\tau_{ij}+\rho\tau_{jk}\frac{\partial U_i}{\partial x_k}+\rho\tau_{ik}\frac{\partial U_j}{\partial x_k}+\rho U_k\frac{\partial\tau_{ij}}{\partial x_k}+\frac{\partial}{\partial x_i}\langle pu_j\rangle+\frac{\partial}{\partial x_j}\langle pu_i\rangle-\left\langle p\frac{\partial u_j}{\partial x_i}\right\rangle-\left\langle p\frac{\partial u_i}{\partial x_j}\right\rangle$$
$$=-\rho\frac{\partial}{\partial x_k}\langle u_iu_ju_k\rangle+\mu\frac{\partial^2}{\partial x_k\partial x_k}\tau_{ij}-2\mu\left\langle\frac{\partial u_i}{\partial x_k}\frac{\partial u_j}{\partial x_k}\right\rangle \quad (2.33)$$

The result of this manipulation, which generated the required equation for τ_{ij}, has introduced additional (unknown) triple correlations of the form $\langle u_iu_ju_k\rangle$. This is again a direct consequence of the nonlinear inertia term in the momentum equation (2.6). At the same time, correlations between velocity components and the pressure field are also introduced. Such correlations were discussed in Section 2.3.3.

The central problem of turbulence—the unresolved indeterminancy of the mean-value equations—is thus clearly brought out.

The straightforward contraction ($i=j$) of Equation (2.33) gives the result

$$\rho\frac{\partial}{\partial t}\left(\frac{q^2}{2}\right)+\rho U_k\frac{\partial}{\partial x_k}\left(\frac{q^2}{2}\right)+\rho\tau_{ik}\frac{\partial U_i}{\partial x_k}+\frac{\partial}{\partial x_k}\left\langle u_k\left(p+\rho\frac{q^2}{2}\right)\right\rangle$$
$$=\mu\frac{\partial^2}{\partial x_k\partial x_k}\left(\frac{q^2}{2}\right)-\mu\left\langle\frac{\partial u_i}{\partial x_k}\frac{\partial u_i}{\partial x_k}\right\rangle \quad (2.34)$$

where $q^2/2\equiv\langle u_1^2+u_2^2+u_3^2\rangle/2$ represents the kinetic energy of the turbulence.

It follows directly from the above development that for each order of moment of the momentum equation, there is a proliferation of higher-order correlations. Hence, to determine the turbulence field exactly would require an infinite set of equations (still unclosed) in terms of an infinite number of correlations. In a practical calculation procedure, it is therefore necessary to resort to empiricism to bring about closure.

There is clearly a complete hierarchy of closure conditions depending upon the level of correlation at which the empiricism is to be applied, and upon the amount of work intended in developing the solution.

The simplest form of closure would approximate the stress tensor directly, and this leads to such ideas as the eddy viscosity or the mixing length. While these are physically attractive, simple ideas they are not capable of describing any but very simple flows to a useful level of approximation.

The eddy viscosity concept makes analogy to fluid molecular viscosity and relates the shear force to the mean velocity gradient in the form

$\langle u_1 u_2 \rangle = \varepsilon\, \partial U_1/\partial x_2$. The eddy viscosity coefficient ε will, in general, be a function of position.

The Reynolds shear stress distribution across a turbulent boundary layer has very much the same general form as the distribution of vorticity across a laminar boundary layer. Hence, the consequences of an eddy viscosity formulation may be of the correct overall form but the physical content is incorrect since it invokes the idea of a turbulent fluid rather than of the turbulent motion of a viscous fluid.

At the next level of approximation, the turbulence kinetic energy equation [Equation (2.25)] is employed. Details of these and other techniques will be treated elsewhere in this book. Launder and Spalding[(34)] and Mellor and Yamada[(35)] have reviewed many aspects of the closure hierarchy. Additional comments are made by Bradshaw[(36)] and by Mellor and Herring.[(37)]

An alternative approach to empirical (or approximation) closure is to adopt a completely numerical solution to the entire problem. Some comments in this respect are made by Emmons[(38)] and by Orszag and Israeli.[(39)] Even in this approach, the closure problem is not absent since much of the energy dissipation occurs in the fine-eddy structure and is beyond the grid resolution.

2.5. The Turbulent Boundary Layer

In many flows of practical interest, it is found that the lateral scale of the viscous region is of smaller order than the streamwise scale. When this occurs (in wall turbulence or in shear layers, for example), it is useful to make the boundary-layer approximation in order to develop equations that are more easily solved than the Navier–Stokes equations. While a large simplification is realized by this technique, there are many unresolved difficulties that arise both formally and practically.

Two approaches are possible in the development of the boundary-layer equations for turbulent flow. The basic equations, (2.5)–(2.7), could be subjected to an analysis on the basis of matched asymptotic expansions followed by an application of the mean-value operator. The alternative procedure would take the mean-value equations [like Equations (2.19) and (2.22)] and then invoke singular perturbation methods to establish the required boundary-layer equations.

In either case certain problems will arise while assessing the magnitude of the fluctuation components. Discussions on the development of the boundary-layer equations in the standard literature usually start with order-of-magnitude statements concerning the quantities concerned (Nash and Patel[(40)]).

2.5.1. Preliminary Remarks

In the present development of the boundary-layer approximation, we choose to treat Equations (2.5)–(2.7) as they stand. This process is initiated by introduction of nondimensional quantities into the equations expressed in terms of a characteristic length and velocity (L and U_∞). Then a new "time" variable $U_\infty t/L$ must be employed. This group has the form of a Strouhal number but does not carry the connotation of a periodic motion within the fluid. Now, if λ is the inverse Reynolds number ν/UL, Equation (2.6) becomes

$$\rho\frac{\partial v_i}{\partial t}+\rho v_j\frac{\partial v_i}{\partial x_j}+\frac{\partial P}{\partial x_i}=\lambda\frac{\partial^2 v_i}{\partial x_j\,\partial x_i} \tag{2.35}$$

where it is understood that all variables are now nondimensional. Equation (2.5) retains the form

$$\frac{\partial v_i}{\partial x_i}=0 \tag{2.36}$$

The procedure is started by writing asymptotic expansions of the form

$$\begin{aligned} v_i &\sim {}_o\varepsilon_1(\lambda)\,{}_ov_i^1+{}_o\varepsilon_2(\lambda)\,{}_ov_i^2+\cdots \\ P &\sim {}_o\gamma_1(\lambda)\,{}_oP^1+{}_o\gamma_2(\lambda)\,{}_oP^2+\cdots \end{aligned} \tag{2.37}$$

where the subprefix o denotes the outer expansion. The expansion functions ${}_ov_i^n$ and ${}_op^n$ are of $O(1)$ (independent of the Reynolds number) while the vector gauge functions ε_n and γ_n contain the Reynolds number dependence of the flow and are subject to the limiting condition

$$\lim_{\lambda\to 0}\varepsilon_n/\varepsilon_{n-1}\to 0$$

on each component.

Introduction of Equation (2.37) into Equations (2.35) and (2.36) produces the result

$$\begin{gathered} \rho\frac{\partial\,{}_ov_i^1}{\partial t}+\rho\,{}_ov_j^1\frac{\partial\,{}_ov_i^1}{\partial x_j}+\frac{\partial\,{}_oP^1}{\partial x_i}=0 \\ \frac{\partial\,{}_ov_i^1}{\partial x_i}=0 \end{gathered} \tag{2.38}$$

in the limit as $\lambda\to 0$. To derive this result (see Van Dyke[41]) appeal must be made to the fact that the functions ${}_ou_i^1$ and ${}_oP^1$ satisfy upstream boundary conditions that are independent of the Reynolds number. This fact implies that each component of ε_1 and γ_1 must satisfy an order relation of the form

$${}_o\varepsilon_1(\lambda)\sim{}_o\gamma_1(\lambda)\sim O(1)$$

Equations (2.38) represent the first-order statement of the outer problem and have somewhat the same content as the potential flow system of equations. It should be realized that Equations (2.38) still pertain to the local flow velocities, and application of the mean-value operator, as in Section 2.3.1, produces the result

$$\rho\frac{\partial\, {}_oU_i^1}{\partial t}+\rho\, {}_oU_j^1\frac{\partial\, {}_oU_i^1}{\partial x_j}+\frac{\partial\, {}_oP^1}{\partial x_i}=-\rho\frac{\partial}{\partial x_j}\, {}_o\langle u_i u_j\rangle^1 \tag{2.38a}$$

The covariance ${}_o\langle u_i u_j\rangle^1$ occurs in this equation for the outer mean flow and adopts the nature of an external force field applied to the motion.

Physically, it is not unrealistic that the term ${}_o\langle u_i u_j\rangle^1$ is present in the outer solution since the fluctuations generated by the shear layer turbulence will propagate into the external flow.* This creates a boundary-value problem for ${}_o\langle u_i u_j\rangle^1$ (Phillips[42]). It would be required, for example, that ${}_o\tau_{ij}^1 \equiv {}_o\langle u_i u_j\rangle^1$ vanish in the far field and match with some specified value at the edge of the boundary layer. If the presence of free-stream turbulence is admitted, then ${}_o\tau_{ij}^1$ would have a prescribed distribution in the far field. Mellor[43] ignores the possible nonzero nature of τ_{ij} in the external flow. An external noise field could also be associated with this term.

The inner matching process may not be possible in a strict singular perturbation sense since the form of the fluctuation field will be significantly different in the two flow regions. It is true that the external surface of a turbulent region is clearly delineated from the "potential" outer flow, so that the fluctuation in the latter would be of a considerably different form. Most certainly, the covariance ${}_o\tau_{ij}^1$ is small in the external flow and negligible for most purposes. The disturbance field in the external flow will be in the nature of aerodynamic noise radiated from the turbulence. Laufer[44] discusses this situation for the flow in a supersonic wind tunnel.

The noise field is more concerned with the pressure fluctuation, which to the same approximation as Equation (2.38), is given by

$$\rho\frac{\partial\, {}_ou_i^1}{\partial t}+\rho\frac{\partial}{\partial x_j}({}_oU_i^1\, {}_ou_j^1+{}_oU_j^1\, {}_ou_i^1)+\frac{\partial\, {}_op^1}{\partial x_j}=-\rho\frac{\partial}{\partial x_j}[{}_ou_i^1\, {}_ou_j^1-{}_o\langle u_i u_j\rangle^1] \tag{2.39}$$

This pressure fluctuation ${}_op^1$ is found to correlate with the wall pressure fluctuation under the turbulent boundary layer.

The above equations relate to the propagation of the noise field, generated by the turbulence, into the external flow (Morse and Ingard[45]). Details of the exact form of the near-field boundary condition to be applied to the solution are still not certain.

* Of course the correlation is zero if the fluctuations are irrotational.

2.5.2. Comments

A formal application of singular perturbation theory, as initiated above, cannot be carried out for a turbulent flow without a good understanding of the turbulence process. The following development is restricted to nonhomogeneous wall turbulence where the viscous region is separated into an outer defect layer and an inner wall layer. The major problems that arise are in assessing the order of the Reynolds stress term in each of these layers and in determining the thickness of the individual layers: The two are, of course, related.

The formulation invoked above is no longer applicable for the inner layers because the turbulence stress is not directly displayed. Without these terms, a match between viscous and inertia terms (in the strained coordinate system) would result in the laminar flow equations. It is vital, therefore, that the Reynolds stress be properly accounted for. Yajnik[(46)] and Mellor[(43)] have made some progress in this direction and have removed many of the heuristic assumptions usually involved in writing the turbulent flow boundary-layer equations.

The subsequent analysis will be based upon the following simple suggestions:

1. The wall region is predominantly conscious of the viscous nature of the flow.

2. The defect region is the domain of the major eddy structure and will be inviscid to first order. The viscous dissipation occurring in the fine eddies would be a higher-order phenomenon.

3. The components of the Reynolds stress tensor would be of the same order throughout the entire shear layer.

4. It is expected that a full statement of the order of magnitude associated with the turbulent structure would involve the closure condition.

The following discussion of the boundary-layer equations will be restricted to a steady, two-dimensional configuration. That the turbulence process is highly three-dimensional (particularly in the smaller scales—Deardorff[(47)]) will not feature in the current treatment.

There are, of course, grave doubts about the physical meaning of an asymptotic expansion for the turbulent stress tensor with the Reynolds number involved in the gauge functions. In crude terms this is equivalent to relating terms at each order to a new scale of eddy size which in turn entertains a different level of energy dissipation.

Usually, in matched asymptotic expansion procedures (Van Dyke[(41)]), the problem is completely determinate in the sense that each successive order of gauge function can be determined by matching either with the boundary conditions or between adjacent layers. In the turbulent shear layer problem, that is not the case since the basic (mean-value) equations are indeterminate.

A complete discussion of the problem cannot be given at this time and the following remarks are intended to be provocative as well as tentative. It is first of all assumed that the turbulent boundary layer can be adequately described by equations derivable by some contraction of the Navier–Stokes equations. This may be a satisfactory assumption for a description of the mean motion but, as already made clear, no closed set of equations is available for a description of the turbulence structure.

The next basic premise is that flow quantities of interest can be expanded by asymptotic sequences with gauge functions that depend only on the Reynolds number of the flow. Suitable orders of magnitude are then selected, along with pertinent scale factors, so that different equations are derived in each region of the flow. It must also be understood, from the structure of the flow, what distinct flow regions need to be recognized.

At first sight it may appear that the buffer layer should be included as part of the viscous sublayer and possibly as a higher-order term. In this way it may be expected that something like Spalding's law of the wall would be recovered. We reject this approach and suppose instead that the buffer layer belongs with the log layer a little further from the wall. The main reason for this is that the lower edge of the buffer layer is the demarcation between the two major regions near the wall—the peak in the turbulence production occurs in this range of nondimensional distance y^+. In the viscous sublayer the flow can be well represented without any knowledge of the Reynolds stress tensor other than its influence upon the wall shear stress. Although the flow is dominated by the viscous nature of the fluid, it should be recognized that the motion is in no sense reminiscent of that in a laminar boundary layer. The presence of a strong streamwise, and unsteady, vorticity component in the sublayer flow is well established.[3]

Above the outer edge of the log layer the flow starts to take on an intermittent character typical of the defect layer. In the present discussion it is supposed that the superlayer separating the defect layer from the main stream should also be treated separately since here the flow is very intermittent with the intermittency factor being of the order of 0.1. This superlayer would again require inclusion of the viscous property of the flow and be able to account for both the entrainment from the free stream into the turbulence structure and the rather strong effect that free-stream turbulence has on the entire flow development.

2.5.3. Further Developments

In the inner, or wall, region, suppose that the flow can be expanded in terms of asymptotic functions and gauge functions dependent upon the

The momentum equation would reduce to

$$\frac{\partial \zeta_{d1}}{\partial x_1}\frac{\partial \psi_{d1}}{\partial y_d}-\frac{\partial \psi_{d1}}{\partial x_1}\frac{\partial \zeta_{d1}}{\partial y_d}=T_{d1} \tag{2.51}$$

and suggests that the Reynolds stresses would contribute to a change in vorticity along streamlines in the defect layer.

The solution would be completed with a matching of the defect-layer solution to the superlayer and hence to the free stream.

In no way are the ideas put forward above complete. They are presented only to indicate the difficulties associated with developing a boundary-layer theory for turbulent flow and indicating the strong need for an empirical input. The calculation procedures laid down in later chapters show some of the ways in which this empirical information can be adopted to create viable calculation methods.

2.6. Final Remarks

This introductory material has attempted to acknowledge past work in turbulence and to direct thinking toward future developments.

Notation

a	Constant
e	Specific internal energy
f	Weight function in definition of mean value
g_i	Arbitrary function
$G(x_i;\xi_i)$	Green's function
L	Characteristic length
P	Instantaneous pressure or probability density function
p	Fluctuating pressure
q^2	Kinetic energy of turbulence
t	Time
T	Temperature or time interval
T	Function of the Reynolds stress tensor $\langle u_1\,\partial\omega'/\partial x_2\rangle$
U_∞	Free-stream velocity
U_i	Mean velocity components
u_i	Fluctuating velocity component ($v_i = U_i + u_i$)
u, v	Velocity components
v_i	Instantaneous velocity components
x, y	Rectangular coordinates
x_i	Coordinate directions
∇^2	Laplacian operator
D/Dt	Substantive derivative
$\langle\,\rangle$	Average value
$(\bar{\ })$	Temporal mean
$\varepsilon_i, \beta_i, \gamma_i$	Gauge functions
$\delta(x)$	The delta function
λ	Inverse Reynolds number
ρ	Density
μ	Viscosity
ψ	Stream function
ζ	Vorticity
σ_{ij}	Stress tensor
τ	Dummy variable for t
ξ_i	Dummy variable for x_i
Ξ	The set (ξ_i, τ)
ω_i	Instantaneous vorticity components
ω_i'	Fluctuating vorticity components
Δ	Length scale
τ_{ij}	Covariance $\langle u_i u_j\rangle$
Ω_i	Mean vorticity
τ_w	Wall shear stress
v^*	Friction velocity

References

1. Jenkins, G. M., and Watts, D. G., *Spectral Analysis and Its Applications*, Holden–Day, San Francisco, California (1968).
2. Gartshore, I. S., An experimental examination of the large-eddy equilibrium hypotheses, *J. Fluid Mech.* **24**, 89–98 (1966).
3. Kline, S. J., Reynolds, W. C., Schraub, F. A., and Runstadler, P. W., The structure of turbulent boundary layers, *J. Fluid Mech.* **30**, 741–773 (1967).
4. Kim, H. T., Kline, S. J., and Reynolds, W. C., The production of turbulence near a smooth wall in a turbulent boundary layer, *J. Fluid Mech.* **50**, 133–160 (1971).
5. Corino, E. R., and Brodkey, R. S., A visual investigation of the wall region in turbulent flow, *J. Fluid Mech.* **37**, 1–30 (1969).
6. Willmarth, W. W., and Lu, S. S., Structure of the Reynolds stress near the wall, *J. Fluid Mech.* **55**, 65–92 (1972).
7. Gupta, A. K., Laufer, J., and Kaplan, R. E., Spatial structure in the viscous sublayer, *J. Fluid Mech.* **50**, 493–512 (1971).
8. Fernholz, H., Three-dimensional disturbances in a two-dimensional incompressible turbulent boundary layer, Aeronautical Research Council, London, R and M No. 3368 (1962).
9. Coles, D., The law of the wake in the turbulent boundary layer, *J. Fluid Mech.* **1**, 191–226 (1956).
10. Fiedler, H., and Head, M. R., Intermittency measurements in the turbulent boundary layer, *J. Fluid Mech.* **25**, 719–735 (1966).
11. Kovasznay, L. S. G., Kibens, V., and Blackwelder, R. F., Large scale motion in the intermittent region of a turbulent boundary layer, *J. Fluid Mech.* **41**, 283–326 (1970).
12. Antonia, R. A., Conditionally sampled measurements near the outer edge of a turbulent boundary layer, *J. Fluid Mech.* **56**, 1–18 (1972).
13. Laufer, J., New trends in experimental turbulence research, in: *Annual Review of Fluid Mechanics*, Vol. 7 (M. Van Dyke, W. G. Vincenti, and J. V. Wehausen, eds.), Annual Review, Inc., Palo Alto, California (1975), pp. 307–326.
14. Davies, P. O. A. L., and Yule , A. J., Coherent structures in turbulence, *J. Fluid Mech.* **69**, 513–537 (1975).
15. Tennekes, H., and Lumley, J. L., *A First Course in Turbulence*, MIT Press, Cambridge, Massachusetts (1972).
16. Shinbrot, M., *Lectures in Fluid Mechanics*, Gordon and Breach, New York (1973).
17. Meyer, R. E., *Introduction to Mathematical Fluid Dynamics*, Wiley-Interscience, New York (1971).
18. Ladyzhenskaya, O. A., *The Mathematical Theory of Viscous Incompressible Flow*, Gordon and Breach, New York (1963).
19. Ladyzhenskaya, O. A., Mathematical analysis of Navier–Stokes equations for incompressible liquids, in: *Annual Review of Fluid Mechanics*, Vol. 7 (M. Van Dyke, W. G. Vincenti, and J. V. Wehausen, eds.), Annual Reviews, Inc., Palo Alto, California (1975), pp. 249–272.
20. Sedov, L. I., *A Course in Continuum Mechanics*, Vols. 1–4, Wolters-Noordhoff, Groningen, The Netherlands (1971).
21. Stewart, R. W., Triple velocity correlations in isotropic turbulence, *Proc. Cambridge Phil. Soc.* **47**, 146–157 (1951).
22. Laufer, J., The structure of turbulence in fully developed pipe flow, NACA report No. TR 1174 (1954).
23. Lilley, G. M., and Hodgson, T. H., On surface pressure fluctuations in turbulent boundary layers, AGARD report No. 276, North Atlantic Treaty Organization, Paris (1960).

24. Willmarth, W. W., and Wooldridge, C. E., Measurements of the fluctuating pressure at the wall beneath a thick turbulent boundary layer, *J. Fluid Mech.* **14**, 187–210 (1962).
25. Wills, J. A. B., Measurement of the wave-number/phase velocity spectrum of wall pressure beneath a turbulent boundary layer, *J. Fluid Mech.* **45**, 65–90 (1971).
26. Willmarth, W. W., Pressure fluctuations beneath turbulent boundary layers, in: *Annual Review of Fluid Mechanics*, Vol. 7 (M. Van Dyke, G. W. Vincenti, and J. V. Wehausen, eds.), Annual Reviews, Inc., Palo Alto, California (1975), pp. 13–38.
27. Reynolds, O., On the dynamical theory of incompressible viscous fluids and the determination of the criterion, *Phil. Trans. R. Soc.* **186**, 123–164 (1883).
28. Monin, A. S., and Yaglom, A. M., *Statistical Fluid Mechanics—Mechanics of Turbulence*, Vol. 1, MIT Press, Cambridge, Massachusetts (1971).
29. Kaplan, R. E., and Laufer, J., The intermittently turbulent region of the boundary layer, in: *Applied Mechanics* (Proceedings of the 12th International Congress of Applied Mechanics) (M. Hetényi and W. G. Vincenti, eds.), Springer-Verlag, Berlin (1969).
30. Panchev, S., *Random Functions and Turbulence*, Pergamon Press, Oxford (1971).
31. Von Neumann, J., Proof of the quasi-ergodic hypothesis, *Proc. Natl. Acad. Sci. USA* **18**, 70–82 (1932).
32. Birkhoff, G. D., Proof of the ergodic theorem, *Proc. Natl. Acad. Sci. USA* **17**, 656–660 (1931).
33. Hinze, J. O., *Turbulence: An Introduction to its Mechanism and Theory*, McGraw-Hill Book Co., New York (1959).
34. Launder, B. E., and Spalding, D. B., *Lectures in Mathematical Models of Turbulence*, Academic Press, New York (1972).
35. Mellor, G. L., and Yamada, T., A hierarchy of turbulent closure models for planetary boundary layers, *J. Atmos. Sci.* **31**, 1791–1806 (1974).
36. Bradshaw, P., The understanding and prediction of turbulent flow, *Aeronaut. J.* **76**, 403–418 (1972).
37. Mellor, G. L., and Herring, H. J., A survey of the mean turbulent field closure models, *AIAA J.* **11**, 590–599 (1973).
38. Emmons, H. W., Critique of numerical modeling of fluid mechanics phenomena, in: *Annual Review of Fluid Mechanics*, Vol. 2 (M. Van Dyke, W. G. Vincenti, and J. V. Wehausen, eds.), Annual Reviews, Inc., Palo Alto, California (1970), pp. 15–36.
39. Orszag, S. A., and Israeli, M., Numerical simulation of viscous incompressible flows, in: *Annual Review of Fluid Mechanics*, Vol. 6 (M. Van Dyke, W. G. Vincenti, and J. V. Wehausen, eds.), Annual Reviews, Inc., Palo Alto, California (1974), pp. 281–318.
40. Nash, J. F., and Patel, V. C., *Three-Dimensional Turbulent Boundary Layers*, SBC Technical Books, Atlanta, Georgia (1972).
41. Van Dyke, M., *Perturbation Methods in Fluid Mechanics*, Academic Press, New York (1964).
42. Phillips, O. M., The maintenance of Reynolds stress in turbulent shear flow, *J. Fluid Mech.* **27**, 131–144 (1967).
43. Mellor, G. L., The large Reynolds number asymptotic theory of turbulent boundary layers, *Int. J. Eng. Sci.* **10**, 851–874 (1972).
44. Laufer, J., Aerodynamic noise in supersonic wind tunnels, *J. Aerosp. Sci.* **28**, 685–692 (1961).
45. Morse, P. M., and Ingard, K. U., *Theoretical Acoustics*, McGraw-Hill Book Co., New York (1968).
46. Yajnik, K. S., Asymptotic theory of turbulent shear flows, *J. Fluid Mech.* **42**, 411–427 (1970).
47. Deardorff, J. W., The use of subgrid transport equations in a three-dimensional model of atmospheric turbulence, *J. Fluids Eng.* **95**, 429–438 (1973).

CHAPTER 3

Statistical Concepts of Turbulence

WALTER FROST and JÜRGEN BITTE

3.1. Basic Physical Model of Turbulence

Statistical models that attempt to describe the physical processes occurring in developed turbulent motion hypothesize the flow to be made up of a tumultuous array of eddies (i.e., disturbances, nonhomogeneities, or, currently, vortex elements) of widely different sizes. The largest eddy sizes are on the order of the dimensions of the expanse of turbulent motion and the smallest are on the order of the dimension across which molecular viscosity can effectively transport momentum and thus shear out velocity gradients.

This wide range of eddy sizes is generated by neighboring-size eddies successively becoming eddies of the next order of smallness by a mechanism of vortex stretching. The process occurs in a cascading fashion with eddies breaking down into smaller eddies and the energy of the over-all flow being transferred to smaller and smaller scales, the smallest scale being reached when the eddies lose energy by the direct action of viscous stresses. It is generally accepted that eddies of significantly different sizes have no direct influence on one another and only eddies of comparable size can exchange energy. Viscosity is found to have little influence on the motion or the structure of the main turbulent motion but does become effective in the final

---

WALTER FROST and JÜRGEN BITTE • The University of Tennessee Space Institute, Tullahoma, Tennessee 37388

dissipation of the turbulent energy, which takes place by the velocity gradients in the small eddies working against viscous stresses.

3.1.1. Vortex Stretching

The concept of vortex stretching is illustrated in Figure 3.1. A fluid element under the influence of a linear strain will be stretched in the direction of the strain and its cross section in a plane perpendicular to the strain will become smaller (Figure 3.1a). If we extend this argument to a vortex element as shown in Figures 3.1b and 3.1c the vortex or eddy in the direction of the strain becomes smaller in cross section while that normal to the rate of strain becomes larger. Tennekes and Lumley[1] have shown that, neglecting viscosity, the time rate of change of the vorticity fluctuations ω_i can be related to the fluctuating rate of strain tensor s_{ij} by

$$\frac{d\omega_i}{dt} = \omega_j s_{ij} \tag{3.1}$$

where

$$s_{ij} = \frac{1}{2}\left(\frac{\partial u_i}{\partial x_j} + \frac{\partial u_j}{\partial x_i}\right) \tag{3.2}$$

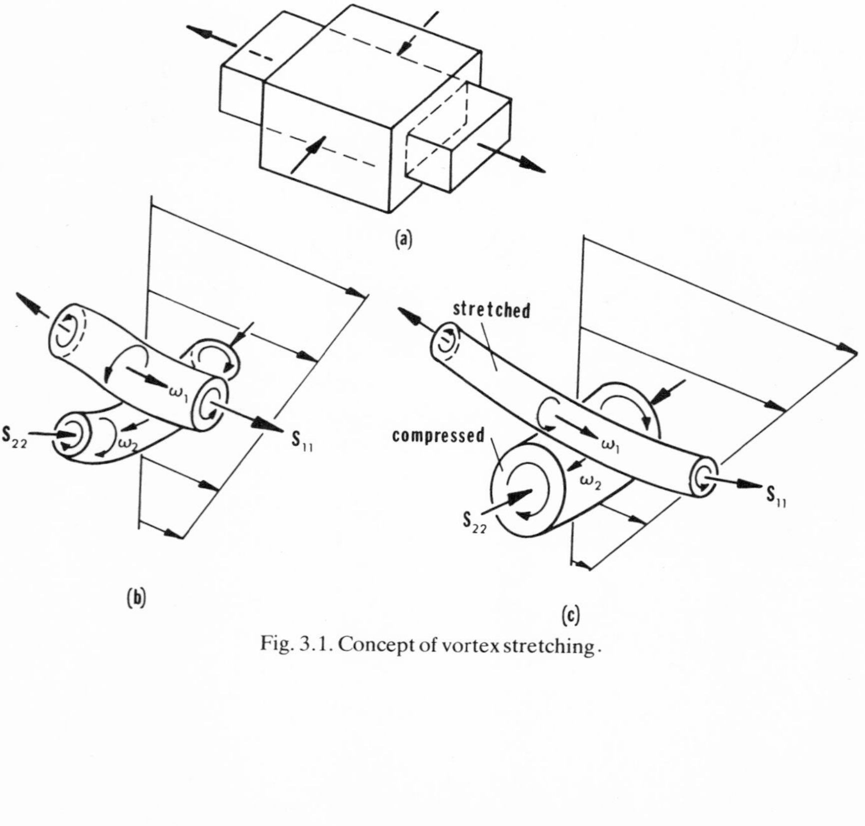

Fig. 3.1. Concept of vortex stretching.

Considering a two-dimensional strain-rate field as depicted in Figure 3.1 and assuming for simplicity $s_{11} = -s_{22} = s$, a constant for all time, and $s_{12} = 0$, then Equation (3.1) becomes

$$\frac{d\omega_1}{dt} = s\omega_1, \qquad \frac{d\omega_2}{dt} = -s\omega_2 \tag{3.3}$$

Integrating gives

$$\omega_1 = \omega_0 e^{st}, \qquad \omega_2 = \omega_0 e^{-st} \tag{3.4}$$

and

$$\omega_1^2 + \omega_2^2 = 2\omega_0^2 \cosh 2st \tag{3.5}$$

The total amount of vorticity thus increases for all positive values of st. The vorticity component in the direction of stretching, ω_1, increases rapidly and that in the direction of compression, ω_2, decreases slowly at large st. This simple calculation illustrates that eddies are stretched at a rapid rate into smaller eddies as previously described while their growth to larger sizes occurs at a much slower rate. Moreover, there are secondary effects associated with the stretching process that counteract the small growth rate of ω_2 as described in the following.

Still neglecting viscosity one can argue from the principle of conservation of angular momentum that the product of the vorticity and the square of the radius must remain constant; or, stated differently, the circulation of the vortex elements must remain constant in the absence of viscous forces during the stretching process. Thus the kinetic energy of rotation increases at the expense of the kinetic energy of the velocity component u_1 that does the stretching. In turn the scale of motion in the x_2, x_3 plane decreases. Therefore an extension in one direction decreases the length scales and increases the velocity components in the other two directions, which stretches other elements of fluid with vorticity components in these directions as illustrated in Figure 3.2. Figure 3.2 shows that two parallel vortices stretched in the x_1 direction result in an increase in u_2 in the positive x_2 direction in an upper plane and in the negative x_2 direction in a lower plane. Thus an increasing strain-rate field is generated, which is experienced by the vortex, ω_2, causing it to stretch. As it stretches, a new strain field is created, stretching in turn other vortices, and so on. This process continues, with the length scale of the augmented motion getting smaller at each stage.

Bradshaw[2] suggests a family tree, shown in Figure 3.3, which demonstrates how the described stretching in the x_1 direction intensifies the motion in the x_2 and x_3 directions, producing smaller-scale stretching in x_2 and x_3 and intensifying the motion in the x_2, x_1, and x_3 directions, respectively. Thus we see qualitatively that the initial stretching in one direction results in

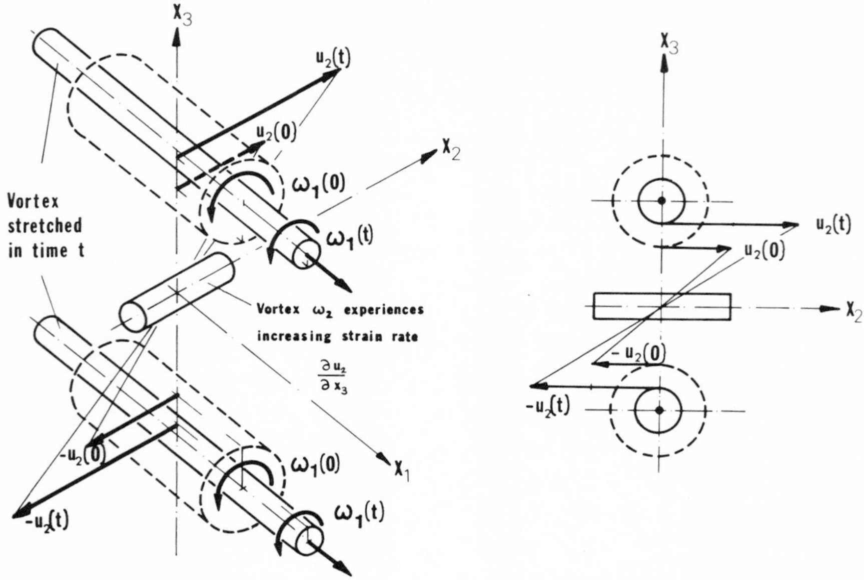

Fig. 3.2. Vortices stretched in the x_1 direction increase the strain rate $\partial u_2/\partial x_3$.

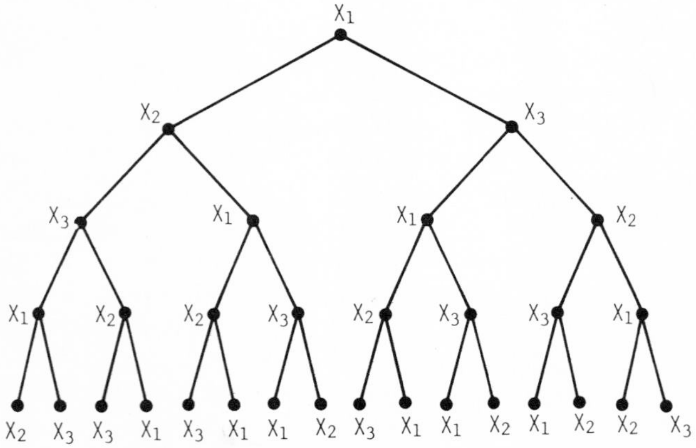

X1	X2	X3
1	0	0
0	1	1
2	1	1
2	3	3
6	5	5
10	11	11
22	21	21

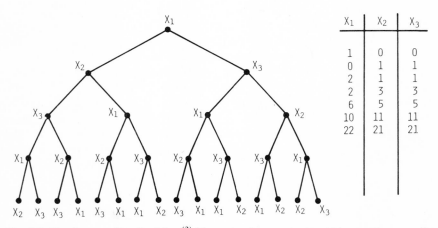

Fig. 3.3. Family tree after Bradshaw[2] illustrating how vortex stretching produces smaller scales. Stretching in the x_1 direction intensifies motion in the x_2 and x_3 directions producing smaller scales, and so on.

progressively more stretching in all three, x_1, x_2, and x_3 directions. Consequently the orienting effect of the mean rate of strain is weakened with each stretching or breaking down of eddies. The small-scale eddies in turbulence therefore tend toward a universal structure that is homogeneous and isotropic despite the fact that the mean flow and large-scale disturbances in any real turbulent flow are nonhomogeneous and anisotropic. Moreover, the characteristic period of the small spatial scales can be expected to be very much smaller than the time of perceptible variation of the mean flow. Hence the regime of fluctuations of sufficiently high order (i.e., sufficiently small spatial scales) will be quasisteady relative to the mean flow. This is the basis for Kolmogorov's theory of a universal equilibrium regime of locally isotropic and statistically steady turbulence.

3.1.2. Energy Cascade

The energy transfer between eddies occurs as the smaller eddies are exposed to the strain-rate field of the larger eddies. The straining increases the vorticity of the smaller eddies with a consequent increase in their energy at the expense of the energy of the larger eddies. Thus a flux of energy from larger to smaller eddies takes place.

The energy increase of smaller eddies during vortex stretching comes from work performed by the strain rate, as can be demonstrated by continuing our simple example from Section 3.1.1 (following Tennekes and Lumley[1]). The amount of energy gained by a disturbance with velocity components u_i and u_j in a strain rate s_{ij} is, from Chapter 2, $-u_iu_js_{ij}$ per unit mass and time. For our strain-rate field of Figure 3.1 the energy exchange rate is

$$T = s(u_2^2 - u_1^2) \tag{3.6}$$

Now as pointed out, when ω_1 increases, u_2 and u_3 increase, and when ω_2 decreases, u_3 and u_1 decrease, with increasing time. Hence T increases from zero at $t=0$ and becomes positive, indicating that the strain rate from the larger eddies performs work to increase the total amount of energy in the smaller-scale vortices.

We indicated in a preceding part of the discussion that only eddies of comparable size exchange energy with one another. To get some feel for the reasoning behind this, we will draw heavily upon an argument presented by Tennekes and Lumley.[1] Consider the energy spectrum illustrated in Figure 3.4. We will later say more about energy spectra; however, at present let us simply state that an eddy of size l is associated with the reciprocal of the wave number κ (i.e., a small wave number indicates a large eddy size and a large wave number, a small eddy size). The energy spectrum gives us the amount of energy contained in a wave-number range from κ to $\kappa - d\kappa$.

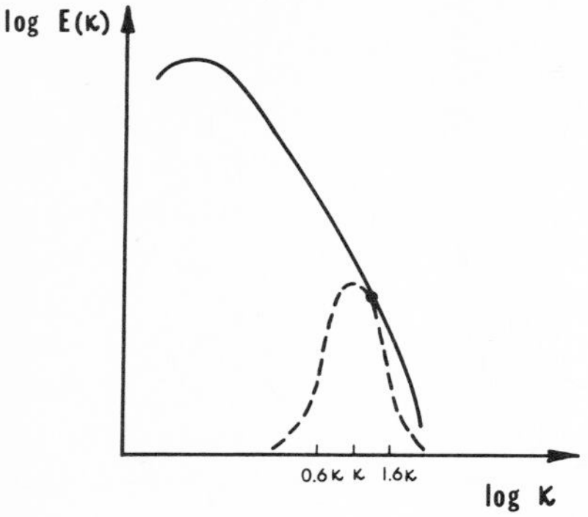

Fig. 3.4. Eddy on the energy spectrum appears as a disturbance containing energy spread over a range of κ.

The eddy on the energy spectrum does not appear as a discrete value of κ but as a disturbance containing energy spread over a range of κ. Say the energy of an eddy of interest is distributed over the range $0.62\text{–}1.62\kappa$ as illustrated in Figure 3.4, which centers the energy around κ on a logarithmic scale. For this arrangement the eddy size l can be shown roughly equal to $2\pi/\kappa$.

The energy associated with the eddy of size $l = 2\pi/\kappa$ is approximately $\kappa E(\kappa)$ and hence a characteristic velocity for the eddy is $[\kappa E(\kappa)]^{1/2}$. The characteristic strain rate of the eddy is then

$$s(\kappa) = [\kappa E(\kappa)]^{1/2}(1/l) = [\kappa^3 E(\kappa)]^{1/2}(1/2\pi) \tag{3.7}$$

It can be shown that $s(\kappa)$ increases with wave number in a manner proportional to $\kappa^{2/3}$.

Imagining the spectrum to be made up of eddies of discrete rather than continuous size, the strain rate imposed on eddies of wave number κ due to eddies of the next larger size (centered about 0.38κ) is on the order of

$$s(0.38\kappa)/s(\kappa) \propto [(0.38\kappa)/\kappa]^{2/3} \simeq 0.5 \tag{3.8}$$

The strain rate of eddies approximately two sizes larger (centered around 0.15κ) is on the order of

$$s(0.15\kappa)/s(\kappa) \simeq 0.25 \tag{3.9}$$

Thus of the total strain rate felt by an eddy of size κ, one-half comes from eddies one size larger, one-quarter from eddies two sizes larger, and so on. This leads one to believe that most of the energy transfer is between eddies of neighboring sizes.

Similar arguments lead to the conclusion that eddies that benefit most during energy transfer from the eddies higher in scale are also immediately neighboring eddies (i.e., on a comparative basis eddies of wave number κ

receive about two-thirds of the total energy from the larger-size eddies, while eddies of the next smaller size receive only one-sixth, etc.).

The cascade of energy of the turbulent motion thus takes place from neighboring eddies to neighboring eddies continuing to smaller and smaller scales (larger and larger velocity gradients) until viscosity finally dissipates the energy received by the smaller eddies. The viscosity, however, does not play an essential part in the stretching process itself.

The preceding discussion suggests that a physical model of turbulence is a tangle of vortex lines that are stretched in a preferred direction by the mean flow and in a random direction by one another. The turbulence is therefore highly three dimensional regardless of the directional dependence of the mean flow. The vortex stretching in all directions transfers the mean-flow kinetic energy to smaller and smaller eddies until it is converted to internal thermal energy by viscous action on the smaller-size eddies.

This chaotic velocity field of elementary vortex lines with energy cascading from one eddy size to another cannot be described by an explicit mathematical relationship because each observation of the phenomenon will be unique. In other words, any given observation of the turbulence will represent only one of many possible results that might have occurred. Thus in analyzing the turbulence we are led to statistical approaches that do not enable us to predict a unique result but do, in some cases, allow us to prescribe the probability of the result occurring. As a matter of fact randomness in space and time is a characteristic often prescribed to turbulent flow. The term randomness carries with it the connotation that statistically distinct average values of the fluid parameters of interest can be discerned even though the motion is highly irregular. The instantaneous velocity of turbulence in its most general form can be said to define a random field, and we shall describe and explain the characteristics of this field in what follows.

3.2. Statistical Definitions

Consider the following hypothetical experiment, which will lead to the establishment of certain required concepts and definitions and finally to the explanation of a random vector field: A flow nozzle with a pressurized stilling chamber as sketched in Figure 3.5 is equipped with M velocity probes, which measure the three orthogonal components of velocity U_1, U_2, and U_3. An experiment is run by opening a valve and allowing the fluid to flow through the nozzle. The probes, placed at various locations, continuously record the fluctuating velocities at these locations. The record of each probe is referred to as a sample function or a sample record. After the flow

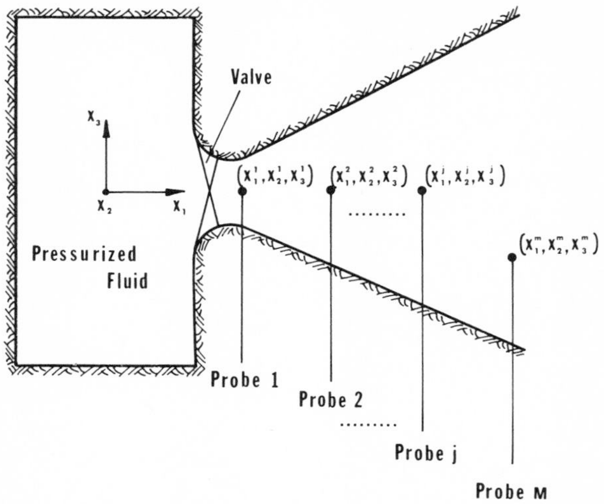

Fig. 3.5. Hypothetical flow nozzle experiment.

has come to rest the flow nozzle is repressurized, the nozzle surroundings are restored to exactly the same conditions as for the previous experiment, and the run is repeated. New sample records or functions are obtained. Comparison of the sample velocity records obtained from any one of the probes during the two runs will reveal that although certain characteristics are similar the details of the record are in general different. If we repeat the process described above a great number of times we will have for each probe a collection of sample records such as illustrated in Figure 3.6. This collection of sample functions is called an ensemble of sample functions. Generally, as illustrated, the plotted data are in the form of digitized values of the analog probe output.

Assume that we now have a very large number, N, of sample functions from all M probes, which were obtained by identically repeating the experiment N times. Ideally we would require N to be infinite, but obviously in practice this is impossible, as is perhaps even the identical repetition of the experiment; however, this will not concern us for the moment.

From our ensemble of sample records lets select for initial examination the record of U_1 generated by probe 1 at position (x_1^1, x_2^1, x_3^1). To aid in visualizing the form of random output, we have compiled Table 3.1, which tabulates 0.1-sec values of random numbers having an χ^2 distribution varying with time. This set of data is artificial but has the characteristic expected of the proposed experiment and serves to illustrate the following

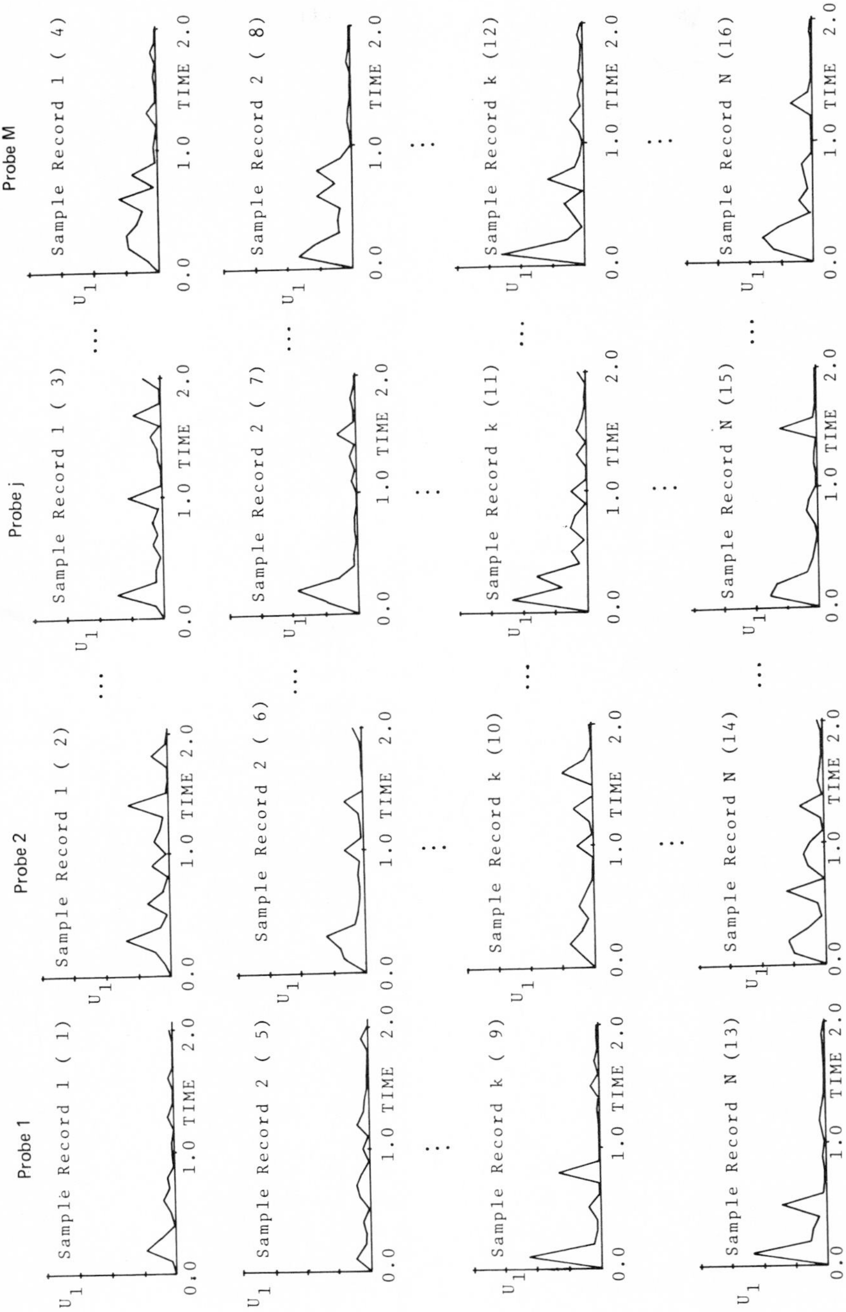

Fig. 3.6. Ensemble of sample functions having a chi-square distribution with time.

Table 3.1. Random Numbers Having a Chi-Square Distribution Varying with Time

SAMPLE RECORD	***TIME*** 0.1000	0.2000	0.3000	0.4000	0.5000	0.6000	0.7000	0.8000	0.9000	1.0000
1	5.2076E-01	5.5514E 00	2.6566E 00	2.5642E-01	9.2306E-01	2.2958E 00	1.0459E 00	1.2767E 00	2.3820E-02	4.6939E-01
2	9.4219E-01	2.8801E 00	8.1132E 00	1.8203E 00	5.7239E-01	4.0793E 00	1.1392E 00	2.6969E-01	3.1097E 00	6.8249E-01
3	1.3849E 00	8.3249E 00	1.3249E 00	1.2924E 00	3.6300E-01	1.8840E 00	7.3791E-01	1.9066E 00	8.2755E-01	6.1888E 00
4	1.9386E 00	5.3970E 00	5.8540E 00	3.7443E 00	2.8171E 00	6.7715E 00	8.4683E-01	4.5666E 00	3.5427E-01	7.1305E-01
5	2.6959E 00	9.9500E-01	7.3321E-01	1.5241E 00	8.0558E-02	1.9543E 00	2.6430E 00	1.4239E 00	4.2462E-03	1.1938E 00
6	3.8079E 00	4.7049E 00	7.2801E 00	1.8486E 00	1.1503E 00	7.3278E-01	7.6108E-01	1.0771E 00	7.8329E-01	3.4887E 00
7	5.6433E 00	1.1011E 01	3.4435E 00	6.4656E-01	6.2555E-01	9.5686E-02	4.6844E-01	5.1161E-01	7.3330E-02	1.6145E-02
8	9.7135E 00	6.8682E 00	2.3580E 00	2.5816E 00	2.2573E 00	6.1076E 00	3.0832E 00	6.1042E 00	1.8187E 00	4.1600E-02
9	1.3484E 01	1.4097E 00	5.2180E-01	6.0189E-01	2.0479E 00	6.0423E-02	1.2130E-02	7.3997E 00	9.1347E-02	5.7543E-03
10	2.3207E 00	4.6261E 00	2.2521E 00	1.1125E 00	2.8248E 00	1.6453E 00	4.6593E-01	5.6978E-02	3.1664E-04	2.8341E 00
11	1.3902E 01	4.5741E 00	9.1630E 00	1.3831E 00	2.9533E 00	5.6804E-01	2.9606E 00	2.0976E 00	1.5004E-01	2.7463E 00
12	1.5154E 01	3.3121E 00	2.9659E-01	1.7928E 00	3.4447E 00	3.1392E-01	6.7629E 00	1.7198E 00	5.0685E-01	9.7509E-02
13	1.4261E 01	3.1245E 00	2.5399E 00	1.5223E 00	8.3091E 00	5.3254E-01	1.4187E-01	2.2876E-01	6.6571E-01	2.0450E-01
14	5.8278E 00	7.0525E 00	3.3539E 00	6.4421E-01	1.4382E 00	6.7554E 00	9.8054E-02	2.8677E 00	3.8744E 00	2.4936E 00
15	9.0871E 00	7.6948E 00	2.0920E 00	1.0430E 00	4.9418E-01	3.3705E-02	3.5913E-01	1.9262E 00	1.5271E 00	4.8611E-02
16	7.4987E 00	9.4645E 00	6.0500E 00	5.6677E-01	2.3932E 00	8.2685E-01	1.5128E 00	2.0907E 00	7.2832E-03	9.1944E-03
17	1.1078E 01	6.2920E-01	4.1767E 00	9.7456E-01	4.9112E-01	3.5682E-01	4.4580E-01	1.8931E 00	5.4343E-01	2.5232E 00
18	2.3391E 01	9.4745E 00	2.6296E 00	8.8822E-01	5.1067E-02	1.3507E 00	1.4659E-03	5.1933E-01	6.3277E-01	8.3397E-01
19	4.9717E 00	3.1893E 00	2.8363E 00	1.0788E-01	4.2669E-01	6.2130E-01	4.8409E 00	1.0261E 00	1.3599E 00	2.0011E 00
20	1.2170E 01	7.2852E 00	4.5462E 00	4.8291E 00	5.2037E 00	4.5157E 00	1.3819E 00	1.6350E 00	7.2737E-01	1.5619E-01
21	8.2291E 00	7.8488E 00	4.9325E 00	1.9677E 01	3.4999E 00	7.3003E-01	3.9593E-01	4.1803E 00	2.4061E 00	1.0911E-01
22	1.3998E 01	7.5089E 00	8.9804E 00	7.0699E-01	1.9540E 00	3.4772E-02	7.4796E-02	6.2186E-05	1.4270E-02	1.5466E 00
23	8.4889E 00	3.4588E 00	4.0058E 00	6.8838E-01	7.3130E-01	1.9494E 00	2.0013E 00	7.5843E-01	1.1341E 00	2.5920E-01
24	4.3788E 00	9.0112E 00	2.6944E 00	1.3229E 00	5.0878E-01	3.2677E 00	1.2003E 00	9.1674E-02	4.0079E-01	1.0907E-01
25	1.1211E 01	6.6254E-01	4.7522E 00	1.8137E 00	1.7935E 00	3.1934E 00	2.5156E 00	1.2715E-01	4.6443E-01	3.7992E-03
26	7.3615E 00	1.3346E 00	8.5335E-01	2.3096E 00	1.5111E 00	7.8130E-01	8.0120E-01	1.5815E 00	2.9629E 00	4.4817E 00
27	1.7046E 01	2.8438E 00	5.2353E 00	4.7781E-01	3.8328E 00	3.2485E 00	1.0106E 00	1.1044E 00	1.5073E-01	2.0809E-01
28	1.3025E 01	9.9921E-01	4.5413E 00	1.9099E 00	8.4500E-01	7.9444E-01	4.3107E 00	1.4838E-01	4.3954E-01	3.0452E-01
29	7.5440E 00	6.5173E 00	1.2299E 00	5.0633E 00	9.9429E-01	3.0915E 00	1.3326E 00	1.0032E 00	2.6676E-01	2.0393E-01
30	1.9115E 01	1.6807E 00	2.5323E 00	3.1524E 00	5.7264E-01	4.8542E-01	5.7615E-01	1.4424E 00	1.0712E-02	1.6991E-01
31	1.3987E 01	1.3998E 01	1.2110E 01	1.2886E 00	6.9178E 00	3.5107E-01	1.6691E 00	1.9952E 00	1.0565E-01	1.5301E 00
32	7.3557E 00	1.0108E 01	4.1331E 00	3.3935E 00	1.6870E 00	2.2596E 00	1.1801E 00	4.5777E-01	1.4321E 00	1.8356E 00
33	8.4378E 00	6.1402E 00	1.7990E 00	7.2446E-01	2.4244E 00	7.1622E-01	3.0516E-01	2.9228E-01	5.4584E-01	2.5763E 00
34	1.1879E 01	9.0737E 00	3.2792E 00	2.0277E 00	1.8204E-01	1.5708E 00	2.4951E-01	1.7873E-01	2.3788E-01	3.2696E 00
35	9.5048E 00	1.9210E 01	3.5586E 00	1.2659E 00	4.4595E-01	4.4219E-01	8.2945E 00	8.9851E-01	5.6441E-01	5.7686E-01
36	1.2222E 01	1.0863E 01	2.4509E 00	1.2440E 00	1.0750E 00	1.8848E-02	2.6425E 00	2.6644E-01	9.4792E-03	3.8360E 00
37	1.2412E 01	5.9582E 00	2.6027E-01	5.3996E 00	2.3826E 00	1.7707E 00	3.3255E-02	1.1299E 00	1.8148E 00	1.2024E 00
38	1.5477E 01	5.3563E 00	2.6849E-01	8.2652E-01	1.1417E 00	2.6360E 00	3.0811E 00	5.3703E-02	1.6181E-01	1.7547E 00
39	1.0871E 01	4.7196E 00	8.1570E 00	1.1474E 00	8.4258E-01	2.4417E 00	1.5823E 00	4.7160E 00	4.2312E-01	1.9799E-01
40	1.4173E 01	5.6636E 00	2.5542E 00	4.4473E 00	3.2004E-01	1.0210E 00	3.5881E-01	1.4570E 00	1.0203E-01	1.2134E 00
41	7.4505E 00	7.0174E-01	1.8926E 00	4.6889E-01	3.1540E 00	8.8177E-03	1.2711E 00	1.0990E 00	1.9918E-01	1.3141E-01
42	8.2278E 00	2.1115E 00	5.1077E 00	6.7873E-01	1.1214E 00	1.9031E 00	1.2481E-02	4.5936E-01	3.0265E 00	2.8500E-01
43	9.3633E 00	8.0184E 00	4.0324E 00	2.0085E-03	1.1405E 00	2.0428E-02	1.1738E 00	1.9333E 00	1.3191E 00	1.0770E-01
44	9.0285E 00	4.8851E 00	9.1247E 00	1.0215E 00	2.7260E 00	5.2541E 00	3.8715E-01	6.1192E-03	6.4335E-02	1.8971E-01
45	7.3276E 00	7.5473E 00	2.8470E 00	3.3359E 00	1.6118E 00	9.9751E-01	8.9944E-01	6.2231E-02	6.0894E-01	1.0984E 00
46	1.0471E 01	6.9099E 00	2.5969E 00	4.0863E 00	1.9493E 00	6.1121E-01	3.3776E-01	6.4302E-01	1.0350E 00	6.5161E-02
47	1.4439E 01	2.8857E 00	1.0285E 00	8.6465E-01	2.8181E 00	2.3625E 00	3.2089E-02	1.9752E 00	1.1675E-01	2.1789E 00
48	1.1473E 01	6.5282E 00	3.4583E 00	5.6770E-01	5.2962E 00	8.4681E-01	1.9980E 00	3.4925E-01	4.6366E-01	2.0341E-01
49	8.8579E 00	9.5009E 00	1.5876E 00	3.3164E 00	2.9844E 00	2.4881E 00	5.4167E-03	4.8422E-05	1.3157E 00	2.0206E-01
50	1.0751E 01	7.0018E 00	7.2902E-01	1.2580E 01	6.7764E-01	2.3359E 00	3.6570E-01	2.5512E 00	6.1661E-02	3.8380E 00

TIME

SAMPLE RECORD	1.1000	1.2000	1.3000	1.4000	1.5000	1.6000	1.7000	1.8000	1.9000	2.0000
1	5.4653E-01	9.5396E-02	1.0460E 00	1.8996E-01	2.3300E-03	8.9816E-01	1.7307E-03	2.2902E-01	9.1123E-02	6.2948E-01
2	2.6958E 00	1.4535E 00	9.6931E-01	7.4221E 00	4.8467E-01	8.7485E-02	2.5964E-05	2.8660E 00	5.0176E-01	9.3214E-04
3	9.4742E-02	2.5327E-03	4.7138E-01	7.5784E-01	1.8308E 00	8.2758E-02	4.6538E 00	5.6916E-02	1.0482E-07	2.9772E 00
4	3.1670E-01	4.3837E-03	1.8242E 00	9.2480E-02	1.1225E-01	5.8825E-01	1.9321E-01	1.0648E 00	9.9956E-02	1.7421E-02
5	1.0436E-01	2.3071E 00	1.0652E 00	7.3000E-01	2.0482E-01	3.0136E-01	7.1846E-02	2.7108E-02	1.2108E 00	1.5157E-02
6	4.8557E-01	5.1624E-01	1.1471E 00	3.5271E 00	2.5068E-02	4.6598E-03	2.1151E-01	9.2790E-02	4.4364E-01	1.8238E 00
7	9.3301E-01	3.0897E-01	1.0947E 00	6.9085E-02	3.2039E 00	3.8175E-01	3.3097E-01	8.0626E-01	9.0278E-02	2.0030E-01
8	5.3675E-02	5.4160E-01	4.2897E-01	7.5686E-02	1.0447E-01	3.3598E-02	6.3162E-01	1.7354E-01	1.0340E-01	2.0289E-02
9	6.4368E-04	2.3909E-02	3.3482E-01	5.9046E-04	1.4843E 00	2.0698E-03	6.8603E-01	2.0775E-03	2.8949E-01	2.1756E-01
10	6.0939E-02	5.5411E-01	3.4099E 00	3.4400E-02	1.2503E-01	5.4703E 00	1.3254E 00	3.6615E-07	2.7942E-02	1.5943E-01
11	1.1983E-02	2.1418E-03	1.7258E 00	1.5161E-05	1.8053E 00	8.2172E-01	5.0760E-02	2.5410E-01	2.2104E-02	1.2772E 00
12	6.1090E-01	2.2418E 00	3.9937E-01	8.4082E-01	9.4422E-02	5.9390E-01	1.3667E-01	1.3674E-01	7.5956E-01	6.9215E-05
13	6.5327E-01	1.1806E 00	7.1345E-01	5.4370E-02	8.8989E-02	4.5819E-02	2.8172E-01	2.9700E-01	5.3963E-01	1.7523E-01
14	1.0044E-01	5.0797E-01	4.0779E 00	8.5253E-03	8.1507E-01	3.4485E-01	2.5689E-03	8.2056E-04	1.2262E-02	6.3052E-01
15	3.0942E-02	5.4562E-01	4.9771E-01	1.3754E-02	6.8512E 00	4.0005E-01	1.2588E-01	7.3456E-05	1.4984E-05	2.3491E-02
16	2.8379E-01	9.2573E-02	3.9597E 00	6.6575E-01	8.5712E-03	2.8305E-02	1.4479E-01	8.8620E-03	3.9658E-01	9.9699E-04
17	2.3802E-02	1.6138E-01	4.3640E-02	2.0298E-01	3.1581E-03	3.2546E 00	3.8294E-01	1.5405E 00	4.6856E-01	4.4941E-01
18	2.4321E 00	3.7332E-01	1.4387E 00	4.4217E-01	4.5725E-03	1.9090E-03	5.5235E-02	8.9480E-04	6.6926E-05	1.2966E 00
19	2.7189E-04	8.9980E-01	6.1738E-01	9.0178E-02	6.4613E-01	8.8723E-04	6.4965E-02	6.6924E-01	2.6793E 00	5.0077E-01
20	3.5703E-03	3.2309E-01	2.6594E-01	4.6566E-01	2.7471E 00	1.1918E-01	2.8532E 00	5.8814E-02	6.2722E-05	1.0923E-01
21	4.6783E-02	5.0817E-01	8.7084E-04	1.8773E-01	2.3170E-02	1.2077E 00	5.3288E-04	2.4668E-01	9.6811E-01	6.8270E-02
22	7.5510E-01	1.8451E-04	1.5838E 00	9.1249E-03	3.7314E-05	2.1181E 00	1.3603E-10	1.6463E 00	2.4949E-02	4.7794E-04
23	3.1393E-03	1.3653E 00	3.7572E-01	8.4938E-01	1.9953E 00	4.0174E 00	9.5349E-04	2.8272E 00	1.9901E-01	3.3394E-03
24	4.2451E-01	5.3194E-01	9.9525E-01	7.6974E 00	5.4194E-03	2.9715E-01	4.8008E-01	3.2604E-01	4.7176E-02	2.5641E-01
25	6.1071E-01	1.7146E-01	2.3775E-01	3.9791E-02	1.0497E 00	8.3479E-01	7.4866E-04	1.2873E-02	9.5820E-04	1.0799E 00
26	2.4401E 00	3.3815E 00	1.3663E-03	6.1939E-01	9.3904E-02	2.8840E-02	6.7187E-02	8.7029E-01	2.5127E-03	2.0070E-92
27	4.1486E 00	1.2538E-02	2.4934E-03	4.4349E-01	9.9152E-03	1.0061E 00	1.0231E-01	4.4665E-02	8.4344E-01	2.7085E)02
28	1.0480E 00	6.9557E-02	2.2041E 00	3.3090E-01	2.6251E-01	5.5229E-03	3.8387E-04	2.0714E-05	1.2017E-03	2.4676E-04
29	8.0384E-01	2.1065E-01	5.4053E-01	1.2532E 00	3.2938E-01	9.7881E-01	6.7352E-02	9.2885E-01	2.6326E-01	1.4249E-01
30	3.2454E-01	4.7105E-03	7.1786E-01	1.2267E 00	3.0917E-01	1.0837E-02	3.4023E-01	1.0925E-01	3.3458E-02	1.2118E 00
31	1.9798E 00	7.9705E-01	1.9067E-01	2.5713E-01	6.4676E-03	6.7619E-04	2.4239E-03	2.1313E-01	1.7073E-01	2.2225E)02
32	2.8035E 00	9.6376E-02	1.9561E-01	1.1594E 00	9.5304E-01	2.8403E-01	5.3474E-03	2.6705E-04	2.8270E-06	9.3840E-03
33	1.3530E 00	8.1127E-01	7.1979E-01	4.1889E-03	1.3404E-02	5.3070E-01	7.9410E-01	2.2115E-02	5.9558E-05	9.8672E-01
34	2.5190E-01	1.0308E-01	5.2331E-03	2.3681E-01	1.4179E-02	1.6220E 00	2.1239E 00	6.3509E-05	7.5372E-04	8.1752E)02
35	4.2043E-01	8.7333E-01	2.1414E-01	2.6847E 00	8.8625E-02	8.9069E-05	1.7429E-01	1.4195E-01	2.6165E-03	1.4053E-02
36	1.2714E-01	1.1871E-01	9.6161E-01	1.1091E-04	1.0942E-01	4.8598E-02	1.4203E 00	2.1226E-01	1.6640E 00	7.7160E-02
37	4.8260E-02	7.6387E-01	8.7838E-02	1.0699E 00	5.3495E 00	4.0337E-01	8.7068E-02	7.0870E-01	4.2668E-01	3.7813R-06
38	2.0600E-01	9.2274E 00	4.2113E-02	1.4227E-01	5.6367E-03	7.2670E-01	8.1283E-01	5.4308E-02	6.8612E-05	1.2674E 00
39	3.4108E-01	2.0319E 00	2.8628E-01	1.8975E-01	1.6731E-04	5.0176E-01	6.8920E-01	9.2881E-02	1.4765E-04	1.5425E 00
40	1.8812E-01	7.1319E-02	1.1934E 00	4.1245E 00	1.5871E 00	5.2345E-02	1.2177E-01	2.0528E-03	1.3174E 00	9.5631E-02
41	3.0618E 00	1.2029E 00	9.8045E-03	3.8251E-01	9.4159E-01	3.7767E 00	4.1519E-04	6.8733E-01	6.5381E-02	2.1022E 00
42	4.8986E-02	1.6405E-02	8.8287E-01	1.0640E 00	1.4355E 00	3.9113E-01	7.1855E-03	1.0478E-03	7.2026E-01	4.0731E-07
43	2.2132E 00	2.8133E 00	2.7910E-03	1.1226E-01	1.2945E-03	5.4310E-04	9.3117E-02	1.6256E 00	4.3204E-01	3.9267E-01
44	2.2361E-01	2.9106E-02	3.0103E 00	4.2466E-02	1.8359E 00	7.1300E-01	7.5163E-01	2.0092E-03	1.1209E-03	1.3028E-06
45	3.9705E-01	1.4533E 00	1.5623E 00	5.0310E-03	1.7026E-01	5.7221E-01	2.7062E-01	1.6151E 00	1.5376E-01	8.5959E-01
46	2.6329E-01	1.6853E-02	4.2596E-01	3.1820E-02	1.5279E 00	1.9489E-01	4.0371E-01	1.8681E 00	2.2188E 00	3.0029E-01
47	1.3847E-06	5.0201E-01	2.5601E 00	7.1191E-02	1.3508E 00	9.3437E-01	2.5906E-01	6.0496E-02	2.8201E-01	1.5982E-01
48	1.6530E-02	1.8534E-02	4.2873E-02	8.0151E-01	3.3899E-02	4.5311E-06	2.2491E-01	1.1927E-02	4.4924E-02	6.3233E-01
49	1.1696E 00	6.4911E-01	3.3298E-02	8.5952E-02	2.4941E-03	3.5780E 00	3.8388E-04	1.1428E 00	3.0688E-01	1.5957E-01
50	1.5055E-01	2.1007E-01	3.0965E-03	3.2132E-01	4.3752E-03	3.9897E-01	6.3182E-02	4.1727E-01	1.0675E-02	1.1331E-02

developments. Figure 3.6 is plotted from these data and illustrates the behavior of 0.1-sec averages (digitized for 0.1-sec intervals). In Figure 3.6 the numbers in parentheses correspond to the numbered entries in Table 3.1. At a given time t_1, measured from the beginning of each experiment, we can find a value of $(U_1)_k$ from each record k and average these to obtain

$$\langle U_1 \rangle = \frac{1}{N} \sum_{k=1}^{N} (U_1)_k \tag{3.10}$$

This is called the ensemble average. More rigorously, the true ensemble average assumes the limit

$$\langle U_1 \rangle = \lim_{N \to \infty} \frac{1}{N} \sum_{k=1}^{N} (U_1)_k \tag{3.11}$$

Note that had we chosen the records from a different probe or had we chosen a different time t, the $\langle U_1 \rangle$ would in general have a different value. Thus $\langle U_1 \rangle$ is a function of position in the flow field and of time and we will designate it as $\langle U_1(\mathbf{r}, t) \rangle$. The symbol $\mathbf{r}$ designates the position vector.

The N sample records allow us also, for a given time and position, to determine a statistical distribution of the variable U_1 $(\mathbf{r}^1, t_1)$ by plotting, as

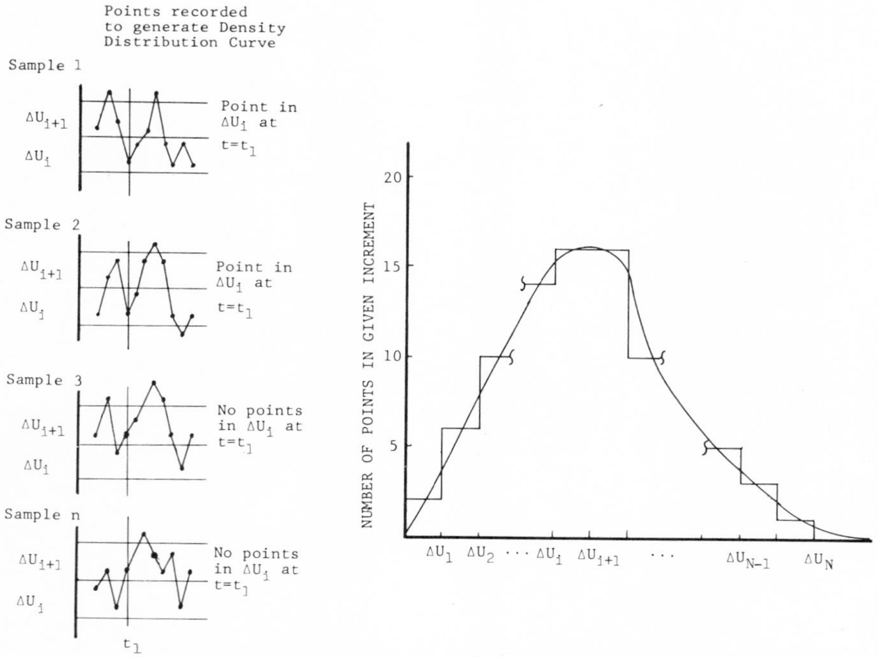

Fig. 3.7. Probability density function.

illustrated in Figure 3.7, the percentage of the N records for which U_1 measured by the probe at $\mathbf{r}^1$ and at time t_1 has a value between $U_1(\mathbf{r}^1, t_1)$ and $U_1(\mathbf{r}^1, t_1)+\Delta U_1$. As long as we consider only one fixed position and one time, the value of $U_1(\mathbf{r}^1, t_1)$, which varies randomly from record to record, behaves like any random variable of classical statistics. Hence $U_1(\mathbf{r}^1, t_1)$ is a random variable and the curve that would result in Figure 3.7 by taking the limit as $\Delta U_1 \to 0$ represents the continuous density distribution function or the probability density denoted by $f(U_1(\mathbf{r}^1, t_1))$. This function allows us to predict the probability of $U_1(\mathbf{r}^1, t_1)$ having a prescribed velocity between $U_1(\mathbf{r}^1, t_1)$ and $U_1(\mathbf{r}^1, t_1)+dU_1$ at the position $\mathbf{r}^1$ and the time t_1.

The probability is expressed as

$$P(U_1(\mathbf{r}^1, t_1) \leq U_1(\mathbf{r}^1, t_1)_p < U_1(\mathbf{r}^1, t_1)+dU_1) = f(U_1(\mathbf{r}^1, t_1))\, dU_1(\mathbf{r}^1, t_1) \tag{3.12}$$

or for calculation purposes normally as

$$P(U(\mathbf{r}^1, t_1) \leq U_1(\mathbf{r}^1, t_1)_p < U_1(\mathbf{r}^1, t_1)+\Delta U_1) = f(U_1(\mathbf{r}^1, t_1))\, \Delta U(\mathbf{r}^1, t_1) \tag{3.13}$$

Example 1. Suppose the probability density for $U_1(\mathbf{r}^1, t_1)$ is given by the normal probability density function

$$f(U_1(\mathbf{r}^1, t_1)) = (\sigma\sqrt{2\pi})^{-1}\, e^{-[U_1(\mathbf{r}^1,t_1)-\langle U_1(\mathbf{r}^1,t_1)\rangle/\sigma]^2/2} \tag{3.14}$$

where the standard deviation is $\sigma = 4$ m/sec and the ensemble average velocity is $\langle U_1(\mathbf{r}^1, t_1)\rangle = 10$ m/sec. We wish to determine the probability that during any one of our experiments the probe at $\mathbf{r}^1$ will read, say, not less than the prescribed value $U_1(\mathbf{r}^1, t_1) = 5$ m/sec, and not more than $U_1(\mathbf{r}^1, t_1) =$ 5.5 m/sec if we examine only those values at time t_1 measured from the beginning of the experiment. The probability is

$$\begin{aligned} P(5\ \text{m/sec} \leq U_1(\mathbf{r}^1, t_1)_p < 5.5\ \text{m/sec}) &\simeq (32\pi)^{-1/2}\, e^{-[(5.25-10.0)/4]^2/2}(0.5) \\ &= 0.0246 \end{aligned} \tag{3.15}$$

There is a 2.5% probability that $5\ \text{m/sec} \leq U(\mathbf{r}^1, t_1) < 5.5\ \text{m/sec}$ [or, stated differently, 2.5 out of every 100 sample records from the probe at $\mathbf{r}^1$ will record $5\ \text{m/sec} \leq U(\mathbf{r}, t) < 5.5\ \text{m/sec}$ at time t_1]. Strictly speaking the above result should be obtained by the integration

$$\begin{aligned} &P(5\ \text{m/sec} \leq U_1(\mathbf{r}^1, t_1) \leq 5.5\ \text{m/sec}) \\ &\quad = (16\pi)^{-1/2} \int_{5.0\ \text{m/sec}}^{5.5\ \text{m/sec}} e^{-[(U_1-10)/4]^2/2}\, dU_1 \end{aligned} \tag{3.16}$$

Tabulated values of this integral are given in standard handbooks.[(3)] The exact value is 0.0247, showing the approximation to be very good.

It is well known that the distribution function of a continuous random variable is given by

$$F(U_1(\mathbf{r}^1, t_1)) = \int_{-\infty}^{U_1(\mathbf{r}^1, t_1)} f(U_1)\, dU_1 \tag{3.17}$$

that the probability of a velocity less than $U_1(\mathbf{r}^1, t_1)$ is

$$P(U_1(\mathbf{r}^1, t_1)_p < U_1(\mathbf{r}^1, t_1)) = F(U_1(\mathbf{r}^1, t_1)) \tag{3.18}$$

and that the integration from $-\infty$ to $+\infty$ must be unity:

$$F(U_1(\mathbf{r}^1, t_1)) = \int_{-\infty}^{\infty} f(U_1)\, dU_1 = 1 \tag{3.19}$$

Summarizing we can state that the velocity at a fixed point in a turbulent flow and at a fixed time in the repeated development of the flow from identical conditions behaves as a random variable with the associated statistical properties.

3.2.1. The Random Process

We have so far considered only a single probe and inspected individually the sample records from this probe at a particular time t_1. Obviously there is a great deal more information in the ensemble of sample records. Let us, however, still consider only probe 1 at $\mathbf{r}^1$, i.e., at a given position in the turbulent flow, but we will now take ensemble averages from the records at several different times beginning with $t = 0$ and increasing in increments of time Δt. A plot of the computed $\bar{U}_1(\mathbf{r}^1, t)$ would behave somewhat as shown in Figure 3.8. In the limit of $\Delta t \to 0$ the ensemble average becomes a continuous function of time. The velocity $U_1(\mathbf{r}^1, t)$ now varies randomly about the average value at any given time t but not necessarily with the same distribution function. This means the probability density function now depends not only on $U_1(\mathbf{r}^1, t_1)$ as before but also on time. If this distribution function were known one could ask the question What is the probability that the velocity recorded by the probe at $\mathbf{r}^1$ is $U_1(\mathbf{r}^1, t) < U_1(\mathbf{r}^1, t)_p < U_1(\mathbf{r}^1, t) + \Delta U_1$, at the particular time t? The probability is given by

$$P(U_1(\mathbf{r}^1, t) \leq U_1(\mathbf{r}^1, t)_p \leq U_1(\mathbf{r}^1, t) + \Delta U_1; t) = f(U_1(\mathbf{r}^1, t), t)\, \Delta U_1 \tag{3.20}$$

In the limit ΔU_1 becomes dU_1.

Example 2. The probability density function for the turbulent flow is assumed to be given by

$$f(U_1(\mathbf{r}^1, t), t) = [2^{1/2t}\Gamma(1/2t)]^{-1} U_1^{(1/2t-1)}\, e^{-U_1/2} \tag{3.21}$$

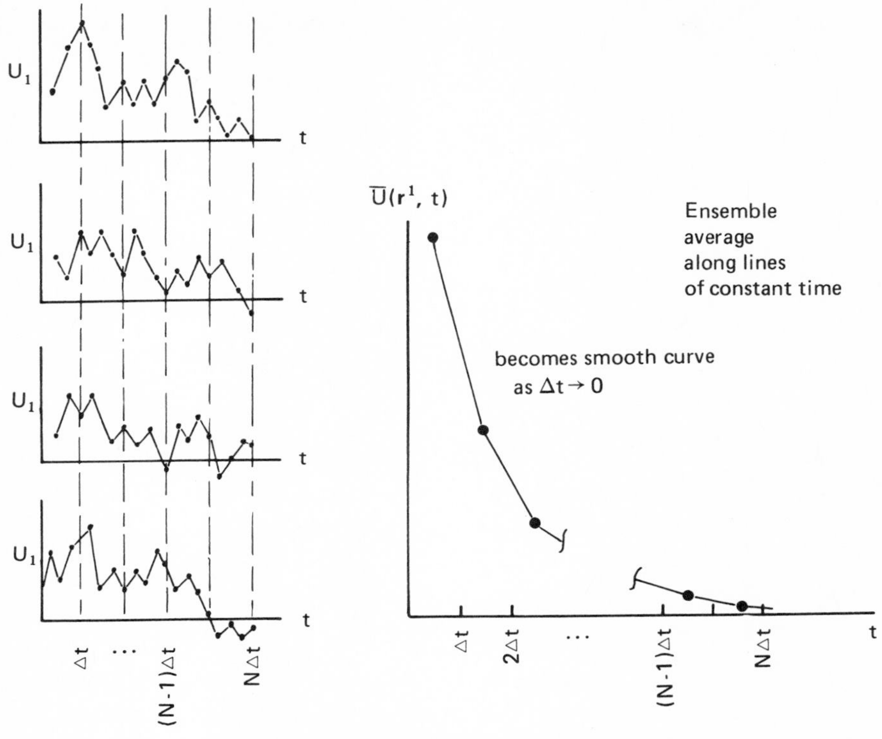

Fig. 3.8. Ensemble average as a function of time.

The probability that the probe $\mathbf{r}^1$ will record $3.75 \leq U_1(\mathbf{r}^1, t)_p < 4.25$ at time $t = 0.1$ is

$$P(3.75 \leq U_1(\mathbf{r}^1, t)_p < 4.25; 0.1) = f(4.00, 0.1)\, \Delta U_1 = 0.052 \quad (3.22)$$

Figure 3.9 shows the probability density of $U_1(\mathbf{r}, t)$ at various times t. Each curve, of which there are an infinite number, represents the distribution of $U(\mathbf{r}, t)$ at that particular time in the flow development. From these curves the probability of finding a given value of velocity in the flow field at position $\mathbf{r}$ can be determined. The ensemble average value is $\langle U(\mathbf{r}, t)\rangle = A/t$, where A is a constant.

The collection or ensemble of sample records of $U_1(\mathbf{r}^1, t)$ from an individual probe such as those from probe 1 in our foregoing discussion is called a random process or a stochastic process. The component of velocity $U_1(\mathbf{r}^1, t)$ is designated as a random function of time. For a fixed instant of

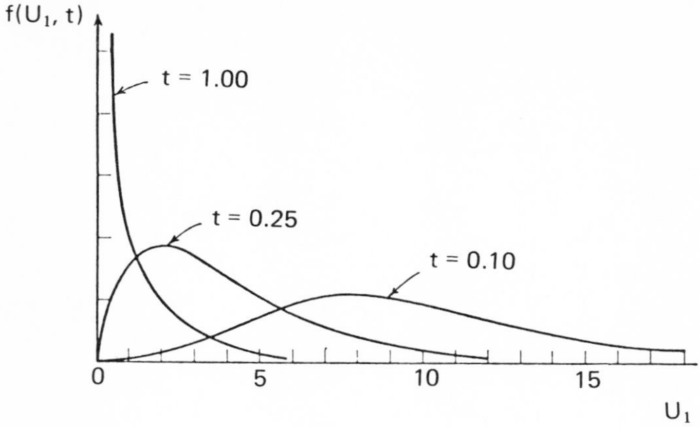

Fig. 3.9. Probability density of $U_1(\mathbf{r}, t)$ at various times.

time, $t = t_1$, as noted before, the random function becomes a random variable in the usual sense of that term.

3.2.2. Stationarity and Ergodicity

There is a classification of random processes that is of importance to us: the class of stationary and ergodic random functions. Random functions whose probability density functions are invariant with respect to the change of the reference reading of time, i.e.,

$$f(U_1(\mathbf{r}^1, t), t) = f(U_1(\mathbf{r}^1, t+\tau), t+\tau) \tag{3.23}$$

are called stationary random functions of a stationary random process. For a more rigorous definition of stationarity see Reference 4.

It is evident from the above that if one takes the average at time t and also at time $t+\tau$ the average at both times will be the same for a stationary random process. The hypothetical experiment we are considering is obviously not a stationary random process since the average velocity initially jumps up and then decays with time. Had we considered an experiment where an identical pressure differential was maintained along N identical wind tunnels that were continuously supplied with fluid at a constant mean rate, the sample records from probes having exactly the same location in each tunnel would appear as sketched in Figure 3.10. The ensemble of records from each probe would constitute a stationary random process. Most stationary random processes have the characteristic called ergodicity. This is defined as follows: Suppose we take the individual sample record k

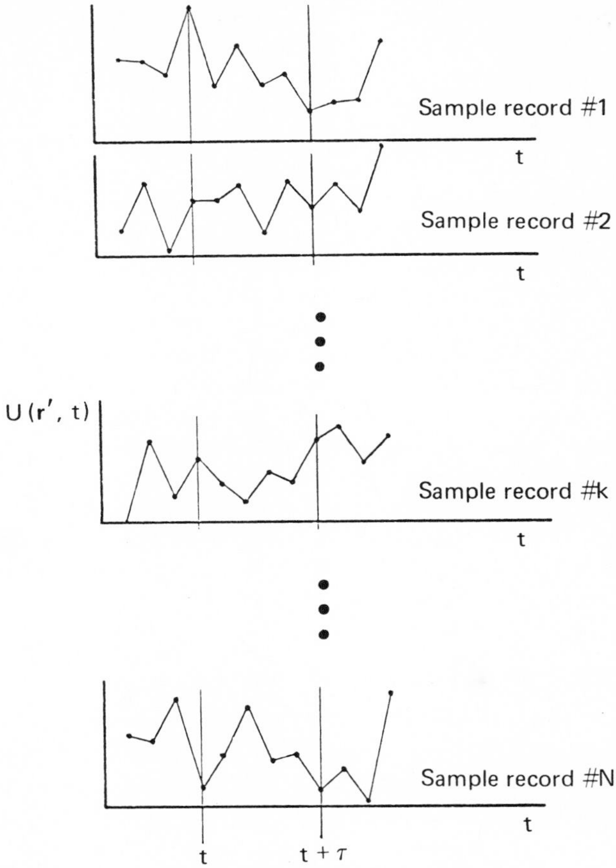

Fig. 3.10. Stationary random process.

and compute the time average for that record:

$$\bar{U}_1^k(\mathbf{r}^1) = \lim_{T\to\infty} \frac{1}{T} \int_0^T U_1^k(\mathbf{r}^1, t)\, dt \tag{3.24}$$

For actual calculation $\bar{U}_1^k(\mathbf{r}^1)$ is computed by

$$\bar{U}_1^k(\mathbf{r}^1, t) = \frac{1}{N} \sum_{i=1}^{N} U_1^k(\mathbf{r}^1, (2i-1)\,\Delta t/2) \tag{3.25}$$

where equal increments of Δt are employed. If the random process is stationary and if $\bar{U}_1^k(\mathbf{r}^1)$ computed from the preceding equation does not differ from the time average computed over any other sample record, say j, i.e.,

$$\bar{U}_1^k(\mathbf{r}^1) = \bar{U}_1^j(\mathbf{r}^1) \tag{3.26}$$

then the random process is said to be ergodic. Physical random processes that are ergodic are extremely convenient to work with since the time-averaged mean value is equal to the corresponding ensemble-averaged value. Thus one experiment conducted over a sufficiently long interval of time is all that is needed to provide the ensemble average. Obviously this is much simpler than our repetitious hypothetical experiment, but remember that our experiment cannot be ergodic since only stationary processes can be ergodic. Fortunately in many physical phenomena the causes that engender the given processes remain constant over an interval of time greater than the duration of the observation required to ensure quasistationary flow. Thus sample records of sufficient length that transitional effects can be disregarded and meaningful time averages can be computed are readily obtainable. The assumption of ergodicity is therefore universally employed in statistically stationary turbulent flows, a classical example being turbulence measurements in wind tunnels. Our hypothetical experiment, however, is not stationary and cannot be assumed ergodic.

3.2.3. Space-Dependent Random Variables

We have seen how the variables of turbulent flow at a fixed position in time and space behave like random variables. If we fix only the position in space and consider the random behavior as it changes with time, we perceive a random function of time, and the collection of sample records of identical experiments from a fixed probe define a random process. There still remains, however, a great deal more information in the ensemble of sample functions from our hypothetical experiment, and further examination of these records will lead to the concept of a random field.

Directing attention again to Figure 3.6, let us evaluate $U_1(\mathbf{r}, t)$ from the kth record of each probe positioned along a line in the x_1 direction (i.e., probes 1, 2, . . . , j, in Figure 3.5) at a fixed time t_1. If this procedure is repeated for all N records and $U_1(\mathbf{r}, t_1)$ is plotted versus x_1 for each record an ensemble of records as sketched in Figure 3.11 will result. Allowing the separation distance between probes to approach zero while the number of probes becomes infinite produces a continuous record of $U(\mathbf{r}, t_1)$ along the direction x_1. Normally, however, we have digitized data averaged over given spatial increments.

An ensemble average at a fixed point x_1^1 can now be generated in the same manner as was done previously for a fixed time t_1. It is apparent that

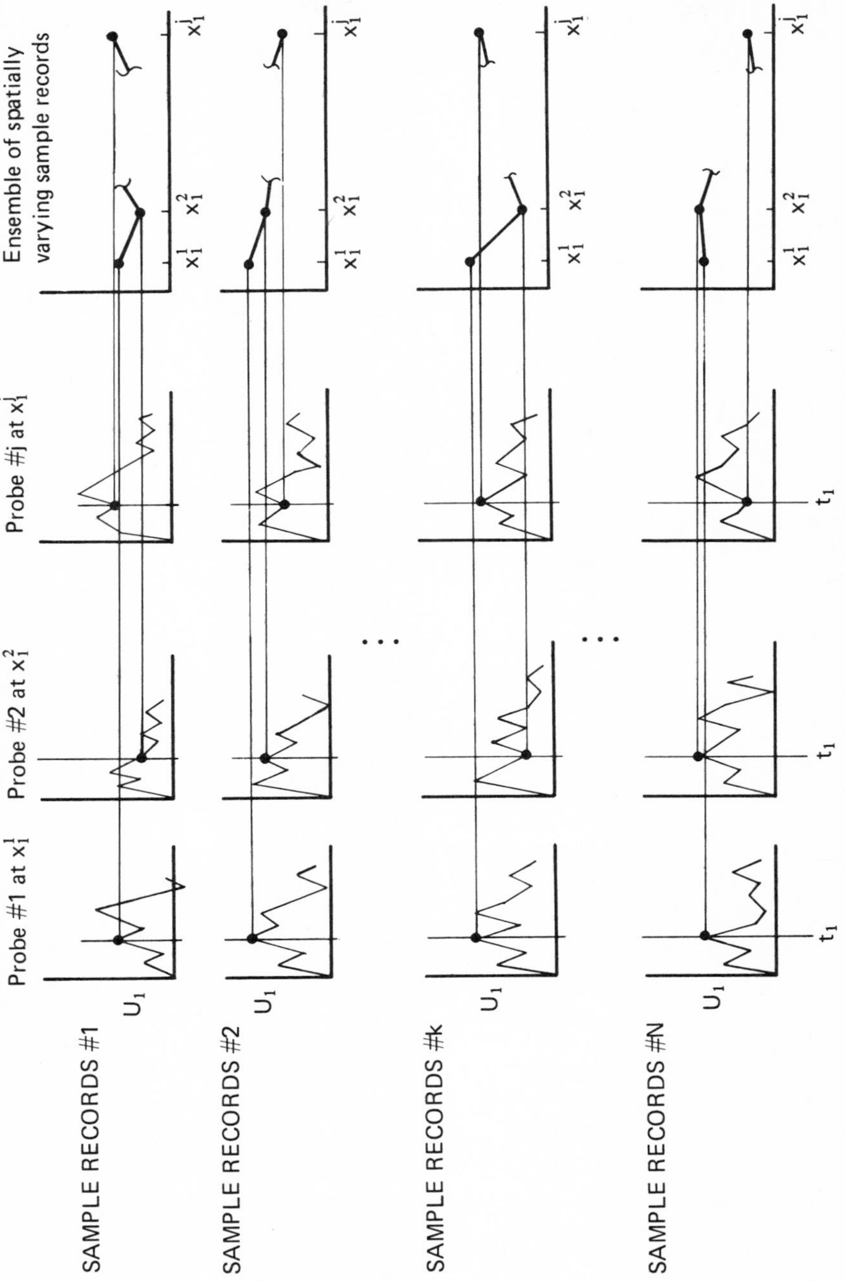

Fig. 3.11. Generation of a spatially dependent random variable from *N* sample records.

these two averages will be the same. Similarly,

$$P\big(U_1(x_1, t_1) \leq U_1(x_1, t_1)_p < [U_1(x_1, t_1) + dU_1]; x_1\big) = f\big(U_1(x_1, t), x_1\big)\, dU_1 \tag{3.27}$$

3.2.4. Homogeneous and Isotropic Turbulence

The question arises also as to whether the space average for, say the kth record,

$$\bar{U}_1^k(t_1) = \lim_{L\to\infty} \frac{1}{L} \int_0^L U_1(x_1, t_1)\, dx_1 \tag{3.28}$$

has a comparable meaning to the time average in a stationary process. The answer lies in the definition of two classes of turbulent flow: homogeneous and isotropic, respectively. Statistically homogeneous flow in the direction x_1 is flow for which

$$f(x_1, t) = f(x_1 + l, t) \tag{3.29}$$

where l is a displacement along the x_1 axis. Statistically homogeneous in a stricter sense is defined more carefully in Reference 4. It is evident that the concept of a homogeneous random function along a straight line is analogous to a stationary random process.

The random function of position is called statistically homogeneous and isotropic if the relationship holds regardless of the direction in which the coordinate axis x_1 is rotated in the flow field. A flow cannot be isotropic unless it is also homogeneous, but the reverse statement does not hold: A flow may be homogeneous but not necessarily isotropic. The concept of isotropy is only applicable to random fields and has no analogy in the theory of random processes.

For homogeneous flow the concept of ergodicity applies, and a space average from a given single sample record is equivalent to the ensemble average. Hence turbulence measurements made from a series of velocity sensors positioned along a given path in homogeneous turbulence or by an aircraft passing very rapidly through homogeneous turbulence can be employed to determine the ensemble average.

3.2.5. The Taylor Hypothesis

The preceding discussion begs the question of whether there are certain classes of turbulence for which the fluctuating velocity pattern with time and the fluctuating velocity pattern in space are related. Taylor's hypothesis states that if $\bar{U}_1 \gg u_1$ the fluctuations at a fixed point of a homogeneous turbulent flow with a constant mean velocity $\bar{U}_1$ in the x_1 direction may behave as if the whole turbulent flow field passes that point with a constant velocity $\bar{U}_1$. The oscillogram of the fluctuating velocity at that point will then be nearly identical with the instantaneous distribution of the velocity along the x_1 axis through that point (the term "frozen turbulence" helps to visualize this concept). Taylor's hypothesis holds if

$$\frac{\partial u_1}{\partial t} = -\bar{U}_1 \frac{\partial u_1}{\partial x_1} \tag{3.30}$$

which follows from the equation of motion when $u_1/\bar{U}_1 \ll 1$. It is apparent then that the gradient of the fluctuation u_1 in time is linearly related to the gradient in position and that the fluctuations occur at a much slower rate, u_1, than the rate of mean motion, $\bar{U}_1$, hence the turbulence is carried past the point before appreciable change in the turbulence pattern can take place.

Taylor's hypothesis, despite its apparent restrictive nature, is frequently used in turbulent flow analyses since it is considerably easier to measure a time variation with a single probe than a spatial variation with several probes. The measured spatial variation, however, is obviously less strongly influenced by the convection velocity of the mean flow and indicates more directly the characteristics of the turbulence. Additional criteria for the applicability of Taylor's hypothesis have been pointed out by Lin.(5) He notes that in shear flows Taylor's hypothesis can be expected to hold only for eddies small enough that velocity gradients across them are negligible compared to the convection velocity. Lumley and Panofsky(6) have observed, however, that the hypothesis is valid for wind shears in excess of the limit prescribed by Lin. The general validity of Taylor's hypothesis is still not clear and must be examined in individual cases.

3.2.6. Random Field

If we go one step further in our preceding argument and measure $U_1(\mathbf{r}, t_1)$ at every location in the flow field rather than just along a line, we obtain a random function of the coordinates x_1, x_2, x_3, which is referred to as a random field. Additionally, letting time vary as described earlier generates a time-dependent random field. In this regard we understand the expression "random function of $\mathbf{r}$ and t" to mean that at each point $(\mathbf{r}, t)$ of the four-dimensional space–time field the value of $U_1(\mathbf{r}, t)$ is a random variable and can only be predicted within a certain probability.

The laws of this probability in principle can be established from the ensemble of $N \times M$ sample records of our N hypothetical runs conducted with M probes, positioned throughout the flow, but of course in a manner undisturbing to the fluid motion. From these data a probability density function can be established and the probability of a value $U_1(\mathbf{r}, t) \leq (U_1)_p < [U_1(\mathbf{r}, t) + dU_1]$ at a position $\mathbf{r}$ and at a time t is given by

$$P\big(U_1(\mathbf{r}, t) \leq U_1(\mathbf{r}, t)_p < [U_1(\mathbf{r}, t) + dU_1]; \mathbf{r}, t\big) = f\big(U_1(\mathbf{r}, t), \mathbf{r}, t\big)\, dU_1 \tag{3.31}$$

Moreover one can conceive of a multidimensional probability function that would enable one to predict the probability of the velocity component U_1 having given values at several different points in the flow. For example a two-dimensional probability density function would allow the computation of the probability of the velocity component $U_1(\mathbf{r}^1, t)$ lying between U_1^1 and $U_1^1 + dU_1^1$ at a position $\mathbf{r}^1$ and lying between U_1^2 and $U_1^2 + dU_1^2$ at a position $\mathbf{r}^2$ at a time t. Further generalizations involve values of $U_1(\mathbf{r}, t)$ observed at more than two points:

$$\begin{aligned} P\big(U_1^1 \leq (U_1^1)_p < U_1^1 + dU_1^1, \ldots, U_1^N \leq (U_1^N)_p \leq U_1^N + dU_1^N; \mathbf{r}^1, \mathbf{r}^2, \ldots, \mathbf{r}^N, t\big) \\ = f(U_1^1 \cdots U_1^N, \mathbf{r}^1 \cdots \mathbf{r}^N, t)\, dU_1^1 \cdots dU_1^N \end{aligned} \tag{3.32}$$

3.2.7. Random Scalar and Vector Fields

The random field for one individual component of the velocity or for any other quantity defined by one random variable, for example temperature or pressure, is a scalar random field. In turbulent flow, however, velocity is one of the primary variables of interest, and it is a vector, known when its three components U_1, U_2, and U_3 are known. The turbulent velocity field is therefore a random vector field. Each one of the velocity components making up the field is a random variable, and to predict the probability of a given velocity at some point in space and time a three-dimensional probability density function is required:

$$P(\mathbf{V} \leq \mathbf{V}_p < \mathbf{V} + d\mathbf{V}; \mathbf{r}, t) = f(U_1, U_2, U_3, x_1, x_2, x_3, t)\, dU_1\, dU_2\, dU_3 \tag{3.33}$$

Extension to a $3n$-dimensional probability enabling one to compute the probability of the three components of velocity at n positions in space is conceptually possible; however, the hopes of ever establishing such probability distributions from any practical experiment have long since lost credibility.

3.3. Statistical Moments

A turbulent flow field is considered known if the multidimensional probability density distribution function described in the preceding is known; however, development of the $3n$-dimensional function, as we have indicated, is not possible in practice. Fortunately, the random field can be adequately characterized for most purposes by statistical moments of different orders that are more readily obtainable from experiment.

3.3.1. Ordinary Moments

Statistical moments in the most general case are defined as

$$B_i^{k_i}, B_j^{k_j}, \ldots, B_p^{k_p}(\mathbf{r}_1, \mathbf{r}_2, \ldots, \mathbf{r}_n, t) = \langle U_i^{k_i}(\mathbf{r}_1, t) U_j^{k_j}(\mathbf{r}_2, t) \cdots U_p^{k_p}(\mathbf{r}_n, t)\rangle \tag{3.34}$$

This is called an n-point statistical moment of the kth order, where $k = k_i + k_j + \cdots + k_p$. The angle brackets denote the probability mean, which again is defined in terms of the probability density function, i.e.,

$$\begin{aligned} &\langle U_i^{k_i}(\mathbf{r}_1, t) U_j^{k_j}(\mathbf{r}_2, t) \cdots U_p^{k_p}(\mathbf{r}_n, t)\rangle \\ &\quad = \int_{-\infty}^{\infty} \int_{-\infty}^{\infty} \cdots \int_{-\infty}^{\infty} U_i^{k_i}(\mathbf{r}_1, t) U_j^{k_j}(\mathbf{r}_2, t) \cdots U_p^{k_p}(\mathbf{r}_n, t) \\ &\qquad \times f(U_i, U_j, \ldots, U_p, \mathbf{r}_1, \mathbf{r}_2, \ldots, \mathbf{r}_n, t)\, dU_i\, dU_j \cdots dU_p \end{aligned} \tag{3.35}$$

Fortunately, calculation of this integral can generally be avoided in turbulent studies.

The two-point moments of second, third, and fourth order have the form shown below:

$$\begin{aligned} B_{i,j}(\mathbf{r}_1, \mathbf{r}_2, t) &= \langle U_i(\mathbf{r}_1, t) U_j(\mathbf{r}_2, t)\rangle \\ B_{ij,k}(\mathbf{r}_1, \mathbf{r}_2, t) &= \langle U_i(\mathbf{r}_1, t) U_j(\mathbf{r}_1, t) U_k(\mathbf{r}_2, t)\rangle \\ B_{ij,km}(\mathbf{r}_1, \mathbf{r}_2, t) &= \langle U_i(\mathbf{r}_1, t) U_j(\mathbf{r}_1, t) U_k(\mathbf{r}_2, t) U_m(\mathbf{r}_2, t)\rangle \end{aligned} \tag{3.36}$$

The comma between the indices indicates that the i component of velocity for the first example and the i and j components of velocity for the second and third examples are measured at position r_1, and the j component for the first example, the k component for the second example, and the k and m components for the third example are measured at position r_2. The moments are said to be symmetric with respect to the different groups of indices. A three-point moment of third order has the form

$$B_{i,j,k}(\mathbf{r}_1, \mathbf{r}_2, \mathbf{r}_3, t) = \langle U_i(\mathbf{r}_1, t) U_j(\mathbf{r}_2, t) U_k(\mathbf{r}_3, t)\rangle \tag{3.37}$$

and so on.

In general most turbulence analyses deal with two-point moments normally of not more than third order. It should be noted that the statistical moments of order n have the characteristics of a tensor of order n.

3.3.2. Central Moments

Moments formed from the fluctuating quantities—the deviations from the mean value—are referred to as central moments and frequently denoted with the lower case symbol, i.e.,

$$\begin{aligned} b_{ij}(\mathbf{r}, t) &= \langle u_i(\mathbf{r}, t) u_j(\mathbf{r}, t)\rangle \\ &= \Big\langle [U_i(\mathbf{r}, t) - \langle U_i(\mathbf{r}, t)\rangle][U_j(\mathbf{r}, t) - \langle U_j(\mathbf{r}, t)\rangle]\Big\rangle \end{aligned} \tag{3.38}$$

The one-point central moments of the components are readily recognized as the Reynolds stresses and often given the symbol $\overline{u_i u_j}$ or $\overline{u_i^2}$. The quantity $\overline{u_i^2}$ is also denoted as σ^2 in much of the literature.

Example 3. In practice a two-point central moment can be calculated from two probes in the flow field with the use of the equations

$$\langle u_i(\mathbf{r}_1, t) u_j(\mathbf{r}_2, t)\rangle = \frac{1}{N} \sum_{k=1}^{N} u_i^{(k)}(\mathbf{r}_1, t) u_j^{(k)}(\mathbf{r}_2, t) \tag{3.39}$$

$$\begin{aligned} \langle u_1(\mathbf{r}_1, t) u_1(\mathbf{r}_2, t)\rangle = \frac{1}{N} \sum_{k=1}^{N} & [U_1^{(k)}(\mathbf{r}_1, t) - \langle U_1(\mathbf{r}_1, t)\rangle] \\ & \times [U_1^{(k)}(\mathbf{r}_2, t) - \langle U_1(\mathbf{r}_2, t)\rangle] \end{aligned} \tag{3.40}$$

Consider that the first 25 lines of Table 3.1 are measured with the probe at $\mathbf{r}_1$ and the remaining 25 lines are measured simultaneously with the probe at $\mathbf{r}_2$. We will calculate the two-point correlation of second order at $t = 0.2$ sec. The ensemble averages are

$$\langle U_1(r_1, t)\rangle = (1/25)[5.55 + 2.88 + 8.32 + \cdots + 9.01 + 0.66] = 5.44 \tag{3.41}$$

and

$$\langle U_1(r_2, t)\rangle = (1/25)[1.33 + 2.84 + 1.00 + \cdots + 0.50 + 7.00] = 6.42 \tag{3.42}$$

The correlation is then

$$\begin{aligned} \langle u_1(\mathbf{r}_1, t) u_1(\mathbf{r}_2, t)\rangle &= (1/25)[(0.11)(-5.09) + (-2.56)(-3.58) \\ &\quad + (2.88)(-5.42) + \cdots + (3.57)(3.08) + (-4.78)(0.58)] \\ &= 0.568 \end{aligned} \tag{3.43}$$

3.3.3. Joint Moments

The moments formed with random variables from several different random fields such as temperature or pressure are called joint moments. The two-point joint moment of pressure and velocity, for example, is

$$B_{p,j}(\mathbf{r}_1, \mathbf{r}_2, t) = \langle p(\mathbf{r}_1, t) U_j(\mathbf{r}_2, t)\rangle \tag{3.44}$$

3.3.4. Space–Time Moments

Moments taken at a given instant of time are normally called space moments. The more general case where an average is taken of the product between random variables at different positions and at different times is a space–time moment, i.e.,

$$B_{i,j}(r_1, r_2, t_1, t_2) = \langle U_i(r_1, t_1) U_j(r_2, t_2) \rangle \tag{3.45}$$

Time moments are the mean values of products of values of the fluid dynamic fields at the same position but at different instants. Thus

$$B_{ij}(\mathbf{r}, t_1, t_2) = \langle U_i(\mathbf{r}, t_1) U_j(\mathbf{r}, t_2) \rangle \tag{3.46}$$

Most often time moments are used when dealing with random processes rather than random fields. In this event the terminology "correlation function" is generally employed. An autocorrelation is then given by

$$B_{ii}(\mathbf{r}, t_1, t_2) = \langle U_i(\mathbf{r}, t_1) U_i(\mathbf{r}, t_2) \tag{3.47}$$

and a cross correlation by

$$B_{ij}(\mathbf{r}, t_1, t_2) = \langle U_i(\mathbf{r}, t_1) U_j(\mathbf{r}, t_2) \rangle \tag{3.48}$$

Example 4. The autocorrelation from the sample random data of Table 3.1 is computed from

$$B_{ii}(\mathbf{r}, t_1, t_2) = \frac{1}{N} \sum_{k=1}^{N} U_i^{(k)}(\mathbf{r}, t_1) U_i^{(k)}(\mathbf{r}, t_2) \tag{3.49}$$

Let $t_1 = 0.2$ and $t_2 = t_1 + \tau$ where $\tau = 0.5$ is the time delay between signals. Hence

$$\begin{aligned} B_{11}(\mathbf{r}, 0.2, 0.7) &= (1/50)[(5.55)(1.04) + (2.88)(1.13) \\ &\quad + \cdots + (9.50)(0.01) + (7.00)(0.37)] \\ &= 9.32 \end{aligned} \tag{3.50}$$

3.3.5. Other Terminology

The term "correlation function" is also used for space moments. Hinze[7] refers to the two-point second-order and third-order moments as double and triple velocity correlations, respectively.

The expression "correlation function" in particular is used for two-point second-order moments. "Covariance" is another term used for such moments.

The correlation coefficient is a nondimensional correlation function defined as

$$R_{i,j}(\mathbf{r}_1, \mathbf{r}_2, t) = \langle U_i(\mathbf{r}_1, t)U_j(\mathbf{r}_2, t)\rangle/[\langle U_i^2(\mathbf{r}_1, t)\rangle\langle U_j^2(\mathbf{r}_2, t)\rangle]^{1/2}$$
$$= B_{i,j}(\mathbf{r}_1, \mathbf{r}_2, t)/[B_{ii}(\mathbf{r}_1, t)B_{jj}(\mathbf{r}_2, t)]^{1/2} \tag{3.51}$$

The term "moments" will generally refer to space moments in this book; however, "correlation," "covariance," and "moments" will be used interchangeably. Also, the notation

$$R_{ij}(\mathbf{r}, t) = \overline{u_i u_j}/\sigma_i\sigma_j \tag{3.52}$$

is frequently used here and in the literature.

Physical arguments will lead to the conclusion that the correlation coefficient must be in the range

$$-1 \leq R_{ij} \leq 1 \tag{3.53}$$

$R_{ij} = 1$ occurs when $U_i(\mathbf{r}, t) = U_j(\mathbf{r}, t)$ and represents a perfect correlation. $R_{ij} = -1$ may occur when a quantity is correlated with its negative. $R_{ij} = 0$ occurs when there is no relationship even of a statistical nature between the fluctuations.

3.3.6. Longitudinal and Lateral Correlations

Two correlations of frequent occurrence in the analysis of turbulence are the longitudinal and lateral correlations. The longitudinal correlation is the two-point moment

$$b_{ll}(\mathbf{r}_1, \mathbf{r}_2, t) = \langle u_l(\mathbf{r}_1, t)u_l(\mathbf{r}_2, t)\rangle \tag{3.54}$$

where l denotes the velocity component along the line joining the points $\mathbf{r}_1$ and $\mathbf{r}_2$ as illustrated in Figure 3.12. The lateral correlation in turn is given by

$$b_{nn}(\mathbf{r}_1, \mathbf{r}_2, t) = \langle u_n(\mathbf{r}_1, t)u_n(\mathbf{r}_2, t)\rangle \tag{3.55}$$

where n denotes the velocity component normal to the line connecting the points, $\mathbf{r}_1$ and $\mathbf{r}_2$.

The symbols $f(r)$ and $g(r)$ are commonly used for the longitudinal and lateral correlation coefficients, respectively, particularly in isotropic turbulence,

$$f(r) = b_{ll}(r, t)/b_{ll}(0, t) \tag{3.56}$$

and

$$g(r) = b_{nn}(r, t)/b_{nn}(0, t) \tag{3.57}$$

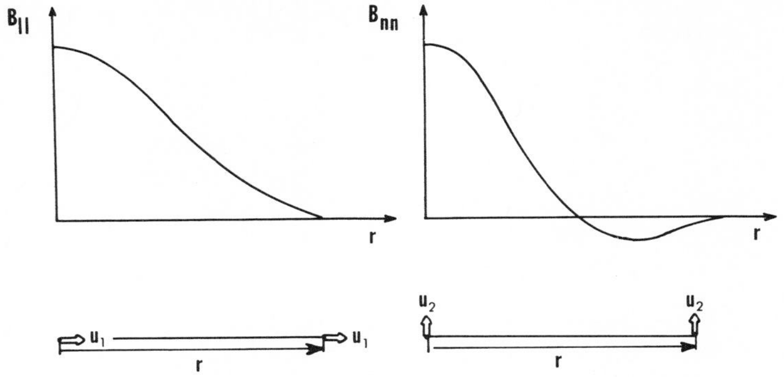

Fig. 3.12. The longitudinal and lateral correlations.

The value of r in isotropic turbulence is simply the distance between the two points in the flow field, and the orientation of the points has no significance.

It should also be mentioned here that for a stationary process the correlation $B_{ij}(\mathbf{r}, t_1, t_2)$ depends only on the time lag $\tau = t_2 - t_1$ and hence can be written as $B_{ij}(\mathbf{r}, \tau)$. Similarly for homogeneous flow the second-order moment $B_{i,j}(\mathbf{r}_1, \mathbf{r}_2, t)$ depends only the separation distance between the two position vectors, i.e., $\mathbf{l} = \mathbf{r}_2 - \mathbf{r}_1$ and is normally written as $B_{i,j}(\mathbf{l}, t)$. For isotropic flow the second-order moment depends only on the magnitude of $\mathbf{l}$ and hence is written $B_{i,j}(l, t)$.

3.3.7. Characteristics of Correlation Coefficients

3.3.7.1. Time Correlations. No generalizations can be made regarding the most general correlation coefficient; however, the characteristics of the time autocorrelation coefficient for statistically stationary and/or statistically homogeneous flow can be described. Figure 3.13 shows the typical behavior of the time autocorrelations coefficient with delay time τ. The autocorrelation is symmetric about $\tau = 0$. Since the turbulent motion is continuous there is a finite delay time before the velocity fluctuation changes, hence the autocorrelation is flat or level at $\tau = 0$.

The autocorrelation is used to define characteristic time scales of the turbulence. The average persistance of the turbulence activity at a point is the integral time scale

$$T_E = \int_0^\infty R(\tau)\, d\tau \tag{3.58}$$

T_E is a measure of the longest connection in the turbulent behavior of $u_i(t)$, for example.

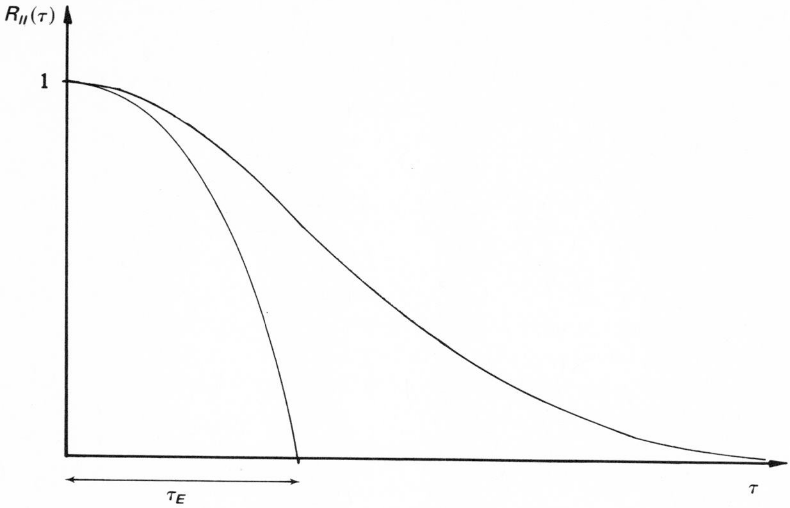

Fig. 3.13. Typical time autocorrelation coefficient.

A micro-time-scale is defined from the correlation as follows. The curvature at the top of the correlation is determined by the small scales of the turbulence. Expanding the correlation in a Taylor's series gives

$$R(\tau) = 1 + \left.\frac{\partial R}{\partial \tau}\right|_0 \tau + \frac{1}{2}\left.\frac{\partial^2 R}{\partial \tau^2}\right|_0 \tau^2 + \cdots \tag{3.59}$$

Owing to symmetry, $\partial R/\partial \tau|_0 = 0$. At small values of τ

$$R(\tau) \simeq 1 - \left(\frac{\tau}{\tau_E}\right)^2_{\tau=0} \tag{3.60}$$

where

$$\tau_E = -\left(\frac{1}{2}\frac{\partial^2 R}{\partial \tau^2}\right)^{-1/2} \tag{3.61}$$

τ_E is called the micro-time-scale of the turbulence. The value of τ_E is given by the interception of the $R(\tau) = 0$ axis by the parabolic curve fitted through $R(\tau)$ at $\tau = 0$ (see Figure 3.13). The micro-time-scale can be shown to be proportional to the ratio of the root-mean-square value of u_i to the root-mean-square value of its derivative, i.e.,

$$\tau_E \propto u_i'[(\partial u_i/\partial t)']^{-1} \tag{3.62}$$

τ_E is a measure of the most rapid changes that occur in the fluctuations of $u_i(t)$. It is important to recognize that time-correlation curves, and the time scales derived from them, depend not only on the structure of the turbulence, but also on the mean velocity convecting the turbulence past the point of measurement.

3.3.7.2. Spatial Correlations. Two-point, spatial autocorrelation curves, unlike time correlations, may take on negative values with increasing separation distance between the points, and also they may be asymmetric with respect to the separation distance. The latter condition occurs when the structure of the turbulence changes substantially over a distance for which the fluctuations have significant correlation. For example, turbulent flow over a plane wall where $\bar{U}_1(x_2)$ is confined to the x_1–x_2 plane will have marked asymmetry in $R(x_2, x_2+l_2)$, slight asymmetry in $R(x_1, x_1+l_1)$, and virtually no asymmetry in $R(x_3, x_3+l_3)$. In homogeneous turbulence the correlations are symmetric in all directions. Additionally, if the turbulence is isotropic the correlations are not only symmetric but they vary the same in all directions.

The integral length scale is defined in terms of the spatial correlations by

$$\Lambda = \int_0^\infty R(x_i, x_i + l)\, dl \tag{3.63}$$

Λ is a measure of the longest connection or correlation distance between the velocities at two points of the flow field. The magnitude of Λ depends on the fluctuations correlated and on the direction of separation. The longitudinal length scale

$$\Lambda_{ll} = \int_0^\infty R_{ll}(x_i, x_i + l)\, dl \tag{3.64}$$

is considerably greater than the lateral length scale

$$\Lambda_{nn} = \int_0^\infty R_{nn}(x_i, x_i + l)\, dl \tag{3.65}$$

for flow near a wall.

As with the time correlation the micro-length-scale is defined as

$$\lambda = -\left(\frac{1}{2}\frac{\partial^2 R}{\partial l^2}\right)_{l=0}^{-1/2} \tag{3.66}$$

provided R is symmetric near $l = 0$. The microscales most commonly used are the longitudinal microscale

$$\lambda_{ll} = -\left(\frac{1}{2}\frac{\partial^2 R_{ll}}{\partial l^2}\right)_{l=0}^{-1/2} \tag{3.67}$$

and the lateral microscale

$$\lambda_{nn} = -\left(\frac{1}{2}\frac{\partial^2 R_{nn}}{\partial l^2}\right)^{-1/2} \tag{3.68}$$

λ_{ll} does not necessarily equal λ_{nn}.

In practice the length scales are normally determined with the application of Taylor's hypothesis. That is,

$$\Lambda_{ll} = \bar{U}T_E \tag{3.69}$$

where $\bar{U}$ is the mean velocity and Λ_{ll} is the correlation of the fluctuations in the direction of the mean flow.

The micro-length-scale is not the smallest scale of turbulence. The dissipation length scale is smaller. Kolmogorov has shown that the smallest scales of motion must be related to the energy dissipation rate per unit mass ε and the kinematic viscosity ν. With these parameters the smallest length scale, η, is equal to $(\nu^3/\varepsilon)^{1/4}$.

Notation

$B(\ )$	Statistical moment
$b(\ )$	Central moment
$E(\kappa)$	Energy spectrum
$F(\ )$	Distribution function
$f(\ )$	Probability density distribution
$f(r)$	Longitudinal correlation coefficient
$g(r)$	Lateral correlation coefficient
l	Separation distance
$\mathbf{l}$	Separation distance vector
l	Scale of eddy size
P	Pressure
$P(\ )$	Probability
$R(\ )$	Correlation coefficient
$\mathbf{r}$	Position vector
S_{ij}	Fluctuating strain tensor
T_E	Integral time scale
t	Time
U	Instantaneous velocity
$\bar{U}$	Mean velocity
u	Fluctuating velocity component
$\mathbf{V}$	Velocity vector
x	Coordinate distance
ω	Vorticity
Λ	Integral length scale
λ	Microscale
κ	Wave number
$\Gamma(\ )$	Gamma function
τ	Lag time
τ_E	Micro-time-scale
$'$	Root-mean-square value
$\overline{(\ )}$	Time average
$\langle\ \rangle$	Ensemble average

References

1. Tennekes, H., and Lumley, J. L., *A First Course in Turbulence*, MIT Press, Cambridge, Massachusetts (1972).
2. Bradshaw, P., *An Introduction to Turbulence and its Measurement*, Pergamon Press, Oxford, England (1971).

3. Hodgman, C. D. (ed.), *Handbook of Chemistry and Physics*, Chemical Rubber Publishing Co., Cleveland (1954).
4. Panchev, S., *Random Functions and Turbulence*, Pergamon Press, Oxford, England (1971).
5. Lin, C. C., *Statistical Theories of Turbulence*, Princeton University Press, Princeton, New Jersey (1961).
6. Lumley, J. L., and Panofsky, H., *The Structure of Atmospheric Turbulence*, Interscience Publishers, New York (1964).
7. Hinze, J. O., *Turbulence*, McGraw-Hill Book Company, New York (1959).

CHAPTER 4

Spectral Theory of Turbulence

WALTER FROST

4.1. Introduction

The statistical correlations described in the preceding chapter are useful in turbulence analyses and are relatively easy to measure. However, another powerful tool for describing turbulence is the method of spectral analysis. The spectral theory and the correlation theory are intimately connected mathematically with one another by the Fourier transformation. There is no additional information contained in the spectra that is not already contained in the correlations, but the two methods of description put different emphases on different aspects of the problem. For example, we discussed earlier the concept of energy transfer between different scales or orders of eddies. Spectral analysis allows us to describe the exchange of kinetic energy associated with different eddy sizes or with different fluctuation frequencies occurring in the turbulence.

There are two types of spectra that are of interest to analyses of turbulence: the frequency spectra and the wave-number spectra. We shall describe the frequency spectra first since they are less complex conceptually. Toward this goal we begin with a review of the theory of harmonic analysis.

WALTER FROST • The University of Tennessee Space Institute, Tullahoma, Tennessee 37388

4.2. Harmonic Analysis

4.2.1. Fourier Series

Consider first the deterministic, periodic function $f(t)$, which can be expanded in a Fourier series

$$f(t)=\frac{a_0}{2}+\sum_{n=1}^{\infty}(a_n\cos\omega_n t+b_n\sin\omega_n t) \tag{4.1}$$

where

$$\omega_n=n\omega'=2\pi n/T$$

and ω' is the fundamental frequency and $T=2\pi/\omega'$ is the period of $f(t)$. The coefficients in the series are given by

$$\begin{aligned} a_n&=\frac{2}{T}\int_{-T/2}^{T/2} f(t)\cos\omega_n t\,dt, \qquad n=0,1,2\ldots \\ b_n&=\frac{2}{T}\int_{-T/2}^{T/2} f(t)\sin\omega_n t\,dt, \qquad n=1,2,3\ldots \end{aligned} \tag{4.2}$$

which are valid representations if

$$\int_{-T/2}^{T/2}|f(t)|\,dt \tag{4.3}$$

is finite.

Although turbulence analyses deal only with real functions, for computational convenience it is often advantageous to express the Fourier series in complex form, i.e.,

$$f(t)=\sum_{n=-\infty}^{\infty}F(n)\,e^{i\omega_n t} \tag{4.4}$$

where

$$F(n)=\tfrac{1}{2}(a_n-ib_n), \qquad n=0,\pm1,\pm2,\ldots \tag{4.5}$$

or

$$F(n)=\frac{1}{T}\int_{-T/2}^{T/2} f(t)\,e^{-i\omega_n t}\,dt, \qquad n=0,\pm1,\pm2,\ldots$$

$F(n)$ is called the complex spectrum of the periodic function $f(t)$; it is a function of the harmonic order n and is a representation of the periodic time function in the frequency domain. $F(n)$ contains full information concerning the harmonic amplitudes and phase angles of the sinusoids into which $f(t)$ is resolved. Because the harmonic order n assumes only discrete values, the spectrum $F(n)$ is a line spectrum.

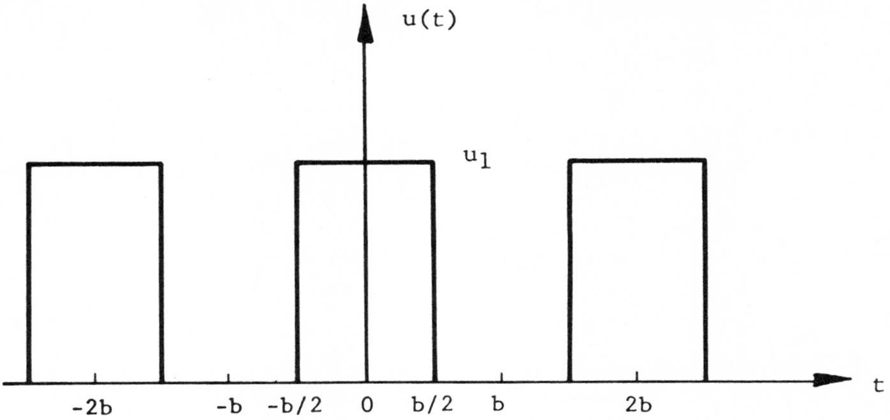

Fig. 4.1. Pseudo-variation velocity made up of an infinite number of sinusoidal eddies.

The amplitude and phase angle characteristic contained in $F(n)$ can be separated out as

$$|F(n)| = \tfrac{1}{2}(a_n^2 + b_n^2)^{1/2} \tag{4.6}$$

which is the amplitude of the nth harmonic and as

$$\theta(n) = \tan^{-1}(-b_n/a_n) \tag{4.7}$$

which is the phase angle of the nth harmonic. Thus the periodic function can be expressed as

$$f(t) = \sum_{n=-\infty}^{\infty} |F(n)|\, e^{i[\omega_n t + \theta(n)]} \tag{4.8}$$

The function $|F(n)|$ is called the amplitude spectrum of $f(t)$, and the function $\theta(n)$ is called the phase spectrum of $f(t)$.

As an example of how the above might relate in a hypothetical manner to turbulent flow, consider that a probe positioned in a flow field measures the longitudinal velocity component u_1 as a periodic wave of the rectangular form shown in Figure 4.1. This is not a particularly realistic velocity fluctuation but it is useful for our discussion.

The complex spectrum of the wave is

$$F(n) = \frac{1}{T}\int_{-b/2}^{b/2} u_m\, e^{-i\omega_n t}\, dt \tag{4.9}$$

$$= \frac{u_m}{2}\frac{\sin(\omega_n b/2)}{\omega_n b/2} \tag{4.10}$$

Expressing the velocity as the real part of the periodic function $f(t)$ gives

$$u_1(t) = \sum_{n=1}^{\infty} \frac{u_m}{2} \frac{\sin(\omega_n b/2)}{\omega_n b/2} \cos \omega_n t \tag{4.11}$$

The turbulent flow can then be considered as made up of an infinite number of eddies each having a cosine wave form. The amplitude of the nth eddy is $[u_m \sin(\omega_n b/2)]/\omega_n b$ and the phase angle of all eddies is zero. The probe measures the algebraic sum of all eddies passing the probe position at a given instant of time. These add together to form the pulse total velocity variation shown in Figure 4.1. If this were a representation of a real flow, we could then separate out each eddy making up the turbulent motion and inspect its behavior.

With this in mind let us investigate the correlation between periodic functions of common fundamental frequency, which is defined as

$$B_{ij}(\tau) = \frac{1}{T} \int_{-T/2}^{T/2} f_i(t) f_j(t+\tau)\, dt \tag{4.12}$$

where τ is a continuous time of displacement independent of t, and

$$f_i(t) = \sum_{n=-\infty}^{\infty} F_i(n)\, e^{i\omega_n t} \tag{4.13}$$

$$f_j(t) = \sum_{n=-\infty}^{\infty} F_j(n)\, e^{i\omega_n t} \tag{4.14}$$

An important feature of the correlation is that its Fourier transform is given by $E_{ij}(n) = F_i^*(n) F_j(n)$, where the asterisk denotes the complex conjugate. To prove this introduce

$$f_j(t+\tau) = \sum_{n=-\infty}^{\infty} F_j(n)\, e^{i\omega_n (t+\tau)} \tag{4.15}$$

into Equation (4.12), which gives

$$B_{ij}(\tau) = \frac{1}{T} \int_{-T/2}^{T/2} f_i(t) \sum_{n=-\infty}^{\infty} F_j(n)\, e^{i\omega_n (t+\tau)}\, dt \tag{4.16}$$

Noting that the integration is independent of τ, and that the order of summation and integration are interchangeable, gives

$$B_{ij}(\tau) = \sum_{n=-\infty}^{\infty} F_j(n)\, e^{i\omega_n \tau} \frac{1}{T} \int_{-T/2}^{T/2} f_i(t)\, e^{i\omega_n t}\, dt \tag{4.17}$$

Since

$$F_i^*(n) = \frac{1}{T} \int_{-T/2}^{T/2} f_i(t)\, e^{i\omega_n t}\, dt \tag{4.18}$$

Equation (4.12) becomes

$$B_{ij}(\tau) = \sum_{n=-\infty}^{\infty} E_{ij}(n)\, e^{i\omega_n \tau} \tag{4.19}$$

Thus $E_{ij}(n)$ is the complex spectrum of $B_{ij}(\tau)$ and in turn the Fourier transform of $B_{ij}(\tau)$ is

$$E_{ij}(n) = \frac{1}{T}\int_{-T/2}^{T/2} B_{ij}(\tau)\, e^{-i\omega_n \tau}\, d\tau \tag{4.20}$$

The autocorrelation of the periodic function $f_i(t)$ can then be expressed as

$$B_{ii}(\tau) = \sum_{n=-\infty}^{\infty} E_{ii}(n)\, e^{i\omega_n \tau} \tag{4.21}$$

where

$$E_{ii}(n) = F_i^*(n) F_i(n) = |F_i(n)|^2 \tag{4.22}$$

For zero displacement $\tau = 0$

$$B_{ii}(0) = \frac{1}{T}\int_{-T/2}^{T/2} f_i^2(t)\, dt \tag{4.23a}$$

$$= \sum_{n=-\infty}^{\infty} |F_i(n)|^2 \tag{4.23b}$$

Equation (4.23) expresses the fact that the mean square value of the function $f_i(t)$ is equal to the sum of the square of the absolute value of the spectrum over the entire range of harmonics, or in our hypothetical case, over the entire range of eddies. If $f(t)$ represents a velocity, then Equation (4.23a) is proportional to the mean kinetic energy of the flow. Moreover in view of the fact that $|F(n)|$ represents the velocity amplitude of the nth harmonic or eddy, then $|F(n)|^2$ is proportional to the kinetic energy contributed to the total kinetic energy by that particular eddy.

The function $E_{ii}(n) = |F_i(n)|^2$ is called the energy spectrum.

An interesting feature of the energy spectrum of the autocorrelation function should be noted here. Since it is the square of the absolute value of the complex spectrum $F_i(n)$ of the given periodic function its phase spectrum is always zero for all harmonics. The autocorrelation function thus retains all harmonics of the given function and contains no new ones, but discards all phase angles. In other words, all periodic functions having the same harmonic amplitudes but differing in their initial phase angles have the same autocorrelation function, which is equivalent to saying that the power spectrum of a periodic function is independent of the phase angles of its harmonics.

Returning to our hypothetical example of a turbulent flow made up of eddies having cosine velocity fluctuations which when summed together register as a periodic velocity variation with time having a rectangular pulse wave form, the autocorrelation behaves as shown in Figure 4.2.

The energy spectrum may be obtained either from direct integration, i.e.,

$$E_{ii}(n)=\frac{1}{T}\int_{-T/2}^{T/2} B_{ii}(\tau)\,e^{-i\omega_n\tau}\,d\tau \tag{4.24}$$

or more simply from

$$E_{ii}(n)=|F_i(n)|^2 \tag{4.25}$$

which for our example is

$$E_{ii}(n)=\left(\frac{u_m}{\omega_n b}\right)^2 \sin^2\frac{\omega_n b}{2} \tag{4.26}$$

The energy spectrum is shown in Figure 4.3. Figure 4.3 gives the amount of kinetic energy associated with each harmonic or eddy making up our imagined turbulent flow. It is observed that those eddies having the smaller frequencies, which are the larger eddies, contain the larger amounts of kinetic energy. Thus in our particular example the major turbulence kinetic energy is contained in the larger eddies. This is to be expected since the flow is periodic.

Unfortunately turbulence is not made up of discrete-size eddies as we have hypothesized but of all size eddies such that the range of eddy sizes is continuous. To handle this situation we will consider one more degree of complexity in our hypothetical flow, which will lead us to the Fourier integral representation of an aperiodic function.

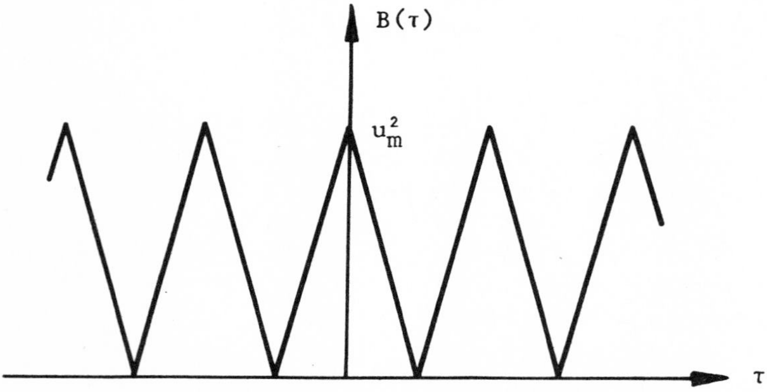

Fig. 4.2. Autocorrelation of rectangular pulse wave.

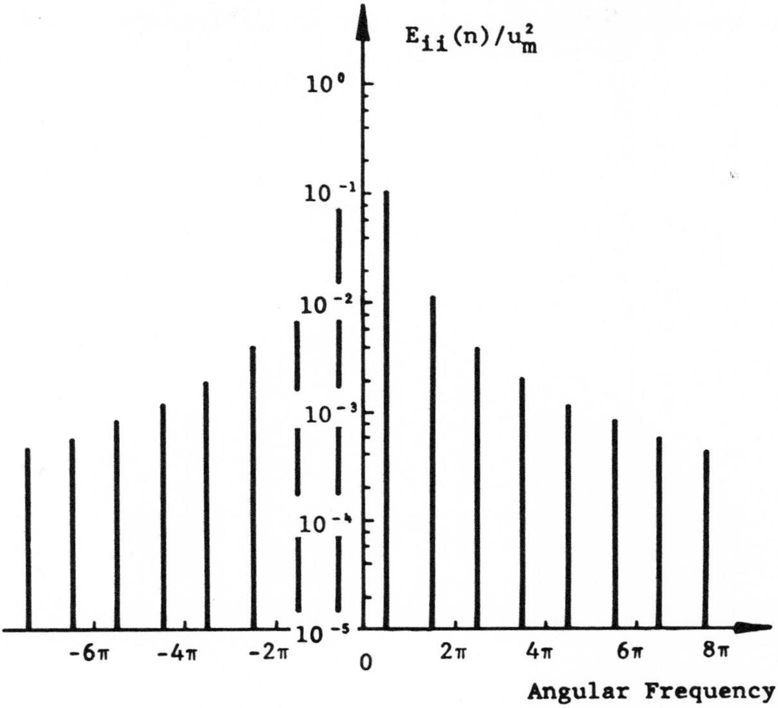

Fig. 4.3. Energy spectrum of rectangular pulse wave.

4.2.2. Fourier Integral

Consider again the periodic pulse function $f(t)$ shown in Figure 4.1. If we allow the period between pulses to become larger and larger then the pulses become more widely separated, as shown in Figure 4.4, until finally at $T \to \infty$ the function reduces to a single pulse.

A heuristic argument as to the behavior of the Fourier series as $T \to \infty$ can be gained by first considering the spectrum for $f(t)$ given by

$$F(n) = \frac{2u_m}{T} \frac{\sin \omega_n}{\omega_n} \tag{4.27}$$

where Equation (4.27) appears different from Equation (4.10) because b has been arbitrarily set equal to 2 and T has been left as a variable. Successive values of the argument ω_n appearing in Equation (4.27) differ by a constant amount, i.e.,

$$\Delta\omega = \frac{2n\pi}{T} - \frac{2(n-1)\pi}{T} = \frac{2\pi}{T} \tag{4.28}$$

We observe that as $T \to \infty$, $\Delta\omega \to 0$. Writing Equation (4.27) as

$$F(n) = \frac{u_m}{\pi} \frac{\sin \omega_n}{\omega_n} \Delta\omega \tag{4.29}$$

we observe that $(\sin \omega_n)/\omega_n$ is a factor that gives the relative scale of the amplitude of the nth harmonic. Plotting this factor at successive values of $\omega_n + \Delta\omega$ beginning at $\omega_n = 0$ results in the curves of Figure 4.5, which show the behavior of the scale factor as T becomes large.

Two features of the curves are evident: (1) The vertical scales of the curves decrease at a rate inversely proportional to T (or directly proportional to $\Delta\omega$); (2) the horizontal interval between ordinates decreases at a rate inversely proportional to T (or equal to $\Delta\omega$). The fact that as $T \to \infty$ (or $\Delta\omega \to 0$) the frequencies of the terms in

$$f(t) = \sum_{n=-\infty}^{\infty} F(n) \cos \omega_n t \tag{4.30}$$

become more and more closely spaced and the coefficients approach zero suggests that the series thought of as a function of T is actually a sum of infinitesimals whose limit is an integral.

Considering now the general complex form of the Fourier series

$$f(t) = \sum_{n=-\infty}^{\infty} F(n)\, e^{i\omega_n t} \tag{4.31}$$

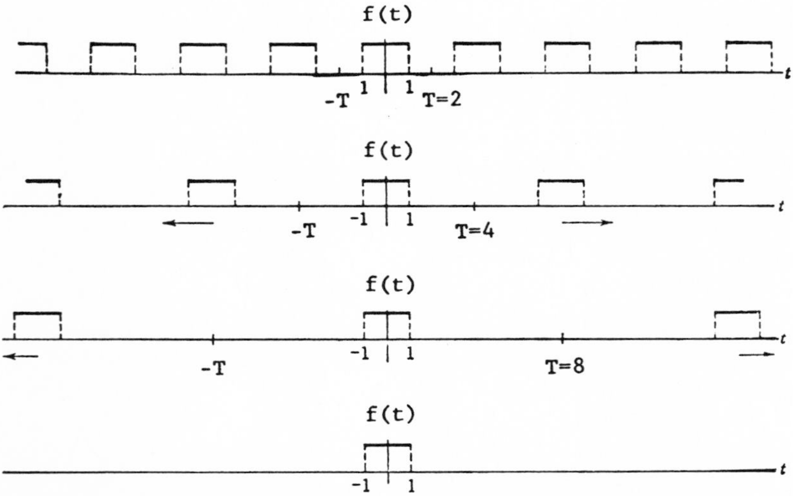

Fig. 4.4. The transition of a periodic pulse wave to an aperiodic pulse as $T \to \infty$.

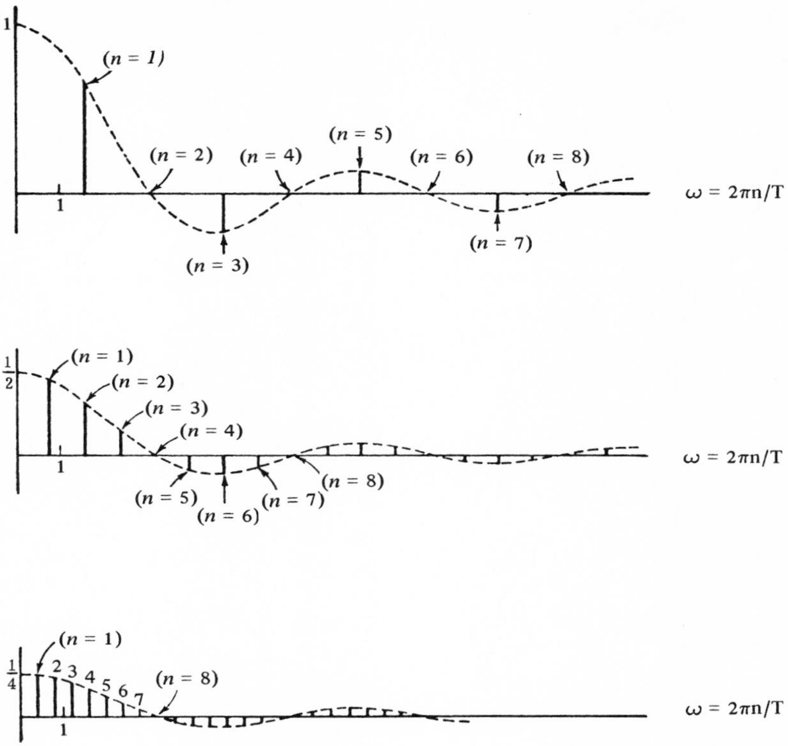

Fig. 4.5. Fourier series coefficients as T increases from upper to lower curves.

where

$$
\begin{aligned}
f(t) &= \sum_{n=-\infty}^{\infty} \frac{1}{T}\left[\int_{-T/2}^{T/2} f(t')\, e^{-i\omega_n t'}\, dt'\right] e^{i\omega_n t} \\
&= \sum_{n=-\infty}^{\infty} \frac{1}{2\pi}\left[\int_{-T/2}^{T/2} f(t')\, e^{i\omega_n t'}\, dt'\right] e^{i\omega_n t}\, \Delta\omega
\end{aligned}
\tag{4.32}
$$

Defining

$$
F(\omega) = \frac{1}{2\pi}\int_{-T/2}^{T/2} f(t)\, e^{i\omega t}\, dt \tag{4.33}
$$

Equation (4.31) becomes

$$
f(t) = \sum_{n=-\infty}^{\infty} F(\omega_n)\, e^{i\omega_n t}\, \Delta\omega \tag{4.34}
$$

where ω_n is the left-hand point in the nth subinterval $\Delta\omega = \omega_n - \omega_{n-1}$. Under very general terms the limit of this form as $\Delta\omega \to 0$ is the integral

$$\int_{-\infty}^{\infty} F(\omega)\, d\omega \tag{4.35}$$

Hence since $T \to \infty$ implies $\Delta\omega \to 0$ it follows that there is good reason to believe that as $T \to \infty$ the nonperiodic limit of $f(t)$ can be written as the integral

$$f(t) = \int_{-\infty}^{\infty} \frac{1}{2\pi} e^{i\omega t} \int_{-\infty}^{\infty} f(t')\, e^{-i\omega t'}\, dt'\, dt \tag{4.36}$$

This is the Fourier integral representation of the aperiodic function $f(t)$, where more rigorous considerations (see Reference 1) show that the validity of the integral requires that

$$\int_{-\infty}^{\infty} |f(t)|\, dt \tag{4.37}$$

be finite.

Thus

$$f(t) = \int_{-\infty}^{\infty} F(\omega)\, e^{i\omega t}\, dt \tag{4.38}$$

and

$$F(\omega) = \frac{1}{2\pi} \int_{-\infty}^{\infty} f(t)\, e^{-i\omega t}\, dt \tag{4.39}$$

are Fourier transform pairs. The function $F(\omega)$, called the complex continuous spectrum of $f(t)$, is a continuous function of the angular frequency ω and is in general complex. Hence

$$F(\omega) = P(\omega) + iQ(\omega) \tag{4.40}$$

where the amplitude density spectrum is given by the absolute value of $F(\omega)$, i.e.,

$$|F(\omega)| = [P^2(\omega) + Q^2(\omega)]^{1/2} \tag{4.41}$$

and the phase density spectrum is given by

$$\theta(\omega) = \tan^{-1}[Q(\omega)/P(\omega)] \tag{4.42}$$

The term "density" is used in view of the fact that $|F(\omega)|$ represents the amplitude per unit angular frequency. That is, in the expression

$$f(t) = \int_{-\infty}^{\infty} [F(\omega)\, d\omega]\, e^{i\omega t} \tag{4.43}$$

the function $f(t)$ is synthesized by an infinite aggregate of sinusoids $e^{i\omega t}$ of all angular frequencies ω in the continuous frequency range from $-\infty$ to ∞, each of which has an infinitesimal amplitude of

$$|F(\omega)|\,d\omega \tag{4.44}$$

It is therefore evident that $|F(\omega)|$ is not the actual amplitude characteristic of $f(t)$, because all amplitudes are of infinitesimal magnitude, but it is a characteristic that shows the relative magnitude of the infinitesimal amplitudes associated with each sinusoid, $e^{i\omega t}$.

The correlation of the aperiodic functions $f_i(t)$ and $f_j(t)$ is given by

$$B_{ij}(\tau) = \int_{-\infty}^{\infty} f_i(t) f_j(t+\tau)\,dt \tag{4.45}$$

which assumes that

$$\int_{-\infty}^{\infty} |f(t)|\,dt \tag{4.46}$$

exists for both functions. The Fourier transform of B_{ij} yields

$$E_{ij}(\omega) = 2\pi F_i^*(\omega) F_j(\omega) \tag{4.47}$$

such that

$$B_{ij}(\tau) = \int_{-\infty}^{\infty} E_{ij}(\omega)\,e^{i\omega t}\,d\omega \tag{4.48}$$

and

$$E_{ij}(\omega) = \int_{\infty}^{\infty} B_{ij}(\tau)\,e^{-i\omega\tau}\,d\tau \tag{4.49}$$

The autocorrelation becomes

$$B_{ii}(\tau) = \int_{-\infty}^{\infty} 2\pi |F_i(\omega)|^2\,e^{i\omega\tau}\,d\tau \tag{4.50}$$

where

$$E_{ii}(\omega) = 2\pi |F_i(\omega)|^2 \tag{4.51}$$

is called the energy density spectrum of $f_i(t)$. At $\tau = 0$

$$B_{ii}(0) = \int_{-\infty}^{\infty} f_i^2(t)\,dt = \int_{-\infty}^{\infty} 2\pi |F_i(\omega)|^2\,d\omega \tag{4.52}$$

If $f(t)$ is a velocity then

$$\int_{-\infty}^{\infty} f_i^2(t)\,dt \tag{4.53}$$

has units of L^2/T, hence $|F_i(\omega)|^2$ must have units of L^2/T per unit angular frequency. In this regard the development for the aperiodic function is different from that of the periodic function, where

$$\frac{1}{T}\int_{-\infty}^{\infty} f_i^2(t)\,dt = \sum_{n=-\infty}^{\infty} |F(n)|^2 \tag{4.54}$$

The units of the integral in this case are L^2/T^2, which are units of kinetic energy, and hence $|F(n)|^2$ can within a constant of proportionality be taken as the kinetic energy associated with each discrete eddy. Thus $|F(n)|^2$ has a clear physical interpretation for the case where $f(t)$ represents, say, a periodic fluid velocity fluctuation, but $2\pi|F(\omega)|^2$, on the other hand, does not have a clear physical meaning in fluid mechanics. For electrical analysis, however, where $f(t)$ is a voltage or current, then $|F(n)|^2$ becomes the average power associated with the nth harmonic and is called the power spectrum and $2\pi|F(\omega)|^2$ is the relative magnitude of the energy associated with each frequency and is called the energy density spectrum. Since electronic equipment is frequently used in turbulence work, care is needed so as not to confuse the terminology when applying it in the different disciplines.

4.2.3. Stationary Random Process

We have now extended the analysis of our hypothetical flow to include all frequencies in the continuous range of frequencies from $-\infty$ to $+\infty$. The model, however, is not yet correct, for we know that the eddies making up our turbulent flow will have random and not deterministic amplitudes. To rectify this, it is necessary to consider the harmonic analysis of random functions. We will restrict our comments to stationary random processes and begin by constructing a stationary random process from a linear combination of two random processes given by $\xi_1 e^{i\omega_1 t}$ and $\xi_2 e^{i\omega_2 t}$. The variables ξ_1 and ξ_2 are complex random variables $\xi_1 = |\xi_1| e^{i\theta_1}$ and $\xi_2 = |\xi_2| e^{i\theta_2}$ with zero mean, and the frequencies ω_1 and ω_2 are real constants. Thus

$$\xi(t) = \xi_1 e^{i\omega_1 t} + \xi_2 e^{i\omega_2 t} \tag{4.55}$$

Since the process is to be stationary, it is necessary for the ensemble mean to be a constant and the autocorrelation to depend only on the lag time $\tau = t_2 - t_1$. (Although this is not a rigorous description of a stationary function it will suffice for our purposes—see Reference 2.) Clearly

$$\xi(t) = \xi_1 e^{i\omega_1 t} + \xi_2 e^{i\omega_2 t} \tag{4.55}$$

which satisfies the first requirement. The autocorrelation of a complex random function is defined in terms of the complex conjugate as $\langle\xi(t)\xi^*(t+\tau)\rangle$. Thus

$$\langle\xi(t)\xi^*(t+\tau)\rangle=\langle|\xi_1|^2\rangle\, e^{i\omega_1\tau}+\langle\xi_1\xi_2^*\rangle\, e^{i(\omega_1-\omega_2)t+i\omega_1\tau}$$
$$+\langle\xi_1^*\xi_2\rangle\, e^{-i(\omega_1-\omega_2)t+i\omega_2\tau}+\langle|\xi_2|^2\rangle\, e^{i\omega_2\tau} \qquad (4.57)$$

from which it is apparent that the autocorrelation will depend only on τ if $\langle\xi_1^*\xi_2\rangle=\langle\xi_2^*\xi_1\rangle=0$. The constructed random process is therefore stationary only if ξ_1 and ξ_2 are uncorrelated random variables with mean zero. The stationary random process, $\xi(t)$, is a superposition of two uncorrelated (possible independent) oscillations of different frequencies and with random amplitudes $|\xi|$ and random phases θ. The autocorrelation is

$$B_{ii}(\tau)=\langle|\xi_1|^2\rangle\, e^{i\omega_1\tau}+\langle|\xi_2|^2\rangle\, e^{i\omega_2\tau} \qquad (4.58)$$

where if $|\xi_1|$ and $|\xi_2|$ are amplitudes of velocity fluctuations then $\langle|\xi_1|^2\rangle$ and $\langle|\xi_2^2|\rangle$ are, within a constant factor, the average kinetic energy of the individual oscillations. Again we observe that the autocorrelation is independent of the phases of the oscillations. Physically $\xi(t)$ must be real for turbulence analyses, which puts more stringent restrictions on the form of $\xi(t)$. For $\xi(t)$ to be real it is necessary that $\omega_2=-\omega_1=-\omega$ and that $\xi_2=\xi_1^*$. Accordingly we set $\xi_1=(\eta-i\zeta)/2$ and $\xi_2=(\eta+i\zeta)/2$ and introduce them into Equation (4.55), which gives, after appropriate algebra,

$$\xi(t)=\eta\cos\omega t+\zeta\sin\omega t \qquad (4.59)$$

If the autocorrelation for the real process

$$B_{ii}(\tau)=\langle(\eta\cos\omega t+\zeta\sin\omega t)[\eta\cos\omega(t+\tau)+\zeta\sin\omega(t+\tau)]\rangle \qquad (4.60)$$

is expanded one obtains

$$\begin{aligned}B_{ii}(\tau)=&\langle\eta^2\rangle\cos\omega\tau(\cos^2\omega t)+\langle\zeta^2\rangle\cos\omega\tau(\sin^2\omega t)\\&-\langle\eta^2\rangle\sin\omega\tau(\sin\omega t\cos\omega t)\\&+\langle\zeta^2\rangle\sin\omega\tau(\sin\omega t\cos\omega t)\\&+\langle\eta\zeta\rangle\cos\omega\tau(\cos\omega t\sin\omega t+\sin\omega t\cos\omega t)\\&+\langle\eta\zeta\rangle\sin\omega\tau(\cos^2\omega t-\sin^2\omega t)\end{aligned} \qquad (4.61)$$

which is stationary only if $\langle\eta^2\rangle=\langle\zeta^2\rangle$ and $\langle\eta\zeta\rangle=0$. The autocorrelation becomes

$$B_{ii}(\tau)=b\cos\omega\tau \qquad (4.62)$$

where

$$b=\langle\eta^2\rangle=\langle\zeta^2\rangle$$

Going a step further, we can construct a more general random process by the superposition of N periodic oscillations of different frequencies, i.e.,

$$\xi(t)=\sum_{k=1}^{N} \xi_k\, e^{i\omega_k t} \tag{4.63}$$

From our preceding arguments it is evident that $\xi(t)$ will be stationary only if

$$\langle \xi_1\rangle=\langle \xi_2\rangle=\cdots=\langle \xi_N\rangle=0 \tag{4.64}$$

and

$$\langle \xi_l \xi_k\rangle=0, \qquad l\neq k \tag{4.65}$$

Then

$$B_{ii}(\tau)=\sum_{k=1}^{N} \langle |\xi_k|^2\rangle\, e^{i\omega_k \tau} \tag{4.66}$$

where $|\xi_k|^2$ is proportional to the average kinetic energy of the separate harmonic oscillations. For $\tau=0$

$$B_{ii}(0)=\sum_{k=1}^{N} |\xi_k|^2 \tag{4.67}$$

which states that in the superposition of uncorrelated periodic oscillations the average kinetic energy of the composite oscillations equals the sum of the average kinetic energies of the separate periodic components.

In order that the above process be real, it is necessary that N be an even number equal to $2M$ and the terms in the sum must separate into M pairs of complex conjugate terms $[\xi_k\, e^{i\omega_k t}, \xi_k^*\, e^{-i\omega_k t}]$. These conditions being satisfied

$$\xi(t)=\sum_{j=1}^{M} (\eta_j \cos \omega_j t+\zeta_j \sin \omega_j t) \tag{4.68}$$

where

$$\begin{aligned} \langle \eta_j \zeta_k\rangle &=0 \qquad \text{for all } j, k \\ \langle \eta_j \eta_k\rangle=\langle \zeta_j \zeta_k\rangle &=0, \qquad k\neq j \\ \langle \eta_j^2\rangle=\langle \zeta_j^2\rangle &=b_j \end{aligned} \tag{4.69}$$

and

$$B_{ii}(\tau)=\sum_{j=1}^{\infty} b_j \cos \omega_j \tau \tag{4.70}$$

if we assure that the quantity

$$\left(\sum_{k=1}^{\infty} \langle |\xi_k|^2\rangle\right)<\infty \tag{4.71}$$

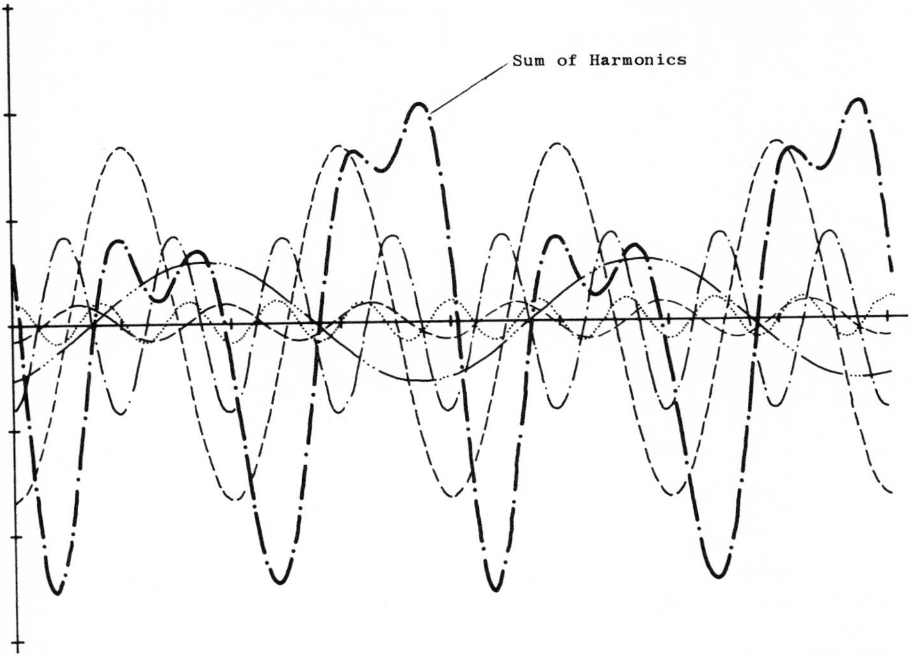

Fig. 4.6. One realization of a stationary real random process generated by the superposition of five sinusoids of random amplitude.

Figure 4.6 illustrates the generation of one realization of a stationary, real random process by the superposition of five sinusoids, i.e.,

$$\xi(t) = \sum_{k=1}^{5} (\eta_k \cos \omega_k t + \zeta_k \sin \omega_k t) \tag{4.72}$$

where for illustration purposes ω_k has been taken as $k\pi/2$ and η_k and ζ_k are assumed normally distributed with zero mean and standard deviations $\sigma_\eta = 2$ and $\sigma_\zeta = 1$, respectively.

4.2.4. Spectral Representation of a Stationary Random Process

We have observed how the representation of a deterministic periodic function with a Fourier series was possible and how the resulting spectrum was discrete. For an aperiodic function a Fourier integral representation was applied that required a continuous range of frequencies from $-\infty$ to $+\infty$. This generated a continuous spectrum. The Fourier integral representation was possible, however, only for functions that vanish sufficiently rapidly at infinity, and it did not have much physical interpretation from a fluid mechanics point of view, although it did illustrate how the transition from discrete to continuous frequencies could be achieved. A stationary random process was then constructed that—in analogy to the Fourier series—was a sum of sinusoids whose coefficients were random variables. This randomness is in accord with the way in which we would expect a velocity record from a turbulent flow to behave, with the exception that turbulence is not periodic. Thus we are led to a random aperiodic process, which suggests a Fourier integral. Such a nonperiodic function will conceptually go on forever and will not vanish at infinity, as is, strictly speaking, required for the Fourier integral. Fortunately, it can be shown that, owing to the nature of random functions, a Fourier expansion with a clear physical interpretation is possible for any stationary random process or homogeneous random field.

To see this, consider the real stationary random process

$$\xi(t) = \sum_{k=1}^{\infty} (\eta_k \cos \omega_k t + \zeta_k \sin \omega_k t) \tag{4.73}$$

Beginning with the first term on the right-hand side (i.e., $\sum_{k=1}^{\infty} \eta_k \cos \omega_k t$) we plot η_k, for one realization, as a function of ω_k, where without loss of generality ω_k is taken equal to $\omega_k = 2k\pi/T$, and T is the period of the sinusoid. Figure 4.7 shows the results of this plot for increasing values of T. We observed that as T becomes large the values of ω_k become closer and closer together and the discrete values of η_k become more and more densely packed. Let us initially focus attention on the interval $[\omega_a, \omega_b]$. Figure 4.7 is similar to the plot of coefficients shown in Figure 4.5 except that the values of η_k do not decrease with increasing T nor does a line drawn through their peaks approach a smooth curve like the dashed line in Figure 4.5. It is obvious that this process will not converge toward the common Riemann integral as it did for the Fourier integral, and therefore before going further in our development we must digress and review the concept of the Stieltjes integral.

The Stieltjes integral is defined as[3]

$$\int_a^b f(x)\, dg(x) = \lim_{|P|\to 0} \sum_{i=0}^{N} f(x_i')[g(x_i) - g(x_{i-1})] \tag{4.74}$$

where f and g are two functions defined on the closed interval $[a, b]$. The interval $[a, b]$ is partitioned, P, into $(x_0, x_1, \ldots, x_N)$ and a set of points $x_1', \ldots, x_N'$ with one in each subinterval of the partition. The mesh fineness $|P|$ is the maximum of the differences $x_1 - x_0$, $x_2 - x_1, \ldots, x_N - x_{N-1}$. The important feature of the Stieltjes integral is that $g(x)$ can be a discontinuous function.

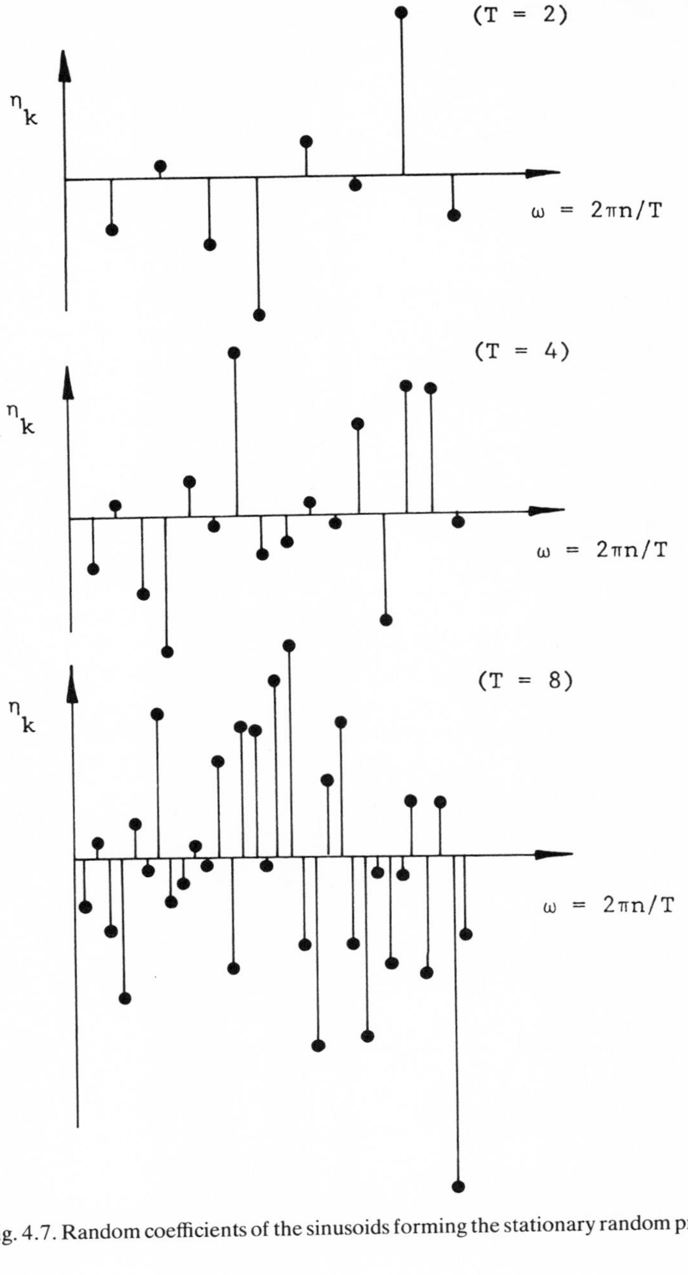

Fig. 4.7. Random coefficients of the sinusoids forming the stationary random process.

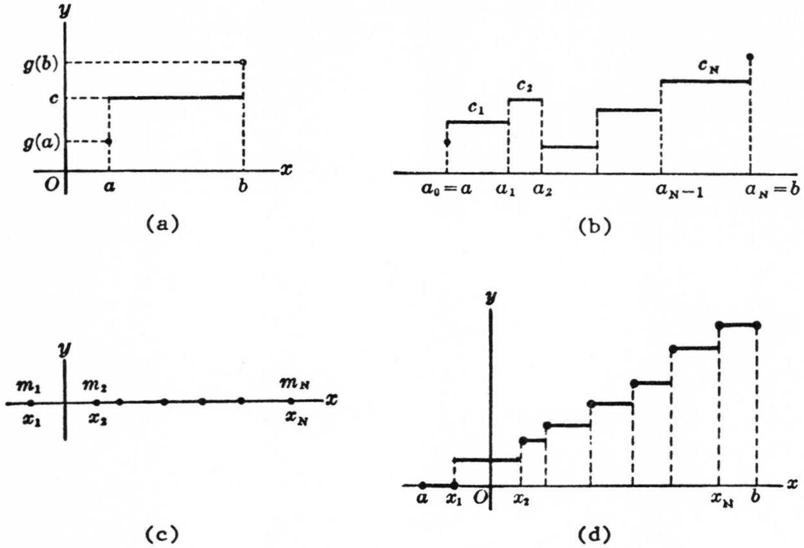

Fig. 4.8. Discontinuous functions $g(x)$ in the Stieltjes integral.

As an example consider the function $g(x)$ shown in Figure 4.8a. Substituting into Equation (4.74) gives

$$\int_a^b f(x)\,dg(x) = \lim_{|P|\to 0} \{f(x_1')[g(x_1)-g(x_0)]+f(x_2')[g(x_2)-g(x_1)]+\cdots$$
$$+f(x_N')[g(x_N)-g(x_{N-1})]\} \tag{4.75}$$

Now

$$g(x_k)-g(x_{k-1})=c-c=0, \qquad k\neq 0 \text{ or } N$$

Thus

$$\int_a^b f(x)\,dg(x) = \lim_{|P|\to 0} f(x_1')[g(x_1)-g(x_0)]+f(x_N')[g(x_n)-g(x_{N-1})] \tag{4.76}$$

which on introducing $g(x_1)=g(x_{N-1})=c$, $g(x_0)=g(a)$, and $g(x_N)=g(b)$ becomes

$$\int_a^b f(x)\,dg(x)=f(a)[c-g(a)]+f(b)[g(b)-c] \tag{4.77}$$

where, since $f(x)$ is continuous,

$$\lim_{|P|\to 0} f(x_0')=f(a)$$

and

$$\lim_{|P|\to 0} f(x'_N) = f(b)$$

A second example is shown in Figure 4.8b, where the interval $[a, b]$ is divided into N parts by the partition $(a_0, a_1, \ldots, a_N)$ and $g(x)$ has the value c_i in the interior of the ith subinterval. The values of $g(x)$ at the point $a, \ldots, a_N$ can be quite independent of the values of $c, \ldots, c_N$. If f is continuous on a, b, the Stieltjes integral becomes

$$\int_a^b f(x)\, dg(x) = f(a)[c_1 - g(a)] + f(a_1)[c_2 - c_1] + \cdots + f(a_{N-1})[c_N - c_{N-1}] + f(b)[g(b) - c_N] \qquad (4.78)$$

Observe that the values of g at $a_1, \ldots, a_{N-1}$ (the interior points of the discontinuity) do not enter into the formula.

A final example, which is useful for our development of the spectral representation of a stationary random process, is to consider the moments of inertia of a number of mass particles distributed along the x axis as shown in Figure 4.8c. The masses $m_1, \ldots, m_N$ are located as indicated with $x_1 < x_2 < \cdots < x_N$. The moment of inertia of this mass system about the y axis is

$$I = m_1 x_1^2 + m_2 x_2^2 + \cdots + m_N x_N^2 \qquad (4.79)$$

which can be written in the form of a Stieltjes integral as follows. Take any closed interval $[a, b]$ containing all points $x_1, \ldots, x_N$. For any x in $[a, b]$ define $g(x) = 0$ if $a \le x \le x_1$ and, if $x_1 < x$, define $g(x)$ as the sum of all masses m_i for which $a \le x_i \le x$. The graph of $g(x)$ appears as in Figure 4.8d. The exact definition of $g(x)$ is

$$g(x) = \begin{cases} 0 & \text{if } a \le x \le x_1 \\ m_1 & \text{if } x_1 < x < x_2 \\ m_1 + m_2 & \text{if } x_2 \le x < x_3 \\ \vdots & \vdots \\ m_1 + m_2 + \cdots + m_N & \text{if } x_N \le x \le b \end{cases} \qquad (4.80)$$

Equation (4.74) gives

$$\int_a^b f(x)\, dg(x) = f(x_1) m_1 + \cdots + f(x_N) m_N \qquad (4.81)$$

in particular if $f(x)=x^2$ then

$$\int_a^b f(x)\,dg(x)=m_1x_1^2+\cdots+m_Nx_N^2=I \tag{4.82}$$

Now returning to our consideration of the plot of η_k versus ω_k, Figure 4.7, we observe that this plot is not unlike the distribution of point masses along the x axis discussed in our third example. In fact we can construct a function $Z_1(\omega)$ in a manner identical to our construction of $g(x)$, i.e.,

$$Z_1(\omega)=\begin{cases} 0, & \omega_a \le \omega < \omega_1 \\ \eta_1, & \omega_1 \le \omega < \omega_2 \\ \eta_1+\eta_2, & \omega_2 \le \omega < \omega_3 \\ \cdot & \cdot \\ \cdot & \cdot \\ \cdot & \cdot \\ \eta_1+\eta_2+\cdots+\eta_{N-1} & \omega_{N-1} \le \omega < \omega_N \\ \eta_1+\eta_2+\cdots+\eta_{N-1}+\eta_N & \omega_N \le \omega \le \omega_b \end{cases} \tag{4.83}$$

A plot of $Z_1(\omega)$ would appear as shown in Figure 4.9. Note that $Z_1(\omega)$ is a random variable and will be finite even if $\omega_a \to -\infty$ and $\omega_b \to \infty$ since there are as many positive η's as there are negative η's making up the sum.

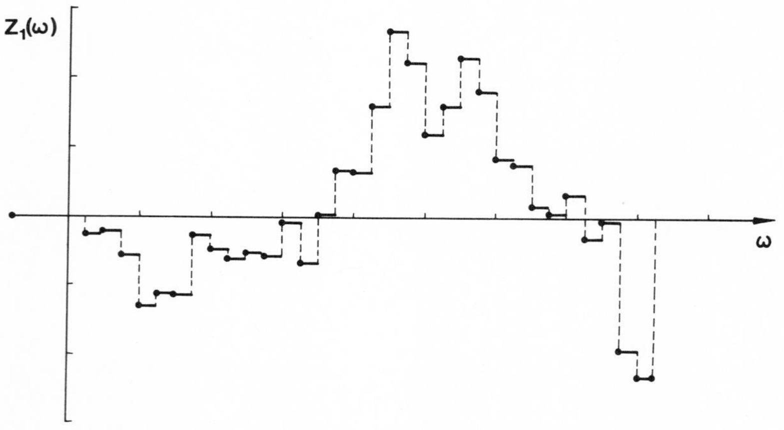

Fig. 4.9. The random variable $Z_1(\omega)$.

Substituting Equation (4.83) into Equation (4.74) gives

$$\int_{\omega_a}^{\omega_b} f(\omega)\,dZ_1(\omega) = \lim_{\max(\omega_k-\omega_{k-1})\to 0} \{f(\omega_1')[Z_1(\omega_1)-Z_1(\omega_0)]+f(\omega_2')$$
$$\times[Z_1(\omega_2)+Z_1(\omega_1)]+\cdots$$
$$+f(\omega')[Z_1(\omega_N)-Z_1(\omega_{N-1})]+f(\omega_{N+1}')$$
$$\times[Z_1(\omega_{N+1})-Z_1(\omega_N)]\} \tag{4.84}$$

If $f(x_k')$ is defined as $\cos\omega_k' t$ and the limit is taken we obtain

$$\int_{\omega_a}^{\omega_b} \cos\omega t\,dZ_1(\omega) = \sum_{k=1}^{N} [Z(\omega_k)-Z(\omega_{k-1})]\cos\omega_k t$$
$$= \sum_{k=1}^{N} \eta_k \cos\omega_k t \tag{4.85}$$

Thus our representation of $\xi(t)$ over the interval $[\omega_a, \omega_b]$ can be expressed in terms of the Stieltjes integral

$$\xi(t) = \int_{\omega_a}^{\omega_b} \cos\omega t\,dZ_1(\omega) + \int_{\omega_a}^{\omega_b} \sin\omega t\,dZ_2(\omega) \tag{4.86}$$

where $Z_2(\omega)$ is defined in a manner identical to $Z_1(\omega)$, with ζ_k replacing η_k.

We have developed the real form of $\xi(t)$ purely for illustrative purposes. The complex development follows directly from the same arguments. In this case $\xi(t)$ becomes

$$\xi(t) = \int_{\omega_a}^{\omega_b} e^{i\omega t}\,dZ(\omega) \tag{4.87}$$

where the integral is the Fourier–Stieltjes integral and $Z(\omega)$ is defined as

$$Z(\omega) = \begin{cases} 0 & \omega \le \omega < \omega_1 \\ \xi_1 & \omega_1 \le \omega < \omega_2 \\ \xi_1+\xi_2 & \omega_2 \le \omega < \omega_3 \\ \cdot & \cdot \\ \cdot & \cdot \\ \cdot & \cdot \\ \xi_1+\xi_2+\cdots+\xi_N & \omega_N \le \omega \le \omega_{N+1} \end{cases} \tag{4.88}$$

In the preceding discussion the frequency has maintained discrete values. Since we desire that $\xi(t)$ will be representative of the random velocity fluctuation encountered in turbulent flows all frequencies must be included such that ω becomes continuous. Figure 4.7 illustrates that as $T\to\infty$,

$\omega_{k+1} \to \omega_k$ and the values of η_k packed in an increment $\Delta\omega$ become infinite and, hence, continuous. This does not destroy the concept of the Stieltjes integral, however. Further, we can allow $\omega_a \to -\infty$ and $\omega_b \to +\infty$ such that Equation (4.87) becomes the improper Fourier–Stieltjes integral,

$$\xi(t) = \int_{-\infty}^{\infty} e^{i\omega t}\, dZ(\omega) \tag{4.89}$$

Equation (4.89) expresses the stationary random process as a sum of sinusoids of all frequencies each having a random amplitude. Thus $\xi(t)$ should display the characteristic expected of a velocity fluctuation measured at a point in a turbulence field.

It is readily apparent that the function $Z(\omega) = Z[-\infty, \omega]$ is a complex random function of ω, called a random point function. It has zero mean,

$$\langle Z(\omega) \rangle = \langle \xi_1 \rangle + \langle \xi_2 \rangle + \langle \xi_3 \rangle + \cdots = 0 \tag{4.90}$$

and

$$\langle [Z(\omega_n) - Z(\omega_m)][Z^*(\omega_l) - Z^*(\omega_k)] \rangle = 0 \tag{4.91}$$

where

$$\omega_m < \omega_n \leq \omega_k < \omega_l \tag{4.92}$$

The latter conditions follow from Equation (4.65), which requires $\langle \xi_k \xi_l^* \rangle = 0$, $k \neq l$, and is immediately evident by expanding Equation (4.91), i.e.,

$$\begin{aligned}
&\langle [\xi_n + \xi_{n-1} + \cdots + \xi_m + \cdots + \xi_1 - \xi_m - \cdots - \xi_1] \\
&\quad \times [\xi_l^* + \cdots + \xi_k^* + \cdots + \xi_n^* + \cdots + \xi_1 - \xi_k^* - \cdots - \xi_1^*] \rangle \\
&= \langle [\xi_n + \cdots + \xi_{m+1}][\xi_l^* + \cdots + \xi_{k+1}^*] \rangle \\
&= \langle \xi_n \xi_l^* \rangle + \cdots + \langle \xi_{m+1} \xi_{k+1}^* \rangle = 0
\end{aligned} \tag{4.93}$$

where the indices are not equal in any of the correlations because of the restriction imposed by Equation (4.92). A more convenient differential form for Equation (4.91) is

$$\langle dZ(\omega)\, dZ^*(\omega_1) \rangle = 0, \qquad \omega \neq \omega_1 \tag{4.94}$$

The representation of $\xi(t)$ as given by Equation (4.88) is called the spectral representation of the stationary random process $\xi(t)$. Since $\xi(t)$ has the characteristic that allows us to model time-dependent velocity fluctuations in turbulent flow we will change the nomenclature $\xi(t)$ to $u(t)$ at this point to give the connotation of a velocity fluctuation. Note, however, that $\xi(t)$ could represent temperature, pressure, or any other random turbulence parameter.

4.2.5. Autocorrelation

The autocorrelation function of a complex random process is given by

$$B_{ii}(\tau) = \langle u_i(t+\tau)u_i^*(t)\rangle \tag{4.95}$$

Introducing $u_i(t+\tau)$ and $u_i^*(t)$ in the form of Equation (4.73) and imposing the conditions of Equation (4.94), we arrive at

$$B_{ii}(\tau) = \int_{-\infty}^{\infty} e^{i\omega\tau}\, dF(\omega) \tag{4.96}$$

where

$$dF(\omega) = \langle |dZ(\omega)|^2\rangle \tag{4.97}$$

For the special case where the absolute value of $B_{ii}(\tau)$ falls off so rapidly with τ that

$$\int_{-\infty}^{\infty} |B(\tau)|\, d\tau < \infty \tag{4.98}$$

it can be shown[4] that Equation (4.96) can be written as

$$B_{ii}(\tau) = \int_{-\infty}^{\infty} e^{i\omega\tau} f(\omega)\, d\omega \tag{4.99}$$

where $f(\omega)$ is called the spectral density of the process $u_i(t)$. $F(\omega)$, called the spectral distribution function, is related to $f(\omega)$ by

$$F(\omega) = \int_{-\infty}^{\omega} f(\omega)\, d\omega \tag{4.100}$$

and obviously

$$f(\omega) = \frac{dF(\omega)}{d\omega} \tag{4.101}$$

so that $F(\omega)$ is everywhere differentiable.

Measurements of $B_{ii}(\tau)$ in turbulent flows justify Equation (4.98) and hence the spectral density function is meaningful in most turbulence applications. This has the particularly attractive feature of enabling us to get along without Stieltjes integrals when working with correlation functions. However, $Z(\omega)$ is in general not differentiable and $Z(\omega)$ cannot be expressed in a form similar to Equation (4.100).

It is useful to investigate the physical significance of the spectral density function, $f(\omega)$. The function $B_{ii}(\tau)$ at $\tau = 0$ reduces to $\langle u_i^2(t)\rangle$, which is proportional to the mean turbulence kinetic energy. Since

$$B_{ii}(0) = \langle u_i^2(t)\rangle = \int_{-\infty}^{\infty} f(\omega)\, d\omega \tag{4.102}$$

it is apparent that $f(\omega)$ is the relative magnitude of the kinetic energy associated with a given frequency ω. The function $f(\omega)$ is therefore called the energy spectrum density function and we shall designate it as $\varepsilon_{ii}(\omega)$.

Both $\varepsilon_{ii}(\omega)$ and $B_{ii}(\tau)$ can be measured experimentally, the former with a spectrum analyzer (a filtering device that transmits harmonic oscillations in a given frequency range and rejects oscillations corresponding to other frequencies) and the latter with a hot-wire anemometer and appropriate electronic equipment. In turn, if one quantity is measured, say $\varepsilon_{ii}(\omega)$, then $B_{ii}(\tau)$ is known from Equation (4.99), or if $B_{ii}(\tau)$ is measured then $\varepsilon_{ii}(\omega)$ can be determined from the inverse Fourier transform, i.e.,

$$\varepsilon_{ii}(\omega)=\frac{1}{2\pi}\int_{-\infty}^{\infty} e^{-i\omega\tau}B_{ii}(\tau)\,d\tau \tag{4.103}$$

In the case of a real process, the spectral distribution function $\varepsilon_{ii}(\omega)$ is an even function of ω and therefore

$$B_{ii}(\tau)=\int_0^{\infty} \cos\omega\tau\, E_{ii}(\omega)\,d\omega \tag{4.104}$$

where

$$E_{ii}(\omega)=2\varepsilon_{ii}(\omega)$$

Also

$$E_{ii}(\omega)=\frac{2}{\pi}\int_0^{\infty} \cos\omega\tau\, B_{ii}(\tau)\,d\tau \tag{4.105}$$

The function $E_{ii}(\omega)$ differs from $\varepsilon_{ii}(\omega)$ only by the factor of 2 and by the fact that it is defined over the interval $[0,\infty]$ rather than $[-\infty,\infty]$. It must be borne in mind that the use of Equations (4.99)–(4.104) requires the condition of Equation (4.98).

Further if $B_{kj}(\tau)=\langle u_k(t)u_j(t+\tau)\rangle$ falls to zero sufficiently rapidly that

$$\int_0^{\infty} |B_{kj}(\tau)|\,d\tau<\infty \tag{4.106}$$

then it can be shown that

$$B_{kj}(\tau)=\int_0^{\infty} E_{kj}(\omega)\cos\omega\tau\,d\omega \tag{4.107}$$

and

$$E_{kj}(\omega)=\frac{2}{\pi}\int_0^{\infty} B_{kj}(\tau)\cos\omega\tau\,d\tau \tag{4.108}$$

The function $E_{kj}(\omega)$ with $k\neq j$ is the cross-spectral density function of the processes $u_k(t)$ and $u_j(t)$.

4.3. Frequency Spectra

Frequency spectra for stationary processes can be determined by measuring the correlation $B_{ij}(\tau)$ and applying either Equation (4.105) or Equation (4.108), or they can be measured directly with a frequency analyzer. If the signal from a probe measuring, say, the u_1 turbulence velocity fluctuations is processed directly through a root-mean-square meter, the reading will be proportional to the mean turbulence kinetic energy $\overline{u_1^2}$ and will include the contributions from the entire frequency range of eddies. However, if the signal is first processed through a spectrum analyzer, which is an arrangement of filters that permits only a small selected band of frequencies $\Delta\omega$ to pass, then the root-mean-square meter reading will be proportional to the turbulence kinetic energy contained by eddies having frequencies only in this small bandwidth. Ideally the value of the energy spectral density function at the midpoint of the band would then be given by

$$E_{ii}(\omega) = \overline{u_i^2(\Delta\omega)}/\Delta\omega \tag{4.109}$$

where $\overline{u_i^2(\Delta\omega)}$ is the square of the root-mean-square reading. This assumes that the filter has ideal transmission characteristics as illustrated in Figure 4.10, whereas in practice the filter will not be ideal and will characteristically transmit as shown also in Figure 4.10.

Therefore some distortion enters the measurement of a turbulence kinetic energy spectrum. However, with quality filters we can measure to a good degree of accuracy the contribution from various frequency bands at selected values of ω over a wide range of frequencies and by applying Equation (4.109) generate an energy spectral density curve.

The energy spectrum curve established by plotting the value of $\overline{u_1^2(\Delta\omega)}$ versus frequency for each frequency band investigated would appear typically as illustrated in Figure 4.11. The function of frequency, $E_{11}(\omega)$, defined by the curve is called in addition to the energy spectrum density the one-dimensional energy spectrum function, the power spectral density function, or simply the spectral density. It follows that

$$\overline{u_1^2} = \int_0^\infty E_{11}(\omega)\, d\omega \tag{4.110}$$

Similarly $\overline{u_2^2}$ and $\overline{u_3^2}$ are related to $E_{22}(\omega)$ and $E_{33}(\omega)$, respectively.

Turbulent fluctuations generally have a broad spectrum; fairly flat peaks can on occasion be distinguished but sharp peaks at discrete frequencies occur only, if at all, during transitional stages from laminar to turbulent flow.

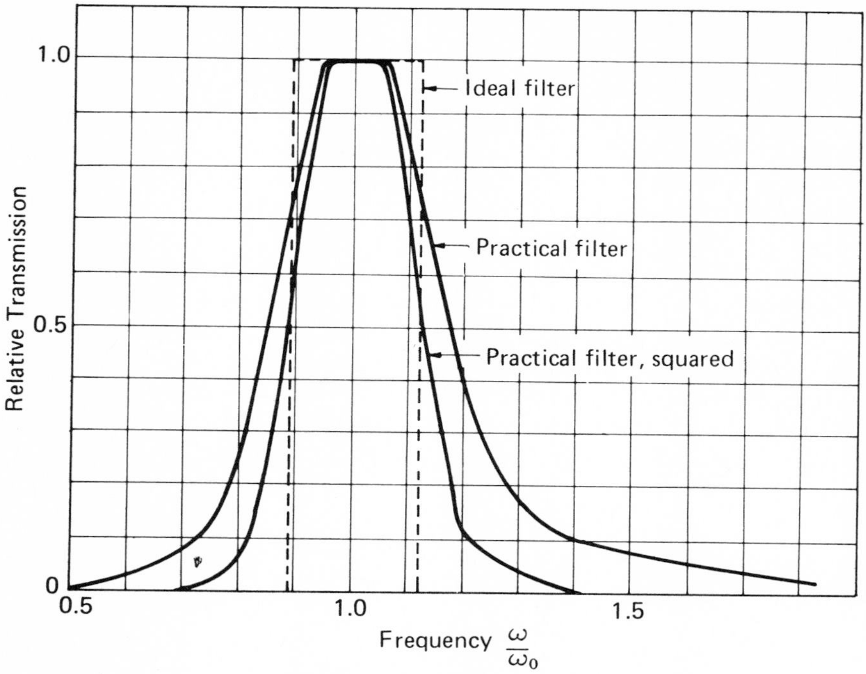

Fig. 4.10. Typical filter transmission.

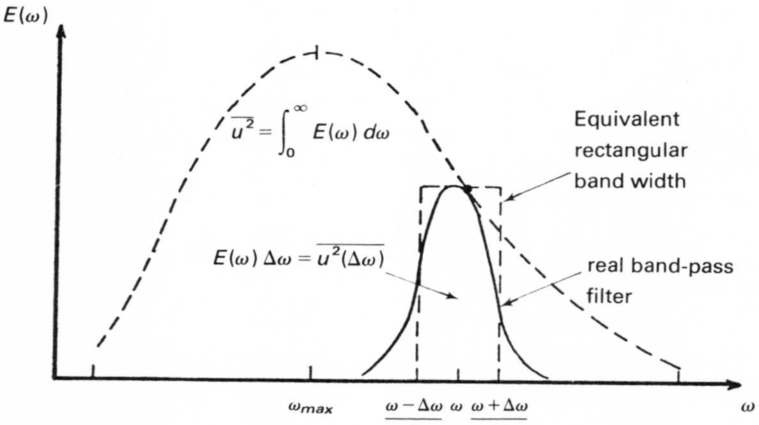

Fig. 4.11. Energy spectrum and a typical frequency band $E(\omega)\,\Delta\omega$.

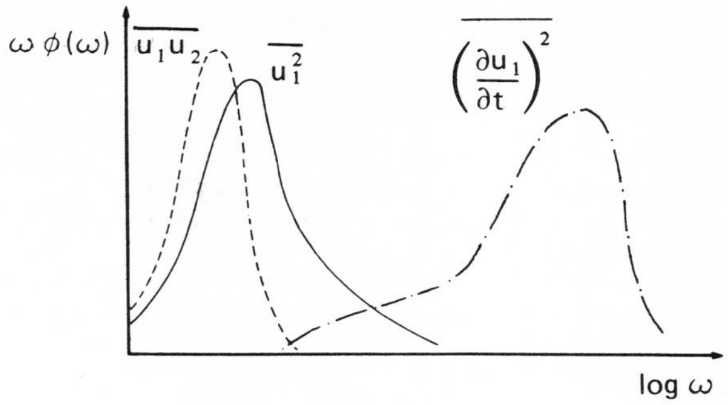

Fig. 4.12. Normalized spectral density functions.

Frequently the energy spectral density is normalized as

$$\phi_{ii}(\omega) = E_{ii}(\omega)/\overline{u_1^2} \tag{4.111}$$

and thus

$$\int_0^\infty \phi_{ii}(\omega)\, d\omega = 1.0 \tag{4.112}$$

The normalized density has the dimensions of time. Common practice also nondimensionalizes the frequency with respect to ω_{max}, the value for which $\phi_{ii}(\omega)$ is a maximum.

Recall that the Reynolds stress tensor has the nine components $\overline{u_i(\mathbf{r})u_j(\mathbf{r})}$. The frequency spectrum tensor in turn is composed of nine components, $E_{ij}(\omega)$, defined by the relationship

$$\overline{u_i(\mathbf{r})u_j(\mathbf{r})} = \int_0^\infty E_{ij}(\omega)\, d\omega \tag{4.113}$$

The energy spectra corresponding to cross correlations are called cross spectra. The integral of the normalized spectrum for $\tau = 0$ over all frequencies is not necessarily unity for the cross spectrum.

Typical plots of the normalized spectral density function for $\overline{u_1^2}$ and $\overline{u_1 u_2}$ are shown in Figure 4.12. Also shown on this plot is $\overline{(\partial u_1/\partial t)^2}$, the significance of which is described later.

In plotting power spectral density distributions, $\omega\phi(\omega)$ is normally plotted versus $\log \omega$. The frequency multiplier gives prominence to high-frequency effects and also enables direct integration under the curve for $\overline{u^2}$, i.e.,

$$\omega\phi(\omega)\, d(\log \omega) = \phi(\omega)\, d\omega \tag{4.114}$$

4.4. Wave-Number Spectra

4.4.1. From Taylor's Hypothesis

Wave-number spectra provide insight into the energy levels associated with eddy sizes rather than frequencies. Wave-number spectra cannot conveniently be measured directly with a commercial spectrum analyzer as are frequency spectra. They are normally measured indirectly either by Fourier transformations of spatial correlation functions or by the application of Taylor's hypothesis to frequency spectra.

With the latter technique the assumption is made that the entire flow field is translated with a uniform velocity $\bar{U}$, say in the x_i direction, and that $\bar{U}$ is large compared with the turbulence velocity fluctuations u_i. The sequences of changes in u_i at a fixed measuring station are then simply those due to the passage of an unchanging pattern of turbulence (Figure 4.13). The variation of u_i in space can then be approximated with the variation in time where the time lag, τ, of the time correlation is related to the separation distance, l, of the spatial correlation in the direction of the flow through the mean velocity $l = \bar{U}\tau$. Therefore, since the velocity fluctuation with time at the point where the probe is located will be nearly identical with the instantaneous distribution of the velocity u_i along the x_i axis passing through that point, the correlation $u_i(t)u_i(t+\tau)$ averaged with respect to time, t, must be nearly identical to the second-order moment $u_i(x_i)u_i(x_i+l_i)$ averaged with respect to position, x_i. Thus

$$B_{i,i}(l) = B_{ii}(\tau) \tag{4.115}$$

Applying Equation (4.104) where ω the radian frequency is related to the wave number κ_i by $\omega/\bar{U}$ gives

$$B_{i,i}(l) = B_{ii}(\tau) = \bar{U}\int_0^\infty E_{ii}(\omega)\cos(\kappa_i l_i)\, d\kappa_i \tag{4.116}$$

Defining $E_{i,i}(\kappa) = \bar{U}E_{ii}(\omega)/2\pi$ we arrive at

$$B_{i,i}(l) = \int_0^\infty E_{i,i}(\kappa)\cos(\kappa_i l_i)\, d\kappa_i \tag{4.117}$$

and similarly

$$B_{i,i}(l) = \int_0^\infty E_{i,i}(\kappa)\, e^{i\kappa_i l_i}\, d\kappa_i \tag{4.118}$$

The above result requires that ω and κ have consistent units, i.e., $\omega \sim$ rad/time and $\kappa \sim$ rad/length. If frequency, n, is measured in cycles per unit time, then the wave number, k, is measured in cycles per unit length. κ and k are related by $\kappa = 2\pi k$. Note that $\bar{U} = \omega/\kappa = 2\pi n/\kappa = n/k$ and $E_{i,i}(k) = \bar{U}E_{ii}(n)/2\pi$.

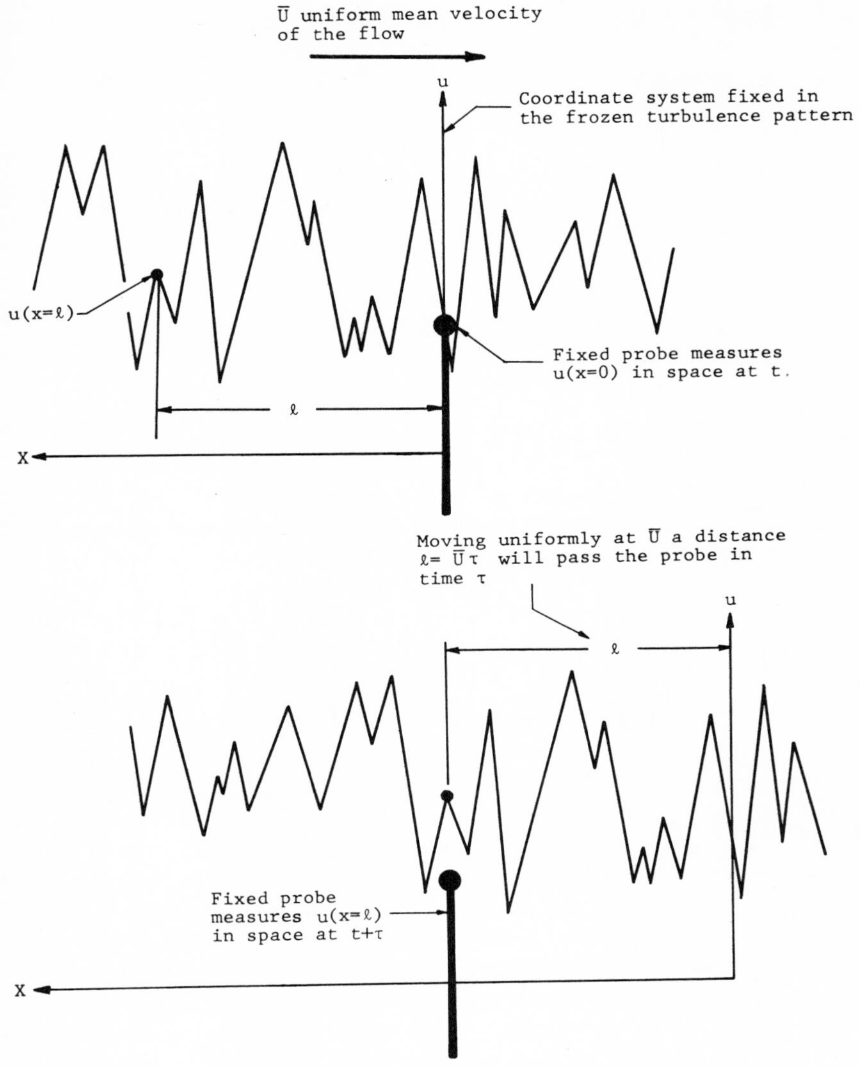

Fig. 4.13. Illustration of Taylor's hypothesis.

The aforementioned application of Taylor's hypothesis is much used in turbulence analyses, being applied to shear flow as well as uniform flow. We have already mentioned some of the proposed conditions for the hypothesis to hold. Experimental and theoretical investigations have also demonstrated that Taylor's hypothesis applies through large regions of inhomogeneous turbulence as well as the weak nearly homogeneous turbulent flows for which it was proposed by Taylor. Observations indicate that it is applicable through approximately 90–95% of the boundary layer, probably through most free flows not having flow reversal and in the cores of channel flows not overly close to the wall.[5]

The one-dimensional, second-order moments most often measured are the longitudinal correlation, $B_{ll}(l)$, and the lateral correlation (also called the transverse correlation), $B_{nn}(l)$. Correspondingly $E_{ll}(\kappa)$ and $E_{nn}(\kappa)$, defined by

$$B_{ll}(l)=\int_{-\infty}^{\infty} E_{ll}(\kappa)\, e^{i\kappa l}\, d\kappa \tag{4.119}$$

and

$$B_{nn}(l)=\int_{-\infty}^{\infty} E_{nn}(\kappa)\, e^{i\kappa l}\, d\kappa \tag{4.120}$$

are called the one-dimensional longitudinal spectrum and the one-dimensional lateral spectrum, which for real functions can be written as

$$E_{ll}(\kappa)=\frac{2}{\pi}\int_{0}^{\infty} B_{ll}(l)\cos\kappa l\, dl \tag{4.121}$$

and

$$E_{nn}(\kappa)=\frac{2}{\pi}\int_{0}^{\infty} B_{nn}(l)\cos\kappa l\, dl \tag{4.122}$$

respectively.

One-dimensional spectra are not particularly appropriate for describing turbulence, however, because they do not give an exact representation of the distribution of energy among the scales or eddies. The cause of this is called aliasing and its mechanism is illustrated in Figure 4.14. If we measure velocity fluctuations along the line x_1 with probes A and B, then an eddy of size R, where $\kappa=\pi/R$, moving with its axis perpendicular to x_1 will create a disturbance equivalent to an eddy of size R_1 moving obliquely to x_1 even though this eddy is much smaller. Note that the smaller eddies have the larger wave number, i.e., $\kappa_1>\kappa$. In the illustrated example both disturbances would be attributed to eddies of wave number κ. Thus a one-dimensional spectrum obtained in three-dimensional turbulence (and recall that all turbulence is three dimensional) contains, at wave number κ, contributions from eddies of all wave numbers greater than κ. For this reason measured one-dimensional spectra generally have a finite value proportional to the integral scale at $\kappa=0$. This finite value of energy at zero wave number is that which has been aliased from the higher wave numbers, and it is not a real quantity.

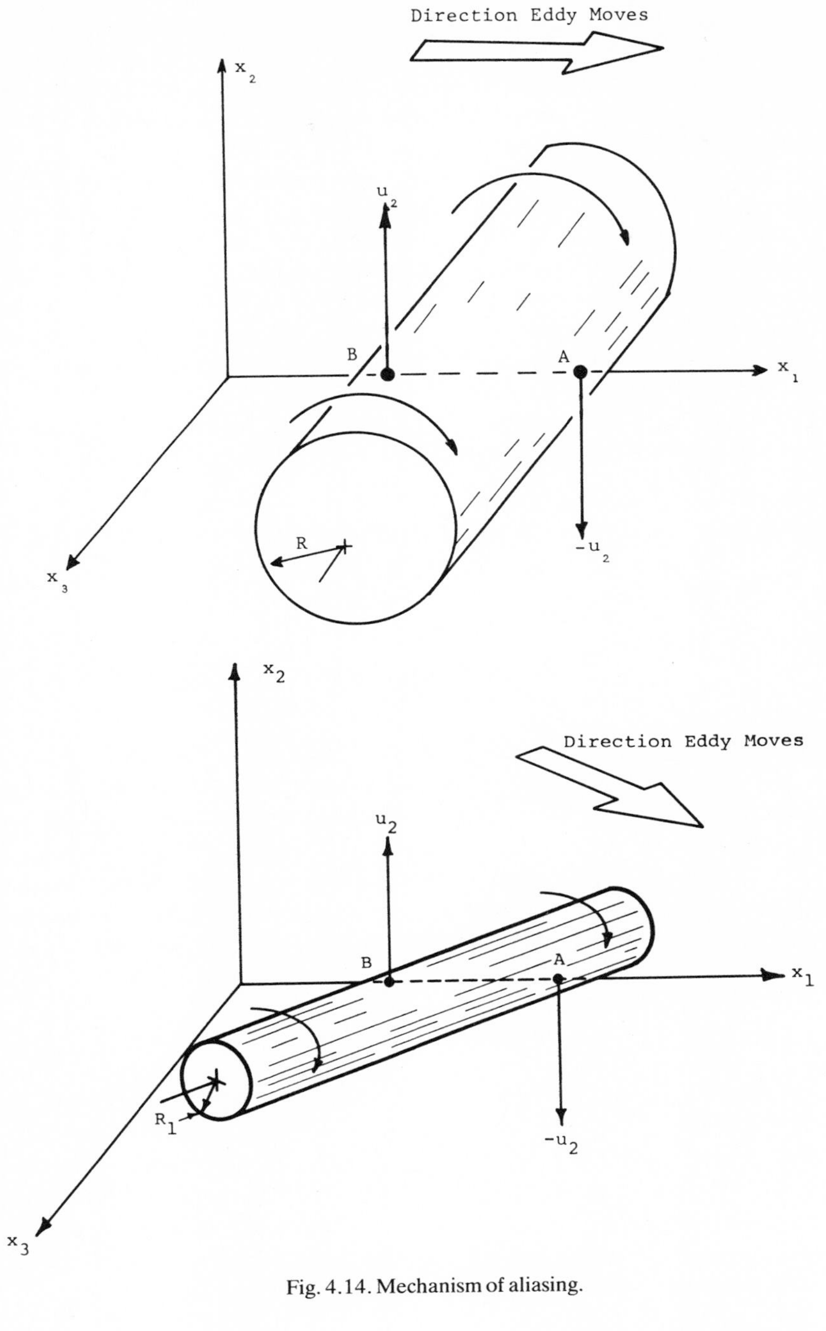

Fig. 4.14. Mechanism of aliasing.

To avoid aliasing problems we must work with measurements in all three directions, from which the necessary spectral data cannot be achieved using Taylor's hypothesis and frequency spectra. Hence we resort to the determination of three-dimensional wave-number spectra from Fourier transforms of statistical moments.

4.4.2. Three-Dimensional Wave-Number Spectra

The three-dimensional wave-number spectra for a random vector field are obtained from the three-dimensional Fourier transformation of the statistical moments. Normally only the two-point second-order moments $B_{i,j}(\mathbf{r}, \mathbf{l})$ are considered in turbulence and we will restrict our discussion to these.

The energy spectrum tensor $E_{i,j}$, which is a function of the wave-number vector $\boldsymbol{\kappa}$ and of the position vector $\mathbf{r}$, is defined by the relationships

$$B_{i,j}(\mathbf{r}, \mathbf{l}) = \iiint_{-\infty}^{\infty} E_{i,j}(\mathbf{r}, \boldsymbol{\kappa})\, e^{i\boldsymbol{\kappa}\cdot\mathbf{l}}\, d\boldsymbol{\kappa} \tag{4.123}$$

and

$$E_{i,j}(\mathbf{r}, \boldsymbol{\kappa}) = \frac{1}{(2\pi)^3} \iiint_{-\infty}^{\infty} B(\mathbf{r}, \mathbf{l})\, e^{-i\boldsymbol{\kappa}\cdot\mathbf{l}}\, d\mathbf{l} \tag{4.124}$$

Both of the above functions are also functions of time in the most general case. The general case, however, is almost intractable, making physical reasoning difficult, and therefore we will further restrict our discussion to statistically homogeneous and stationary turbulence.

The concept of a homogeneous random field can be developed in direct analogy to our previous development of a stationary process. Thus it is possible to express the spatially dependent random variable $u(\mathbf{l})$ in terms of a Fourier–Stieltjes integral as

$$u(\mathbf{l}) = \int e^{i\boldsymbol{\kappa}\cdot\mathbf{l}}\, dZ(\boldsymbol{\kappa}) \tag{4.125}$$

The harmonic oscillations $e^{i\omega t}$ in Equation (4.89) have been replaced by the plane waves $e^{i\boldsymbol{\kappa}\cdot\mathbf{l}}$ and the random amplitude functions $dZ(\omega) = [Z(\omega_k) - Z(\omega_{k-1})]$ have been replaced by $dZ(\boldsymbol{\kappa}) = [Z(\boldsymbol{\kappa}_k) - Z(\boldsymbol{\kappa}_{k-1})]$.

As in Section 4.1.5 it follows that Equation (4.125) can be written as

$$B_{i,j}(\mathbf{l}) = \iiint_{-\infty}^{\infty} e^{i\boldsymbol{\kappa}\cdot\mathbf{l}} E_{i,j}(\boldsymbol{\kappa})\, d\boldsymbol{\kappa} \tag{4.126}$$

if

$$\iiint_{-\infty}^{\infty} B_{i,j}(\mathbf{l})\, d\mathbf{l} < \infty \tag{4.127}$$

Equation (4.127) assures the continuity of the energy distribution of the random field. The condition of Equation (4.127) is normally satisfied in turbulent flow because the physical transfer of energy associated with the cascading of energy between eddies rapidly redistributes the energy at any point in the spectrum and creates a continuous spectrum.

For the homogeneous turbulence under consideration the statistical moments are a function of $\mathbf{l}$ only, and assuming Equation (4.127) is satisfied we obtain the Fourier transform pair

$$B_{i,j}(\mathbf{l}) = \iiint_{-\infty}^{\infty} E_{i,j}(\boldsymbol{\kappa})\, e^{i\boldsymbol{\kappa}\cdot\mathbf{l}}\, d\boldsymbol{\kappa} \tag{4.128}$$

and

$$E_{i,j}(\boldsymbol{\kappa}) = \frac{1}{(2\pi)^3} \iiint_{\infty}^{\infty} B_{i,j}(\mathbf{l})\, e^{-i\boldsymbol{\kappa}\cdot\mathbf{l}}\, d\mathbf{l} \tag{4.129}$$

where, as before, $E_{i,j}(\boldsymbol{\kappa})$ is the energy spectrum tensor. The vector notation $\boldsymbol{\kappa}\cdot\mathbf{l}$ and $d\boldsymbol{\kappa}$ signify $\kappa_1 l_1 + \kappa_2 l_2 + \kappa_3 l_3$ and $d\kappa_1\, d\kappa_2\, d\kappa_3$, respectively. Similarly $d\mathbf{l} = dl_1\, dl_2\, dl_3$.

Equation (4.129) can be considerably simplified for the case of an isotropic field. Under these conditions $B_{i,j}(\mathbf{l})$ is independent of direction and depends only on the magnitude l of the vector $\mathbf{l}$. Introducing spherical coordinates, $d\mathbf{l} = l^2 \sin\theta\, dl\, d\phi\, d\theta$ and $\boldsymbol{\kappa}\cdot\mathbf{l} = \kappa l \cos\theta$, Equation (4.129) becomes

$$E_{i,i}(\kappa) = \frac{1}{(2\pi)^3} \int_0^{\infty} \int_0^{2\pi} \int_0^{\pi} l^2 B_{i,i}(l) \sin\theta\, e^{-i\kappa l \cos\theta}\, dl\, d\phi\, d\theta \tag{4.130}$$

Carrying out the integration with respect to ϕ and θ gives

$$E_{i,i}(\kappa) = \frac{1}{2\pi^2} \int_0^{\infty} \frac{\sin \kappa l}{\kappa l} B_{i,i}(l) l^2\, dl \tag{4.131}$$

Analogously

$$B_{i,i}(l) = \int_0^{\infty} \frac{\sin \kappa l}{\kappa l} E_{i,i}(\kappa)\kappa^2\, d\kappa \tag{4.132}$$

a three-dimensional energy spectrum function $E(\kappa)$ is then defined as

$$E(\kappa) = 2\pi\kappa^2 E_{i,i}(\kappa) \tag{4.133}$$

such that from Equation (4.132)

$$B_{i,i}(0)=3\overline{u^2}=4\pi\int_0^\infty E_{i,i}(\kappa)\kappa^2\,d\kappa \tag{4.134}$$

and thus

$$\int_0^\infty E(\kappa)\,d\kappa=\frac{3}{2}\overline{u^2} \tag{4.135}$$

Note that for isotropic turbulence $u_1=u_2=u_3=u$ and $B_{i,j}(0)=0,\ i\neq j$.

The form of Equations (4.131) and (4.132) applies only for isotropic turbulence.

Batchelor,[6] however, suggests that for nonisotropic homogeneous turbulence the statistical moments and the spectrum function be expressed in terms of a single scaler distance l and wave number κ, respectively, by taking mean values of the functions over spherical surfaces given by $l=$ const in physical space and $\kappa=$ const in wave-number space. Then

$$[E_{i,j}(\kappa)]_{\rm av}=\frac{1}{4\pi\kappa^2}\int E_{i,j}(\kappa)4\pi\kappa^2\,d\kappa \tag{4.136}$$

and

$$[B_{i,j}(l)]_{\rm av}=\frac{1}{4\pi l^2}\int B_{i,j}(l)4\pi l^2\,dl \tag{4.137}$$

The quantity $B_{i,j}(l)_{\rm av}$ is the average statistical moment tensor for two points a distance l apart, while $[E_{i,j}(\kappa)]_{\rm av}\,d\kappa$ is the contribution to the energy tensor $\overline{u_iu_j}$ from wave numbers whose magnitudes lie between κ and $\kappa+d\kappa$. The relationship between $[B_{i,j}(l)]_{\rm av}$ and $[E_{i,j}(\kappa)]_{\rm av}$ becomes[7]

$$[B_{i,j}(l)]_{\rm av}=4\pi\int_0^\infty [E_{i,j}(\kappa)]_{\rm av}\kappa^2\frac{\sin\kappa l}{\kappa l}\,d\kappa \tag{4.138}$$

Equation (4.138) does not uniquely define the function $B_{i,j}(l)$, however, because the energy of plane waves with wave-number vectors in the range $(\kappa,\kappa+d\kappa)$ can be distributed differently between waves of different orientation. The total energy of the field, however, is contained in

$$\sum_{i=1}^{3} B_{ii}(0)=\sum_{i=1}^{3}\overline{u_i^2}=4\pi\int_0^\infty [E_{i,j}(\kappa)]_{\rm av}\kappa^2\,d\kappa \tag{4.139}$$

4.5. Characteristics of Energy Spectra

4.5.1. Three-Dimensional Energy Spectra

In this section a brief description of the characteristics of energy spectra is given. More lengthy and complete discussions can be found in References 7–9. Figure 4.15 shows the typical form of a three-dimensional energy spectrum function $E(\kappa, t)$. Physical argument coupled with dimensional analysis has provided basic insight into the behavior of $E(\kappa, t)$ with κ.

In Section 4.1 it was argued that the larger turbulence eddies associated with the mean motion were the energy-bearing anisotropic eddies. The length scales of these eddies become smaller and smaller owing to vortex stretching, and the energy contained in the eddies cascades to the smaller scales, where it is finally dissipated by viscosity. The small scale eddies lose the preferred orientation of the mean rate of strain, taking on a universal structure or isotropy referred to as local isotropy by Kolmogorov.

For local isotropy to exist the turbulence Reynolds number $\mathrm{Re}_l = u'l/\nu$ must be large, on the order of 100 (see Reference 8), where l is the average size of the energy-containing eddies.

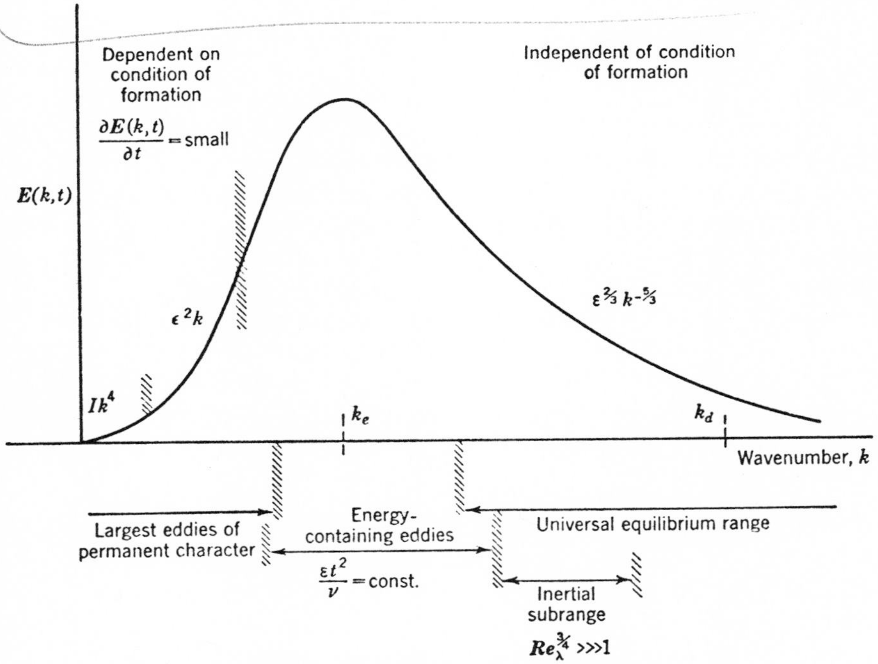

Fig. 4.15. Typical three-dimensional energy spectrum function.

In the part of the spectrum where local isotropy exists time scales are short compared to those of the mean flow and thus the smaller eddies respond sufficiently fast relative to changes in the mean flow that they are essentially in equilibrium with the local mean flow conditions. The range of wavelengths or corresponding eddy length scales over which equilibrium prevails is indicated in Figure 4.15.

The equilibrium range coincides with length scales where the energy cascading from larger eddies is dissipated. In view of this fact and the local equilibrium the character of the turbulence is determined by $\varepsilon(t)$, the total energy supply, and ν, the kinematic viscosity. The characteristic velocity and length scale of the equilibrium region are, from dimensional reasoning,

length scale: $\eta = (\nu^3/\varepsilon)^{1/4}$

velocity scale: $v = (\nu\varepsilon)^{1/4}$

The form of the energy spectrum at high wave numbers can be established from dimensional arguments. That is, $E(\kappa, t)\,[L^3T^{-2}]$ must be a function of $\kappa[L^{-1}]$, $\varepsilon(t)\,[L^2T^{-3}]$, and $\nu\,[L^2T^{-1}]$, and hence

$$E(\kappa, t) = \nu^{5/4}\varepsilon^{1/4}E^*(\kappa\eta)$$
$$[L^3T^2] \sim [L^{5/2}T^{-5/4}][L^{1/2}T^{-3/4}] \tag{4.140}$$

where the function $E^*(\kappa\eta)$ is to be found from experiment.

The size of eddies that provide the main contribution to the total dissipation is roughly associated with the wave number κ_d, where the dissipation curve $\kappa^2E(\kappa, t)$ has a maximum (see Figure 4.12) [the correlation $(\partial u/\partial t)^2$ can be shown to be proportional to the dissipation $\kappa^2E(\kappa, t)$]. The value of κ_d is expected to be on the order $1/\eta$; however, experiments[7] have found $0.09 \leq \eta\kappa_d \leq 0.5$.

At the low-wave-number range of the energy spectrum, a peak occurs at a value of κ_e. The range of wave numbers around κ_e where the eddies make the major contribution to the total kinetic energy of the turbulence is called the range of the energy-containing eddies (Figure 4.15). The maximum kinetic energy is contained in these eddies. At still lower wave numbers lie the largest more permanent eddies. These contain on the order of 20% of the total kinetic energy.

The value of κ_e serves to define the length scale l described earlier as the average size of the energy-containing eddies, i.e., $\kappa_e = 1/l$.

Since the energy $\varepsilon(t)$ dissipated at high wave numbers is the same energy supplied by the larger eddies, the form of the energy spectrum function in the range of energy-containing eddies will also depend on the parameter $\varepsilon(t)$. A second parameter that becomes important in this region is time, t, because the relative rate of change of the total kinetic energy is the same order as the time scale of the large eddies. Finally, $E(\kappa, t)$ should depend on ν in the energy-containing eddy range if the Reynolds number does not have an extremely high value. If the Reynolds number does have an extremely high value there exists an inertial subrange between the equilibrium range and the energy-containing eddy range. This range is described later. In the energy-containing eddy range the energy spectrum $E(\kappa, t)$ should depend on κ, ε, t, and ν.

Since any two of the three parameters will render $E(\kappa, t)$ dimensionless the three parameters cannot be dependent and hence must form a dimensionless group, $\varepsilon t^2/\nu$.

From dimensional arguments the form of $E(\kappa, t)$ for the energy-containing eddy range becomes

$$E(\kappa, t) = \nu^{5/4}\varepsilon^{1/4}E^*(\kappa\eta\varepsilon t^2/\nu) \tag{4.141}$$

Kolmogorov hypothesized, and experiment has confirmed, that at very high turbulence Reynolds numbers a range of wave numbers exists where the effects of viscous dissipation is negligibly small compared with the flux of energy transferred by inertial effects. This range is called the inertial subrange and the energy spectrum for it is solely dependent on the parameter $\varepsilon(t)$. Thus

$$E(\kappa, t) = \text{const} \times \varepsilon(t)^{2/3}\kappa^{-5/3} \tag{4.142}$$

By a limiting process Tennekes and Lumley[8] have established a map of the region of existence of the inertial subrange (Figure 4.16).

The magnitude of the turbulence Reynolds number $u'l/\nu$ necessary for an inertial subrange to exist is observed to be of the order of 10^5. Therefore, inertial subranges are not normally found in laboratory flows but are frequently observed in geophysical flows.

The form of $E(\kappa, t)$ in the range of the largest eddies of permanent character is shown to have the form of

$$E(\kappa, t) = \epsilon^{8/3}I^{-1/3}E^*[\kappa(I/\epsilon^2)^{1/3}] \tag{4.143}$$

for the very low wave numbers and

$$E(\kappa, t) = \epsilon^{5/4}\varepsilon^{1/4}E^*[\kappa(\epsilon^3/\varepsilon)^{1/4}] \tag{4.144}$$

for the wave numbers intermediate to the very low wave numbers and those on the lower end of the energy-containing eddy range. The parameter ϵ is an eddy viscosity and I is the Loitsianskii integral, both of which are fully described in Reference 7.

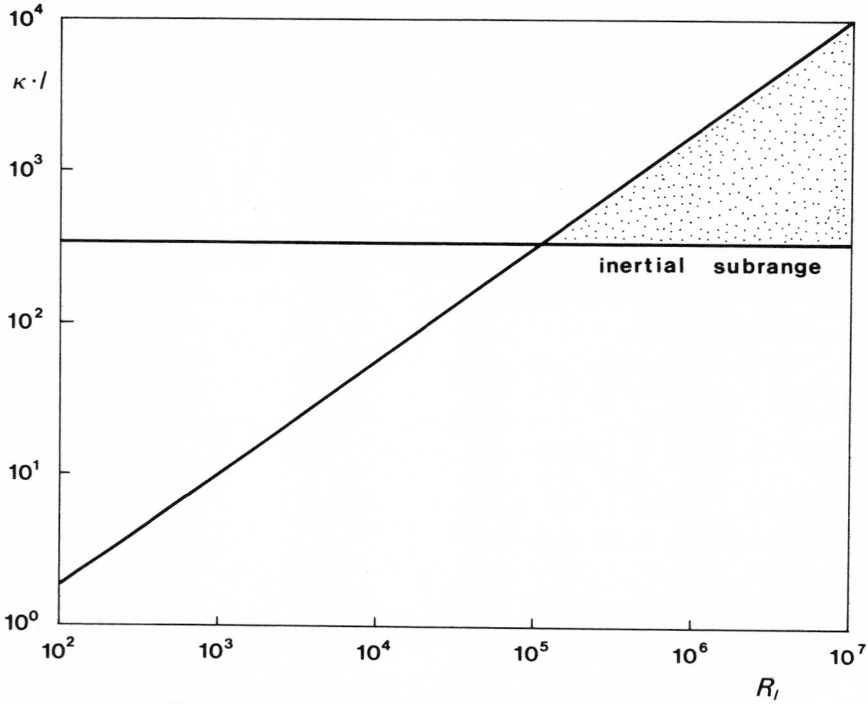

Fig. 4.16. Region in which an inertial subrange exists.[8]

4.5.2. One-Dimensional Energy Spectra

The most common spectra measured are one-dimensional. These are defined in terms of the longitudinal correlation and the lateral or transverse correlation, Equations (4.119) and (4.120).

The relation between E_{ll}, E_{nn}, and $E_{i,j}$, the three-dimensional spectrum tensor, can be determined with the application of Equation (4.128):

$$B_{ll}(r, 0, 0) = \int_{-\infty}^{\infty} e^{i\kappa_1 r}\left[\iint_{-\infty}^{\infty} E_{1,1}(\kappa)\, d\kappa_2\, d\kappa_3\right] d\kappa_1 \tag{4.145}$$

Comparison with Equation (4.119) shows

$$E_{ll}(\kappa_1) = \int_{-\infty}^{\infty}\int_{-\infty}^{\infty} E_{1,1}(\kappa)\, d\kappa_2\, d\kappa_3 \tag{4.146}$$

This again illustrates the aliasing effect mentioned earlier. For example, with

homogeneous isotropic turbulence the kinetic energy associated with eddies of wave number κ is contained within the spherical shell

$$E(\kappa) = 4\pi \iint E_{i,i}(\kappa)\kappa^2\, d\kappa \qquad (4.147)$$

as illustrated in Figure 4.17. From a one-dimensional spectrum, however, one is led to believe that the energy associated with $\kappa = \kappa_1$ is the energy contained in the slice extending to infinity in both the κ_2 and κ_3 directions. Thus energy from higher wave numbers than κ_1 is aliased to κ_1. It is apparent from Figure 4.17 that at $\kappa_1 = 0$ the energy is zero whereas the one-dimensional spectra would clearly show the amount contained within the slice formed from the plane $(0, \kappa_2, \kappa_3)$ and $d\kappa_1$.

One-dimensional spectra have the characteristic illustrated in Figure 4.18.[8] Longitudinal correlation functions normally do not have negative

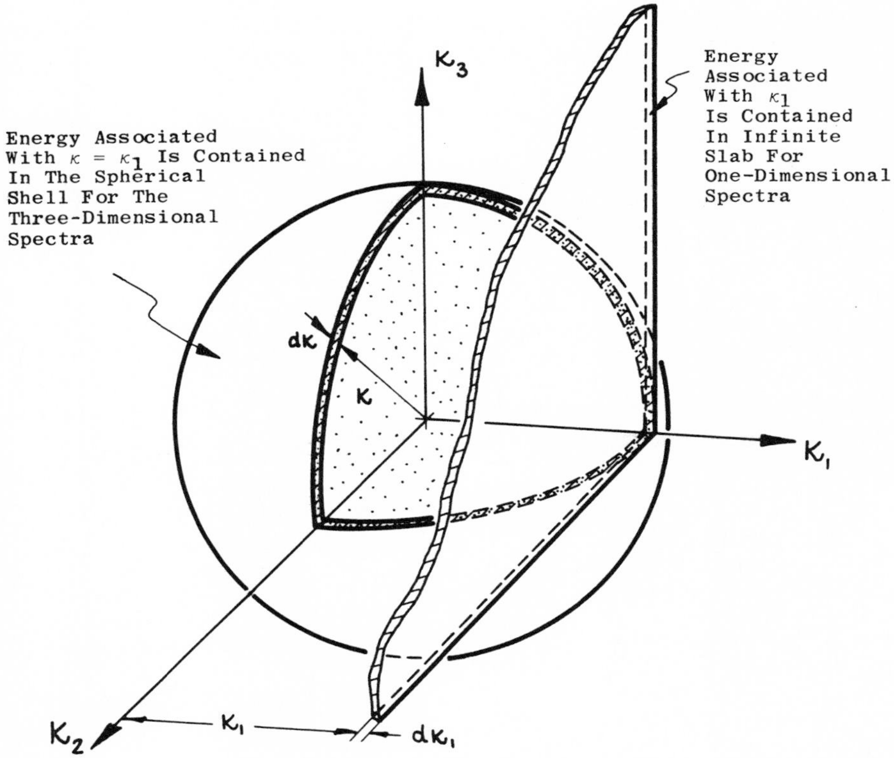

Fig. 4.17. One-dimensional and three-dimensional spectra.

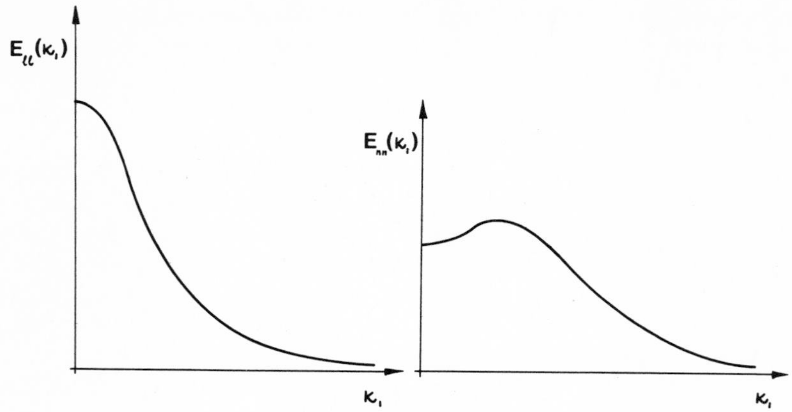

Fig. 4.18. Characteristics of one-dimensional spectra.

values, hence the peak in the energy spectrum function E_{ll} occurs at the origin. This peak is proportional to the integral length scale

$$\Lambda = \int_{-\infty}^{\infty} R(r, 0, 0)\, dr \tag{4.148}$$

Unlike the longitudinal correlation the lateral correlation does become negative and the corresponding spectrum is likely to have a peak away from the origin.

Notation

a_n, b_n	Fourier series coefficients
$B(\ )$	Correlation or statistical moment
$E_{i,j}$	Energy spectrum tensor
$E(\ )$	Energy spectrum
ε	Total energy transferred between scales
ε	Spectral distribution function
$F(\ )$	Complex spectrum
$f(\ )$	Function of
$g(\ )$	Discrete function of
L	Length scale
l	Separation distance
$\mathbf{l}$	Separation distance vector
n	Frequency in cycles per second
$\mathbf{r}$	Radius vector
$R(\ )$	Correlation coefficient
T	Time scale
T	Period
t	Time
$\bar{U}$	Mean velocity
u_m	Amplitude of rectangular wave
v	Velocity scale
Z	Complex random variable
ξ	Complex random variable
n	Kolmogorov length scale
η	Real part of complex random variable
ζ	Imaginary part of complex random variable
θ	Phase angle
$\phi(\ )$	Normalized energy spectral density
ω	Frequency in radians

κ	Wave number	ν	Kinematic viscosity
τ	Lag time	ϵ	Eddy viscosity
Λ	Integral length scale	*	Complex conjugate
σ	Standard deviation	′	Root-mean-square value

References

1. Churchill, R. V., *Fourier Series and Boundary Value Problems*, McGraw-Hill Book Co., New York (1941).
2. Yaglom, A. M., *An Introduction to the Theory of Stationary Random Functions*, translated and edited by R. A. Silverman, Prentice-Hall Inc., New Jersey (1962).
3. Taylor, A. E., *Advanced Calculus*, Ginn and Company, New York (1955).
4. Gnedenko, B. V., *The Theory of Probability*, translated by B. D. Sickler, Chelsea Publishing Co., New York (1962).
5. Reynolds, A. J., *Turbulent Flows in Engineering*, John Wiley and Sons, New York (1974).
6. Batchelor, G. K., *Homogeneous Turbulence*, Cambridge University Press, London (1967).
7. Hinze, J. O., *Turbulence*, McGraw-Hill Book Co., New York (1957).
8. Tennekes, H., and J. L. Lumley, *A First Course in Turbulence*, MIT Press, Cambridge, Massachusetts (1972).
9. Panchev, S., *Random Functions and Turbulence*, Pergamon Press, Oxford, England (1971).

CHAPTER 5

Turbulence: Diffusion, Statistics, Spectral Dynamics

H. TENNEKES

5.1. Introduction

It is probably a bit misleading to say that turbulent flow is "random"; even if the birth of an eddy out of an instability somewhere in the flow field is a matter of "chance," its subsequent evolution must be governed by the Navier–Stokes equations. Being careful, we shall say no more than that the turbulence is *chaotic*, and that for many practical purposes it is sufficient to know something about the statistical properties of the flow field. However, there are circumstances in which a statistical treatment will not do. For example, in weather forecasting, we want to predict at what time a particular eddy will pass over a certain area, and at which stage in its life cycle.

Our approach will not consider such questions; it is more like a theory of climate or a mean-motion model of the atmosphere's general circulation. In effect, we accept *a priori* that the predictability of turbulent flows is limited to relatively short time intervals, no matter how well the initial conditions are known. If we are interested in space and time intervals beyond the limits of predictability for individual realizations, we *have* to

H. TENNEKES • Dept. of Aerospace Engineering, Pennsylvania State University, University Park, PA 16802. Present address: Royal Netherlands Meteorological Institute, de Bilt, the Netherlands adopt stochastic methods. There are profound issues here—as always when one gets involved in statistical mechanics—but we shall have to let those rest.

5.2. Turbulent Diffusion

To get a feeling for the techniques that are commonly used in the analysis of turbulent flow, let us have a look at a simple diffusion problem. Take an infinitely large box, and fill it with stationary and homogeneous turbulence (homogeneous turbulence cannot be stationary, and vice versa, but we can ignore that difficulty). At the coordinate origin we release fluid "particles" that are tagged in some way (say, with radioactive isotopes). How do those particles diffuse away from the origin? The very meaning of the verb "diffuse" shows that we are not interested in the trajectories of individual particles, but only in certain averaged properties.

The position of the Lagrangian fluid point at time t is (we shall work with the y direction only)

$$y(t) = \int_0^t v(t')\,dt' \tag{5.1}$$

Averaging this over many realizations (i.e., a long series of releases of particles), we find that the mean value of y is zero if the velocity fluctuations satisfy certain reasonable conditions (Tennekes and Lumley,[1] Chapter 7). In other words, the center of gravity of a cloud of particles does not drift away from the origin (statistically speaking). A more interesting equation is obtained by multiplying Equation (5.1) by $v(t)$:

$$y(t)v(t) = y(t)\frac{dy}{dt} = \frac{d}{dt}\left(\frac{1}{2}y^2\right) = \int_0^t v(t)v(t')\,dt' \tag{5.2}$$

Taking averages, we find that

$$\frac{d}{dt}\left(\overline{\frac{1}{2}y^2}\right) = \int_0^t \overline{v(t)v(t')}\,dt' \tag{5.3}$$

Both sides of this equation have the dimensions of a diffusivity ($m^2\,sec^{-1}$); in fact, Equation (5.3) *defines* the eddy diffusivity of the turbulence in which the tagged particles are released.

This line of thinking was developed by G. I. Taylor; he patterned his theory of diffusion after Einstein's theory of Brownian motion. If we want to solve Equation (5.3) we need to know more about the Lagrangian autocorrelation that is involved. If the turbulent motion is stationary, the correlation is an even function of the time interval $\tau = t - t'$; we put

$$\overline{v(t)v(t')} = R_L(\tau) = \overline{v^2}\rho_L(\tau) \tag{5.4}$$

Here, $\rho_L(0) = 1$ and $\rho_L(\tau) = \rho_L(-\tau)$. The *shape* of the correlation function presumably is related to the dynamics of turbulence; we shall address that issue later. We expect $\rho_L(\tau)$ to be well behaved, so that we can define the *Lagrangian integral time scale* T_L by

$$T_L = \int_0^\infty \rho_L(\tau)\, d\tau \tag{5.5}$$

We assume that $\rho_L(\tau) \to 0$ as $\tau \to \infty$; in practice, one often finds that $\rho_L(\tau) < 0.01$ as soon as $\tau > 3T_L$.

Substitution of Equation (5.4) into Equation (5.3) yields

$$\frac{d}{dt}\left(\overline{\frac{1}{2}y^2}\right) = \int_0^t R_L(\tau)\, d\tau = \overline{v^2}\int_0^t \rho_L(\tau)\, d\tau \tag{5.6}$$

For times $t \gg T_L$, this may be approximated by

$$\frac{d}{dt}\left(\overline{\frac{1}{2}y^2}\right) = \overline{v^2}\int_0^\infty \rho_L(\tau)\, d\tau = \overline{v^2}T_L \tag{5.7}$$

Upon integration, this yields

$$\overline{y^2(t)} = 2\overline{v^2}T_L t \tag{5.8}$$

This confirms the parabolic nature of diffusion over time intervals that are long compared to the integral scale.

5.3. Fourier Transforms

In a box with stationary, homogeneous turbulence the velocity of a fluid particle is a stationary function of time. Functions of that kind often are studied with the aid of Fourier transforms. Fourier transforms of the random functions themselves involve certain mathematical problems, but the correlation function $R_L(\tau)$ is so well behaved that simple cosine transforms can be used (Monin and Yaglom,[2] Chapter 6):

$$R_L(\tau) = \int_{-\infty}^{\infty} e^{i\omega\tau}F_L(\omega)\, d\omega = \int_0^\infty (\cos \omega\tau)E_L(\omega)\, d\omega \tag{5.9}$$

and conversely

$$F_L(\omega) = \frac{1}{2\pi}\int_{-\infty}^{\infty} e^{-i\omega\tau} R_L(\tau)\, d\tau \tag{5.10a}$$

or

$$E_L(\omega) = \frac{2}{\pi}\int_{0}^{\infty} (\cos \omega\tau) R_L(\tau)\, d\tau \tag{5.10b}$$

For $\tau = 0$, Equation (5.9) gives

$$R_L(0) = \overline{v^2} = \int_0^{\infty} E_L(\omega)\, d\omega \tag{5.11}$$

This is one of the reasons why $E_L(\omega)$ is called a *power spectral density*, a *power spectrum*, or simply a *spectrum*. In fact, it is possible to show (I will not do it here) that $E_L(\omega)\, d\omega$ is the contribution to $\overline{v^2}$ made by fluctuation components with frequencies in an interval of width $d\omega$ centered at ω. For the analysis of turbulent flow, is this really useful information? In what way does it make sense to speak of the "frequency" of an eddy?

For $\omega = 0$, Equation (5.10b) becomes

$$E_L(0) = \frac{2}{\pi}\int_0^{\infty} R_L(\tau)\, d\tau = \frac{2}{\pi}\overline{v^2} T_L \tag{5.12}$$

The product of $\overline{v^2}$ and T_L was encountered earlier, in the diffusion equations (5.7) and (5.8). Apparently, the value of the spectrum at the origin is (apart from a numerical factor) the eddy diffusivity that is involved in the description of diffusion over large time intervals.

Another function of interest is the Lagrangian *structure function* $D_L(\tau)$, which is defined by

$$D_L(\tau) = \overline{[v(t) - v(t+\tau)]^2} \tag{5.13}$$

The structure function is related to $R_L(\tau)$ and $E_L(\omega)$ by

$$D_L(\tau) = 2\overline{v^2} - 2R_L(\tau) \tag{5.14}$$

$$= 2\int_0^{\infty} (1 - \cos \omega\tau) E_L(\omega)\, d\omega \tag{5.15}$$

For small time intervals, the structure function is a reasonably good filter, one that selects mainly the high-frequency components of $E_L(\omega)$. We shall use that feature later.

5.4. Particle Diffusion

Taylor's approach to turbulent diffusion is easily generalized to the motion of particles whose inertia is not negligible. Under favorable circumstances, the velocity $w(t)$ of a finite particle (a speck of dust, a water droplet) is governed by a simple differential equation of the type

$$\Lambda\frac{dw}{dt}+w(t)=v(t) \tag{5.16}$$

where Λ is a time constant related to the particle's inertia, and $v(t)$ represents the fluid velocity encountered by the particle along its trajectory (Monin and Yaglom,[3] Chapter 5). If $w(t)$ and $v(t)$ are stationary random functions, it makes sense to construct an equation relating their correlation functions. Denoting the particle acceleration dw/dt by $a(t)$, we obtain from Equation (5.16)

$$\Lambda^2\overline{a(t)a(t+\tau)}+\overline{w(t)w(t+\tau)}=\overline{v(t)v(t+\tau)} \tag{5.17}$$

We need to introduce the correlation functions involved here:

$$\overline{w(t)w(t+\tau)}=R_p(\tau)=\overline{w^2}\rho_p(\tau) \tag{5.18}$$

$$\overline{a(t)a(t+\tau)}=-\frac{d^2R_p}{d\tau^2}=-\overline{w^2}\frac{d^2\rho_p}{d\tau^2} \tag{5.19}$$

$$\overline{v(t)v(t+\tau)}=R_*(\tau)=\overline{v^2}\rho_*(\tau) \tag{5.20}$$

Note that $\rho_*(\tau)$ generally differs from the correlation coefficient $\rho_L(\tau)$, because a finite particle does not follow the same trajectory as a Lagrangian fluid point.

Substitution of Equations (5.18)–(5.20) into Equation (5.17) yields

$$-\Lambda^2\frac{d^2R_p}{d\tau^2}+R_p(\tau)=R_*(\tau) \tag{5.21}$$

Taking Fourier transforms, we obtain

$$(\omega^2\Lambda^2+1)E_p(\omega)=E_*(\omega) \tag{5.22}$$

The simplicity of this kind of relation between "input" and "output" spectra is a major point in favor of the use of Fourier transforms. However, this simplicity is lost if the basic equation is nonlinear.

Integrating Equation (5.21) from $\tau = 0$ to infinity, we find that the first term disappears because $dR_P/d\tau = 0$ at $\tau = 0$ and $\tau \to \infty$. There results

$$\int_0^\infty R_p(\tau)\,d\tau = \int_0^\infty R_*(\tau)\,d\tau \tag{5.23}$$

or, in view of Equations (5.12) and (5.22)

$$\overline{w^2}T_p = \overline{v^2}T_* \tag{5.24}$$

$$E_p(0) = E_*(0) \tag{5.25}$$

The "output" diffusivity apparently equals the "input" diffusivity, and it is independent of the time constant (Λ) of the particles involved. This property is of great help in applications of diffusion equations such as Equation (5.8). The eddy diffusivity needed to compute the diffusion of particles is *not* a function of the particle size.

At $\tau = 0$, Equation (5.21) leads to

$$(2\Lambda^2/\lambda_p^2 + 1)\overline{w^2} = \overline{v^2} \tag{5.26}$$

where the particle *microscale* λ_p is defined by

$$2\frac{\overline{w^2}}{\lambda_p^2} = -\left(\frac{d^2R_p}{d\tau^2}\right)_{\tau=0} = \overline{\left(\frac{dw}{dt}\right)^2} = \int_0^\infty \omega^2 E_p(\omega)\,d\omega \tag{5.27}$$

The microscale thus relates to the root-mean-square value of the acceleration fluctuations, to the curvature of the correlation function at the origin, and to the second moment of the spectrum. If $E_p(\omega)$ decreases as ω^{-3} or slower at high frequencies, the integral does not converge, so that $\lambda_p = 0$.

The mean-square particle velocity is given by

$$\overline{w^2} = \int_0^\infty E_p(\omega)\,d\omega = \int_0^\infty \frac{E_*(\omega)\,d\omega}{1+\omega^2\Lambda^2} \tag{5.28}$$

If the correlation function R_* is approximated by

$$R_*(\tau) = \overline{v^2}\exp(-\tau/T_*) \tag{5.29}$$

then

$$E_*(\omega) = \frac{2}{\pi}\frac{\overline{v^2}T_*}{1+\omega^2T_*^2} \tag{5.30}$$

In the same spirit, the Fourier transform of the filter $F(\omega) = (1+\omega^2\Lambda^2)^{-1}$ is

$$R_F(\tau) = \frac{\pi}{2\Lambda}\exp\left(\frac{-\tau}{\Lambda}\right) \tag{5.31}$$

An application of Parseval's relation gives

$$\overline{w^2} = \int_0^\infty E_*(\omega)F(\omega)\,d\omega = \frac{2}{\pi}\int_0^\infty R_*(\tau)R_F(\tau)\,d\tau$$

$$= \frac{2\overline{v^2}}{\pi}\frac{\pi}{2\Lambda}\int_0^\infty \exp\left(-\frac{\tau}{T_*}-\frac{\tau}{\Lambda}\right)d\tau$$

$$= \overline{v^2}T_*(\Lambda+T_*)^{-1} \tag{5.32}$$

When $\Lambda\to 0$, this gives $\overline{w^2}\to\overline{v^2}$, in agreement with Equation (5.26). When $T_*\ll\Lambda$, this gives $\overline{w^2}\approx\overline{v^2}T_*\Lambda^{-1}$, which indicates [see Equation (5.24)] that $T_p\approx\Lambda$ if T_* is small.

Substitution of Equation (5.32) into Equation (5.26) yields

$$(2\Lambda^2/\lambda_p^2+1)T_* = \Lambda + T_* \tag{5.33}$$

This leads to

$$\lambda_p^2 = 2T_*\Lambda \tag{5.34}$$

For small, fixed values of T_* the microscale is much smaller than the time constant Λ, and vice versa. Also, for fixed values of Λ or T_*, λ_p is proportional to the square root of the other time constant. Similar relations occur in turbulence dynamics.

5.5. Another Look at Fourier Transforms

A Fourier transform is a mathematical operation that maps the problem to be studied from one space to another. After the equations have been transformed, they may not be any easier to interpret. Turbulence dynamics in Fourier space (wave-number space or frequency space) often seems like an attempt to fit eddies into the straightjacket of wave mechanics.

A Fourier coefficient registers the *average* amplitude of velocity components at a particular wave number or frequency over the *entire* space filled with turbulence (a very large box, say). A Fourier coefficient in wave-number space has no sense of position in coordinate space, and a Fourier coefficient in frequency space has no sense of time. Conversely, we have the same problem: The velocity at a point in coordinate space has no sense of the wave numbers of the components that contribute to it, and the velocity at a point in time knows nothing about frequencies.

There are certain advantages to this. The Fourier coefficient at a particular wave number k incorporates contributions from *all* eddies that have appreciable energy content at or near k, wherever the eddies may be in the box of turbulence. A Fourier coefficient is an average, and often it is easier to work with averages than with local properties. But there are disadvantages, too: When an eddy of a certain size and position breaks up into a number of smaller ones, the "daughters" probably will be found at positions near the "mother" (Kraichnan[4] should be given credit for inventing this terminology). The process of breaking up relates to an energy cascade in wave-number space, but the Fourier coefficients involved in the cascade do not know *where* the individual events creating the cascade happen.

We conclude that a Fourier coefficient should not be confused with an eddy. Fourier coefficients are associated with an *ensemble* of eddies, and when one analyzes the dynamics of turbulence in wave-number space one should use the plural only: (the collective of) eddies at or near a certain wave number are involved.

In order to stimulate discussion, let me put forward some offbeat ideas. If turbulence is made of eddies, we should take those eddies to be reasonably *local*, both in coordinate space and in wave-number space. This means that an eddy is not a wave, but similar to a wave *packet*. We have to be a bit careful, because eddies do not propagate in the way most wave-type motions do; however, since we will not attempt Fourier transforms in space-time, the use of the word *wave* packet should not lead to confusion. My choice of words is borrowed from quantum mechanics; I do that on purpose.

A wave (Fourier coefficient) carries no information on position at all, but a wave packet *does* have a position, even if its edges are rather blurred. This is consistent with our concept of an eddy: It is hard to say where one eddy ends and its neighbor takes over. Obviously, we need something similar to an uncertainty principle: The spatial "fuzzyness" and the wave-number "blurring" of an eddy should be complementary. Dimensional considerations and a reference to the formal nature of Fourier transforms suggest that we should put

$$\Delta x \cdot \Delta k \sim 2\pi \qquad (5.35)$$

This says that a large eddy occupies a small volume in wave-number space, and that a small eddy incorporates contributions from many Fourier coefficients. Neighboring Fourier coefficients can belong to the *same* eddy, and their interactions are likely to be quite different from those with Fourier coefficients that are several Δk away.

If we think of an eddy as a vortex ring (shaped roughly like a doughnut), then it is clear that its size is comparable to the dominant wavelength of its motion. So, the smallest sensible value of Δx is

$$\Delta x \approx \lambda \qquad (5.36)$$

Since we determine the wavelength λ from such a short wave packet (one that contains only one cycle), the corresponding blurring in wave-number space is quite severe. The wave number k is related to λ by $k = 2\pi/\lambda$. This means that Equation (5.36) can be written as

$$\Delta x \cdot k \sim 2\pi \tag{5.37}$$

With the aid of the "uncertainty" relation, Equation (5.35), we obtain

$$\Delta k/k \sim 1 \tag{5.38}$$

This relation will turn out to be quite useful in our discussion of the spectral energy cascade. It states that, in wave-number space, an eddy (or group of eddies) is about an octave wide. One may visualize this by saying that a mother eddy breaks up into at least two daughters, which are about half as big as the mother. The evolution of and the interaction between eddies involve nonlinear "mixing," both in coordinate space and in wave-number space.

For the sake of completeness, I should mention that the spectral energy cascade appears to involve increasing "intermittency" at increasing wave numbers, with smaller and smaller eddies occupying a smaller and smaller fraction of the total fluid volume. This would seem to imply that equations such as (5.36) and (5.35) are violated in some subtle way, corresponding to increasing mixing ($\Delta k/k$ increasing with k) in wave-number space.

5.6. On the Interpretation of Frequency

As soon as we wish to consider possible relations between time and frequency in turbulence, we depart from kinematical aspects and enter into a discussion of dynamical features. It is easy to hypothesize that, in analogy with Equation (5.35), we must have

$$\Delta t \cdot \Delta\omega \sim 2\pi \tag{5.39}$$

but that does not state anything about the way in which the frequency ω should be defined. The methods of wave mechanics are of no use, because there is no propagation speed involved, and there is no dispersion relation between wave number and frequency. Knowing that "vortex stretching" is important in turbulence, and that vorticity has the dimensions of frequency, we now assume that the vorticity ω of an eddy is the frequency-like variable involved in Equation (5.39).

Equation (5.39) states that the "lack of focus" on the value of the vorticity in an eddy is inversely proportional to the "life time" of the eddy. A more specific interpretation needs to be developed, but I have not found it yet. As it stands Equation (5.39) is merely a formal statement.

The temporal equivalent of Equation (5.36) is

$$\Delta t \sim T \tag{5.40}$$

where T is the "period" of the eddy, defined by $T = 2\pi/\omega$. Equation (5.40) is a very specific assumption about the dynamics of eddies in turbulence. It states that the lifetime of an eddy is comparable to its period of revolution. To be somewhat more specific, we may think of Δt as the "time constant" of the evolution of an eddy, so that Equation (5.40) implies that any given eddy will have completely disappeared in a few revolutions. This assumption finds general support in many aspects of turbulence theory.

The statement contained in Equation (5.40) leads to the temporal equivalent of Equation (5.37):

$$\Delta t \cdot \omega \sim 2\pi \tag{5.41}$$

and to the equivalent of Equation (5.38) in "frequency" space:

$$\Delta\omega/\omega \sim 1 \tag{5.42}$$

Note that this particular frequency space is three dimensional, because vorticity is a vector. Equation (5.41) shows that high-frequency eddies break up rapidly, and that low-frequency eddies require more time to evolve. All of this seems quite reasonable.

5.7. Strong Interactions

The hypotheses proposed above are inspired in part by a few passages in Kadomtsev's book on plasma turbulence.(5) Allow me to quote:

> . . . in a stationary turbulent state the damping of each mode is compensated by the input due to the beat interaction. The wave k, ω then exists for a time $\sim 1/\gamma$ (γ is the growth rate) and occupies a region in space of size $\omega/\gamma k$. During a period of the order $1/\gamma$, an individual wave disappears completely and is replaced by another wave, which originates from the beat interaction and consequently cannot be correlated with the first. Although the mode may initially be localized, it in general spreads out as time passes, to fill a region of space whose characteristic size L will be of the order of the distance through which the wave propagates during its lifetime, i.e., $L \sim (1/\gamma)\, d\omega/dk \sim \omega/\gamma k$.
>
> Thus the state of turbulent motion of a continuous medium must be regarded as a system of many wave packets. For $\gamma/\omega \ll 1$ these packets exist for a long time and are very extended, so that one can describe them as waves which are almost completely unlocalized in space. But as γ/ω increases, we must *explicitly* [italics mine—HT] consider that the elements of the turbulent motion are not the Fourier components, but the wave packets. In other words, for finite γ/ω, nearby Fourier components can no longer be considered weakly correlated.

In wave mechanics, $\gamma/\omega \sim 1$ [which is identical to our Equation (5.41)] means that we have to construct a theory of "strong interactions." Attempts at linearization, such as Kraichnan's direct-interaction approximations, may not be appropriate to the task at hand.

5.8. Vorticity and Velocity

Since eddies do not propagate in the way that waves do, we need a relation that serves the function of a dispersion relation but does not carry the same interpretation. The vorticity vector is defined as the curl of the velocity vector:

$$\omega_i = \varepsilon_{ijk}\, \partial u_k/\partial x_j \tag{5.43}$$

We shall assume that the "frequency" ω of an eddy with velocity $\hat{u}$ and wave number k can be estimated with the relation

$$\omega \sim \hat{u}k \tag{5.44}$$

This implies that, for a wave packet or eddy of "size" Δx and "time constant" Δt,

$$\Delta x/\Delta t \sim \hat{u} \tag{5.45}$$

This seems altogether reasonable.

5.9. The "First Law" of Turbulence

Let us now consider the decay of turbulence behind a grid in a wind tunnel. The kinetic energy of the turbulence decays according to (we shall use a frame of reference moving with the mean velocity)

$$\frac{\partial}{\partial t}\overline{(\tfrac{1}{2}u_i u_i)} = -\nu\overline{\omega_i \omega_i} \tag{5.46}$$

The right-hand side represents the rate at which kinetic energy is converted into thermal energy as small eddies are strained by viscous stresses. This *dissipation rate* is so important to turbulence that a special symbol is used:

$$\varepsilon \equiv \nu\overline{\omega_i \omega_i} \tag{5.47}$$

Now, if the most energetic eddies created in the wake of the grid have a velocity $\hat{u}$ that is comparable to the turbulence intensity q ($q^2 = \overline{u_i u_i}$), and if their wavelength is l, then Equations (5.41) and (5.44) imply that the time constant of decay is

$$\Delta t \sim \frac{2\pi}{\omega} \sim \frac{2\pi}{\hat{u}k} \sim \frac{l}{\hat{u}} \sim \frac{l}{q} \tag{5.48}$$

Therefore,

$$\varepsilon = -\frac{\partial}{\partial t}\overline{(\tfrac{1}{2}u_i u_i)} \sim \frac{q^2}{\Delta t} \sim \frac{q^3}{l} \tag{5.49}$$

This relation is so important to turbulence theory that I am tempted to call it the "first law" of turbulence. There is no firm theoretical basis for Equation (5.49), because there is no satisfactory theory of the statistical mechanics of nonlinear nonequilibrium ensembles. However, Equation (5.49) and its consequences have been solidly confirmed by experimental data. In essence, Equation (5.49) states that the rate at which turbulence dissipates its energy is determined by the "turnover" time of the most energetic eddies. The dissipation rate is *independent of viscosity*. In turbulence, it does not help to choose a fluid with a small viscosity; the losses of mechanical energy will remain the same.

5.10. The Energy Cascade

Equation (5.49) determines the rate at which the most energetic eddies of turbulence lose their kinetic energy. That energy is not lost immediately to viscosity, but it is transferred to smaller eddies (typically with a wave number about an octave higher than the first), which in turn transfer energy to smaller daughters, and so on, until the offspring have become so small in size (so large in wave number) that viscosity can annihilate them almost immediately (for more details, see Tennekes and Lumley,[1] Chapter 8).

As this spectral energy transfer proceeds, the characteristic times of the eddies involved decrease. At wave numbers much smaller than those of the eddies containing most of the energy, the energy transfer times are quite small compared to the turnover time of the large eddies. This suggests that the statistical properties of the motion at high wave numbers might not depend on time explicity, but only on the *present* rate at which energy is transferred down the spectrum. If, moreover, we restrict ourselves to wave numbers that are small compared to those at which the actual dissipation of energy by viscous stresses takes place, we may postulate that the expression exhibited in Equation (5.49) should be valid at all wave numbers:

$$\varepsilon \sim \hat{u}^3 k \tag{5.50}$$

Here, $\hat{u}$ represents the average amplitude of the Fourier coefficients in an octave band around k.

Kolmogorov applied this idea first to the spatial structure function

$$D(r) = \overline{[v(x) - v(x+r)]^2} \tag{5.51}$$

This is a fairly good filter at small r [see the text following Equation (5.13)]; contributions to Equation (5.50) are made mainly by eddies of size r, which have a kinetic energy—see Equation (5.50)—proportional to $\varepsilon^{2/3}r^{2/3}$. Therefore, Kolmogorov hypothesized that

$$D(r) = c\varepsilon^{2/3}r^{2/3} \tag{5.52}$$

The spectral equivalent of this expression requires the introduction of the "three-dimensional" spectrum $E(k)$, which is a function of the modulus of the wave number (it represents energy averaged over spherical wave-number shells), and integrates over all k to the total kinetic energy:

$$\tfrac{1}{2}\overline{u_i u_i} = \tfrac{1}{2}q^2 = \int_0^\infty E(k)\, dk \tag{5.53}$$

According to Equation (5.38), the kinetic energy of all eddies with a wave number near k may be represented as $\Delta k \cdot E(k) \sim kE$. Substitution of this into Equation (5.50) yields

$$\varepsilon \sim (kE)^{3/2}k \tag{5.54}$$

which gives the famous Kolmogorov spectrum in the inertial subrange:

$$E(k) = \alpha\varepsilon^{2/3}k^{-5/3} \tag{5.55}$$

These expressions imply that the vorticity ω and the strain rate s of an eddy of wave number k vary as

$$\omega(k) \sim s(k) \sim (k^3 E)^{1/2} \sim \varepsilon^{1/3}k^{2/3} \tag{5.56}$$

These increase as the wave number increases, so that—see Equation (5.41)—the characteristic lifetime of small eddies is short. This, of course, supports Kolmogorov's equilibrium hypothesis.

The energy transfer to higher wave numbers comes to an end when the viscous frequency νk^2 becomes comparable to the inertial frequencies given in Equation (5.56). The *smallest* scale of motion thus corresponds to $\nu k^2 \sim \varepsilon^{1/3}k^{2/3}$; accordingly, the *Kolmogorov microscale* is defined by

$$\eta = (\nu^3/\varepsilon)^{1/4} \tag{5.57}$$

In this line of reasoning, the smallest scale of motion adjusts itself to the rate of energy supply. If the dissipation rate is large, η is small, and vice versa. At moderate Reynolds numbers the microstructure of turbulence is relatively coarse grained, at very high Reynolds numbers the microstructure is fine grained.

5.11. Some Enlightening Errors

The Kolmogorov spectrum, Equation (5.55), is often "explained" with the aid of spectral adaptations of eddy-viscosity and mixing-length models. Spectral energy transfer arises when the Reynolds stresses of smaller eddies perform work against the strain rates of larger ones. If this deformation work is reasonably "local" in wave-number space, we can write

$$\varepsilon \sim \hat{u}^2 \cdot k\hat{u} \sim k^{5/2}E^{3/2} \tag{5.58}$$

which leads to the Kolmogorov spectrum.

This kind of reasoning does not always produce correct spectral forms. For example, Tchen (see Hinze,[6] p. 268) made the mistake of assuming that the mean shear ($\partial U/\partial y = S$) in the bottom part of a turbulent boundary layer is so large that the mean-flow contribution to the strain rate dominates all strain-rate fluctuations. Therefore, he took $\varepsilon \sim \hat{u}^2 S \sim kES$, which gave

$$E(k) = c_T \varepsilon k^{-1} S^{-1} \tag{5.59}$$

Since this decreases more slowly than the Kolmogorov spectrum, it indicates that the mean strain rate is a rather ineffective spectral transfer agent for kinetic energy.

A much more interesting mistake was made by Kraichnan in his *direct-interaction approximation*. This is a weak-coupling model in Fourier space. It bears some resemblance to eddy-viscosity models, and it can be explained in those terms. Energy transfer is like a stress times a strain rate; stress can be written as the product of an eddy viscosity and a strain rate; and an eddy viscosity is like a energy times a time constant. Then

$$\varepsilon \sim \tau \cdot \hat{u}^2 \cdot (k\hat{u})^2 \tag{5.60}$$

In the direct-interaction approximation, the time constant that dominates the energy transfer is the time it takes for an eddy of size λ (wave number k) to pass a fixed point. This time is of order $1/qk$ (q is defined as the root-mean-square turbulence velocity). Therefore, Equation (5.60) becomes:

$$\varepsilon \sim (qk)^{-1} \cdot kE \cdot k^3 E \tag{5.61}$$

which yields (see Kraichnan[7])

$$E(k) = c(\varepsilon q)^{1/2} k^{-3/2} \tag{5.62}$$

What is at issue here? The eddy diffusivity at wave number k is the product of a time integral scale (correlation time) times the average energy in the eddies involved [see Equation (5.7), for example]. In the direct-interaction approximation, the correlation time is selected on the basis of the so-called "adiabatic interaction" (Kadomtsev[5]) between inertial-range eddies and the most energetic eddies. This "adiabatic interaction" represents the advection of small eddies by large eddies past a fixed observation point; it has no *dynamical* consequences to speak of because the energy transfer associated with the straining of the small eddies by the large ones is negligible.

Indeed, because adiabatic interactions have negligible consequences for energy transfer in wave-number space, they have to be filtered out of the direct-interaction equations. When this is done, one retrieves the Kolmogorov form of the inertial subrange.

Adiabatic interactions are of immediate relevance in certain specific cases. The frequency spectrum seen at a fixed observation point in a box of turbulence without a mean flow is determined by the advection of small eddies by large eddies. As motions with velocities of order q sweep eddies of wave number k past the observation point, frequencies of order qk (the reciprocal of the adiabatic decorrelation time) are generated, so that the Kolmogorov spectrum, Equation (5.55), converts to (Tennekes[8])

$$\Phi_E(\omega)=\alpha'(\varepsilon q)^{2/3}\omega^{-5/3} \tag{5.63}$$

The Eulerian frequency spectrum is different from its Lagrangian counterpart (one in which adiabatic interactions are absent by definition). The latter reads

$$\Phi_L(\omega)=\beta'\varepsilon\omega^{-2} \tag{5.64}$$

The form of this spectrum may be rationalized by noting that $\varepsilon \sim u^2/\tau \sim u^2\omega$. Because u^2 is estimated as $\omega\Phi_L(\omega)$, Equation (5.64) follows. The corresponding Lagrangian structure function is

$$D_L(\tau)=\overline{[v(t)-v(t+\tau)]^2}=\beta_L\varepsilon\tau \tag{5.65}$$

which is valid for time intervals long compared to the viscous time scale $(\nu/\varepsilon)^{1/2}$ and short compared to the inertial time scale q^2/ε.

5.12. Other Inertial Ranges

These Kolmogorov scaling concepts can be extended easily to the spectra of other variables. The spectrum of pressure fluctuations, for example, should be related to dynamic-head fluctuations. For an octave band around wave number k, we obtain

$$p(k)\sim\tfrac{1}{2}\rho\hat{u}^2(k)\sim\rho kE\sim\rho\varepsilon^{2/3}k^{-2/3} \tag{5.66}$$

Squaring this, we find

$$kS_p(k)\sim p^2(k)\sim\rho^2\varepsilon^{4/3}k^{-4/3} \tag{5.67}$$

so that

$$S_p(k)=\alpha_p\rho^2\varepsilon^{4/3}k^{-7/3} \tag{5.68}$$

The corresponding spatial structure function is

$$\overline{\{p(x)-p(x+r)\}^2}\sim\rho^2\varepsilon^{4/3}r^{4/3} \tag{5.69}$$

The spectrum of pressure-gradient fluctuations, based on Equation (5.68), is proportional to the minus one-third power of k. In other words,

$$\overline{\left(\frac{\partial p}{\partial x}\right)^2_k}\sim k^2p^2(k)\sim\rho^2\varepsilon^{4/3}k^{2/3} \tag{5.70}$$

These gradients become steeper as the wave number increases. Clearly, the largest contributions to $\partial p/\partial x$ are fluctuations made near the Kolmogorov microscale η. This means that the spectrally averaged variance of $\partial p/\partial x$ may be estimated as (Batchelor[9])

$$\overline{\left(\frac{1}{\rho}\frac{\partial p}{\partial x}\right)^2}\sim\varepsilon^{4/3}\eta^{-2/3}\sim(\varepsilon^3/\nu)^{1/2} \tag{5.71}$$

Another spectrum of interest is that of temperature fluctuations. Temperature here serves as the prototype for all scalar additives or contaminants that are mixed or homogenized by turbulent motion. Spectral transfer of kinetic energy is accompanied by spectral transfer of contaminant fluctuations, because vortex stretching makes long, thin ribbons out of scalar concentration inhomogeneities. Let us denote the rate of molecular destruction of temperature variance by the symbol χ:

$$\chi\equiv\gamma\overline{\frac{\partial\theta}{\partial x_i}\frac{\partial\theta}{\partial x_i}} \tag{5.72}$$

where γ is the thermal diffusivity of the fluid. Equation (5.72) is the counterpart for temperature variance of the definition of ε given in Equation (5.47). Note that χ is the rate at which $\frac{1}{2}\overline{\theta^2}$ is destroyed, and ε is the rate at which $\frac{1}{2}\overline{u_iu_i}$ is destroyed.

In the inertial range, we assume that the effects of molecular diffusion are negligible (this now includes ν and γ, so that not only the Reynolds number has to be large, but also the Péclet number). The spectral transfer of temperature variance behaves like a heat flux interacting with a temperature gradient:

$$\chi\sim[\hat{u}(k)\theta(k)]\cdot k\theta(k) \tag{5.73}$$

which leads to

$$kS_\theta(k)\sim\theta^2(k)\sim\chi/k\hat{u}(k) \tag{5.74}$$

In the inertial range, $\hat{u} \sim \varepsilon^{1/3} k^{-1/3}$, so that we obtain

$$S_\theta(k) = \alpha_\theta \chi \varepsilon^{-1/3} k^{-5/3} \tag{5.75}$$

The way in which ε occurs here deserves some comment. The spectral density of temperature fluctuations decreases when ε increases because it is impossible to maintain the same level of temperature variance if the turbulent motion homogenizes the temperature inhomogeneities at a faster pace. We can see this in another way be estimating the spectrally averaged temperature variance from Equation (5.74), under the assumption that the temperature fluctuations are introduced at scales comparable to the integral scale (l, say) of the turbulent motion. This yields

$$\overline{\theta^2} \sim \chi \tau \tag{5.76}$$

or

$$\chi \sim \overline{\theta^2}/\tau \sim \overline{\theta^2} \cdot q/l \tag{5.77}$$

This shows that, for the same value of χ, the temperature variance is *less* if the mixing rate q/l is faster. Conversely, the spectral transfer rate of temperature variance is proportional to the variance itself, divided by the "time constant" of spectral transfer. In this sense, Equation (5.77) is similar to Equation (5.49)—recall that the latter states that the spectral transfer rate of velocity variance is proportional to the variance involved, divided by the "time constant" of the spectral flux.

Another interesting special case occurs in fluids with large Prandtl numbers. If the Prandtl number is large the viscosity is much larger than the thermal diffusivity. Straining motions in the turbulence then can create very thin filaments in the temperature microstructure. The problem is similar to that of the mixing of different colors of paint. For large Prandtl numbers, the smallest temperature inhomogeneities are much smaller than the smallest eddies in the velocity field, so that there will be a temperature subrange at wave numbers so large that the velocity fluctuations have been dissipated by viscosity, while the temperature fluctuations are not yet being attacked by the thermal diffusivity. In this range, the spectral rate of transfer, χ, must be related to the spectrally local variance of temperature as $\chi \sim \theta^2(k) s$, where s is an appropriate strain rate. In other words,

$$k S_\theta(k) \sim \theta^2(k) \sim \chi/s \tag{5.78}$$

Turbulent motion generates its largest strain rates at its smallest scales. At scales comparable to the Kolmogorov microscale η, the value of s is of order $(\varepsilon/\nu)^{1/2}$, so that we obtain

$$S_\theta(k) = \beta_\theta \chi (\nu/\varepsilon)^{1/2} k^{-1} \tag{5.79}$$

Further details on this subrange and on several others can be found in Tennekes and Lumley,[1] Chapter 8.

5.13. Turbulent Diffusion Revisited

In the inertial subrange of frequencies, the Lagrangian structure function is given by Equation (5.65). The corresponding correlation function is—see Equation (5.14)—

$$R_L(\tau) = \overline{v^2} - \tfrac{1}{2}\beta_L \varepsilon \tau \qquad (5.80)$$

Therefore, we can integrate the dispersion rate equation, Equation (5.69), over time intervals small compared to the Lagrangian integral scale:

$$\frac{d}{dt}\overline{\left(\frac{1}{2}y^2\right)} = \int_0^t R_L(\tau)\, d\tau = \overline{v^2}t - \frac{1}{4}\beta_L \varepsilon \tau^2 \qquad (5.81)$$

This is not altogether fair, because Equation (5.80) is an approximation that takes the curvature of R_L at $\tau = 0$ to be infinite. At large Reynolds numbers the errors involved are small, however. Further integration yields

$$\overline{y^2} = \overline{v^2}t^2 - \tfrac{1}{6}\beta_L \varepsilon t^3 \qquad (5.82)$$

Because of the decreasing correlation, the dispersion $\overline{y^2}$ is less than the value that would be obtained if the velocity fluctuations were perfectly persistent. Here we meet the consequences of adiabatic interactions again. The *initial* dispersion away from a fixed point is determined completely by the intensity of the velocity fluctuations at that point, because tagged particles are advected away from the origin. For time intervals so short that the velocity does not change, we have $y = vt$, so that $\overline{y^2} = \overline{v^2}t^2$. But as soon as the velocity fluctuations begin to lose their correlation, the dispersion rate begins to slow down.

In that perspective, it is not hard to estimate how turbulent diffusion over small time intervals behaves when the initial advection effects are removed. We do this by releasing particles *in pairs*, in such a way that their initial separation is small. As long as the velocity fluctuations remain perfectly correlated the separation Δy of adjacent particles will not change, but as soon as the decorrelation begins, we expect—on the basis of Equation (5.82)—that

$$\overline{(\Delta y)^2} \sim \varepsilon t^3 \qquad (5.83)$$

We dare to do this because it is the very process of decorrelation that allows Δy to grow as time proceeds.

The eddy diffusivity associated with relative diffusion is

$$\frac{1}{2}\frac{d}{dt}\overline{(\Delta y)^2} \sim \varepsilon t^2 \tag{5.84}$$

Equation (5.83) permits us to rewrite this in such a way that the explicit dependence on time is removed:

$$\frac{1}{2}\frac{d}{dt}\overline{(\Delta y)^2} \sim \varepsilon^{1/3} r^{4/3} \tag{5.85}$$

where $r^2 = \overline{(\Delta y)^2}$. This is Richardson's famous four-thirds law, often written as

$$K(r) = c\varepsilon^{1/3} r^{4/3} \tag{5.86}$$

In the inertial subrange, the eddy diffusivity thus varies quite rapidly with the size of the diffusing cloud of particles.

5.14. Conclusions

Coming to the end of this excursion into the statistical dynamics of turbulence, we realize—not without some despair—that we have not made any substantial use of the Navier–Stokes equations. That may not be typical of the state in which turbulence theory finds itself, but I am afraid I have to confess that the absence of a sound theory is one of the most disturbing aspects of the turbulence syndrome. We have yet a long way to go.

It should be mentioned that I have not made any attempts to give due credit at every point in the analysis. Complete references can be found in Monin and Yaglom[(2,3)]; a list of all major texts and monographs on turbulence is given in Tennekes and Lumley.[(1)]

Notation

a	Particle acceleration
c	Constant
D_L	Lagrangian structure function
E_L	Power spectral density
F_L	Fourier transform of R_L
k	Wave number
l	Wavelength
R_L	Correlation function
S	Mean shear
S_p	Spectrum of pressure fluctuations
S_θ	Spectrum of temperature fluctuations
t, t'	Time
T_L	Lagrangian integral time scale
u_i	Velocity components
$\hat{u}$	Eddy velocity
v	Velocity

w	Particle velocity
y	Distance
γ	Thermal diffusivity
ρ_L	Lagrangian autocorrelation function
η	Kolmogorov microscale
ε	Dissipation rate of kinetic energy
Φ	Frequency spectrum
Λ	Time constant
ν	Kinematic viscosity
ω	Frequency, vorticity
ω_i	Vorticity components
χ	Dissipation rate of temperature variance
τ	Time interval

References

1. Tennekes, H., and Lumley, J. L., *A First Course in Turbulence*, MIT Press, Cambridge, Massachusetts (1972).
2. Monin, A. S., and Yaglom, A. M., *Statistical Fluid Mechanics*, Volume 2, MIT Press, Cambridge, Massachusetts (1975).
3. Monin, A. S., and Yaglom, A. M., *Statistical Fluid Mechanics*, Volume 1, MIT Press, Cambridge, Massachusetts (1971).
4. Kraichnan, R. H., On Kolmogorov's inertial-range theories, *J. Fluid Mech.* **62**, 305–330 (1974).
5. Kadomtsev, B. B., *Plasma Turbulence*, Academic Press, New York (1965).
6. Hinze, J. O., *Turbulence*, McGraw-Hill, New York (1959).
7. Kraichnan, R. H., The structure of isotropic turbulence at high Reynolds numbers, *J. Fluid Mech.* **5**, 497–543 (1959).
8. Tennekes, H., Eulerian and Lagrangian time microscales in isotropic turbulence, *J. Fluid Mech.* **67**, 561–567 (1975).
9. Batchelor, G. K., *The Theory of Homogeneous Turbulence*, Cambridge University Press, New York (1953).

CHAPTER 6

Transition

R. BETCHOV

6.1. Introduction

Transition is the most complex problem of fluid mechanics, and perhaps of modern physics. A century of effort has produced very limited insight and only a few guidelines for the benefit of design engineers. Transition is the process by which a laminar flow becomes turbulent. While the definition of the word "turbulent" still leaves some uncertainty, there seems to be less doubt about what we call "transition." In a laminar flow, it is reasonable to try to describe the motion by specifying the velocity at any point and any time. The theoretician will use analytic expressions, the numerologist will interpolate between values known at certain grid points. In a turbulent flow, this objective becomes meaningless. A full description of the flow would be of little use, since interest centers on averaged quantities, and especially on the Reynolds stress. Thus, statistical information becomes all-important.

It is because transition straddles the fence that it presents great difficulties. It no longer deals with a well-ordered system and not yet with a fully developed disorder.

From another point of view, we can always offer a first approximation for a laminar flow: Blasius layer, Poiseuille profile, etc. We can approximate a turbulent flow by a mean profile plus Gaussian fluctuations, which may be simply correlated. But no one, so far, has offered a useful "first approximation" for flows undergoing transition.

R. BETCHOV • The University of Notre Dame, Notre Dame, Indiana 46556

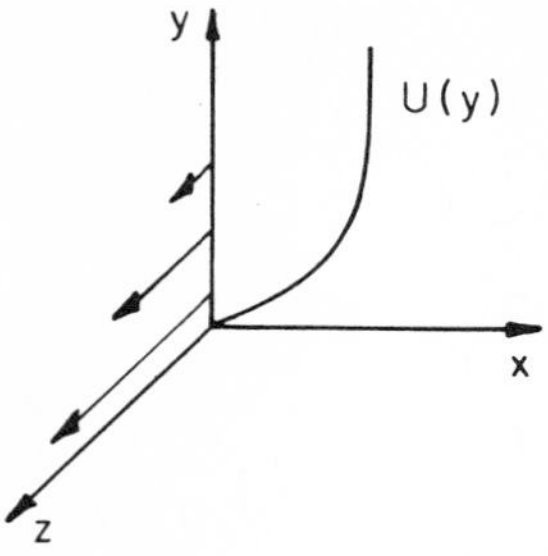

Fig. 6.1. Vortex lines in a laminar boundary layer.

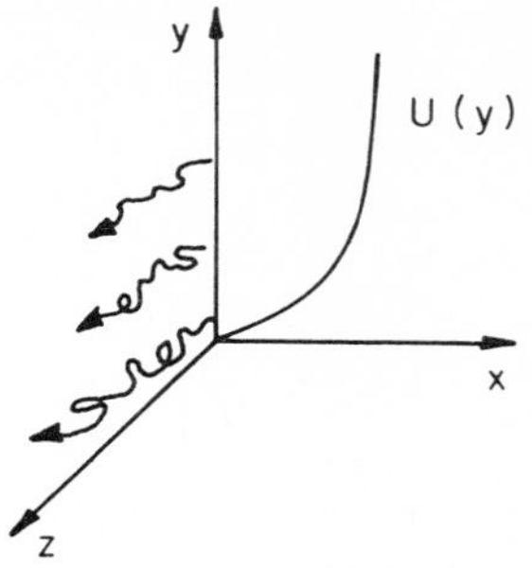

Fig. 6.2. Vortex lines in a turbulent boundary layer.

Confronted with this situation, those interested in transition should always begin by taking a candid look at various movies showing the pattern formed by smoke or ink injected into flows undergoing transition. In particular, the movies made by F. N. M. Brown are most instructive.*

In interpreting this kind of experimental data, one must be careful. We are only looking at streaklines, that is, lines formed by various tagged particles, released sequentially from one common origin. As shown by Hama,[1] a flow can have a very simple Eulerian velocity field, and yet it can produce very contorted streaklines. The streaklines seem to suggest growing oscillations and separation into rotating clusters of points, but this does not necessarily mean that the flow is really involved in any such things. The pattern formed at one instant by various particles, which entered the flow at various times, or various points, depends upon the history of the flow. This has the effect that simple variations in velocity can produce very intricate displacements.

It is not the mere presence of vorticity that characterizes turbulence. It is the complexity of the vorticity field. In a laminar boundary layer, the vortex lines are parallel (see Figure 6.1) and stacked near the wall, like uncooked spaghetti. In the turbulent layer (see Figure 6.2), the vortex lines are constantly changing and twisting. Near the wall, major entanglements appear, and the vortex lines may develop knots and crossover points. The spaghetti is cooked.

The history of our efforts to understand transition seems to fall nicely into three phases. First, the research centered on the stability of simple disturbances of two-dimensional flows: The spaghetti started to move, but remained rigid. In a second phase, one or two simple disturbances were

* Archives, Department of Aerospace Engineering, University of Notre Dame, Notre Dame, Indiana.

considered: The spaghetti began to wiggle. This included cases where the mean flow is already modified by the unstable process. This brings us to the threshold of the third phase, which deals with very localized but strong phenomena. This third phase is characterized by frequent reliance on numerical methods, and a plurality of efforts.

6.2. Weak Oscillations of Simple Flow

After Reynolds pointed out that transition in pipe flow does not occur unless the viscous forces have dropped below a level amounting to less than 0.1% of the inertial forces, the search for an explanation started. How can such weak viscous forces control transition?

At the turn of the century, it became clear that, in the absence of viscous forces, the classic flows along walls are very stable. Instability does not appear unless the velocity profile has an inflection. This means (see Figures 6.3 and 6.4), that the vorticity must have a clear maximum. The instability caused by inflection was first studied by Helmholtz,[2] in the limiting case where two parallel flows are separated by a very thin shear layer. Rosenhead[3] showed that the vortex layer tends to roll up, something that linearized theories do not show clearly. As shown in Figure 6.5, once a vortex filament is displaced, the influence of its close neighbors aggravates the disturbance, and an irreversible process is started. The vorticity tends to concentrate in large clusters. The relation between vorticity, velocity fluctuations, and pressure fluctuations is illustrated in Figure 6.6, which shows the potential flow perturbations. For a detailed analysis, see Reference 4. This

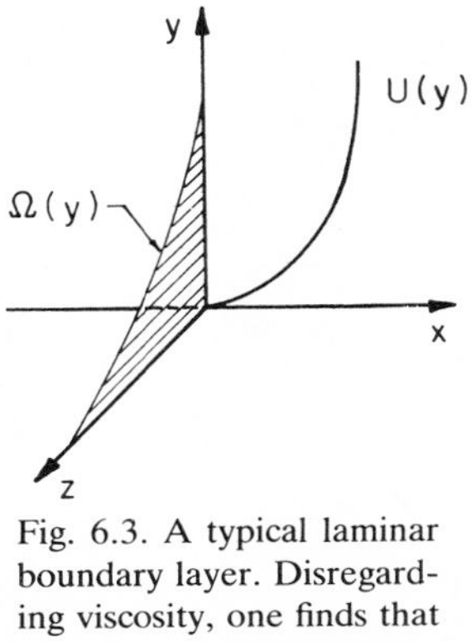

Fig. 6.3. A typical laminar boundary layer. Disregarding viscosity, one finds that this flow is stable to small perturbations, because the velocity profile has no inflections.

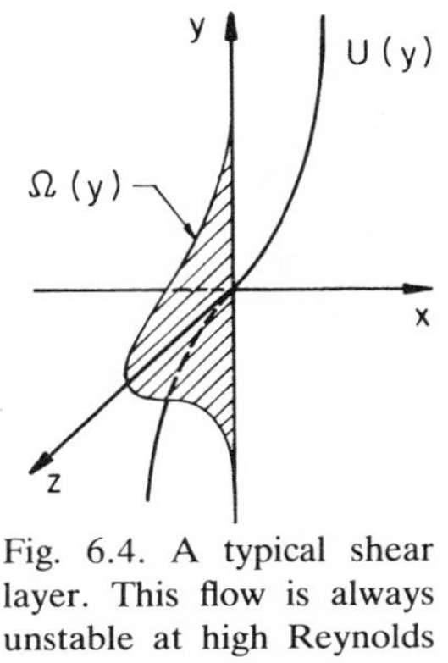

Fig. 6.4. A typical shear layer. This flow is always unstable at high Reynolds numbers, because of the presence of an inflection in the velocity profile.

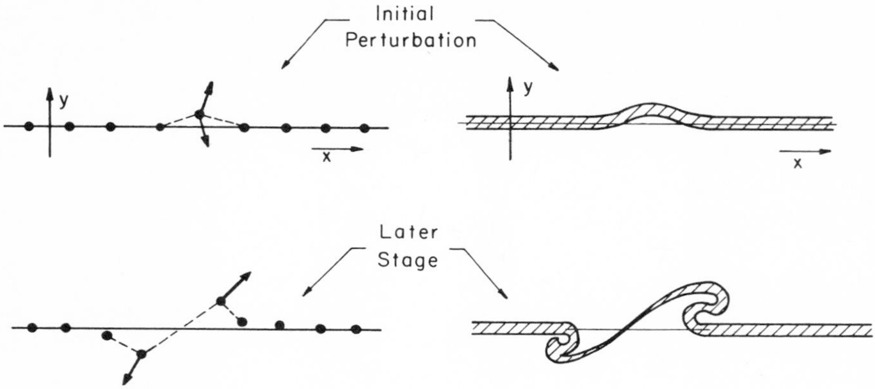

Fig. 6.5. Instability of a vortex row or vortex sheet. Once a vortex filament moves out of line, an irreversible roll-up process takes place.

process is limited to wavelengths large in comparison with the width of the shear layer, or vortex sheet. Viscosity always tends to damp the instability, and has very little influence[(5)] if the Reynolds number of the shear layer is larger than 20. For short wavelength, the oscillation becomes strongly damped—not because of viscosity but because of a remarkable kind of damping, which will be discussed below—and bears a strong analogy to the Landau damping well known in plasma physics.

The stability of a shear layer is governed by three basic mechanisms, which decide the location of the neutral line. The classic stability diagram is shown in Figure 6.7. Let us now examine the Landau damping, since it affects the short wavelengths and represents the major obstacle to transition.

Basically, the vorticity fluctuations ω are transported by the mean flow. They induce a velocity fluctuation v, which disturbs the mean flow and generates new vorticity fluctuations. The basic equations (see Reference 4) read

$$\frac{\partial \omega(x, y, t)}{\partial t} + U(y)\frac{\partial \omega}{\partial x} + \frac{d^2 U}{dy^2} v = 0 \tag{6.1}$$

$$\frac{\partial^2 v}{\partial x^2} + \frac{\partial^2 v}{\partial y^2} = -\frac{\partial \omega}{\partial x} \tag{6.2}$$

For a fluctuation periodic in x, and with the appropriate boundary conditions, the integration gives

$$v(x, y, t) = \frac{i}{2}\int_{-\infty}^{+\infty} \omega(\eta, t)\exp(-\alpha|y-\eta| + i\alpha x)\, d\eta \tag{6.3}$$

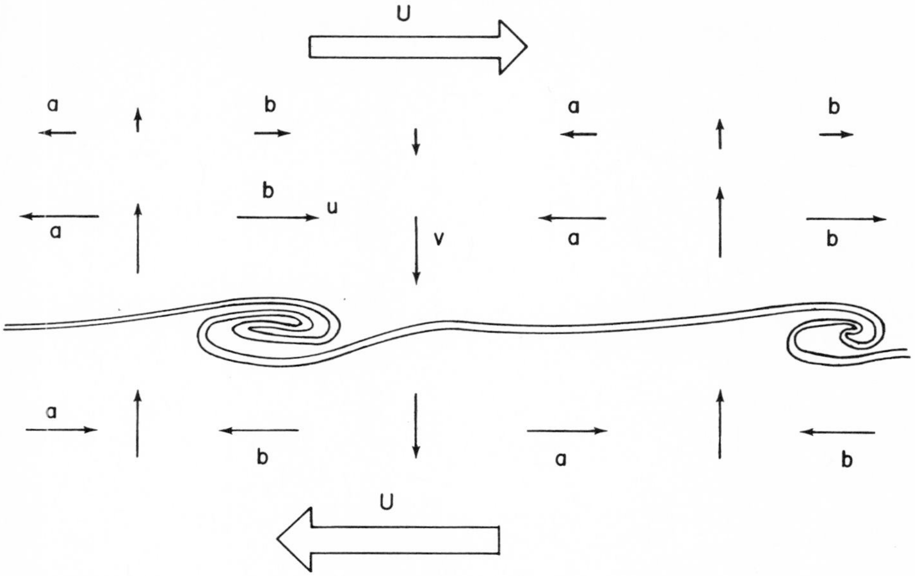

Fig. 6.6. Potential perturbations induced by a rolling-up vortex sheet. The thin arrows indicate velocity fluctuations, superposed on the main flow. The pressure is high at points marked a and low at points marked b (Bernoulli theorem).

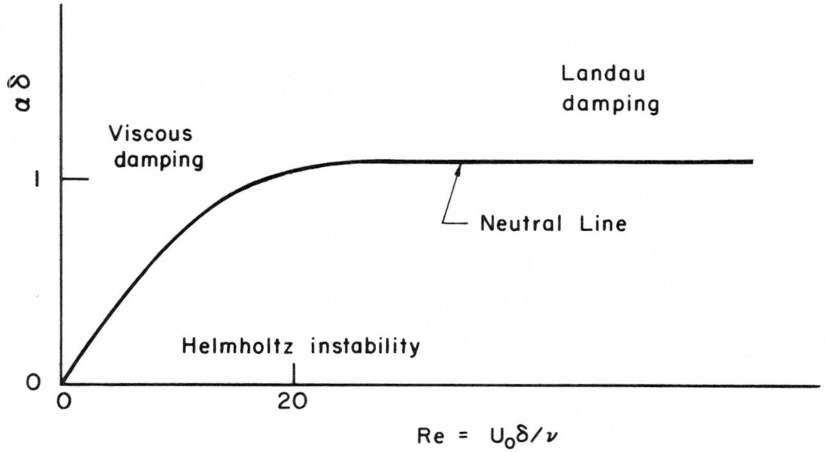

Fig. 6.7. Stability diagram for a shear layer of thickness δ. In the upper-right-hand corner, the oscillations are damped by an inviscid process similar to the Landau damping encountered in plasma physics.

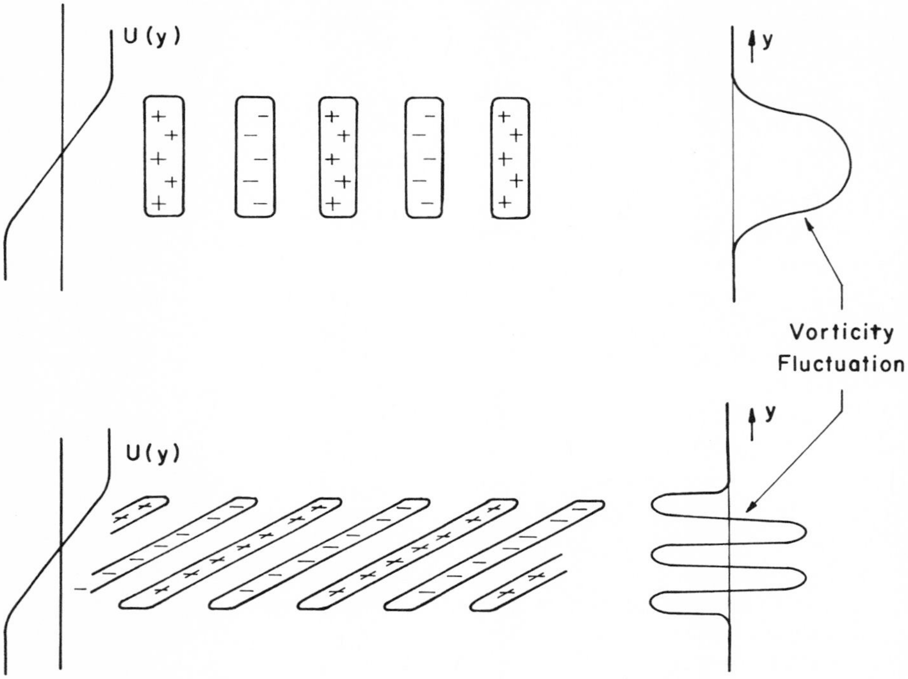

Fig. 6.8. Landau damping in a shear layer. Above: Initial perturbation of the vorticity. Below: Later distribution, disregarding the production term in Equation (6.1). The induced potential flow is rapidly attenuated by the "shredding" of the vorticity. Before vanishing, it recreates a new initial perturbation of the type illustrated here.

Let us now consider perturbation in the shear layer $U = U_0 \tanh(y/\delta)$. The mean flow transforms any initial periodic vorticity fluctuation ω_0 into a "shredded" pattern (see Figure 6.8). This attenuates the induced velocity field so rapidly that it becomes practically null after some characteristic time $t_1 \approx \lambda/U_0$. But, before vanishing, the induced velocity, corresponding to potential oscillations outside the shear layer, disturbs the mean flow vorticity and recreates new vorticity fluctuations, similar to those shown in Figure 6.8.

Let us assume $v_0 \approx \omega_0 \delta$. Then, in the time interval t_1 this produces a second generation of vorticity fluctuations of the order of $\omega_1 \approx v_0(U/\delta^2)t_1$. The new vorticity persists before it is again shredded by the mean flow. The process repeats itself, and at time $t = nt_1$ we have roughly

$$v_n \approx \omega_n \delta \approx v_0 \left(\frac{Ut_1}{\delta}\right)^{t/t_1} = v_0 \exp\left[\frac{U_0 t}{\lambda} \ln\left(\frac{\lambda}{\delta}\right)\right] \tag{6.4}$$

In terms of the wave number $\alpha = 2\pi/\lambda$ and of the imaginary part of the phase velocity, this leads to

$$v = v_0 \exp(\alpha c_i t)$$

with

$$c_i \approx -U_0 \ln(\alpha\delta) \tag{6.5}$$

Thus, for $\alpha\delta = 1$ the oscillation is neutrally stable. For $\alpha\delta > 1$, it is damped. This result was predicted theoretically by Tatsumi and Gotoh[6] and clarified by Gotoh.[7] We found with a computer that, in the limit of small viscosity, this damping indeed appears, but that the eigenfunctions become discontinuous. Incidentally, it seems that any inviscid process involving shredding of the vorticity does not lend itself to ordinary Fourier analysis. This may be related to the fact that, without viscosity, the distance between the zeros of the vorticity tends to vanish at $1/t$.

A basically similar mathematical problem is encountered in plasma, where the electron density obeys the Vlasov equation. It induces an electric field, which, by interaction with a mean profile, generates new electron disturbances. The problem is somewhat simpler to handle, and Landau[8,9] used a detour in the complex plane to obtain an analytic solution.

Turning now to boundary layers, we can understand that, at high Reynolds numbers, they can be very stable (Figure 6.9). They have no inflections, and the Landau damping is dominant.

At the beginning of the century, Orr and Sommerfeld established that nothing short of a certain equation, containing both the viscous terms and

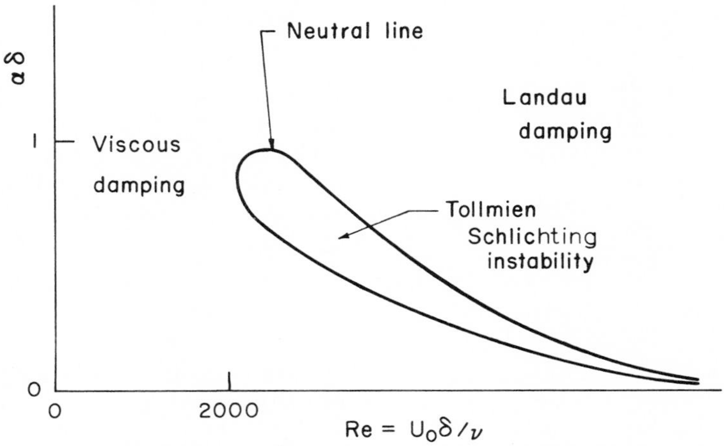

Fig. 6.9. Stability diagram for a boundary layer. The resistive instability is confined to the indicated region.

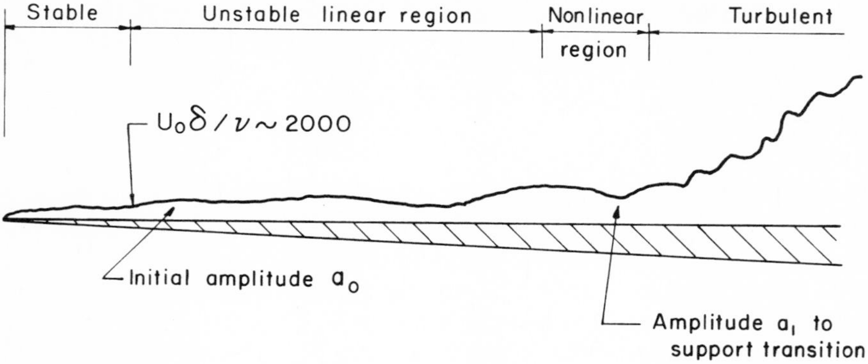

Fig. 6.10. Unstable boundary layer along a flat plate. The linear two-dimensional process prepares the flow for turbulent transition.

the curvature of the mean profile, could yield any explanation of Reynolds criteria for transition. Only in the mid-twenties did Tollmien and later Schlichting produce eigenvalues and eigensolutions for this formidable equation. They found that, if the viscosity is neither too large nor too small, the boundary layer has unstable oscillations. The experiments of Schubauer confirmed the theoretical predictions. For a review and general bibliography the reader is referred to Reference 4.

A Tollmien–Schlichting wave is really a resistive instability that depends essentially upon viscous effects. Viscosity plays a crucial role only in a thin portion of the fluid moving with the phase velocity of the oscillation. It redistributes the vorticity, which otherwise would become discontinuous. The vorticity relocated by the diffusion process induces a special type of velocity fluctuation. The viscosity, in this way, creates a contribution to the motion that is out of phase with the main oscillation, and this renders the boundary layer unstable. The out-of-phase vorticity induces special velocity fluctuations and thereby creates Reynolds stress. These feed energy from the mean flow to the perturbation. The critical layer is very thin, and the growth or decay of the wave is very sensitive to the status of this layer. It is probably this part of the entire boundary layer that is most sensitive to random perturbations. It is located away from the wall. It is within this critical layer that nonlinear effects are most likely to develop, but little is known about this topic.

The stability analysis was extended to other flows, but, in the case of the Poiseuille two-dimensional channel flow, it caused a serious controversy that could not be resolved until the computer became available.

The experimentors also pointed out that these oscillations are of such low frequency and long wavelength that they bear no resemblance to turbulence. Resistive instability, beginning at a Reynolds number of about 10^3, only prepares the flow for a more severe type of accident (Figure 6.10).

Thus it became clear that the Orr–Sommerfeld equation does not cover transition, and that the enigma was still intact.

6.3. Multiple Perturbations of Laminar Flow

The experimental work, initiated to verify Tollmien's results, was carried further. Klebanoff *et al.*[(10)] found that, when the resistive oscillations gain amplitude, they begin to show the first symptoms of turbulence. It begins with isolated and sharp spikes in the velocity recordings. Further downstream, the spikes appear in pairs, and soon in larger groups, which gradually become turbulent bursts (see Figure 6.11). Emmons,[(11)] observing transition in a layer of water flowing down an inclined table, had already noticed that turbulence begins in spots and spreads laterally (see Figure 6.12).

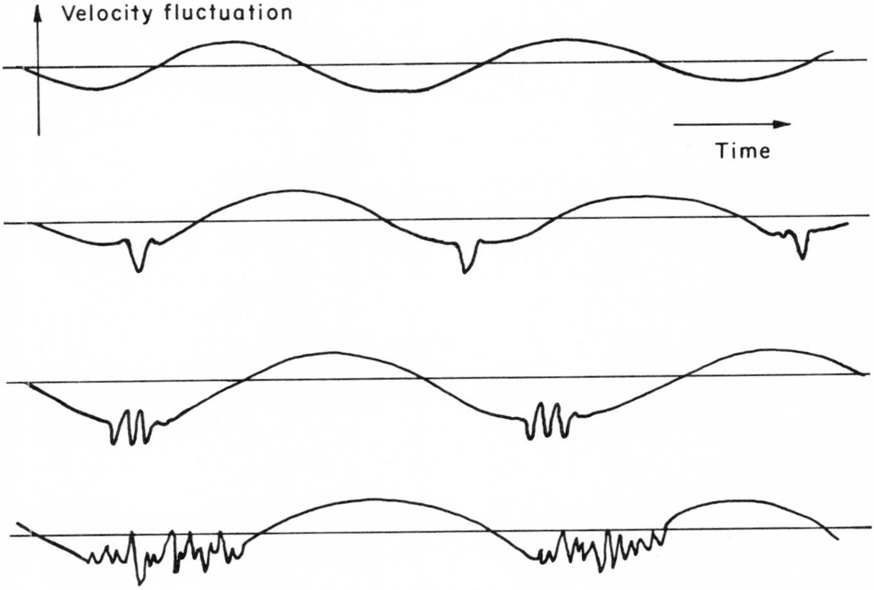

Fig. 6.11. Velocity fluctuations in an oscillating boundary layer. Above: Upstream of the transition region. Middle: First symptoms of approaching transition. Below: Inside the transition region.

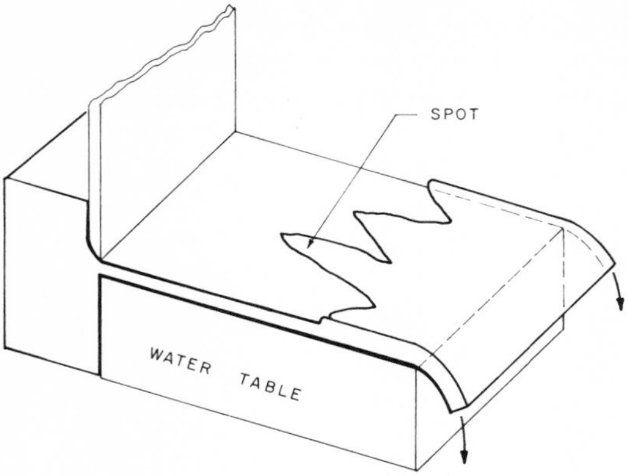

Fig. 6.12. Transition in a laminar boundary layer. Emmons observed that the free surface of a water flow reveals the irregular nature of the boundary separating laminar and turbulent regions, and noted the importance of special spots.

This behavior was unexpected, since it did not fit with the prevailing views. Richardson had described turbulence as a cascade process with big whirls producing progressively smaller whirls, until viscosity stopped the process. Carried over by the tremendous success of linear methods, the mathematicians could not approach the problem without starting with Fourier transform and following by other expansions, in the hope that, in the end, everything would converge and lead to a valid description. A typical example of this unwarranted confidence in superposition can be found in Landau and Lifschitz[(12)]:

> When the Reynolds number increases still further, more and more new periods appear in succession . . . The new flows themselves are on a smaller scale . . .

Thus the theoretical work went along, extending to oblique oscillations (first studied by Squire[(13)] and later by Criminale and Kovasznay,[(14)] who showed how an initially localized perturbation produces a gently spreading wave packet). Görtler,[(15)] Benney and Lin,[(16)] and others studied other situations, along curved walls or with streamwise vortices. Meanwhile, the experimenters found that a variety of mean flow modifications could create the conditions favorable to the appearance of the first spikes. A comprehensive review was given by Tani.[(17)]

It seems that the laminar flow must acquire some three-dimensional structure, and that its profile must develop, locally, a point of inflection.

Using precisely controlled "single-shot" perturbations of an unstable Blasius layer, Kovasznay *et al.*[(18)] showed that a large concentration of vorticity develops, some distance away from the wall. The vortex lines are strongly curved, and the vorticity disturbances become larger than the mean flow vorticity. The spaghetti strands are no longer rigid, merely shifting to and from the wall; they are now bending and forming pronounced loops.

Thus it became clear that Fourier analysis was not a suitable tool to approach this type of phenomenon: The breakdown is localized, nonlinear, and sharp. It cannot be described by a few sinusoidal components; instead it must implicate several components, with strong phase coupling.

Unfortunately, just when the theoretician lost his favorite tool, the experimentor reached the limits of his instruments. Every hot-wire anemometer perturbs the flow, and it is not practical to introduce one probe downstream of another.

6.4. Amplification of Initial Perturbations

An important sideline must now be examined. A. M. O. Smith[19] studied the total amplification supplied by Tollmien–Schlichting waves to simple perturbations, between the point of neutral stability and the region where the flow becomes fully turbulent. A study of various flows, backed by extended stability calculations for various velocity profiles, showed that an initial perturbation has the opportunity to grow by a factor e^9, or about 8000. The uncertainty in the exponent seems to be 9 ± 0.5. This criterion seems sufficiently accurate to be of service to design engineers.

This discovery raises the question of initial perturbations. It is well known that a wide variety of disturbances can cause transition: free-stream turbulence, acoustic waves, wall roughness, vibrations, entropy fluctuations, etc.

The free-stream turbulence is directly related to vorticity fluctuations. Acoustic waves are basically irrotational; however, whenever they are reflected by solid walls, the friction creates a special boundary layer and thereby injects additional vorticity into the flow. The entropy fluctuations are generally the effect of heat transfer at the wall. When both the pressure and the entropy fluctuate, the acceleration created by pressure gradients is no longer the gradient of a function and vorticity is created.

We still do not know how to compare the effectiveness of two different kinds of perturbations in producing transition. Perhaps the various possible perturbations should all be expressed in terms of vorticity since this seems to be the key factor.

The fact that Smith found a roughly constant amount of total amplification suggests that, when reasonable precautions have eliminated the main disturbances, the net effect of all remaining perturbations is equivalent to a sort of uniform background of possible initial perturbation. Since most of the amplification follows an exponential process, and since turbulence breaks out when the fluctuations have reached a given level a_1 (around 1% of free stream vorticity), the basic relation is of the type

$$a_0 \exp G = a_1 \tag{6.6}$$

where a_0 is the initial perturbation and G the overall growth. Note that a reduction of a_0 by a factor of 2 should require an increase of G from 9 to 9.7 to yield the same a_1. This suggests that a_0 has a rather well-defined magnitude.

Could thermal agitation provide a constant level of disturbances? In a gas every molecule has a random velocity of the order of c, the speed of sound. A small cloud of N molecules has a random collective motion with random velocity fluctuations of order $cN^{-1/2}$. Indeed the cloud has a Brownian agitation, with kinetic energy kT per degree of freedom. Thus, at all scales, a gas has a certain thermal agitation. The same property exists in a liquid, although the inventory of the degrees of freedom is more delicate. The spectrum of these velocity fluctuations has been given by Landau and Lifshitz(12) and later evaluated by Betchov.(20) The omnipresent turbulence level, for fluid particles formed of N molecules, is roughly $(c/U)N^{-1/2}$. This activity includes vorticity fluctuations.

With $G=9$, and $a_1 = 10^{-2}$, we need $a_0 = 10^{-6}$. Note that the ratio U/c is simply the Mach number and let us take the value $U/c = 0.03$. This leads to $N = 10^{14}$, or about 10^{-9} mole. In air, this is a cube with a side of 0.03 cm, or about 10 mils. This is about the thickness of the critical layer of a Tollmien–Schlichting wave, that is, of that part of the boundary layer where viscosity redistributes the fluctuating vorticity.

Can Brownian motion really trigger the Tollmien–Schlichting waves, and hence determine transition? At first, it seems that the thermal agitation is too weak to drive an essentially two-dimensional process. It takes too many of the above small cubes to fill the critical layer. As long as we have not integrated the inhomogeneous Orr–Sommerfeld equation, and analyzed the delicate relation between a noise and a slowly released oscillator, the question cannot be dismissed easily.

A detailed study of this difficult subject leading to comparison with experimental results has been pursued by Tsuge.(21) An approximate study of the transition process for a shear layer gave a Reynolds number of 150 (Reference 5).

6.5. Strong Disturbances of Simple Flows

We are only at the threshold of this phase of the work. It is perhaps too early to characterize it, beyond some general statements. This phase seems

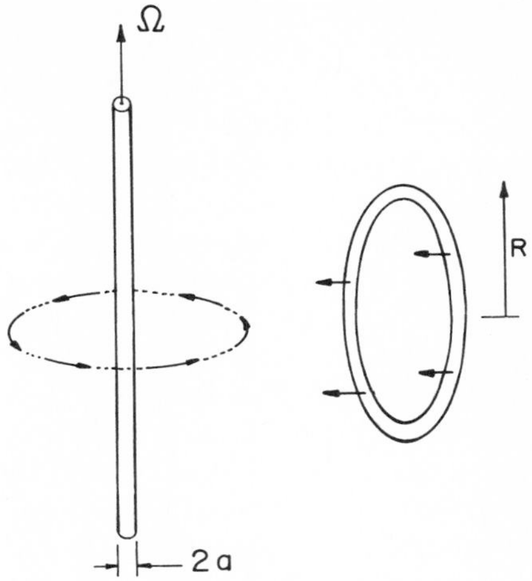

Fig. 6.13. The straight and the circular vortex filaments. The thin arrows indicate induced velocities.

to rely primarily on the use of computers. No doubt, the day will come when the hot wires, or the laser, will again allow further progress. As for numerical modeling of the details of a flow during the transition process, the task may demand as many grid points and programming skills as meteorological studies.

Some workers have focused their attention on simple flows with highly concentrated vorticity. The relation between vorticity and velocity has been studied long ago. If the vorticity is confined to a single line, the velocity is given by a Biot–Savart integral of the type

$$v = c_0 \int \frac{(\boldsymbol{\omega} \times \mathbf{r})}{r^3} \cdot d\mathbf{r} \qquad (6.7)$$

Unless the vortex line is straight, the calculations require the use of a computer. The case of a straight line is shown in Figure 6.13. The vortex line must be treated as a filament of finite radius a, since otherwise both the velocity and the kinetic energy diverge at the center. Under the effect of its own vorticity, the straight filament spins, but it does not travel.

If the vortex filament is circular, with a radius R, the induced velocity transports the vorticity. This is the familiar smoke ring. The translational velocity was given by Prandtl:

$$v = \frac{a^2\Omega}{2R} \ln\left(\frac{6R}{a}\right) \qquad (6.8)$$

Note that, if $a \to 0$, $v \to \infty$. The reason for this effect is that the motion of the filament is controlled by those portions of the vortex line that are close but already out of alignment. For a vortex filament of arbitrary shape, Arms (unpublished communications) proposed to use Equation (6.8) with the local radius of curvature. This implies that circular and helicoidal vortex lines retain their shapes while traveling, but that all others will undergo various contortions. A vortex filament, initially confined to a plane surface but with variable curvature, will develop torsion. This process was studied by Hama.[22] Later Betchov[23] examined the intrinsic relations between the torsion and the curvature and Hasimoto[24] showed that curvature and torsion can be combined to form a complex function governed by a single equation similar to the Schrödinger equation for a self-attracting particle. This nonlinear equation had already attracted the attention of the high-energy physicist (Korte–De Vries equation). It has the remarkable property that certain solutions propagate along an intially straight filament without changing shape. In quantum mechanics, these solutions are called "solitons," as they might represent single stable particles.

In hydrodynamics this means that, starting with a weak disturbance of a straight vortex filament, we can expect two helicoidal disturbances running away, on either side. It is not difficult to observe such solitons on a numerical solution, but severe computational instabilities stand in the way of more advanced work (see Figure 6.14).

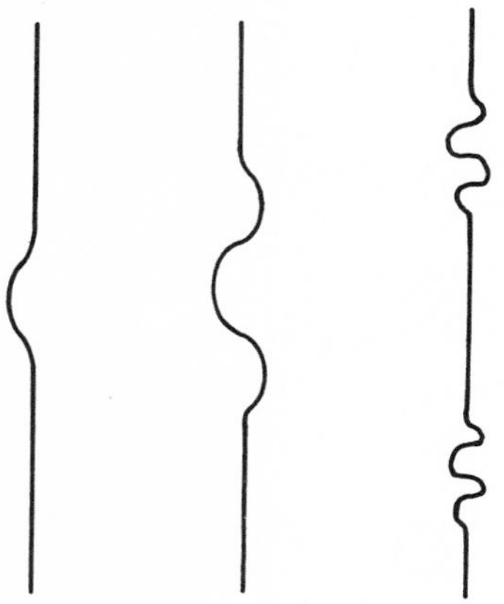

Fig. 6.14. Small perturbations on a straight vortex filament. A simple initial bump (left) without torsion, produces two helicoidal disturbances that wiggle away along the filament.

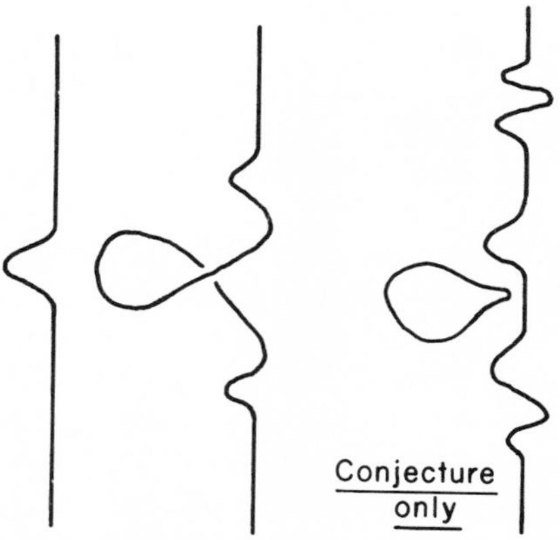

Fig. 6.15. Large perturbation on a straight vortex filament. It is possible, but not certain, that the filament may cut itself and that a viscous process may produce a separate loop, free to wander away.

What happens when the initial disturbance is strong? Can it create a pair of helicoidal disturbances and change into a stable solution? Can the vortex line, during this initial process, intersect itself? In the event that two distinct portions of the filament would run across each other, the viscous effects would again play a decisive role. Leonard[25] recently studied such cases and found that two vortex filaments can perform a crisscross operation. Thus, it is conceivable that a vortex ring can detach itself from the initial filament and drift far away (see Figure 6.15).

Fluid mechanics, through the ages, has provided many basic concepts that guided the developments of other fields of physics. The words "density," "current," and "waves," which now permeate electrodynamics and quantum mechanics, have one common origin: hydrodynamic. Perhaps the vortex ring and the soliton will continue this tradition.

Transition in compressible flows is an even more complex problem. With high Mach numbers comes heat transfer and entropy gradients. When the entropy and the pressure vary together, a new source of vorticity enters the picture. Furthermore, in a compressible flow energy can be transferred by acoustic waves, with the delays inherent to wave propagation. This greatly increases the possibilities of unstable oscillations, as shown for example by Mack.[26] A comprehensive review of the subject by Morkovin[27] not only shows the difficulty of interpreting a mass of data, but it also suggests a basic unity. Even in the hypersonic regime, the Reynolds number of transition remains within the familiar range. This suggests that the traditional sources of vorticity or instability do not profoundly modify the situation.

6.6. Statistical Models

Transition can also be approached from another direction. Nee and Kovasznay[28] proposed a theory for two-dimensional turbulent flows based on a phenomenological treatment of the eddy viscosity. They proposed an equation accounting for several effects: how the eddy viscosity diffuses itself and how it grows in a shear flow or decays like isotropic turbulence. This model was very successful when applied to fully developed turbulent boundary layers.

Recently Nee (private communication) applied this method to transition flow. Starting with a Blasius profile, with critical Reynolds number and zero-eddy viscosity except for a small initial perturbation, he used a numerical method to study the consequences. The layer is most sensitive to disturbances introduced in the vicinity of the critical layer. The eddy viscosity builds up starting from an arbitrary initial level of 10% of the molecular viscosity. It grows most rapidly near the wall and soon alters the mean flow. The effect then shifts away from the wall, and the velocity profile evolves towards the classical turbulent shape. This method has one noteworthy feature: It creates a sharp interface between turbulent and laminar flow. This is often called the superlayer. Across this superlayer, the eddy viscosity collapses.

The laminar and turbulent situations are shown in Figure 6.16 corresponding to an increase of the eddy viscosity up to 60 times the molecular viscosity.

The turbulent curves correspond to a location where, if the layer had remained laminar, it would have reached a Reynolds number of about 4000. Further downstream, the eddy viscosity would keep growing, up to a level of about 200 times the molecular viscosity. Note the presence of a wall region, where the eddy viscosity increases linearly.

This model has no spikes or spots, because it is inherently two dimensional. Before extending it to a three-dimensional process, the eddy viscosity equation should be expanded to account for transversal effects. This is not a simple task, since the spread of turbulence is not sufficiently understood, and some experimental data would be necessary. Perhaps the study of simple disturbances carried by one or several vortex lines will provide some clues.

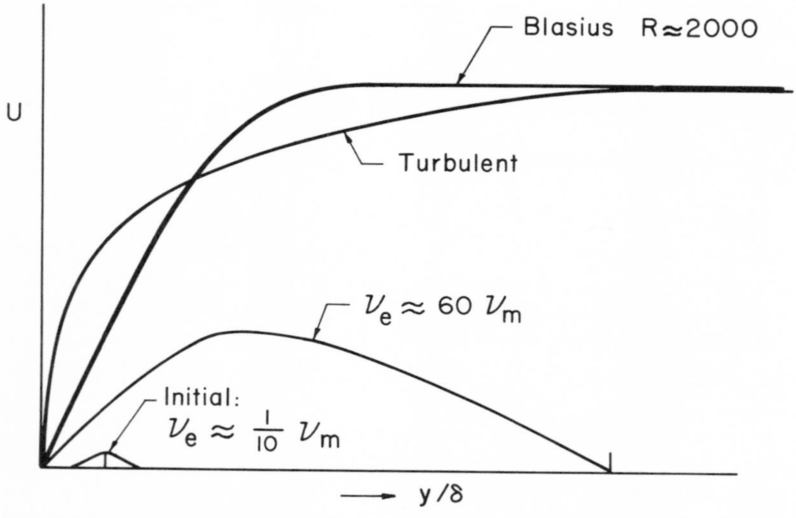

Fig. 6.16. Two-dimensional model of the transition process (V. W. Nee). A rate equation is postulated for the eddy viscosity ν_e. The initially laminar flow promptly changes to a turbulent profile, with rapid growth of the eddy viscosity.

6.7. Comment

Slowly, but surely, one by one, the pieces of the transition puzzle come together. At times, it seems that progress has stopped. But these interludes often cover the incubation of new techniques, or the erosion of obsolete ideas. It took us a century to reach the present degree of understanding. There is no doubt that more discoveries are awaiting us.

Notation

a	Radius of vortex filament
a_0	Initial perturbation
c	Sound speed
G	Growth factor
N	Number of molecules
R	Radius of circular vortex
t	Time
U	Mean velocity
u, v	Velocity components
x, y	Rectangular coordinates
α	Wave number
δ	Boundary-layer thickness
λ	$2\pi/\alpha$
ω	Vorticity fluctuation
Ω	Vorticity

References

1. Hama, F. R., Streaklines in a perturbed flow, *Phys. Fluids* **5**, 644–650 (1962).
2. Helmholtz, H. von, Über discontinuierliche Flüssigkeitsbewegungen, *Akad. Wiss. Berlin* April, 215 (1868).
3. Rosenhead, L., The formation of vortices from a surface of discontinuity, *Proc. R. Soc. London, A* **134**, 170–192 (1932).
4. Betchov, R., and Criminale, W. O., *Stability of Parallel Flow*, Academic Press, New York (1967).
5. Betchov, R., and Szewezyk, A., Stability of a shear layer between parallel streams, *Phys. Fluids* **6**, 1391–1396 (1963).
6. Tatsumi, T., and Gotoh, D., The stability of free boundary layers between two uniform streams, *J. Fluid Mech.* **7**, 433–441 (1960).
7. Gotoh, D., The damping of the large wave number disturbances in a free boundary layer flow, *J. Phys. Soc. Japan* **20**, 164–169 (1965).
8. Landau, L. D., Oscillation of an electron plasma, *Sov. Phys.—JETP* **10** (1946).
9. Stix, T. H., *The Theory of Plasma Waves*, McGraw-Hill, New York (1962).
10. Klebanoff, P. S., Tidstrom, K. D., and Sargent, L. M., The three-dimensional nature of boundary layer instability, *J. Fluid Mech.* **12**, 1–34 (1962).
11. Emmons, H. W., The laminar–turbulent transiton in a boundary layer—Part I, *J. Aeronaut. Sci.* **18**, 490–498 (1951).
12. Landau, L. D., and Lifshitz, E. M., *Theoretical Physics*, Vol. 6: *Fluid Mechanics*, Addison-Wesley Publishing Co., Reading, Massachusetts (1959).
13. Squire, H. B., On the stability for three-dimensional disturbances of viscous fluid flow between parallel walls, *Proc. R. Soc. London, A* **142**, 621–628 (1933).
14. Criminale, W. O., and Kovasznay, L. S. G., The growth of localized disturbances in a laminar boundary layer, *J. Fluid Mech.* **14**, 59–80 (1962).

15. Gortler, H., Über den Einfluss der Wandkrümmung auf die Entstehung der Turbulenz, *Z. Angew. Math. Mech.* **20**, 138–147 (1940).
16. Benney, D. J., and Lin, C. C., On the secondary motion induced by oscillations in a shear flow, *Phys. Fluids* **3**, 656–657 (1960).
17. Tani, I., Boundary layer transition, in: *Annual Review of Fluid Mechanics*, Vol. 1, W. R. Sears and M. Van Dyke (eds.), Annual Reviews, Inc., Palo Alto, California (1969), pp. 169–196.
18. Kovasznay, L. S. G., Komoda, H., and Vasudeva, B. R., Detailed flow field in transition, in: *Proceedings of the 1962 Heat Transfer and Fluid Mechanics Institute* (F. E. Ehlers, J. J. Kauzlarich, C. A. Sleicher, and R. E. Street, eds.), Stanford University Press, Stanford, California (1962), pp. 1–26.
19. Smith, A. M. O., Transition, pressure gradient and stability theory, *Proceedings of the Ninth International Congress on Applied Mechanics*, Published by the Congress (1956).
20. Betchov, R., Thermal Agitation and Turbulence, in: *Proceedings of the Second International Symposium on Rarefied Gas Dynamics* (L. Talbot, ed.), Academic Press, New York (1961), pp. 307–321.
21. Tsugé, S., Approach to the origin of turbulence on the basis of two-point kinetic theory, *Phys. Fluids* **17**, 22–23 (1974).
22. Hama, F. R., Progressive deformation of a perturbed line vortex filament, *Phys. Fluids* **6**, 526–534 (1963).
23. Betchov, R., On the curvature and torsion of an isolated vortex filament, *J. Fluid Mech.* **22**, 471–479 (1965).
24. Hasimoto, H., A soliton on a vortex filament, *J. Fluid Mech.* **51**, 477–485 (1972).
25. Leonard, A., Numerical simulation of interacting three-dimensional vortex filaments, *Proceedings of the Fourth International Conference on Numerical Methods in Fluid Dynamics* (R. D. Richtmeyer, ed.), Springer, Berlin (1975).
26. Mack, L. M., Boundary layer stability theory, Jet Propulsion Laboratory, California, Institute of Technology report No. 900–277 (1969).
27. Morkovin, M., On the many faces of transition, in: *Symposium on Viscous Drag Reduction* (S. Wells, ed.), Plenum Press, New York (1969), pp. 1–31.
28. Nee, V. W., and Kovasznay, L. S. G., Simple phenomenological theory of turbulent shear flows, *Phys. Fluids* **12**, 473–484 (1969).

CHAPTER 7

Turbulence Processes and Simple Closure Schemes

R. G. DEISSLER

7.1. Introduction

The origin of the closure problem in turbulence was discussed in some of the earlier papers in this volume. Herein we briefly review the closure problem and introduce simple closure schemes in order to obtain solutions for some simple flows. These solutions will be used to illustrate the processes occurring in turbulence. Closure by specification of initial conditions will then be considered. Finally, practical closure schemes for more complicated flows, such as boundary layers and pipe flows, will be discussed.

7.2. Theoretical Development

In order to illustrate the important processes of turbulence dissipation and turbulence transfer between wave numbers or eddy sizes we first consider a statistically homogeneous turbulent field without mean gradients. Such a turbulence will decay with time since no energy is added to the system, and so we must consider an initial-value problem. The turbulence must be generated initially by some means, such as by flow through a grid.

R. G. DEISSLER • Lewis Research Center, National Aeronautics and Space Administration, Cleveland, Ohio 44135

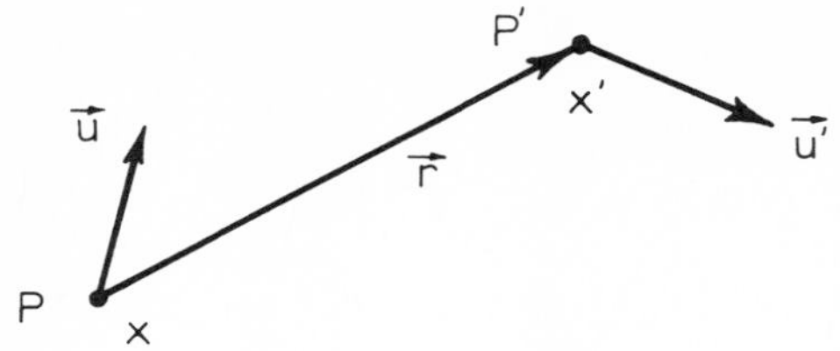

$$\left[\frac{\partial u_i}{\partial t} + \frac{\partial (u_i u_k)}{\partial x_k} = -\frac{1}{\rho}\frac{\partial p}{\partial x_i} + \nu \frac{\partial^2 u_i}{\partial x_k \partial x_k}\right] u_j'$$

$$\left[\frac{\partial u_j'}{\partial t} + \frac{\partial (u_j' u_k')}{\partial x_k'} = -\frac{1}{\rho}\frac{\partial p'}{\partial x_j'} + \nu \frac{\partial^2 u_j'}{\partial x_k' \partial x_k'}\right] u_i$$

$$\frac{\partial \overline{u_i u_j'}}{\partial t} + \frac{\partial}{\partial r_k}\left[\overline{u_i u_j' u_k'} - \overline{u_i u_j' u_k}\right] = -\frac{1}{\rho}\left[\frac{\partial \overline{p' u_i}}{\partial r_j} - \frac{\partial \overline{p u_j'}}{\partial r_i}\right] + 2\nu \frac{\partial^2 \overline{u_i u_j'}}{\partial r_k \partial r_k}$$

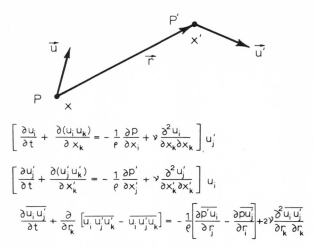

Fig. 7.1. Two-point correlation equation.

We start the analysis by writing the Navier–Stokes equations at two arbitrary points P and P' in the turbulent fluid, separated by the vector $\mathbf{r}$. (See Figure 7.1, where the subscripts in the equations can take on the values 1, 2, or 3, and a repeated subscript in a term indicates a summation. The quantities u_i and u_j are instantaneous velocity components, x_i is a space coordinate, t is the time, ρ is the density, ν is the kinematic viscosity, and p is the instantaneous pressure.) Then we multiply the equation at point P through by a velocity component at P', and that at P' by a velocity component at P, add the equations, take average values, and finally arrive at the equation involving correlations between quantities at points P and P' shown at the bottom of Figure 7.1.[1] Besides velocity–velocity correlations, the equation contains pressure–velocity correlations and triple-velocity correlations. The equation can be converted to spectral form by taking its three-dimensional Fourier transform and setting $i = j$, as shown at the top of Figure 7.2, where κ is the wave number, E is the energy spectrum function, and W is the energy transfer term due to triple correlations.[1] Note that this operation also converts the partial differential equation into an ordinary one. The various terms in the spectral equation can be interpreted by multiplying the equation through by a wave-number band of width $d\kappa$ and referring to the sketch in Figure 7.2. The energy spectrum function E gives contributions from various eddy sizes to the total turbulent energy. The area under the spectrum curve is thus equal to the total turbulent energy $\overline{u_i u_i}/2$. The first term in the spectral equation represents the rate of change of turbulent energy in the cross-hatched wave-number band. The next term, $W\,d\kappa$, represents the net rate of transfer of energy into the wave-number band by nonlinear effects, that is, by the triple correlation term in Figure 7.1. Finally, the last term, $2\nu\kappa^2 E\, d\kappa$, represents the rate at which turbulent energy is dissipated within the wave-number band by viscous action. The dissipation term is always negative, in contrast to the transfer term. Note that a spectral equivalent of the pressure-velocity correlations is absent in this equation for $E\, d\kappa$, the total turbulent energy in a wave-number band. This indicates that the pressure–velocity correlations do not contribute to the rate of change of the turbulent energy but they can transfer energy between its directional components.

Unfortunately the spectral equation contains two unknowns, E and W, so that we cannot, in general, obtain a solution without more information. That is, of course, a manifestation of the closure problem in turbulence and is a consequence of the nonlinearity of the original Navier–Stokes equations. However, if the turbulence is sufficiently weak, the inertia or transfer term W will be negligible, and we can obtain, subject to particular initial conditions, the remarkably simple solution shown at the bottom of Figure 7.2.(1) J_0 and t_0 are constants that depend on initial conditions. Although this solution applies only to very low Reynolds numbers, it can be used to study the viscous dissipation, the only turbulence process accounted for in its derivation. If we set $\partial E/\partial\kappa = 0$, we get $1/[\nu(t-t_0)]^{1/2}$ for κ_m, the wave number at which E is a maximum. Thus as time increases, the turbulent energy shifts to lower wave numbers, or to larger eddy sizes. The physical interpretation of this shift is that the smaller eddies die out faster than the larger ones because of the larger velocity gradients (larger shear stresses) between the smaller eddies. The essence of the turbulence dissipation process appears to be that the dissipation is always negative and that it affects mainly the smaller eddies.

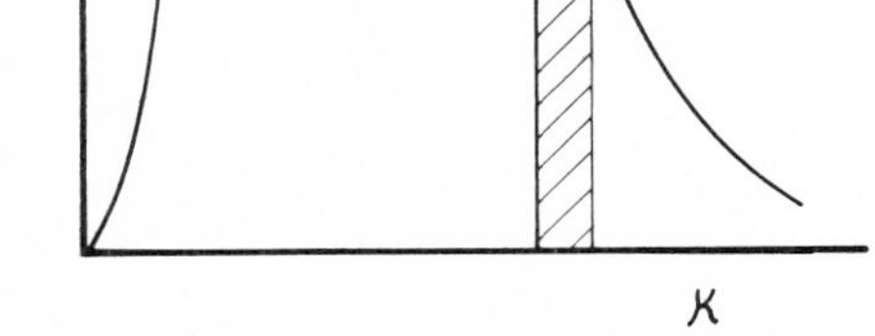

$$\frac{\partial E}{\partial t} = W - 2\nu K^2 E$$

$$\frac{\partial (E\, dK)}{\partial t} = W\, dK - 2\nu K^2 E\, dK$$

$$\int_0^\infty E\, dK = \frac{\overline{u_i u_i}}{2} = \frac{1}{2}\left[\overline{u_1^2} + \overline{u_2^2} + \overline{u_3^2}\right]$$

FOR WEAK TURBULENCE:

$$E = J_0 K^4 \exp\left[-2\nu K^2 (t - t_0)\right]$$

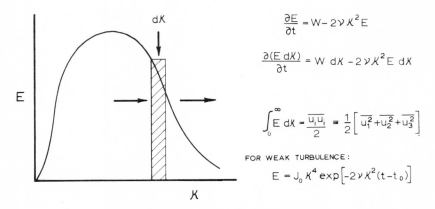

Fig. 7.2. Spectral equations and energy spectrum.

Unless the turbulence level is very low, as in the final period of decay, the inertia or transfer effects will not be negligible ($W \neq 0$), and so we would like to be able to take them into account in some way. A large number of proposals for calculating the transfer function W have been given including those of Heisenberg,[2] Kovasznay,[3] and Kraichnan.[4] There are in fact almost as many proposals for the transfer function as there are workers in turbulence. For our purposes it will be sufficient to consider a simple deductive approach which is essentially a perturbation on the solution for the final period of decay.[5,6]

In carrying out the analysis we consider, in addition to the two-point correlations of Figure 7.1, three, and sometimes four or more points. Thus, we can write the Navier–Stokes equations at three and at four arbitrary points in the turbulent fluid, as in Figure 7.3. We can then construct three-point correlation equations involving triple and quadruple correlations, and four-point equations involving quadruple and quintuple correlations, as shown in the figure. However, there are still more unknowns than there are equations, again as a result of the nonlinearity of the Navier–Stokes equations. To make the system of equations determinate, we use an operation similar to that used for the final period of decay, but instead of assuming that the turbulence is weak enough to neglect the inertia term in the two-point equation, we assume only that it is weak enough to neglect the inertia term in the highest-order equation considered. Herein, in order to give the simplest possible representation of the transfer or inertia process, we use only two- and three-point equations,[5] although the analysis has also been carried out for four points.[6] Thus, neglecting the quadruple correlations in the three-point equation gives a determinate set of equations that should be applicable at times somewhat before the final period of decay.

The resulting energy transfer spectra are plotted in Figure 7.4.[5] They are negative at small wave numbers and positive at higher ones, indicating that energy is transferred out of the low-wave-number region and into the

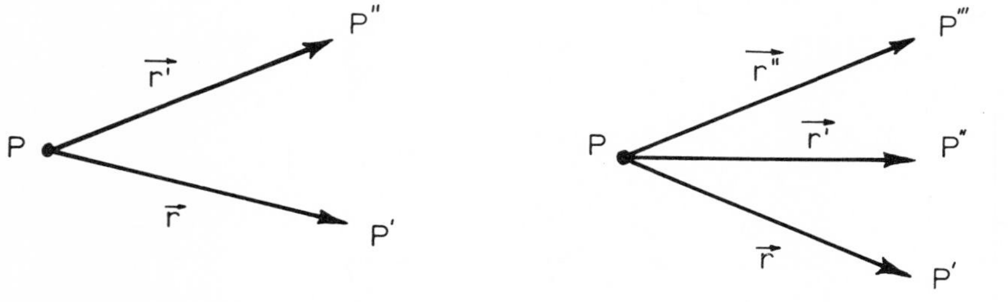

$$\frac{\partial}{\partial t}(\overline{u_i u_j' u_k''}) = \frac{\partial}{\partial r_l}(\overline{u_i u_j' u_k'' u_l}) + \cdots \qquad \frac{\partial}{\partial t}(\overline{u_i u_j' u_k'' u_l'''}) = \frac{\partial}{\partial r_m}(\overline{u_i u_m u_j' u_k'' u_l'''}) + \cdots$$

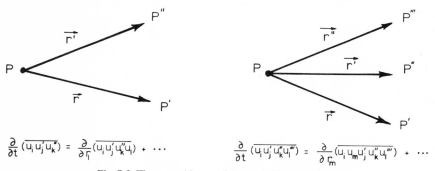

Fig. 7.3. Three- and four-point correlation equations.

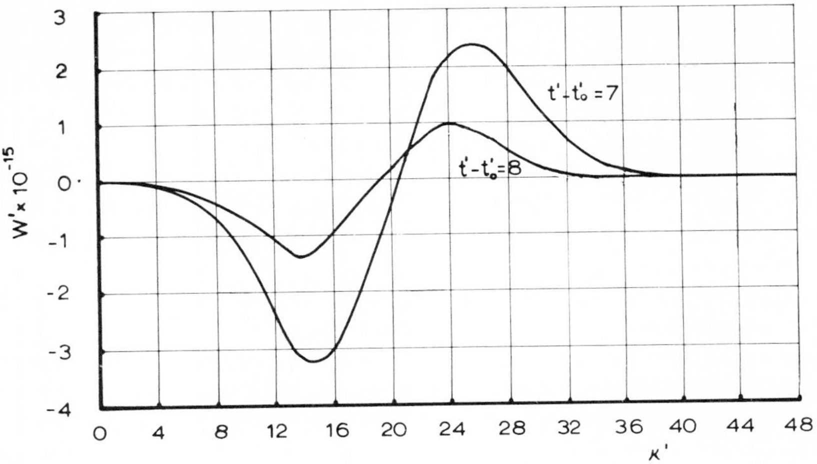

Fig. 7.4. Dimensionless energy transfer spectra. $W' = J_0 W/\nu^5$; $\kappa' = J_0^{1/3}\kappa/\nu^{2/3}$; $t' = \nu^{7/3}t/J_0^{2/3} \times 10^3$; $t_0' = -6.33 \times 10^{-3}$. J_0 and β_0 are constants of the initial condition.

higher-wave-number region. The net areas under the curves are zero, since the term makes no contribution to the rate of change of the total turbulent energy. The transfer of energy to the high-wave-number or small-eddy region can be thought of as a breakup of big eddies into smaller ones, or alternatively as a stretching of vortex filaments (to smaller diameters).

Calculated energy spectra are plotted for various times in Figure 7.5.[5] The curves show how the spectrum changes shape with time and approaches

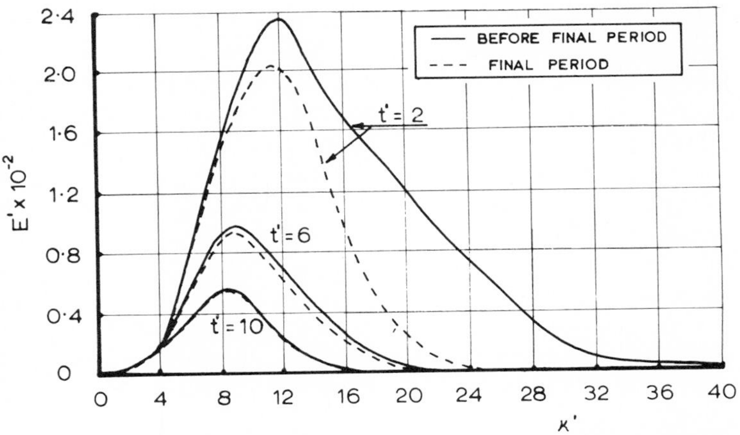

Fig. 7.5. Comparison of dimensionless spectra before final period with those for final period of decay. $E' = J_0^{1/3}E/\nu^{8/3}$ (see Fig. 7.4 for definitions of other quantities).

that for the final period. For small times the inertia or transfer terms transfer energy into the high-wave-number or small-eddy region and cause the slopes on the high-wave-number sides of the spectra to be less steep than they are at larger times, when the spectrum assumes a more or less symmetrical shape. Thus, the function of the inertia terms in the equations is to excite the higher-wave-number or small-eddy regions of the spectrum by transferring energy into those regions. If it were not for inertia effects, those regions of the spectra would be absent, as they are in the final period of decay.

Inertia and dissipation tend to shift the energy in opposite directions on the wave-number scale. However, the mechanisms for the two effects appear to be different. Whereas inertia tends to transfer the energy to higher wave numbers by a breakup of large eddies into smaller ones (or by a stretching of vortex filaments), dissipation tends to shift the energy to smaller wave numbers by selective annihilation of eddies, the small eddies being the first to go. As the turbulence decays, the dissipation effects must, of course, eventually win out, since the inertia effects become negligible at the low Reynolds numbers occurring at large times.

Before going on to the effects of mean gradients on turbulence, I should like to briefly consider closure by specification of initial conditions. In the closure methods just considered, we note that as we use more points in the turbulent field, or higher-order correlation equations, we have to specify more initial conditions to solve the set of equations (since there are more dependent variables). This is in addition to the closure assumption that must be made for the highest-order correlation. If we are willing to specify multiple initial conditions, as indeed we must for a complete specification of the initial turbulence,[1] there is an alternative way of looking at the problem that does not require an assumption for the highest-order correlation. It is not hard to show that, given the initial multipoint correlations and their correlation equations, all of the time derivatives of the turbulent energy tensor and of other pertinent turbulence quantities can be calculated.[7] These time derivatives can then be used in a series to calculate the evolution of the turbulent energy tensor (or of the equivalent energy spectrum tensor) and of other turbulence quantities.

When the turbulence is treated in this way, we no longer have the usual problem of closing an infinite set of correlation or spectral equations. The correlation equations are used only to relate the correlations at an initial time to their derivatives, and those correlations must be given in order to have a complete specification of the turbulence at that time. Of course, in practice only a small number of the correlations, and thus of their time derivatives, will ordinarily be available, but a sufficient number may be known to give a reasonably good representation. In the theory of Reference

7 it was found that by specifying n spectra at an initial time, where n is an odd integer greater than or equal to 3, the evolution in time of those n spectra was predictable. Good agreement with experimental data was obtained for $n = 3$ or 5.[7,8]

It may be that the nature of the problem is such that three or more spectra have to be specified initially in order to calculate the evolution of any of them (except for weak turbulence). However, particularly in an applied problem, three or more initial spectra will often not be available. In that case possible courses of action are, first, the required initial spectra might be estimated from previous experimental or analytical results, or, second, the introduction of physical or mathematical hypotheses into the theory might be allowed. In Reference 9 the latter course of action is followed by assuming that the energy transfer is a function of the initial conditions and of the energy at κ. By allowing that hypothesis it is found that by specifying only two initial spectra, E and W, the evolutions of those spectra are predictable. The results are compared with experiments in Figures 7.6–7.8, where A is a constant for the initial conditions. As in the theory of Reference 7 the present theory[9] contains no adjustable constants or functions.

We turn now to the important case of turbulent shear flow. In order to consider a solvable problem in which we can study the turbulence processes associated with the mean shear, we use for our model a homogeneous turbulent field with a uniform mean shearing velocity gradient. Since we have already considered the energy transfer process, it is assumed in this

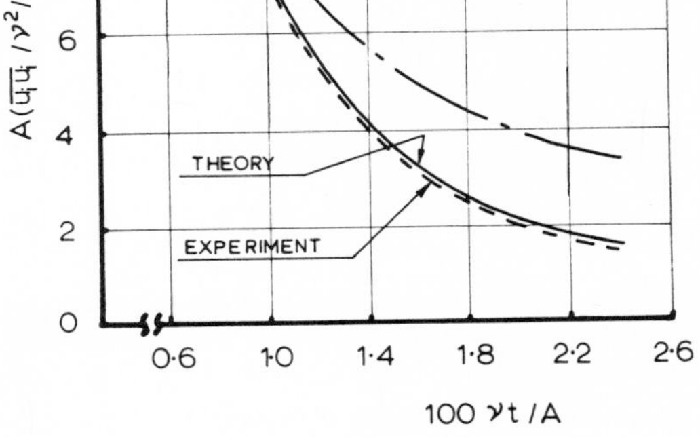

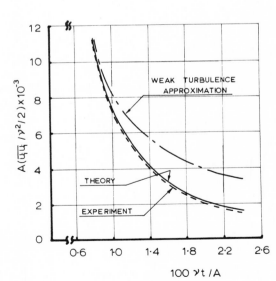

Fig. 7.6. Comparison of theory and experiment for decay of turbulent energy. Data from Ling and Huang.[23]

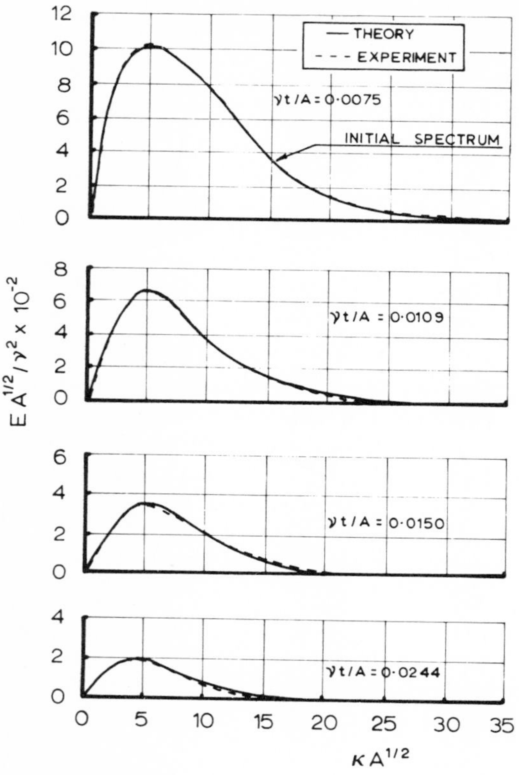

Fig. 7.7. Comparison of theory and experiment for decay of three-dimensional turbulent-energy spectra. Data from Ling and Huang.[23]

case that the interaction between the turbulence and the mean gradient is large compared with the turbulence self-interaction. Thus, we need only consider the two-point correlation equations, since we can close them if we neglect the self-interaction or triple correlation terms by comparison with the mean gradient terms. The analysis is similar to that just described, except that we break the velocity into mean and fluctuating components. In that way we obtain equations for the $\overline{u_i u_j}$ and their Fourier transforms in the presence of a uniform velocity gradient.[10]

The resulting equation for the energy spectrum is

$$\frac{\partial E}{\partial t} = P(\kappa)\frac{\partial U_1}{\partial x_2} + T(\kappa)\frac{\partial U_1}{\partial x_2} - 2\nu\kappa^2 E$$

As in the equation at the top of Figure 7.2, the first term gives the rate of change of the energy spectrum function E at wave number κ. The next term is proportional to the mean shearing velocity gradient, and $P(\kappa)$ in that term

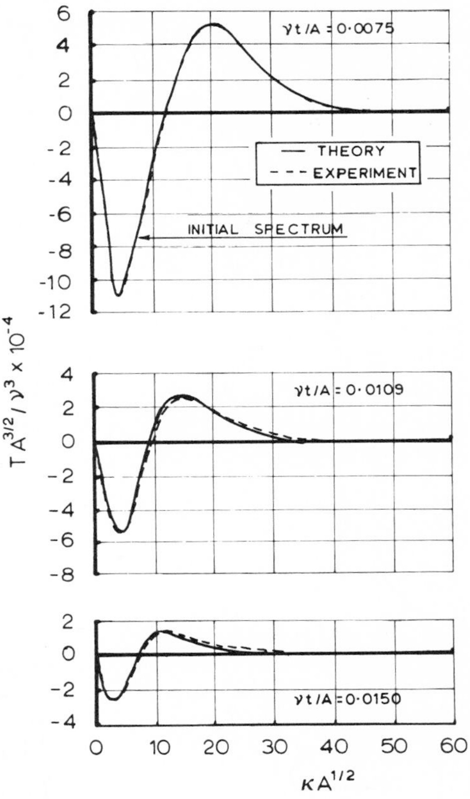

Fig. 7.8. Comparison of theory and experiment for decay of energy transfer spectra. Data from Ling and Huang.[23]

is proportional to a spectral component of the turbulent shear stress. That term is thus interpreted as a production term by which energy is produced at wave number κ by work done by the velocity gradient on a spectral component of the turbulent shear stress.

The third term is also proportional to the mean velocity gradient, but we interpret it as a transfer term rather than as a production term. That is because if we integrate the term over κ from 0 to ∞, we get the result zero.[10] The term gives zero contribution to the rate of change of the total turbulent energy, but it can transfer energy between wave numbers. Thus, although we neglect triple correlations, we still get an energy transfer T that appears similar to that produced by triple correlations. The difference between the two cases is that, whereas in the case of triple correlations the transfer is produced by the nonlinear action of the turbulence on itself, in the present case it is due to the external action of the mean velocity gradient on the turbulence. However, the results in both cases are similar, as shown in

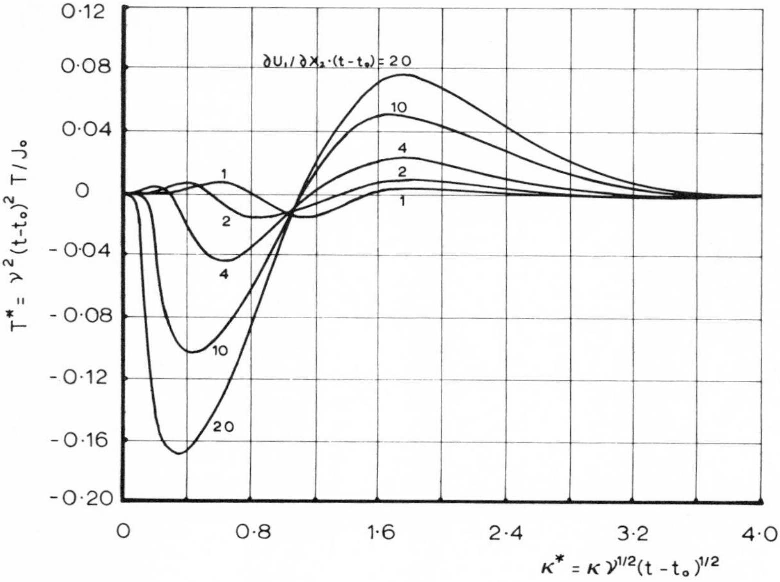

Fig. 7.9. Spectra of transfer term due to mean velocity gradient.

Figure 7.9 for a particular initial condition.(10) Since the dimensionless transfer term is primarily negative at low wave numbers and positive at higher ones (net area = 0), the energy transfer is mostly from small to large wave numbers, as is the case for energy transfer by triple correlations. The reason the dimensionless T^* can increase with time is that T^* itself contains time.(10)

Finally, the last term in the spectral equation is again the dissipation term.

Plots of the energy spectrum, production term, and dissipation term for a particular time are shown (normalized to the same height) in Figure 7.10.(10) From these we can summarize the history of the turbulent energy as follows. The energy comes into the turbulent field from the mean velocity field mainly through the large eddies. This energy is then transferred from the big-eddy region to the small-eddy region by the transfer term just discussed. Physically this transfer might be thought of as being produced by the stretching of turbulent vortex filaments by the mean velocity gradient. Finally the energy is dissipated in the small-eddy region by viscous action. It is physically reasonable that the dissipation should occur mainly in the small-eddy region since the shear stress should be greater between the small eddies than between the larger ones. The energy resides in a region between

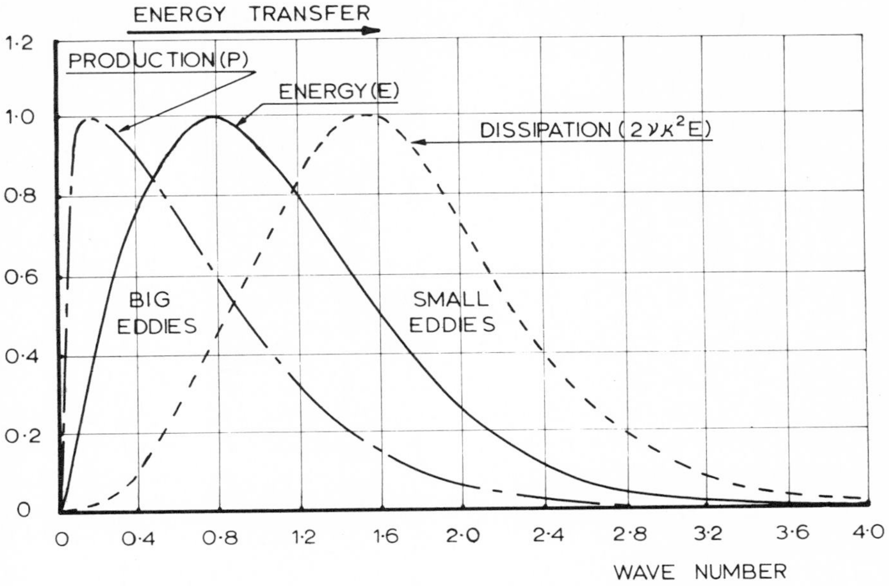

Fig. 7.10. Production, energy-containing, and dissipation regions.

the production and dissipation regions, as shown by the plot of the energy spectrum.

By using realistic initial conditions the analysis has been compared with experiments for uniformly sheared turbulence in Reference 11. Some of the results are shown in Figures 7.11 and 7.12. The agreement is good, including a negative region in one of the two-point correlation curves.

Although the agreement between theory and experiment is good for the data shown in Figures 7.11 and 7.12, it should be pointed out that the results are not directly applicable to fully developed inhomogeneous flows (e.g., pipe flows), since all of the predicted components of the turbulence decay at large times. This is shown by the dashed curves in Figure 7.13.[12] The decay apparently occurs because there are no production terms in the full equations for $\overline{u_2^2}$ and $\overline{u_3^2}$, the transverse turbulence components. In order to test that hypothesis, $\overline{u_2^2}$ and $\overline{u_3^2}$ were arbitrarily set equal to $\frac{1}{2}\overline{u_1^2}$. The solid curves in Figure 7.13 indicate that when this is done, all of the components grow at large times.

Thus, for simulating a sustained turbulence it is important that the energy remain distributed among the various directional components of the turbulence. One way of maintaining this directional distribution (other than by forcing it as in Figure 7.13) is by stretching the turbulence in the direction of flow as it is being sheared ($\partial U_1/\partial x_1 > 0$). This action tends to put energy into the transverse components, so that the turbulence is maintained at large times. This is shown by the solid curves in Figure 7.14,[12] where all of the turbulence components grow at large times.

In a fully developed pipe flow, where longitudinal straining is absent, the distribution of the turbulent energy between its directional components might be maintained by an interaction of triple correlations with the pressure–velocity correlations. That effect was neglected in the analysis considered here, although the effect of mean gradients on the pressure–velocity correlations was included. Inhomogeneities in the turbulent field may also have a directional redistribution effect on the turbulent energy.[13]

This discussion leads us to a consideration of more complicated turbulent flows than those considered thus far, such as those in boundary layers and pipes. A fully developed turbulent pipe flow may seem simple, but in reality it is exceedingly complicated when approached from a fundamental standpoint. A complete solution for an inhomogeneous turbulent shear flow has not been obtained, although some progress has been made in Reference

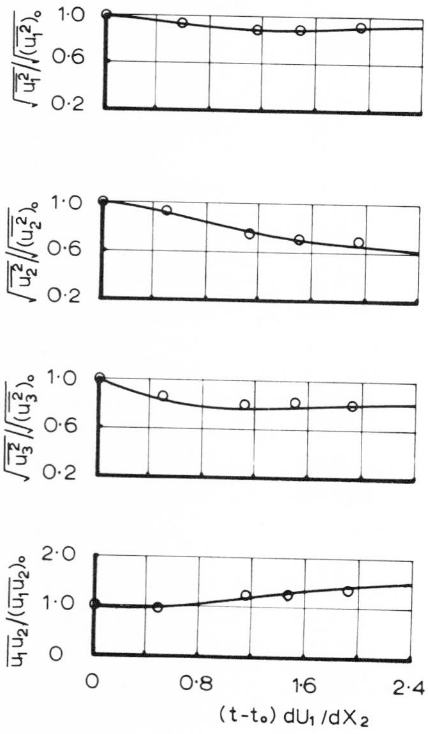

Fig. 7.11. Comparison of theory (solid curves) and experiment (circles) for evolution of one-point turbulence components with uniform shear. Data from Champagne *et al.*[22]

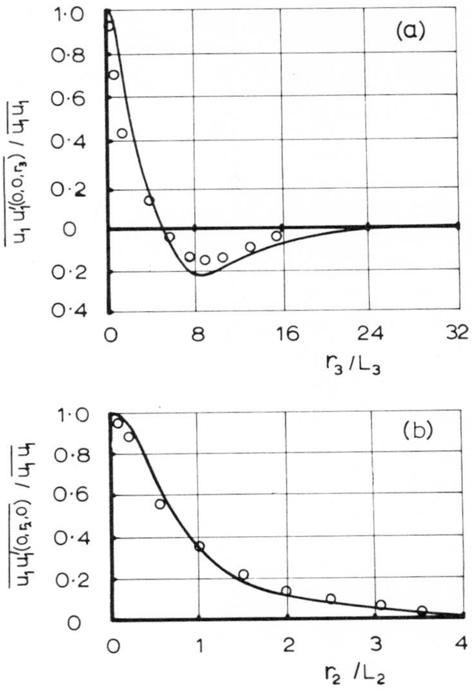

Fig. 7.12. Comparison of theory (solid curves) and experiment (circles) for two-point velocity correlations. $(dU_1/dx_2)(t-t_0)=2.27$. Data from Champagne *et al.*[22] Conditions are as follows: (a) Two points separated along direction normal to both flow and velocity gradient. (b) Two points separated along direction of velocity gradient.

13. Work to obtain a more complete solution is currently in progress, but it is too early to say whether it will be successful.

The complete two-point correlation equations for the double-velocity and pressure–velocity correlations for an inhomogeneous flow with mean velocity gradients are given in Reference 10. The principal process occurring in inhomogeneous turbulence that has not been considered thus far is turbulence diffusion between regions of higher and lower turbulence intensity. We have not been able to obtain solutions to illustrate the diffusion process, except for viscous diffusion.[10] Since the two-point equations for inhomogeneous turbulence are extremely complex, we content ourselves here with considering only the one-point equations, the equations for the mean flow. Those equations are[10]

$$\frac{\partial U_i}{\partial t}+U_k\frac{\partial U_i}{\partial x_k}=-\frac{1}{\rho}\frac{\partial p}{\partial x_i}+\frac{\partial}{\partial x_k}\left(\nu\frac{\partial U_i}{\partial x_k}-\overline{u_iu_k}\right) \tag{7.1}$$

where the overbars and the U_i designate mean values. These equations were obtained from the Navier–Stokes equations by breaking the instantaneous

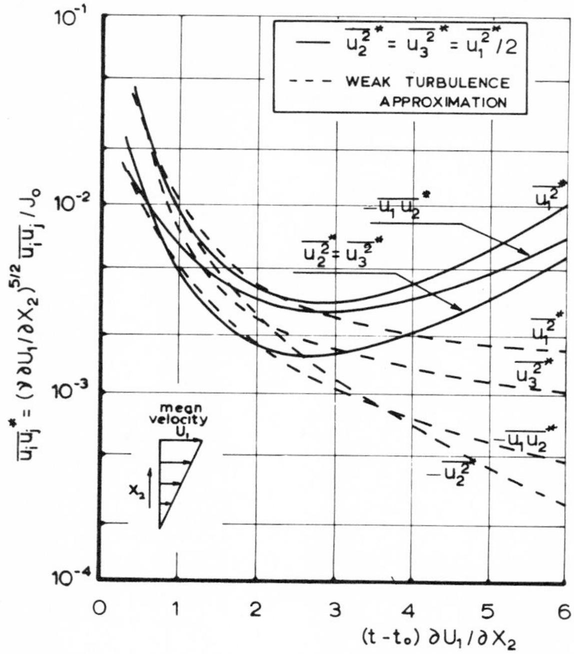

Fig. 7.13. Growth due to shear of weak homogeneous turbulence when $\overline{u_2^2} = \overline{u_3^2} = \frac{1}{2}\overline{u_1^2}$ and comparison with case where components are all calculated from weak turbulence approximation.

velocities and pressures into mean and fluctuating components. The system of equations is closed in this case by introducing assumptions for the $\overline{u_i u_j}$.

It is an interesting observation that as we have gone from comparatively simple to more complex systems we have been forced to consider fewer points in the turbulent fluid, and to use lower-order closures. Thus for homogeneous turbulence without mean gradients we were able to obtain a solution by closing the three- or four-point correlation equations. When uniform mean gradients were added, closure was obtained at the two-point level. Finally for inhomogeneous turbulence we will consider only the one-point equation [Equation (7.1)]. Many workers have also used higher-order one-point equations, which are somewhat easier to solve than the higher-order multipoint equations. Some of those methods will be considered in later papers in this volume. However, it should be pointed out that when higher-order one-point equations are introduced, the number of unknowns goes up faster than the number of equations, so that the amount of empirical information that must be supplied is greatly increased. Nevertheless, such equations have been found useful in some cases.

If we restrict our attention to flows for which the boundary-layer assumptions are applicable, the only component $\overline{u_i u_k}$ in Equation (7.1) that we need consider is $\overline{u_1 u_2}$, the Reynolds shear stress, where the shear velocity gradient is $\partial U_1/\partial x_2$. The above system of equations then reduces to a single equation. One flow for which that equation can be easily closed is that in a moderately short boundary layer with a severe pressure gradient. It is shown in Reference 14 that for this case $\overline{u_1 u_2}$ can be considered as frozen at its initial given values as it is convected along stream lines. Theoretical velocity profiles and Stanton numbers (dimensionless heat transfer coefficients) are compared with experiment in Figures 7.15 and 7.16.[14,15] The quantity τ_w is the shear stress at the wall. For obtaining the heat transfer results, an energy equation similar to the above equation for the mean flow [Equation (7.1)] was used in addition to that equation. The agreement between theory and experiment is considered good.

The above analysis is an example of a case in which closure can be obtained in a simple manner at the Reynolds stress level. For longer boundary layers, or for less severe pressure gradients, the simplification used breaks down, and so other less deductive methods have to be used. The

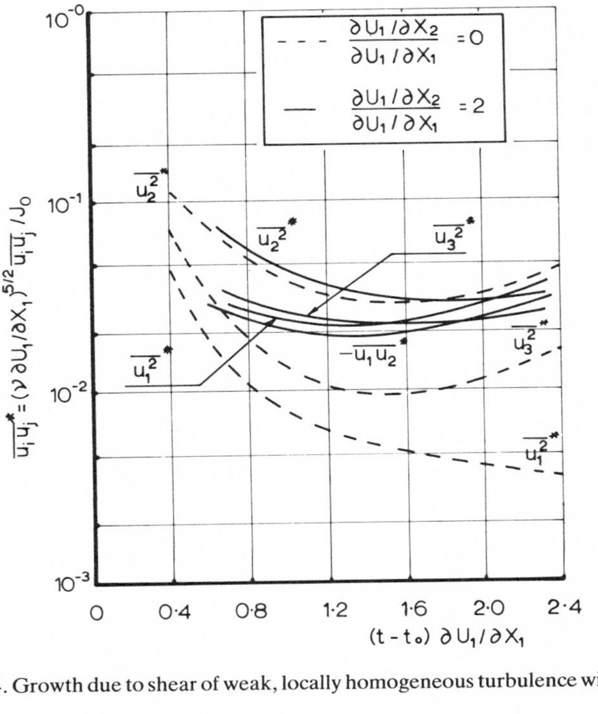

Fig. 7.14. Growth due to shear of weak, locally homogeneous turbulence with normal strain.

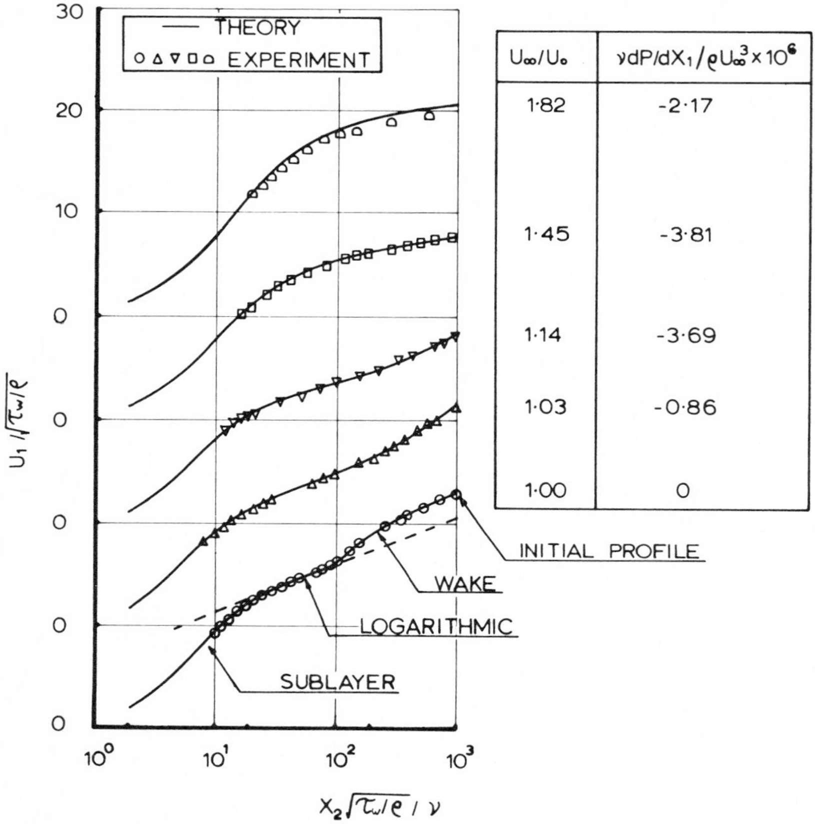

Fig. 7.15. Semilogarithmic law of the wall plot of theoretical velocity profiles for severe favorable pressure gradients and comparison with experiment of Patel and Head.[21] (Note shifted vertical scales.)

reason a solution could be obtained is that, for the case considered, the given initial conditions were all-important. At the other extreme we have fully developed flows, where initial conditions have no effect at all on the turbulence. For that case and for other intermediate cases, we generally employ simple models based on physical reasoning and dimensional analysis. As mentioned above, if we make the boundary-layer assumptions, we need consider only the $\overline{u_1u_2}$ component of the Reynolds stress $\overline{u_iu_j}$ in the one-point Equation (7.1).

There are at least three types of closure schemes we can use. First, we can make an assumption for $\overline{u_1u_2}$ directly (Reynolds shear-stress closure). Second, by analogy with the molecular viscosity, we can introduce an eddy

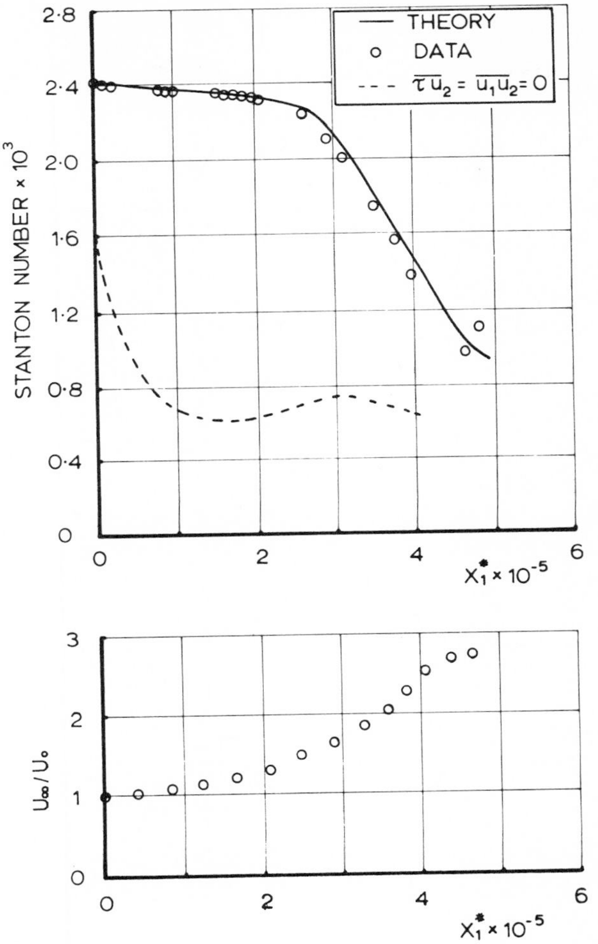

Fig. 7.16. Comparison of theory and experiment for evolution of Stanton number in moderately short highly accelerated turbulent boundary layers. Data from Moretti and Kays.[20]

viscosity ε given by

$$\overline{u_1 u_2} \equiv -\varepsilon\, \partial U_1 / \partial x_2 \tag{7.2}$$

and make an assumption for ε. Finally, we can introduce a mixing length for the turbulence (Prandtl's mixing length[16]) given by

$$\overline{u_1 u_2} \equiv -l^2 (\partial U_1 / \partial x_2)^2 \tag{7.3}$$

and make an assumption for l. Which of these three closure schemes is used is to some extent a matter of taste, although one will sometimes be more convenient than another. The eddy viscosity or the mixing-length method is often preferred because those methods ensure that the Reynolds shear stress is zero for zero velocity gradient (for finite ε or l). That is usually the case, although there can be exceptions for certain asymmetric flows. Some workers prefer to use the mixing length, because it seems to them easier to make an assumption for a length than for an eddy viscosity or for a Reynolds shear stress. However, it should be pointed out that the choice of Equation (7.3) as the definition of the mixing length is itself an assumption, since that definition is by no means the only one that might be given. For example we might have used Taylor's mixing length.[16] Alternatively, a mixing length l' might be given by

$$\overline{u_1 u_2} \equiv -l' U_1 \, \partial U_1 / \partial x_2 \tag{7.4}$$

and an assumption made for l'.

Probably the most reasonable expression for the region away from walls is von Kármán's similarity expression.[17] That expression is most easily obtained by assuming that, away from boundaries, the turbulence at a point is a function only of conditions in the vicinity of the point, in particular of the first and second velocity derivatives at the point. In this case it is equally convenient to use Reynolds stress, eddy viscosity, or mixing length [Equation (7.3)], but Equation (7.4) is not used because its use would imply that the velocity relative to the wall is important here. If we start from the Reynolds stress and apply dimensional analysis we get

$$\overline{u_1 u_2} = \overline{u_1 u_2}\left(\frac{\partial U_1}{\partial x_2}, \frac{\partial^2 U_1}{\partial x_2^2}\right) = -K^2 \frac{(\partial U_1/\partial x_2)^4}{(\partial^2 U_1/\partial x_2^2)^2} \tag{7.5}$$

where K is the von Kármán constant. If we had started from the eddy viscosity [Equation (7.2)], we would have obtained, by dimensional analysis,

$$\varepsilon = \varepsilon\left(\frac{\partial U_1}{\partial x_2}, \frac{\partial^2 U_1}{\partial x_2^2}\right) = K^2 \frac{(\partial U_1/\partial x_2)^3}{(\partial^2 U_1/\partial x_2^2)^2} \tag{7.6}$$

Finally, starting from the mixing length, Equation (7.3) would give

$$l = l\left(\frac{\partial U_1}{\partial x_2}, \frac{\partial^2 U_1}{\partial x_2^2}\right) = K \frac{\partial U_1/\partial x_2}{\partial^2 U_1/\partial x_2^2} \tag{7.7}$$

Clearly, all three of these starting points give the same result for $\overline{u_1 u_2}$.

von Kármán's hypothesis is applicable in the region away from a wall. Close to a wall we assume that the eddy viscosity, or the mixing length given by Equation (7.4), is a function only of quantities measured relative to a wall,

U_1 and x_2, and of ν.[17] The simplest assumption consistent with dimensional analysis and the requirement that the effect of ν should become small for large x_2 is then

$$\varepsilon = \varepsilon(U_1, x_2, \nu) = n^2 U_1 x_2 (1 - \exp(-n^2 U_1 x_2/\nu)) \tag{7.8}$$

where n is an experimental constant ($n = 0.124$). If we had started from Equation (7.4), rather than from Equation (7.2), we would have obtained the equivalent expression

$$l' = l'(U_1, x_2, \nu) = n^2 x_2 (1 - \exp(-n^2 U_1 x_2/\nu)) \tag{7.9}$$

There is a third region, the so-called wake region, where ε tends to be approximately constant. That is important in free turbulent flows, but it also occurs near the outer edge of a boundary layer and near the center of a pipe. However, we can often neglect it in the latter two cases, particularly if we use 0.36 rather than 0.4 for the experimental constant K. Equations (7.6) and (7.8) give results for flow and heat (or mass) transfer in tubes and boundary layers that are in good agreement with experiment (Figures 7.17 and 7.18).[17,18]

The closure assumptions used here are by no means the only ones that can be made. For instance Prandtl has assumed that $l = Kx_2$, where x_2 is the distance from a wall, in Equation (7.3).[16] Van Driest[19] has modified that assumption by introducing a damping factor:

$$l = Kx_2(1 - \exp[-x_2(\tau_w/\rho)^{1/2}/A]) \tag{7.10}$$

where A is another experimental constant. Equation (7.10) appears to be reasonably applicable to the regions both close to and away from a wall and

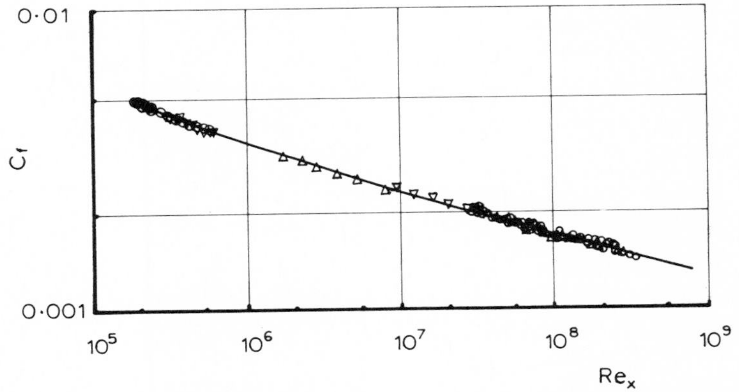

Fig. 7.17. Comparison of theory and experiment for flat-plate skin-friction coefficient for low-speed flows.

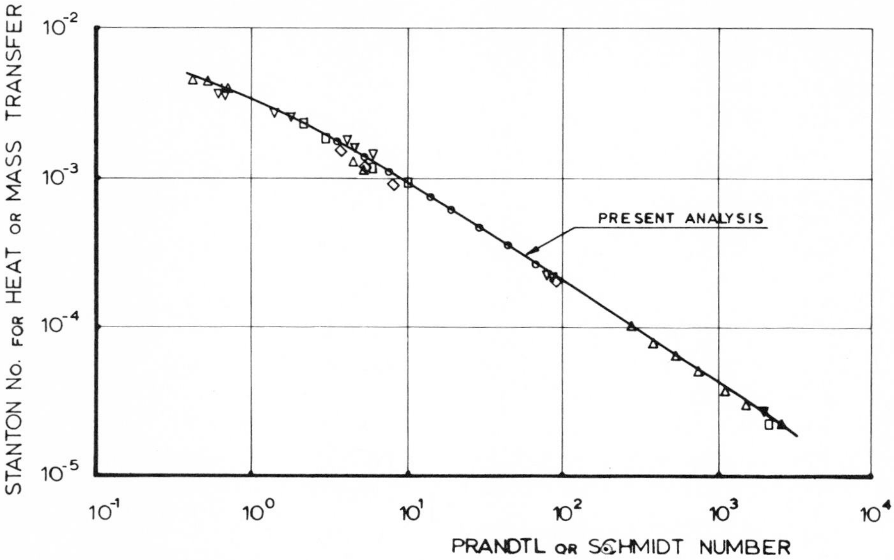

Fig. 7.18. Comparison of theory and experiment for fully developed heat and mass transfer. Reynolds number, 10,000.

so is sometimes considered to be more convenient to use than Equations (7.6) and (7.8). However, there is basically no reason why one equation should apply to both regions, since the turbulence mechanism close to a wall differs from that away from a wall. Thus the use of two equations might be considered reasonable.

7.3. Final Remarks

This discussion has attempted to show how turbulence solutions can be obtained by introducing simple closures into the turbulence equations. It is hoped that it has also provided some insight into the processes occurring in turbulence.

Notation

A	Constant depending upon initial conditions
E	Energy spectrum function
J_0	Constant depending upon initial conditions
n	Constant
p	Pressure
P, P'	Points in space
$\mathbf{r}$	Space vector
T	Energy transfer

t	Time	U_i	Mean velocity components
t_0	Constant depending upon initial conditions	W	Energy transfer term
u_i, u_i'	Instantaneous velocity components	x_i	Space coordinates
		κ	Wave number
		ν	Kinematic viscosity

References

1. Batchelor, G. K., *The Theory of Homogeneous Turbulence*, Cambridge University Press, New York (1953).
2. Heisenberg, W., Zur statistischen Theorie der Turbulenz, *Z. Phys.* **124**, 628–657 (1948).
3. Kovasznay, L. S. G., Spectrum of locally isotropic turbulence, *J. Aeronaut. Sci.* **15**, 745–753 (1948).
4. Kraichnan, R. H., Lagrangian-history closure approximation for turbulence, *Phys. Fluids* **8**, 575–598 (1965).
5. Deissler, R. G., On the decay of homogeneous turbulence before the final period, *Phys. Fluids* **1**, 111–121 (1958).
6. Deissler, R. G., A theory of decaying homogeneous turbulence, *Phys. Fluids* **3**, 176–187 (1960).
7. Deissler, R. G., Decay of homogeneous turbulence from a given initial state, *Phys. Fluids* **14**, 1629–1638 (1971).
8. Deissler, R. G., Further comparison of theory and experiment for decay of homogeneous turbulence, *Phys. Fluids* **15**, 1353–1355 (1972).
9. Deissler, R. G., Remarks on the decay of homogeneous turbulence from a given state, *Phys. Fluids* **17**, 652–653 (1974).
10. Deissler, R. G., Effects of inhomogeneity and of shear flow in weak turbulent fields, *Phys. Fluids* **4**, 1187–1198 (1961).
11. Deissler, R. G., Comparison of theory and experiment for homogeneous turbulence with shear, *Phys. Fluids* **18**, 1237–1240 (1975).
12. Deissler, R. G., Growth of turbulence in the presence of shear, *Phys. Fluids* **15**, 1918–1920 (1972).
13. Deissler, R. G., Problem of steady-state shear-flow turbulence, *Phys. Fluids* **8**, 391–398 (1965).
14. Deissler, R. G., Evolution of a moderately short turbulent boundary layer in a severe pressure gradient, *J. Fluid Mech.* **64**, 763–774 (1974).
15. Deissler, R. G., Evolution of the heat transfer and flow in moderately short turbulent boundary layers in severe pressure gradients, *Int. J. Heat Mass Transfer* **17**, 1079–1085 (1974).
16. Hinze, J. O., *Turbulence*, McGraw-Hill Book Company, New York (1959).
17. Deissler, R. G., Turbulent heat transfer and friction in smooth passages, in: *Turbulent Flows and Heat Transfer* (C. C. Lin, ed.), Princeton University Press, Princeton, New Jersey, 288–313 (1959).
18. Deissler, R. G., and Loeffler, A. L., Analysis of turbulent flow and heat transfer on a flat plate at high Mach numbers with variable fluid properties, NASA report No. TR-17, Washington, D. C. (1959).
19. Van Driest, E. R., On turbulent flow near a wall, *J. Aeronaut. Sci.* **23**, 1007–1011 (1956).
20. Moretti, P. M., and Kays, W. M., Heat transfer to a turbulent boundary layer with varying free-stream velocity and varying surface temperature—An experimental study, *Int. J. Heat Mass Transfer* **8**, 1187–1202 (1965).

21. Patel, V. C., and Head, M. R., Reversion of turbulent to laminar flow, *J. Fluid Mech.* **34**, 371–392 (1968).
22. Champagne, F. H., Harris, V. G., and Corrsin, S., Experiments on nearly homogeneous turbulent shear flow, *J. Fluid Mech.* **41**, 81–139 (1970).
23. Ling, S. C., and Huang, T. T., Decay of weak turbulence, *Phys. Fluids* **13**, 2912–2924 (1970).

CHAPTER 8

Kinetic Energy Methods

P. T. HARSHA

8.1. Introduction

Problems in which turbulent flow fields dominate form a major portion of engineering fluid mechanics and heat transfer work. In principle, the calculation of these flows involves the solution of the time-dependent Navier–Stokes equations. But these equations cannot be solved without recourse to numerical methods, which must divide the flow field into a finite number of calculation points. The fundamental problem in the computation of turbulent flows then becomes the fact that turbulence introduces motions on a scale far smaller than the distances between the calculation points on the smallest practical numerical solution grid. Indeed, even if it were possible to compute the velocity field in a turbulent flow down to the smallest scale of motion of interest, another problem would be encountered. Because the velocity field in a turbulent flow fluctuates randomly, the variables of engineering interest in the flow are in general time or ensemble averages of the fluctuating quantities. In order to predict these averages, it would be necessary to repeat a detailed computation a great number of times, each with a slightly different initial condition, and ensemble-average the results. For these reasons, a direct assault on the problem of the computation of turbulent flows is impractical.

An alternate approach to the computation of turbulent flow fields is to devise a model for the flow that represents as closely as possible its average behavior. All current techniques represent models for the flow field. These

P. T. HARSHA • R & D Associates, Santa Monica, California. Present address: Science Applications, Inc., Woodland Hills, California models differ in the complexity of the technique, in the detail that is computed, and in their success in providing a general description of at least a class of turbulent flows, but it must always be remembered that the solutions these techniques supply represent the behavior of a model flow field, which to a greater or lesser extent mimics the behavior of the true turbulent flow.

The models of turbulence described in this article all are based on the Reynolds-averaged equations of motion. Fundamentally the Reynolds-averaged equations are themselves models of the turbulent flow, for the averaging process may mask some characteristic patterns in the flow field. Large-scale, ordered motion has been shown to underlie at least some turbulent flows (see, e.g., Reference 1). Because of this, it has been proposed[(2)] that a three-level averaging procedure be used in which the flow field is divided into a mean motion, a large-scale turbulent motion, and a fine-scale turbulent motion. The large-scale motion may be modeled by a deterministic model representing the large eddies in a turbulent flow, with a degree of randomness introduced at this level by postulating a random occurrence in space and time of these structures. The fine-scale turbulent motion is then thought of as an essentially isotropic turbulence carried along by the mean motion and by the large eddies. This model has been used with some success to calculate the noise radiated by turbulent jets.[(3)] However, turbulent jet noise depends on the details of the turbulence structure to a much greater degree than other quantities of engineering interest. The success obtained using the Reynolds-averaging approach to compute turbulent mixing rates and heat-transfer phenomena indicates that the two-level Reynolds-averaged model is appropriate for the calculation of phenomena that do not depend strongly on the detailed turbulence structure.

The mean momentum equation obtained after Reynolds-averaging can be written

$$\frac{\partial U_j}{\partial t}+\frac{\partial}{\partial x_k}(U_jU_k+\langle u_ku_j\rangle)=\frac{\partial}{\partial x_k}(-\delta_{jk}P+T_{kj}) \tag{8.1}$$

where

$$T_{kj}\equiv\nu\left(\frac{\partial U_k}{\partial x_j}+\frac{\partial U_j}{\partial x_k}\right) \tag{8.2}$$

represents the viscous stress tensor. For the incompressible flow problem, this equation and the mean continuity equation form a closed set of equations that can be solved, subject to appropriate boundary conditions, provided that the term $\langle u_ku_j\rangle$ can in some way be modeled in terms of either known or calculable quantities. This term, the Reynolds stress tensor, represents the turbulent shear stress in the flow. The subject of turbulence modeling is thus in essence the subject of providing a model for $\langle u_ku_j\rangle$, i.e., obtaining "closure" of the Reynolds-averaged equations.

Mellor and Herring,[4] in their excellent survey paper on mean turbulent field closure, divide turbulent field models into mean velocity field (MVF) closure and mean turbulent field (MTF) closure. In MVF calculations, rather simple (and in many circumstances successful) models are used to obtain an estimate of the turbulent shear stress. These models predict only the mean velocity field in addition to the turbulent shear stress. On the other hand, MTF closure models attempt to supply some information on the mean turbulence field as well as a prediction of the mean velocity field. Mellor and Herring [4] further subdivide the mean turbulent field closure models on the basis of the additional turbulence information that is computed: In mean Reynolds stress (MRS) closure an attempt is made to compute the full Reynolds stress tensor $\langle u_k u_j \rangle$, while mean turbulent energy (MTE) models rely on a computation of the turbulent kinetic energy $e \equiv \frac{1}{2}\langle u_k u_k \rangle$. The latter group of models will form the primary subject of this article.

Mean velocity field (MVF) closure schemes are represented by a number of different models. Eddy viscosity (or mixing length) models form the simplest class of MVF closures, examples of which are the Prandtl mixing length[5] and eddy viscosity[6] formulations; more recent eddy viscosity models have attempted to account for the effects of density variations on the eddy viscosity in a compressible flow.[7–9] An alternate approach is that taken by Nee and Kovasznay[10] and Saffman.[11] In these approaches the eddy viscosity is assumed to be a transportable scalar, and transport equations for it are written by analogy to the equations for the turbulent transport of a scalar which can be rigorously obtained from the Navier–Stokes equations. Mellor and Herring[4] also include in this group the model devised by Bradshaw *et al.*,[12] since, although these investigators solve an equation for the turbulent shear stress, it is not the turbulent shear-stress equation rigorously derived from the Navier–Stokes equations. Conceptually, then, the model fits in between the eddy viscosity transport equation approach and the MTE and MRS models that form the MTF group in Mellor and Herring's classification. All of the eddy viscosity approaches have been applied to wall boundary layers and free flows (remote from walls). The models of Nee and Kovasznay[10] and Saffman[11] have primarily been applied to wall boundary problems (the latter, e.g., by Saffman and Wilcox[13]). The method of Bradshaw *et al.*[12] is restricted to planar flows. It has, however, been applied to two-dimensional free flows by Morel *et al.*[14] and by Morel and Torda.[15]

Quite a substantial number of papers have appeared describing the application of different forms of MTF closure to boundary-layer problems. Among these are the applications to wall boundary layers described by Glushko,[16] Beckwith and Bushnell,[17] Mellor and Herring,[18] and Ng and Spalding,[19] all of which incorporate the semiheuristic models of Kolmogorov[20] and Prandtl and Wieghardt[21] relating the eddy viscosity to the turbulent kinetic energy and a turbulence length scale. In the first three of these approaches[16–18] the turbulence length scale is specified directly, as a function of the boundary layer width, while Ng and Spalding[19] use a differential equation for the spatial variation of the length scale. This equation is based upon a scale transport equation derived by Rotta[22] from the Reynolds-averaged Navier–Stokes equations. Another two-equation MTE closure for wall boundary layers is described by Jones and Launder[23]; the two equations comprise an equation for the turbulent kinetic energy and one for the turbulent kinetic energy dissipation rate, ε. The dissipation rate ε can be in turn related to the turbulence length scale, although it is not necessary in the Jones and Launder formulation[23] to do this explicitly.

Free turbulent flows (i.e., flows remote from walls) have also been extensively investigated using MTE closure. Approaches that use the Prandtl–Kolmogorov eddy viscosity hypothesis have been described by Rodi,[24] Rodi and Spalding,[25] and Launder *et al.*[26]; in all of these a form of turbulence length scale differential equation is utilized. Lee and Harsha[27] and Harsha[28,29] have described the application of MTE models in which the relation between the shear stress and kinetic energy used by Bradshaw *et al.*[12] is assumed, but the turbulent kinetic energy transport equation is solved instead of a model equation for the transport of shear stress. The Lee and Harsha[27] and Harsha[28,29] models all use an algebraic definition of the turbulent length scale, rather than a scale transport equation.

All of the MVF and MTF/MTE models so far mentioned are based on field-point (i.e., finite-difference) calculations of the turbulent flow field. Several integral methods have also been developed, for example, for wall boundary layer flows by Patel and Head,[30] based on the MVF model of Bradshaw *et al.*,[12] and for free flows by Peters and Phares,[31] based on the MTF/MTE model of Harsha.[28] In general, integral methods are looked upon as allowing an increase in speed of calculation when compared to a finite-difference computation, but they also have another advantage: By introducing known data into the equations they can, properly used, compensate for a lack of knowledge of other variables. An example of this is the model described by Peters and Phares,[31] in which the introduction of a set of empirical profiles for the turbulent kinetic energy obviates the need for a model for turbulent kinetic energy diffusion.

A much more detailed analysis of turbulent flows is contemplated in the mean Reynolds stress (MRS) class of closure models, for in this class, equations are derived for all of the components of the Reynolds stress tensor, generally in conjunction with equations for the several length scales that enter the problem. Taking a boundary-layer incompressible flow as an example, this entails writing equations for the turbulent shear stress $\langle uv \rangle$, equations for each of the intensities u'^2, v'^2, and w'^2 as well as a length-scale equation. Examples of MRS models are given by Chou,[32] Rotta, [33] and Daly and Harlow,[34] calculations made with a MRS model for boundary-layer and free flows are presented by Donaldson[35] and by Lewellen *et al.*[36] A hybrid MTE/MRS model has been utilized by Hanjalic and Launder[37] and by Launder *et al.*,[26] applied to channel flows and free flows; in this model equations are written for the turbulent kinetic energy $e = \frac{1}{2}(u'^2 + v'^2 + w'^2)$, for the turbulent shear stress $\langle uv \rangle$, and for the turbulent kinetic energy dissipation rate, ε.

It is, of course, possible to obtain turbulence models without recourse to the Reynolds-averaged equations, for example by appealing to the apparent large-scale deterministic structures in some types of turbulent flows (see, e.g., Kovasznay[2]). Reynolds[38] examines in detail the implications of averaging techniques, leading to a conclusion that such an approach may indeed be necessary, if a fully consistent closure is sought. An alternate approach is to proceed from the kinetic theory and obtain new moment equations for the turbulent correlation terms directly from the Boltzmann equations. The Navier–Stokes equations are retained for the mean velocity, but new equations are derived to obtain the Reynolds stresses. This approach has been described by Yen[39]; for low-speed compressible flow it results in a closed set of 40 equations for 40 unknowns. No solutions have been obtained.

As this brief review indicates, a great variety of techniques now exist for the calculation of turbulent flows. In this article, the emphasis is on those models that incorporate the turbulent kinetic energy equation or an analogous equation (such as for the turbulent shear stress or the turbulent eddy viscosity). A selection of these models will be described in the following sections. The Nee and Kovasznay[10] and Saffman[11] MVF models will be described first. Although Mellor and Herring[4] include the model developed by Bradshaw *et al.*[12] in the MVF closure group, the basic ideas that lead to this model are similar to other models in the MTF closure, and thus this model along with the similar model developed by Morel *et al.*[14] will be discussed with the turbulent kinetic energy models, which include the models described by Harsha,[29] Launder *et al.*,[26] Ng and Spalding,[19] and Rodi.[24] Comparisons between the predictions of these finite-difference models and the integral models developed by Patel and Head[30] and Peters and Phares[31] will also be presented.

8.2. Eddy Viscosity Transport Models

Since it is well known that equilibrium boundary layers are correctly predicted by eddy viscosity (or mixing length) concepts, a natural extension of the eddy viscosity concept is the proposal that in a nonequilibrium boundary layer, the eddy viscosity can be described by a rate equation. In this way, the locally dependent nature of a classical eddy viscosity approach can be avoided, while retaining the good profile prediction characteristics of eddy viscosity models.

In general, for boundary-layer flows, the eddy viscosity is assumed to be a scalar quantity. Nee and Kovasznay[10] note that any transportable scalar quantity F subject to the conservation laws is transported according to the equation

$$\frac{\partial F}{\partial t}+U_j\frac{\partial F}{\partial x_j}=\frac{\partial \phi_{F,j}}{\partial x_j}+\text{production}-\text{decay} \tag{8.3}$$

where $\phi_{F,j}$ represents the flux of F due to diffusion. Assuming that the total viscosity (i.e., the sum of the molecular viscosity ν and the turbulent eddy viscosity ν_T), $n=\nu+\nu_T$, is a transportable scalar quantity, and that turbulence is a self-diffusive process, so that the diffusion coefficient for n is n, Equation (8.3) becomes

$$\frac{\partial n}{\partial t}+U_j\frac{\partial n}{\partial x_j}=\frac{\partial}{\partial x_j}\left(n\frac{\partial n}{\partial x_j}\right)+G-D \tag{8.4}$$

Expressions for the production term G and the decay term D were devised by Nee and Kovasznay, in part by analogy to the production and decay terms in the turbulent kinetic energy transport equation. Thus,

$$G=A(n-\nu)\left|\frac{\partial U}{\partial y}\right|, \qquad D=\frac{B}{L^2}n(n-\nu) \tag{8.5}$$

for a two-dimensional boundary-layer flow, where A and B are expected to be universal constants and L is a length scale.

In the application of this method to nonequilibrium boundary-layer predictions,[40] Nee and Kovasznay postulated an additional term

$$-C\frac{n(n-\nu)}{U_\infty^2}\frac{dU_\infty}{dx}\left|\frac{\partial U}{\partial y}\right| \tag{8.6}$$

to account for the effect on the eddy viscosity of the straining of the turbulence by the acceleration due to the mean velocity gradient. Thus, the eddy viscosity rate equation becomes, for a two-dimensional, boundary-layer flow

$$U\frac{\partial n}{\partial x}+V\frac{\partial n}{\partial y}=\frac{\partial}{\partial y}\left(n\frac{\partial n}{\partial y}\right)+A(n-\nu)\left|\frac{\partial U}{\partial y}\right|-\frac{Bn(n-\nu)}{L^2}$$
$$-\frac{Cn(n-\nu)}{U_\infty^2}\frac{dU_\infty}{dx}\left|\frac{\partial U}{\partial y}\right| \tag{8.7}$$

In the majority of the flow fields computed with this model, Nee and Kovasznay chose $A = 0.133$, $B = 0.8$, and $C = 0$; however, an improved level of agreement with the data for some of the flows included in the 1968 Stanford Conference[41] was found when the constants were revised to $A = 0.10$, $B = 1.0$, and $C = 1.0$.

A somewhat similar model for turbulent flows has been described by Saffman.[11] In this case, the model equations have been developed from the hypothesis that a turbulent flow can be described by an energy density ε^* and a vorticity density ω^*, both of which are described by model transport equations. The eddy viscosity, ν_T, is then given by the relation

$$\nu_T = \gamma\varepsilon^*/\omega^* \tag{8.8}$$

where γ is expected to be a universal constant. Roughly, ω^* is defined by Saffman as the mean square vorticity of the "energy-containing eddies" and ε^* is the kinetic energy of the motion induced by this vorticity; plausible transport equations for ε^* and ω^* are developed by analogy with the exact turbulent kinetic energy and mean square vorticity equations. The resulting equations are

$$\frac{\partial \varepsilon^*}{\partial t} + U_i \frac{\partial \varepsilon^*}{\partial x_i} = \alpha^* \varepsilon^* (2S_{ij}^2)^{1/2} - \beta^* \varepsilon^* \omega^* + \frac{\partial}{\partial x_i}\left(\sigma^* \nu_T \frac{\partial \varepsilon^*}{\partial x_i}\right) \tag{8.9}$$

$$\frac{\partial \omega^{*2}}{\partial t} + U_i \frac{\partial \omega^{*2}}{\partial x_i} = \alpha \omega^{*2}\left[\left(\frac{\partial U_i}{\partial x_j}\right)^2\right]^{1/2} - \beta \omega^{*3} + \frac{\partial}{\partial x_i}\left(\sigma \nu_T \frac{\partial \omega^{*2}}{\partial x_i}\right) \tag{8.10}$$

in which all constants are expected to be universal. The term $S_{ij} = \frac{1}{2}(\partial U_i/\partial x_j + \partial U_j/\partial x_i)$. Saffman[11] shows, by means of general arguments regarding various limiting turbulent flows, that $\beta^* = 1$, $\gamma = 1$, $\frac{5}{3} < \beta < 2$, $\sigma = \sigma^* = \frac{1}{2}$, $\alpha^* = 3$, and $(\alpha^*/2)^{1/2} < \alpha < (\alpha^*/2)$, so that, without recourse to direct comparison with experimental data, the values of the constants in Equations (8.9) and (8.10) can be established within fairly narrow limits.

These model equations have been applied to the prediction of a compressible flat-plate boundary layer by Saffman and Wilcox.[13] A major problem in the application of this formulation to the flat-plate boundary-layer problem is the specification of the boundary condition on the vorticity near a solid boundary. Saffman and Wilcox[13] show that an appropriate boundary condition can be obtained. Further, the van Driest form of the compressible law of the wall is deduced from the model equations, showing that this law is a proper boundary condition on the Saffman model in the sense of matched asymptotic expansions. This, of course, allows a drastic decrease in the computation time for a given wall-boundary-layer computation.

Neither the Saffman[11] model nor the Nee and Kovasznay[10] model has been extensively used for computations of a variety of turbulent flows. Nee and Kovasznay[40] presented their model at the 1968 Stanford Conference[41]; the results of that conference show that the eddy viscosity transport hypothesis does not produce results as accurate as the turbulent kinetic energy models also presented. Perhaps the most fundamental objection to both the Saffman and the Nee and Kovasznay approaches is that in these models the modeling effort is not applied to a physically real quantity; the eddy viscosity in the Nee and Kovasznay model and the energy and vorticity densities in the Saffman model are themselves models for a turbulent flow, rather than fundamental physical quantities. On the other hand, all developments utilizing the Reynolds-averaged Navier–Stokes equations also relate to what is fundamentally a model for the flow, and thus the only important distinction between models should be their relative success in predicting turbulent flow fields. On this basis, the turbulent kinetic energy models described in the next section are demonstrably superior.

8.3. Turbulent Kinetic Energy Models

Turbulent kinetic energy models for turbulent flows can be defined as those models that (1) relate the local turbulent shear stress to the local turbulent kinetic energy, either by means of a model or through a coupled set of partial differential equations, and (2) solve the turbulent kinetic energy equation (or an equation derived from the turbulent kinetic energy equation) to obtain a spatial shear stress distribution. Turbulent kinetic energy methods may involve the solution of other equations as well, for example, equations for a turbulence length scale or the turbulent kinetic energy dissipation rate; however, models that involve the solution of equations for all components of the Reynolds stress tensor (MRS models in the terminology of Mellor and Herring[4]) are excluded from this group.

Turbulent kinetic energy models can also be conveniently categorized in terms of the model for the shear-stress–kinetic-energy relation used. One group uses the relation proposed by, among others, Nevzgljadov[42] and Dryden[43]:

$$\tau = a_1 \rho e \tag{8.11}$$

—we shall call these "ND" models, while the other group ("PK" models) uses the relation proposed by Prandtl[21] and Kolmogorov[20]:

$$\tau = \rho \nu_T \frac{\partial U}{\partial y} \tag{8.12}$$

where

$$\nu_T = C_\mu e^{1/2} l_k \tag{8.13}$$

in which l_k represents a turbulent length scale. Both a_1 and C_μ, are, it is hoped, "universal" constants whose values may be specified *a priori* for all flows. However, it is immediately clear that Equation (8.11) cannot be applied without some modification to flows in which the shear stress changes sign or to flows such as circular jets, in which the shear stress is zero on the centerline but the turbulent kinetic energy is not. For such flows, Lee and Harsha[27] proposed writing

$$a_1 = a_1^* \frac{(\partial U/\partial y)}{|\partial U/\partial y|_{\max}} \tag{8.14}$$

where $|\partial U/\partial y|_{\max}$ represents the absolute value of the mean velocity gradient at the point in the shear profile at which the maximum (absolute) value of the shear stress is achieved. This effectively converts Equation (8.11) into a form such as Equation (8.12), with

$$\nu_T = a_1^* e / |\partial U/\partial y|_{\max} \tag{8.15}$$

The models developed by Bradshaw *et al.*[12] for two-dimensional boundary layers, by Lee and Harsha[27] and by Harsha[28,29] for two-dimensional and axisymmetric free flows, by Morel *et al.*[14] and by Morel and Torda[15] for two-dimensional free flows, by Patel and Head[30] for boundary layers, and by Peters and Phares[31] for two- and three-dimensional free flows, all use the ND model. Additional work using the ND model has been reported by Lee *et al.*,[44] by Mikatarian and Benefield,[45] and by Rhodes *et al.*,[46] with emphasis in these more recent papers on the modeling of two-gas and reacting flow fields.

PK models have been described by Launder *et al.*[26] for two-dimensional and axisymmetric free flows, by Ng and Spalding[19] for two-dimensional boundary layers, and by Rodi[24] for two-dimensional and axisymmetric free flows. A variety of versions of these models exist, differing primarily in the variables chosen for use in the auxiliary equation (for length scale or turbulent kinetic energy dissipation rate), and a variety of applications of the PK models has been described. These include predictions of low-Reynolds-number two-dimensional boundary layers,[23] and predictions of the elliptic flow field downstream of a sudden enlargement in a pipe.[47]

8.3.1. ND Models I: Bradshaw et al.[12]

The earliest practical calculation technique using the Nevzgljadov–Dryden turbulence model [Equation (8.11)] is that described by Bradshaw *et al.*[12] in 1967. This model is summarized in a paper by Bradshaw and Ferriss[48] in which additional applications of the model are described; as stated in Reference 48, the method

> has been successfully applied to boundary layers with compressibility, heat transfer, three dimensionality or unsteady flow, with an empirical input obtained almost entirely for low-speed flow at constant pressure. With numerical changes in the empirical input the method has been extended to duct flows and free shear layers.

Since the essential problem in the prediction of turbulent flows is the specification of an expression for the Reynolds stress, it is natural to consider the transport equation for the rate of change of the Reynolds stress along a streamline. For a two-dimensional incompressible boundary layer flow the equation for the Reynolds shear stress

$$\tau = -\rho\overline{uv}$$

is

$$\underbrace{U\frac{\partial\tau}{\partial x}+V\frac{\partial\tau}{\partial y}}_{\text{I}}=\underbrace{\rho\overline{v^2}\frac{\partial U}{\partial y}}_{\text{II}}-\underbrace{\overline{p'\left(\frac{\partial v}{\partial x}+\frac{\partial u}{\partial y}\right)}}_{\text{III}}+\underbrace{\frac{\partial}{\partial y}(\overline{p'v}+\overline{\rho v^2})}_{\text{IV}} \\ \underbrace{-\mu(\overline{u\nabla^2 v}+\overline{v\nabla^2 u})}_{\text{V}} \tag{8.16}$$

where the various terms may be interpreted as

I: transport by mean flow
II: generation
III: destruction by pressure forces
IV: transport by turbulence
V: viscous term

Equation (8.16) can be exactly derived from the Navier–Stokes equations, but as with all of the Reynolds-averaged turbulent flow equations, the terms appearing in it must be modeled in terms of known or calculable quantities before the equation is usable in a calculation procedure. This process requires experimental information, but such information is unavailable, since there is as yet no generally-accepted way of measuring the pressure fluctuation terms appearing in terms III and IV. A similar, but more tractable, equation is the turbulent kinetic energy equation for the quantity

$$\rho e = \tfrac{1}{2}\rho(u^2+v^2+w^2)$$

which may be written

$$\underbrace{U\frac{\partial \rho e}{\partial x}+V\frac{\partial \rho e}{\partial y}}_{\text{I}}=\underbrace{\tau\frac{\partial U}{\partial y}}_{\text{II}}-\underbrace{\frac{\partial}{\partial y}(\overline{p'v}+\overline{\rho e v})}_{\text{III}}-\underbrace{\rho\varepsilon}_{\text{IV}} \tag{8.17}$$

in which the various terms are interpreted as

I: transport by mean flow (advection)
II: generation
III: diffusion
IV: viscous dissipation

Equation (8.17) still contains a pressure fluctuation term, but measurements of the other terms in this equation indicate that this term is in general fairly small and in any case forms a part of the same process of diffusion as does the term $\rho\overline{ev}$. The other terms in Equation (8.17) can be measured, and their behavior, at least for constant-pressure incompressible flow, is qualitatively well understood.

In order to use Equation (8.17) in a calculation scheme, Bradshaw and Ferriss, inspired by the work of Townsend,[49] devised the hypothesis that "for any given shear stress profile $\tau(y)$ there is one and only one possible profile for any one of the other turbulence quantities."[48] Thus, to close Equation (8.17) they write

$$a_1 \equiv \frac{\tau}{\rho e} \tag{8.18}$$

$$L \equiv \frac{(\tau/\rho)^{3/2}}{\varepsilon} \tag{8.19}$$

$$G = \frac{\overline{p'v}/\rho + \overline{ev}}{(\tau/\rho)(\tau_{max}/\rho)^{1/2}} \tag{8.20}$$

where a_1 and G are dimensionless and L is a length (the dissipation length). τ_{max} is defined as the maximum value of the shear stress in the outer part of the boundary layer.

Equations (8.18)–(8.20) are just definitions; for them to be useful, values must be assigned to a_1, L, and G that are applicable over a wide range of flows. Bradshaw and Ferriss[48] note that experience has shown that general empirical functions may be obtained for a_1, L, and G; these are reproduced here as Figure 8.1.

It is of considerable interest to observe that with the diffusion term defined as in Equation (8.20), the set of equations [continuity, momentum, and Equation (8.17) converted to an equation for the turbulent shear stress using Equation (8.18)] is *hyperbolic*. This allows the method of characteristics to be used to obtain a numerical solution; it also has the physical interpretation that a small disturbance introduced at a point in the flow is felt only within a region of influence bounded by the characteristics passing through that point.

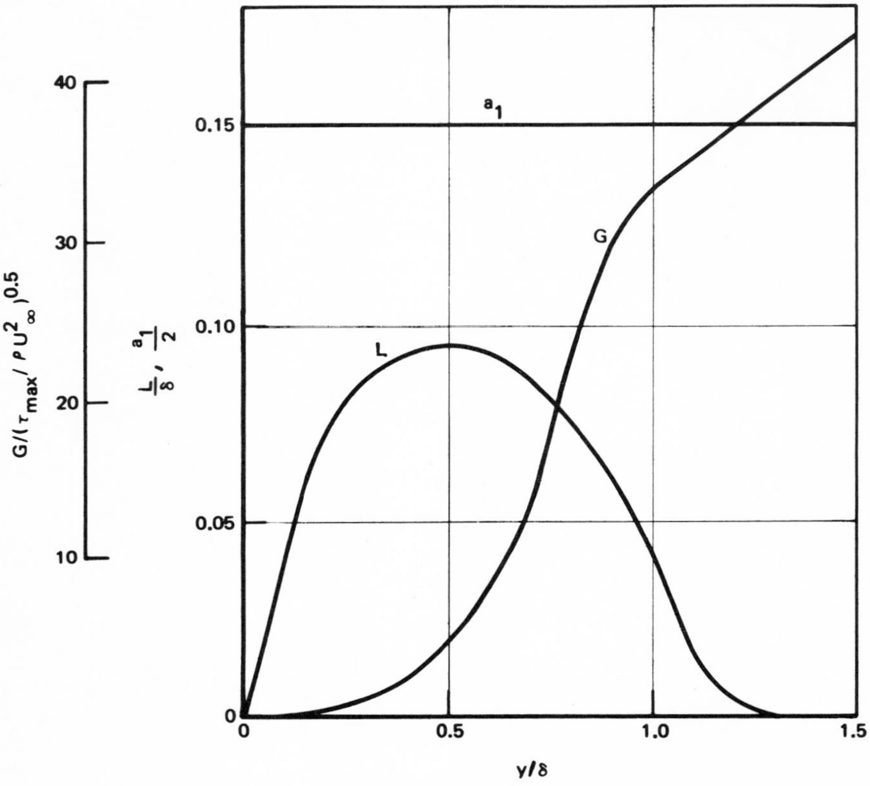

Fig. 8.1. Empirical functions used by Bradshaw *et al.*[12]

The accuracy of this model has been demonstrated for a great variety of boundary-layer flows, e.g., in References 12 and 48 and in the results of the 1968 Stanford Conference.[41] It has been extended to the prediction of compressible flow on adiabatic walls,[50] and to incompressible heat or pollutant transfer,[51] by developing an equation for the transport of a passive scalar contaminant in analogy to Equation (8.17), and to the prediction of three-dimensional flow,[52] by developing an equation for the second component of the shear stress—$\rho\overline{wv}$ (v is the velocity component normal to the surface). Developments of the basic method have also been applied to the prediction of two-dimensional duct flows[53] and free shear flows.[14,15,27–29]

An interesting variation of Bradshaw's approach is the half-integral technique developed by Patel and Head.[30] In this version the momentum integral equation

$$\frac{d\theta}{dx}+(H+2)\frac{\theta}{U_\infty}\frac{dU_\infty}{dx}=\frac{\tau_w}{\rho U_\infty^2} \tag{8.21}$$

in which θ is the boundary-layer momentum thickness and H is the boundary-layer shape factor, is used with Bradshaw's shear-stress equation in the form

$$U\frac{\partial}{\partial x}\left(\frac{\tau}{2a_1\rho}\right)+V\frac{\partial}{\partial y}\left(\frac{\tau}{2a_1\rho}\right)=\frac{\tau}{\rho}\frac{\partial U}{\partial y}-\left(\frac{\tau_{\max}}{\rho}\right)^{1/2}\frac{\partial}{\partial y}\left(G\frac{\tau}{\rho}\right)-\frac{(\tau/\rho)^{3/2}}{L} \tag{8.22}$$

where a_1, G, and L are as shown in Figure 8.1, and a half-integral momentum equation in the form

$$\tau(y)=\tau_w+\int_0^y\frac{dP}{dx}dy+\int_0^y 2\rho U\frac{\partial U}{\partial x}dy-\rho U\int_0^y\frac{\partial U}{\partial x}dy \tag{8.23}$$

where the upper limit on the integrals in Equation (8.23) is $y=0.5\delta$. Thompson's two-parameter family of velocity profiles and associated wall shear-stress law[54] is used in Equation (8.23); these velocity profiles and the skin friction coefficient are functions of the integral shape parameters θ and H.

The procedure used by Patel and Head to solve the set of Equations (8.21)–(8.23) is then as follows: At a downstream station $(i+1)$ the value of H is assumed known. Equation (8.21) then defines θ, given the external pressure gradient. The profile family for the appropriate θ and H is then used to obtain the shear stress at $y=0.5\delta$ at the axial station $(i+\frac{1}{2})$. Finally, Equation (8.22) is solved. This solution is used to determine whether the increment in shear stress along $y=0.5\delta$ from the axial station $(i-\frac{1}{2})$ to station $(i+\frac{1}{2})$ obtained from the profile family satisfies Equation (8.22). In general, it will not, and the procedure is iterated, using a new value of H, until Equations (8.21) and (8.22) are both satisfied.

This approach is computationally somewhat faster than the method of characteristics used by Bradshaw and Ferriss,[48] but, of more interest, it is also appreciably more accurate. A comparison of predictions of equilibrium boundary layer data is shown in Figure 8.2; for this flow the predictions are virtually indistinguishable and in excellent agreement with the data. However, for nonequilibrium boundary layers, the half-integral method is considerably more accurate (particularly for predictions of the shape factor H), as shown in Figure 8.3. This is despite the fact that the original data were obtained by Bradshaw[55]; disagreement of a particular investigator's theory with his own experimental data is relatively uncommon.

Also shown in Figure 8.3 are predictions made in unpublished work by Harsha and Glassman. These predictions were made using a boundary-layer

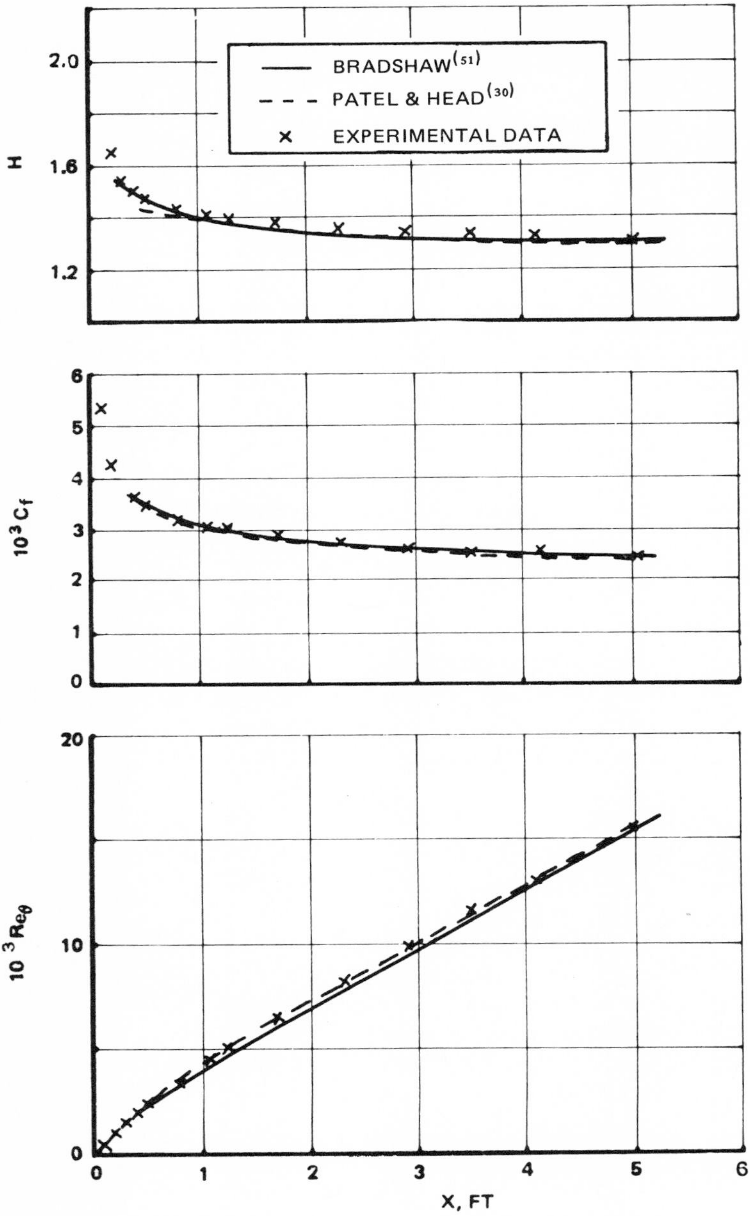

Fig. 8.2. Comparison of integral and finite-difference TKE models for equilibrium boundary layers. Data: Wieghardt and Tillman, Stanford[41] ident 1400; figure after Patel and Head.[30]

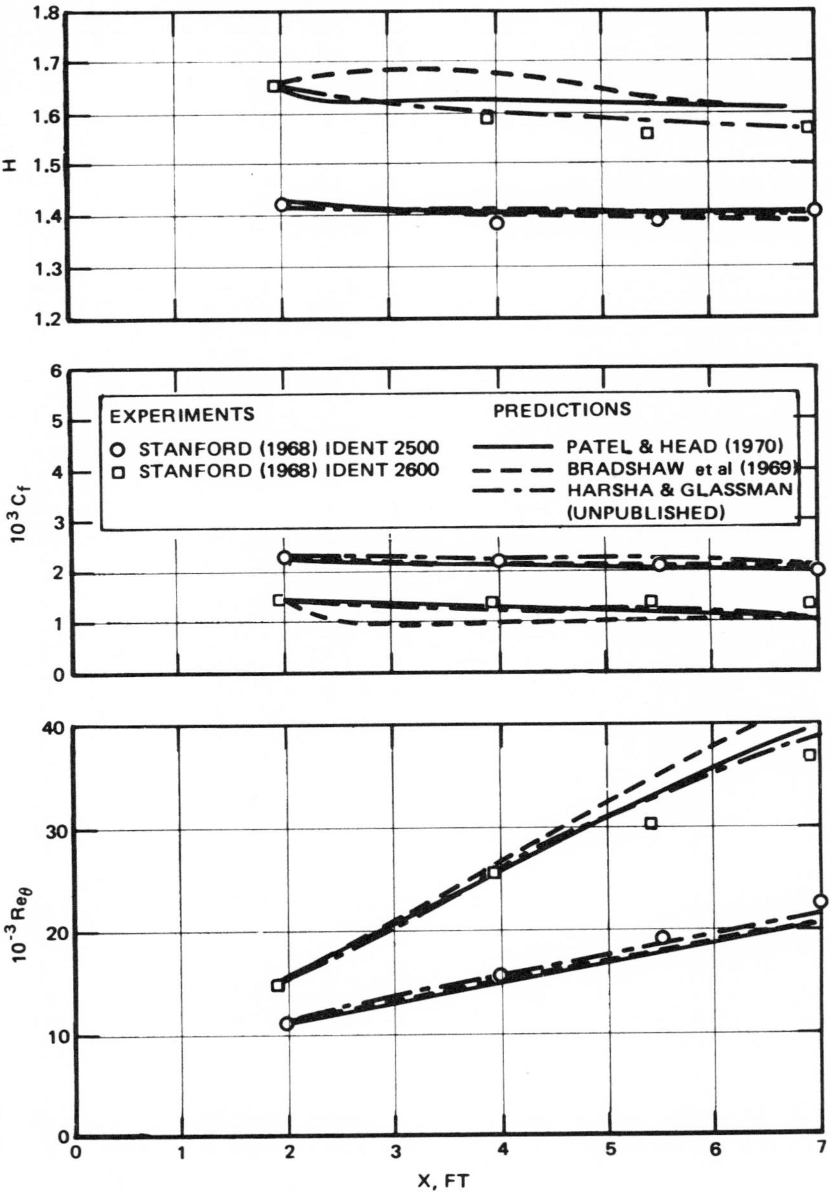

Fig. 8.3. Comparison of integral and finite-difference TKE models with plane boundary-layer data.

version of the model originally developed by Lee and Harsha[27] for free-mixing flow. Since, as described below, the Lee and Harsha model is based on the Bradshaw model (albeit with a gradient diffusion term in the turbulent kinetic energy equation), the predictions shown in Figure 8.3 represent a rather satisfactory "closing of the circle."

8.3.2. ND Models II: Morel et al.[14]

The application of Bradshaw's method to two-dimensional free turbulent jets, wakes, and shear layers represents another interesting variation of the basic method. It rests on the concept of "interaction," i.e., that turbulent shear flows in which the shear-stress profile changes sign can be represented by the interaction of two opposite boundary-layer-like flows. Figure 8.4, after Reference 15, illustrates the hypothesis. Two "simple" layers, in each of which the shear stress does not change sign and in which only one set of large eddies exists, interact in the "region of small-scale mixing." Each layer feels the existence of the other only through the mean velocity profile, which is in turn determined by the algebraic sum of the two simple-layer shear-stress fields. In this way, a continuous shear-stress profile is generated in a flow field in which the sign of the shear stress varies.

This concept allows the use of the empiricism developed for the two-dimensional wall boundary layer in a free shear flow, although the concept becomes somewhat problematical when an axisymmetric flow is considered. (An infinite number of interacting layers would have to be postulated.) It also turns out that boundary-layer values of a_1, L, and G cannot be directly applied to free shear flows. However, with some varia-

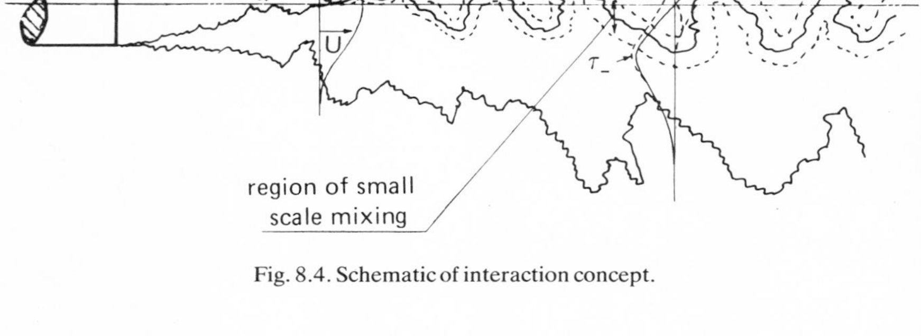

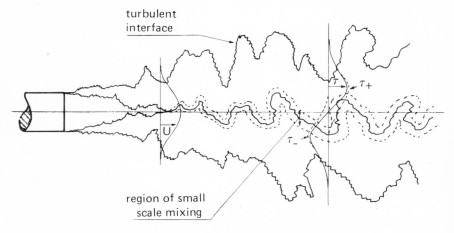

Fig. 8.4. Schematic of interaction concept.

tions in the numerical values of a_1, L, and G (compared to the two-dimensional boundary-layer case), this approach has been applied to a variety of free flows; some results will be shown (in comparison with other approaches) below.

8.3.3. ND Models III: Lee and Harsha[27]

The model chosen by Lee and Harsha[27] for free turbulent flows differs from the Bradshaw model in several major ways. Instead of making the assumption embodied in Equation (8.20), Lee and Harsha assumed a gradient diffusion model for the turbulent kinematic energy, so that, for a two-dimensional flow, for example,

$$(\overline{p'v}+\overline{\rho u v^2}) \equiv \frac{\rho \nu_T}{\sigma_k}\frac{\partial e}{\partial y} \tag{8.24}$$

where σ_k represents a "Prandtl number" for turbulent kinetic energy, assumed to be a constant in a given flow. The dissipation term in Equation (8.17) is also modeled somewhat differently by Lee and Harsha than by Bradshaw *et al.*; it is assumed to take on the form given by Kolmogorov[20]:

$$\rho\varepsilon \equiv \frac{a_2 \rho\, e^{3/2}}{l_k} \tag{8.25}$$

where a_2 is a constant and l_k represents a dissipation length scale, constant across the flow field and algebraically related to the local shear region width. With these assumptions, Equation (8.17) becomes, for a general variable density flow,

$$\rho U\frac{\partial e}{\partial x}+\rho V\frac{\partial e}{\partial y}=\frac{1}{y^\alpha}\frac{\partial}{\partial y}\left(\frac{\rho\nu_T y^\alpha}{\sigma_k}\frac{\partial e}{\partial y}\right)+\rho\nu_T\left(\frac{\partial U}{\partial y}\right)^2-\frac{a_2\rho e^{3/2}}{l_k} \tag{8.26}$$

where Equations (8.11) and (8.12) have been used in writing the generation term. α is zero for a plane flow and unity for axisymmetric flow. This equation is *parabolic* and thus a major difference between the Lee and Harsha[27] model and the Bradshaw[12,48] model is that the former results in a parabolic set of governing equations rather than a hyperbolic set. As fundamental as this difference appears to be, it is found to have relatively little effect when results of the two models are compared. This can be explained by the observation that, for boundary-layer-type turbulent flows at least, the diffusion of turbulent kinetic energy is a relatively small term in Equation (8.26).

To use Equation (8.26) in the computation of free turbulent flows, some fairly extensive modeling of the parameters a_1, a_2, and the length scale l_k is

Table 8.1. Definition of $f_1(y)$ in Lee and Harsha[27,29] Model[a]

Flow regime	Range	Expression	Remarks
Two-dimensional shear layer, and first regime of jets	$0.05 \le \dfrac{U-U_e}{U_j-U_e} \le 0.95$	$f_1(y) = \dfrac{\partial u}{\partial y} \Big/ \left\lvert \dfrac{\partial u}{\partial y} \right\rvert$	Provides sign of shear stress
	$0 \le \dfrac{U-U_e}{U_j-U_e} < 0.05$	none	ν_T held constant at value obtained at $(U-U_e)/(U_j-U_e) = 0.05$
	$0.95 < \dfrac{U-U_e}{U_j-U_e} \le 1.0$	none	ν_T held constant at value obtained at $(U-U_e)/(U_j-U_e) = 0.95$
	$U_c > 0.09\, U_j$	$f_1(y) = \dfrac{\partial u}{\partial y} \Big/ \left\lvert \dfrac{\partial u}{\partial y} \right\rvert$	Provides sign of shear stress
Second regime of axisymmetric and planar jets; wakes	$\left.\begin{matrix} U_c < 0.09\, U_k \\ y \le y_{max} \end{matrix}\right\}$	$f_1(y) = \dfrac{\partial u}{\partial y} \Big/ \left\lvert \dfrac{\partial u}{\partial y} \right\rvert_{max}$	y_{max} is value of lateral coordinate at which $\lvert \partial u/\partial y \rvert$ is a maximum
	$\left.\begin{matrix} U_c > 0.09\, U_j \\ y > y_{max} \end{matrix}\right\}$	$f_1(y) = \dfrac{\partial u}{\partial y} \Big/ \left\lvert \dfrac{\partial u}{\partial y} \right\rvert$	

[a] For zero-momentum flow, where two local maxima in $\lvert \partial u/\partial y \rvert$ occur, the $\partial u/\partial y / \lvert \partial u/\partial y \rvert_{max}$ model for $f_1(y)$ is applied using the appropriate maximum in each part of the flow.

required. This has been described in detail in References 28 and 29. In general, the parameter a_1 may be written

$$a_1 = 0.3 f_1(y) \tag{8.27}$$

where the particular form of $f_1(y)$ varies depending on the flow under consideration, as summarized in Table 8.1. The flow regimes described in Table 8.1 are shown schematically in Figure 8.5; Figure 8.6 shows evidence that for at least some flows the admittedly heuristic form of $f_1(y)$ provides a close approximation to the experimental variation of a_1. The turbulent length scale l_k is also defined differently in different flow regimes, as summarized in Table 8.2. In general, the different definitions are used because it is not possible to define a general length scale that could be applied to many different flow fields.

The turbulent kinetic energy production and dissipation terms are the dominant terms in Equation (8.26) in most flow fields of interest. Because the term representing turbulent kinetic energy production is exact, and the value of 0.3 for $a_1/f_1(y)$ is strongly supported by experimental data,[56] it is clear that modeling of the dissipation term will dominate attempts to improve the generality of the Lee and Harsha model. The 1972 NASA–Langley Free Turbulent Shear Flows Conference[57] encompassed a wide variety of free-mixing flow fields, and the need to extend the Lee and Harsha model (as well as the related Peters and Phares[31] integral technique) to the calculation of as many of these flows as possible led to an extensive modeling

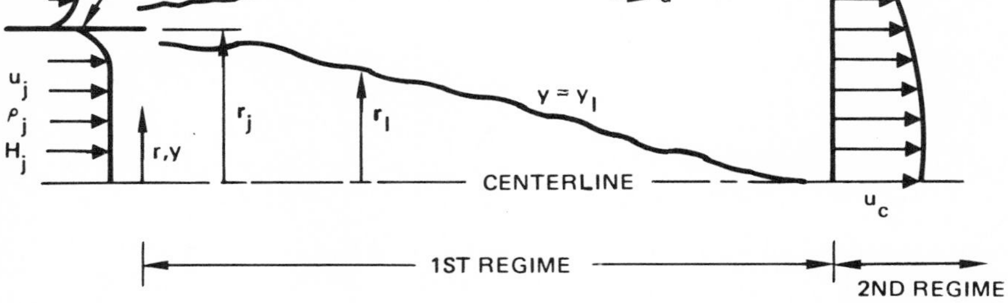

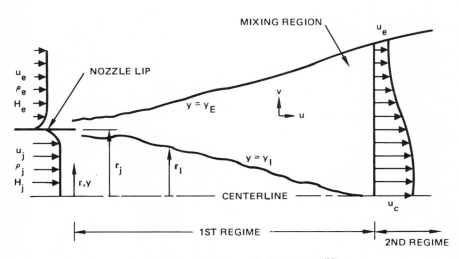

Fig. 8.5. Schematic of free-mixing flow field.[29]

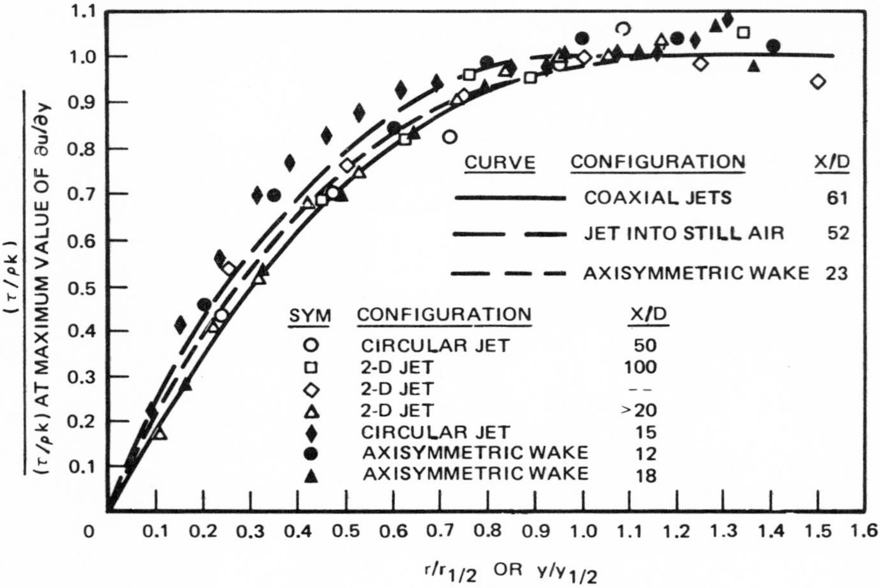

Fig. 8.6. Comparison of calculated and experimental variation of ratio of shear stress to kinetic energy.[29]

Table 8.2. Definition of Length Scale l_k in Lee and Harsha[27,29] Model

Flow regime	Length scale
Two-dimensional shear layer or first regime of jets, with non-fully-developed profiles of velocity	$l_k = y_2 - y_1$, where at y_2: $(U-U_e)/(U_j-U_e)=0.05$ and at y_1: $(U-U_e)/(U_j-U_e)=0.95$
As above, with fully developed profiles	$l_k = 1.57\ (U_j-U_e)/(\partial U/\partial y)_m$ where m refers to point in velocity profile where the shear stress is a maximum[a]
Two-dimensional wake	$l_k = y_2 - y_1$; where at y_2: $(U-U_e)/(U_j-U_e)=0.05$ and $y_1=0$
Second regime of jet	$l_k = 2r_{1/2}$, where at $r_{1/2}$, $(U-U_e)/(U_c-U_e)=0.5$

[a] Note: this definition used when $(y_2-y_1) \le l_k \le 1.57\ (y_2-y_1)$.

Table 8.3. Definition of $f_2(R_T)$ in Harsha[29] Model

$$f_2(R_T)=\begin{cases} 1/C_1 & \text{for} \quad 0<R_T<185 \\ (0.46+0.00762\,R_T)/1.69\,C_1 & \text{for} \quad 185<R_T<360 \\ 3.20/1.69\,C_1 & \text{for} \quad R_T>360 \end{cases}$$

where

$$R_T=\Delta u l_k/\nu_{TM}$$

$$\nu_{TM}=(\tau_m/\rho_m)/(\partial u/\partial y)_m$$

Δu represents the velocity difference across the viscous region and m refers to the point of maximum shear at a given axial station in the flow. The factor C_1 is used to correct for the influence of density variations. It is given by

$$C_1=\begin{cases} 0.984+0.016\,\rho_{e1}/\rho_{j1} & \text{for} \quad (\rho_{e1}/\rho_{j1})>1 \\ 0.95+0.05\,R_eT_e/R_jT_j & \text{for} \quad (\rho_{e1}/\rho_{j1})<1 \end{cases}$$

where the subscript 1 refers to conditions at the origin of mixing, R_e is the gas constant appropriate to the outer flow, R_j the gas constant for the jet, T_e the total temperature of the outer flow, and T_j the jet total temperature.

---

of the dissipation rate parameter a_2. This effort, largely carried out using the Peters and Phares[31] approach, evolved the definition

$$a_2=1.69f_2(R_T) \tag{8.28}$$

where R_T represents a turbulent Reynolds number. The parameter $f_2(R_T)$ depends also on the initial density ratio between the mixing streams, and the overall variation of f_2 is given in Table 8.3. Although complex, the variation summarized in Table 8.3 provides good results over a variety of free-mixing flow fields[28,31] and significantly improves the utility of the model.

As in the case of the Patel and Head[30] integral formulation of the Bradshaw[12] model, an integral formulation of the Lee and Harsha[27,29] model exists.[31] Also as in the case of the Patel and Head version of Bradshaw's model, the inclusion of appropriate empirical information in the integral model has resulted in the development of a model with significantly better predictive capability than the corresponding finite-difference model in some flows.

In the Peters and Phares[31] approach, the describing equations for a compressible axisymmetric flow are integrated, so that the resulting equation set comprises the following:

(1) A continuity equation for the entire flow

$$\int_0^{r_w}\frac{\partial}{\partial x}(\rho U)r\,dr=-\rho_eU_rr_w\frac{dr_w}{dx} \tag{8.29}$$

where r_w is the radius of a frictionless duct which it is assumed encloses the flow, and the subscript e refers to the velocity at the outer edge of the flow (i.e., at the duct wall).

(2) An overall momentum equation

$$\int_0^{r_w} \frac{\partial}{\partial x}(\rho U^2) r\,dr = -\frac{r_w^2}{2}\frac{dP_w}{dx} - \rho_e U_e^2 r_w \frac{dr_w}{dx} \tag{8.30}$$

(3) A half-radius momentum equation

$$\int_0^{r_m} \frac{\partial}{\partial x}(\rho U^2) r\,dr - U_m \int_0^{r_m} \frac{\partial}{\partial x}(\rho U) r\,dr = \tau_m r_m - \frac{r_m^2}{2}\frac{dP_w}{dx} \tag{8.31}$$

where r_m is the radius at which $U_m = \frac{1}{2}(U_c + U_e)$ is the value of the axial velocity at the flow centerline.

(4) A jet species conservation equation

$$\int_0^{r_w} \frac{\partial}{\partial x}(\rho U C) r\,dr = 0 \tag{8.32}$$

and

(5) An integral turbulent kinetic energy equation

$$\int_{r_i}^{r_i+b} \frac{\partial}{\partial x}(\rho U_e) r\,dr = \int_{r_i}^{r_i+b} \tau \frac{\partial U}{\partial r} r\,dr - \frac{a_2}{b}\int_{r_i}^{r_i+b} \rho\, e^{3/2} r\,dr \tag{8.33}$$

In the integrated turbulent kinetic energy equation, the integrals are taken across the viscous region, whose inner edge is at r_i and whose width is b. The integration has eliminated the diffusion term, and, by noting that the production of turbulent kinetic energy equals the negative of the dissipation of mean flow mechanical energy

$$\int_{r_i}^{r_i+b} \tau \frac{\partial U}{\partial r} r\,dr = -\int_{r_i}^{r_i+b} U \frac{\partial}{\partial r}(\tau r)\,dr = \frac{1}{2}\int_0^{r_i+b} \frac{\partial}{\partial x}(\rho U^3) r\,dr$$
$$+\frac{1}{2}U_e^2 \int_0^{r_i+b} \frac{\partial}{\partial x}(\rho U) r\,dr - \frac{dP_w}{dx}\int_0^{r_i+b} U r\,dr \tag{8.34}$$

Equation (8.33) can be written in a form in which the turbulent shear stress τ does not explicitly appear.

A family of profiles is now specified for U, for C, and for e; a unity Lewis number assumption relates the local flow enthalpy directly to C. The profile families are functions of the local characteristic values of the dependent variables and the spatial position and width of the viscous region. All profile families are in the form of cosine functions, and different families are used for the first and second regimes of the flow, with an empirical transition function derived for the turbulent kinetic energy profiles. The empirical data on which these profile families are based are obtained in general from incompressible flow experiments.

By introducing the profile functions into Equations (8.30)–(8.33), a set of ordinary differential equations for P_w, r_i, b, and e_m is obtained in the first regime. In the second regime, the dependent variables are P_w, U_c, b, and e_m, where in both regimes e_m is the value of the kinetic energy at the half-velocity point r_m. Two important features result from this procedure: As has already been noted, the diffusion term drops out of the turbulent kinetic energy equation (diffusion is, of course, implicit in the profiles assumed for the turbulent kinetic energy), and the relationship between the shear stress and the kinetic energy need only be applied at the half-velocity point in the flow, where

$$|\tau|/\rho e = 0.3$$

is an excellent approximation.[(56)]

Both finite-difference[(28)] and integral[(31)] forms of the Lee and Harsha[(27)] model were presented at the 1972 NASA–Langley Conference,[(57)] and thus direct comparisons may be made between the two versions of the basic model. A comparison of predictions for a mildly compressible circular jet is shown in Figure 8.7. For this case, there is little difference between the predictions. However, as Figure 8.8 shows, the integral model provides a substantially better prediction of the supersonic circular jet. Here the inaccuracy of the finite-difference model in the transition region between the first and second regimes of the jet (Figure 8.5) affects the

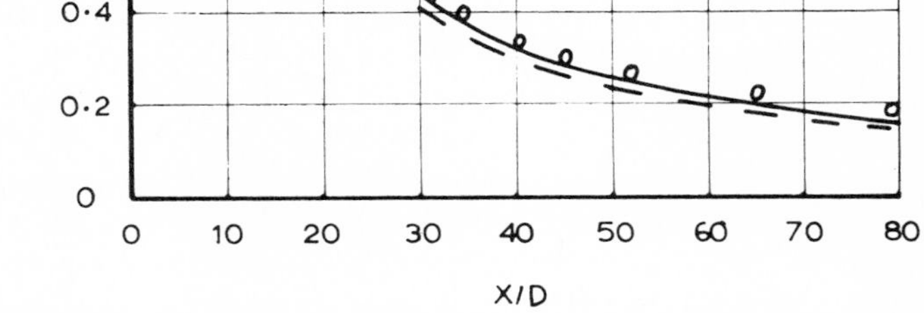

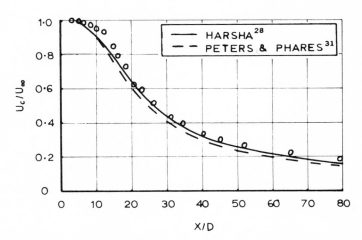

Fig. 8.7. Comparison of integral and finite-difference models $M_0 = 0.64$ circular jet, NASA case 6.[(53)]

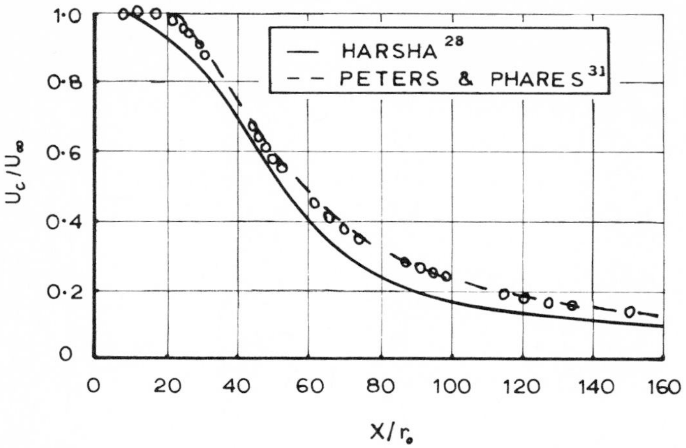

Fig. 8.8. Comparison of integral and finite-difference models $M_0 = 2.2$ circular jet, NASA case 7.(53)

prediction for a substantial distance downstream. This inaccuracy in the finite-difference model is a result of the definitions of $f_1(y)$ and l_k shown in Tables 8.1 and 8.2; these functions change rather abruptly as the flow regimes change.

The effects of the abrupt change in definitions of $f_1(y)$ and l_k are not nearly as apparent for the supersonic hydrogen–air jet as they are for the supersonic air jet case. A comparison of integral and finite-difference model predictions for this case is shown in Figure 8.9. Here the relatively small inaccuracies in the centerline concentration profile prediction exhibited by the integral model are probably caused by the restriction of the integral model to unity Prandtl and Schmidt numbers. This restriction introduces considerably more error into the prediction of the subsonic mixing of hydrogen and air, as shown in Figure 8.10; for this flow, the finite-difference model, which allows nonunity values of the Prandtl and Schmidt numbers, is clearly superior.

Two further comparisons are of interest. The first of them involves the incompressible axisymmetric wake, for which predictions are shown in Figure 8.11. The parameter $W = 1 - U_c/U_e$. Although neither of these predictions is especially accurate (note though that the method of plotting emphasizes small inaccuracies), the shape of the integral model prediction appears closer to the trend of the data than that of the finite-difference model prediction. This comparison becomes clearer in the case of the asymptotic jet, for which predictions are shown in Figure 8.12. Similarity considerations show that the asymptotic rate of decay of the incompressible circular jet should be proportional to x^{-1}. It is clear from this figure that

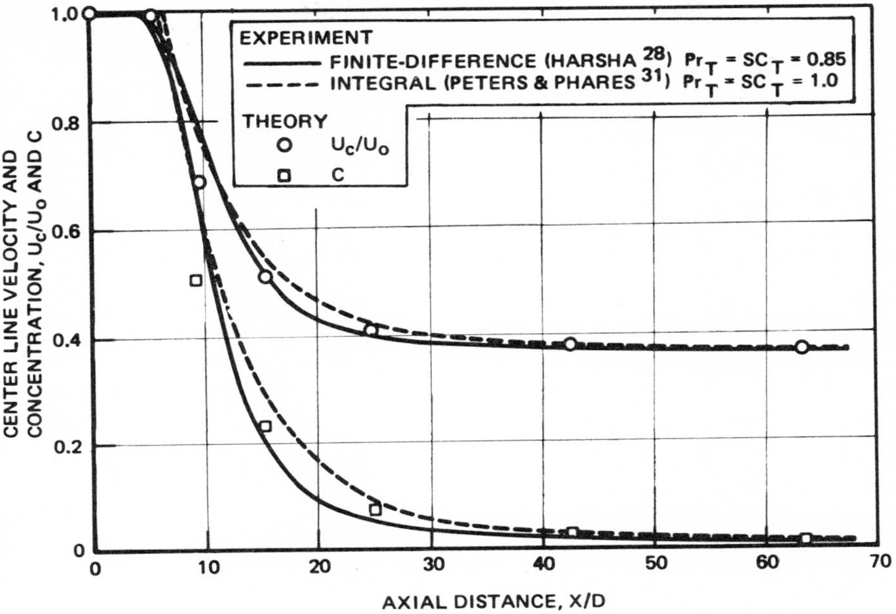

Fig. 8.9. Comparison of integral and finite-difference TKE models; H_2 jet in moving stream, $M_e = 1.33$, NASA case 12.[53]

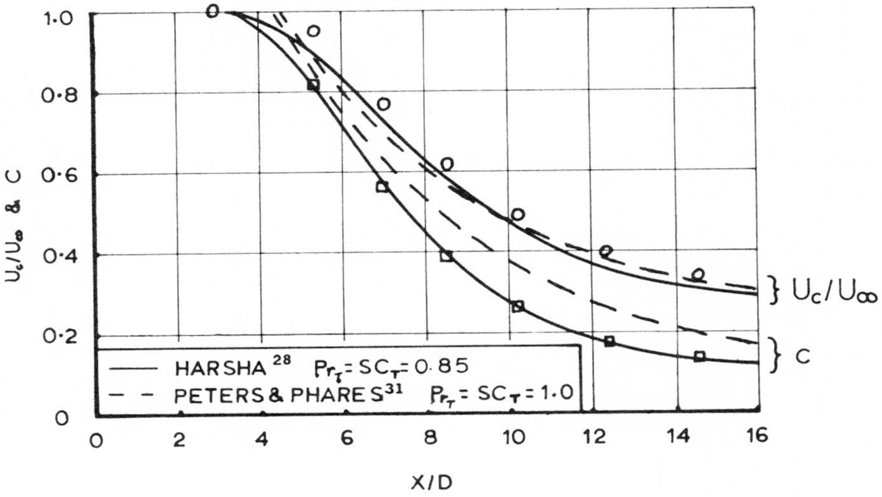

Fig. 8.10. Comparison of integral and finite-difference TKE models; H_2 jet in moving stream, $U_e/U_0 = 0.16$, NASA case 10.[53]

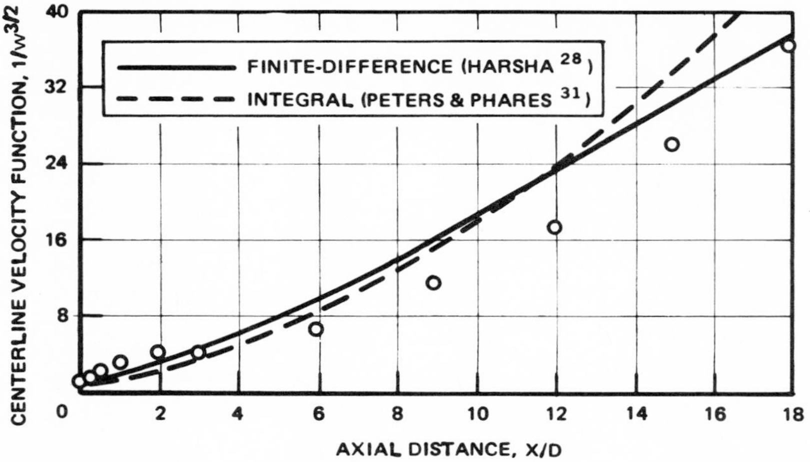

Fig. 8.11. Comparison of integral and finite-difference TKE models; incompressible axisymmetric wake, NASA case 15.[53]

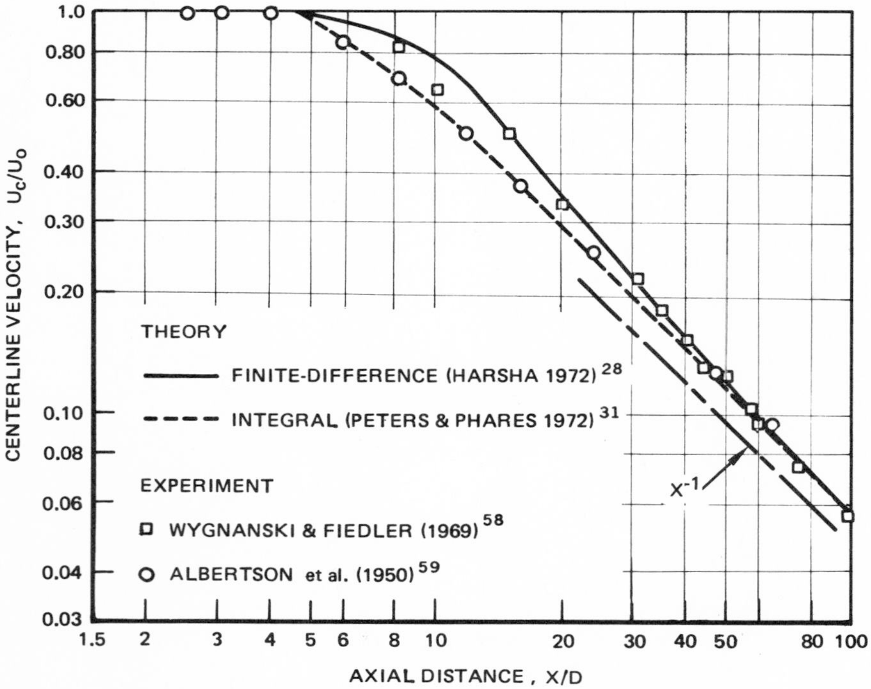

Fig. 8.12. Comparison of integral and finite-difference TKE models; fully developed axisymmetric jet, NASA case 18.[53]

although the finite-difference model predictions closely follow the data of Wygnanski and Fiedler,[58] neither these data nor the finite-difference model predictions achieve an asymptotic x^{-1} decay rate. On the other hand, both the data of Albertson *et al.*[59] and the integral model predictions do achieve the proper asymptotic decay rate.

In both the wake and jet flows the reason for the failure of the finite-difference model to correctly predict the asymptotic behavior is related to the model used for the turbulent kinetic energy diffusion. Although detailed data are not available for the far asymptotic region of a circular wake, turbulent kinetic energy profiles for a circular jet in the asymptotic region clearly indicate that a countergradient diffusion mechanism has operated. Such diffusion cannot be modeled by Equation (8.24), but this diffusion term does not appear in the integral formulation. Since empirical turbulent kinetic energy profiles for the asymptotic jet have been included in the development of the profile models used in the integral approach, a form of countergradient diffusion *is* implicitly included in the integral model. Thus, the integral model is capable of a more accurate prediction of these asymptotic flows than the finite-difference formulation.

8.3.4. PK Models I: Ng and Spalding[19]; Rodi and Spalding[25]

The second widely used kinetic energy model of turbulence is the Prandtl–Kolmogorov (PK) model,[20,21] in which the turbulent eddy viscosity is related directly to the product of the square root of the turbulent length scale, Equation (8.13). Although the available experimental data have been shown to support the ND model,[56] and little direct evidence favors the PK model, the PK model does have the advantage of being explicitly an eddy viscosity formulation. Since we are in any case discussing *models* for a turbulent flow, it is not absolutely necessary that they reproduce all features of the details of the flow field; the overall comparison of the results of the model as compared to real turbulent flows is, in the final analysis, the most important criterion for a turbulence model.

PK turbulence models are generally used with a transport equation for the turbulent length scale transport equation (indeed, it has been reported[26] that a PK model without a length-scale transport equation produces results that, for wall boundary layers, are little better than those obtained with mixing length models), and it is in the details of the length-scale equation formulation that the various PK models differ. In the planar boundary-layer computation reported by Ng and Spalding, the product el_k is used as the dependent variable of the length-scale equation. Another widely used approach involves

$$\varepsilon = e^{3/2} l_k^{-1}$$

as the dependent variable of the auxiliary equation. In general, length-scale equations are written for a variable of the form

$$e^a l_k^b$$

and the same basic transport equation can be used to obtain the spatial variation of

$$z = e^a l_k^b$$

for any a and b.[60] However, in at least some flow fields the use of a transport equation for the dissipation rate ε is most appropriate, if only because it turns out after modeling to contain the fewest terms.[60]

In the model described by Ng and Spalding, the transport equations to be solved to obtain the turbulent eddy viscosity are the turbulent kinetic energy equation, in the form proposed by Kolmogorov,[20] for an incompressible flow:

$$U\frac{\partial e}{\partial x} + V\frac{\partial e}{\partial y} = \frac{1}{\sigma_e}\frac{\partial}{\partial y}\left(e^{1/2} l_k \frac{\partial e}{\partial y}\right) + e^{1/2} l_k \left(\frac{\partial U}{\partial y}\right)^2 - \frac{C_D e^{3/2}}{l_k} \tag{8.35}$$

where σ_e is the turbulent Prandtl number, and both σ_e and C_D are taken to be constants for fully turbulent flows, and an equation for the transport of length scale in the form

$$U\frac{\partial e l_k}{\partial x} + V\frac{\partial e l_k}{\partial y} = \frac{\partial}{\partial y}\left(\frac{e^{3/2} l_k}{\sigma_1}\frac{\partial l_k}{\partial y} + \frac{e^{1/2} l_k^2}{\sigma_2}\frac{\partial e}{\partial y}\right) + C_p e^{1/2} l_k^2 \left(\frac{\partial U}{\partial y}\right)^2 - C'_M e^{3/2} \tag{8.36}$$

In Equation (8.36), which is similar to an equation for the spatial variation of the turbulent length scale derived from the Navier–Stokes equations by Rotta,[22] the quantities σ_1, σ_2, C_p and C'_M are assumed to be universal constants or functions of local properties. It turns out that adequate predictions of wall boundary-layer flows can only be obtained with a variation in C'_M according to the equation

$$C'_M = C_M + C_W (l_k/y)^q \tag{8.37}$$

in which C_M, C_W, and q are constants. Equation (8.37) is without theoretical foundation, but, again keeping in mind that it is a *model* of turbulence that is being developed, the heuristic nature of Equation (8.37) is not a serious difficulty in the use of this technique.

The constants that appear in Equations (8.35)–(8.37) must be obtained by a comparison of flow field predictions with experimental data. Certain of them may be determined by an appeal to those few especially simple turbulent flows for which an analytical solution of the governing equations exists, while the optimization of the remaining constants requires numerous

Table 8.4. Comparison of Model Constants for Boundary Layers[19] and Free Flows[25]

Model	C_D	C_P	C_M	C_W	σ_e	σ_1	σ_2	q
Ng and Spalding[19]	0.1	0.84	0.055	22	2	1.2	2	4
Rodi and Spalding[25]	0.09	1	0.057	—	1	0.3	−0.43	—

computations of other test flows. The constants were obtained by Ng and Spalding by comparison of computations with a set of four test experiments; the results are shown in Table 8.4 in comparison with those obtained for prediction of free shear layers by Rodi and Spalding[25] using a similar model. The constants C_w and q do not appear in the modeling of free flows.

Ng and Spalding[19] report computations for a variety of flows, including flat-plate boundary layers, fully developed pipe flow, plane-channel flow, and the plane wall jet. Satisfactory results are obtained in these computations, although some discrepancy is shown in the case of the plane wall jet. In this flow, experimental evidence indicates that the point in the velocity profile at which $\tau = -\rho\overline{uv} = 0$ does *not* correspond to the position at which $\partial U/\partial y = 0$; an eddy viscosity model such as the PK model cannot model this behavior. Ng and Spalding do not report direct comparisons with the data included in the 1968 Stanford Conference,[41] so that comparison with the other models for the turbulent boundary layers discussed in this article is not possible.

The constants obtained by Rodi and Spalding[25] in modeling free shear flows using the PK model and Equations (8.24) and (8.36) are shown in Table 8.4. Large discrepancies can be seen in the values of the constants required in the diffusion terms of these equations, i.e., σ_e, σ_1, and σ_2. This problem is the basic difficulty with the $e - el_k$ model exemplified by the two techniques discussed in this section. Although reasonably accurate predictions of both wall boundary layers and free shear flows may be obtained with this model, major changes in the constants are required. The lack of generality thus observed has led to the development of the models that use ε as the dependent variable in the scale equation, which will now be described.

8.3.5. PK Models II: Launder et al.[26]

The most generally successful version of the PK model is that in which the dependent variable of the auxiliary equation is the turbulent kinetic energy dissipation rate $\varepsilon = e^{3/2} l_k^{-1}$. Two variations of this model are described in the paper by Launder *et al.*,[26] applied to free shear flows; the basic model has also been used in the calculation of low-Reynolds-number wall boundary layers.[23,61] For planar or axisymmetric compressible boundary layer flows, the equations that govern the distribution of ν_T (at high Reynolds numbers) are

$$\rho U\frac{\partial e}{\partial x}+\rho V\frac{\partial e}{\partial y}=\frac{1}{y^\alpha}\frac{\partial}{\partial y}\left(\frac{\rho \nu_T y^\alpha}{\sigma_k}\frac{\partial e}{\partial y}\right)+\rho\nu_T\left(\frac{\partial U}{\partial y}\right)^2-\rho\varepsilon \tag{8.38}$$

and

$$\rho U\frac{\partial \varepsilon}{\partial x}+\rho V\frac{\partial \varepsilon}{\partial y}=\frac{1}{y^\alpha}\frac{\partial}{\partial y}\left(\frac{\rho \nu_T y^\alpha}{\sigma_\varepsilon}\frac{\partial \varepsilon}{\partial y}\right)+C_{\varepsilon 1}\frac{\varepsilon}{e}\rho\nu_T\left(\frac{\partial U}{\partial y}\right)^2-C_{\varepsilon 2}\frac{\rho\varepsilon^2}{e} \tag{8.39}$$

in which $\alpha = 0$ for plane flow, $\alpha = 1$ for an axisymmetric flow, and σ_k, σ_ε, $C_{\varepsilon 1}$, and $C_{\varepsilon 2}$ are constants. The eddy viscosity relation is given by

$$\nu_T = C_\mu e^2/\varepsilon \tag{8.40}$$

where C_μ is an additional constant. As was the case for the Harsha[28] and Peters and Phares[31] versions of the ND model, an extensive evaluation for the models described by Launder *et al.*[26] was carried out for the 1972 NASA–Langley Conference[57] resulting in the development of two versions of the so-called "$k\varepsilon$" models, which differ in the constants chosen and, to some extent, in their range of applicability.

In the "$k\varepsilon 1$" model, Equations (8.38)–(8.40) are solved with the constants assigned the following values:

C_μ	$C_{\varepsilon 2}$	$C_{\varepsilon 1}$	σ_k	σ_ε
0.09	1.92	1.43	1.0	1.13

These constants were found to produce satisfactory results for a variety of turbulent flows but they do not produce good agreement with data for the axisymmetric jet or, to a lesser extent, with coaxial jet data. To improve the axisymmetric jet predictions, Launder *et al.*,[25] following Rodi,[24] proposed the following modifications:

For axisymmetric flows—as for planar flows, except

$$C_\mu = 0.09 - 0.04f \tag{8.41}$$

$$C_{\varepsilon 2} = 1.92 - 0.0667f \tag{8.42}$$

$$f \equiv \left|\frac{Y_g}{2\Delta U}\left(\frac{dU_{c1}}{dx}-\left|\frac{dU_{c1}}{dx}\right|\right)\right| \tag{8.43}$$

In Equation (8.43) Y_g represents the width of the viscous region (shear layer) and ΔU represents the characteristic velocity difference across the viscous region. These terms are defined in Table 8.5; U_{c1} represents the mean velocity on the centerline of the flow, so that in the first regime of the jet (Figure 8.5) where $U_{c1} = \text{const}$, $f \equiv 0$. With this modification, the $k\varepsilon 1$

Table 8.5. Definition of Y_g Used in Launder et al.[26]

1. For a monotonically increasing or decreasing velocity profile

$$Y_g = y_2 - y_1$$

where

$$\frac{U - U_I}{U_E - U_I} = \begin{cases} 0.1 & \text{at } y_1 \\ 0.9 & \text{at } y_2 \end{cases}$$

where

U_I = axial velocity at inner boundary
U_E = axial velocity at outer boundary

2. For velocity profiles without a maximum or minimum at either boundary

$$Y_g = y_2 - y_1$$

where at y_1: $\dfrac{U - U_I}{U_E - U_I} = 0.1$ (for region of flow between the point of minimum or maximum velocity U_{max})

or at y_1: $\dfrac{U - U_{max}}{U_E - U_{max}} = 0.9$ (for outer region of flow)

and at y_2: $U = U_{max}$

model produces reasonably good predictions of both planar and axisymmetric flows. However, a further difficulty then appears: For weak shear flows, i.e., for cases in which the velocity excess or defect is a small fraction of the velocity of the external stream, the $k\varepsilon 1$ model fails to predict properly the asymptotic velocity decay rate. This problem is not unique to the $k\varepsilon 1$ model; both the Harsha[28,29] and Peters and Phares[31] models also encounter some difficulties in such flows.

In order to improve the prediction of weak shear flows, Rodi[24] proposed to make the constant C_μ a function of some suitable turbulence parameter. The ratio of production of turbulence energy (by the straining action of the mean flow) to the dissipation of energy into heat, G/ε appears to be a suitable parameter; G/ε can also be thought of as a measure of how close the flow is to local equilibrium, for which $G/\varepsilon = 1$. Rodi[24] thus chose to make C_μ a function of G/ε; but rather than relate C_μ directly to the local value of G/ε, he chose to assign a single value to a given axial station, so that $C_\mu = 0.09\, g(\overline{G/\varepsilon})$, where

$$\overline{G/\varepsilon} = \int_{Y_I}^{Y_E} \tau \frac{G}{\varepsilon} y^\alpha \, dy \Big/ \int_{Y_I}^{Y_E} \tau y^\alpha \, dy \tag{8.44}$$

and Y_I and Y_E represent the inner and outer edges of the viscous region,

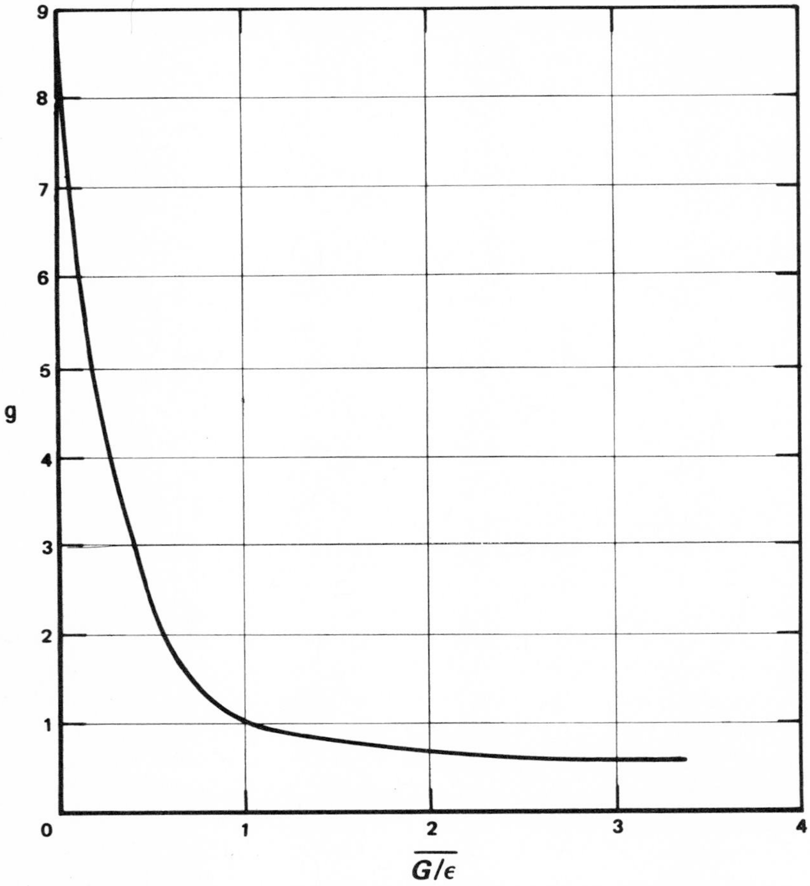

Fig. 8.13. Variation of g with $\overline{G/\varepsilon}$ for $k\varepsilon 2$ model.

respectively. The shear-weighting is used in Equation (8.44) to ensure that regions with large shear stress have a dominating influence on the value of $\overline{G/\varepsilon}$. The function $g(\overline{G/\varepsilon})$ is defined graphically in Figure 8.13, taken from Reference 26.

The model presented by Launder *et al.*[26] incorporating Equation (8.44) is termed the "$k\varepsilon 2$" model. The constants chosen by Launder for this model are

C_μ	$C_{\varepsilon 1}$	$C_{\varepsilon 2}$	σ_k	σ_ε
$0.09\, g(\overline{G/\varepsilon})$	1.40	1.94	1.0	1.0

for planar flows. For axisymmetric flows, a modification of these constants is again required, so that

$$C_\mu = 0.09\, g(\overline{G/\varepsilon}) - 0.0543 f \tag{8.45}$$

$$C_{\varepsilon 2} = 1.94 - 0.1336 f \tag{8.46}$$

with f defined as in Equation (8.43).

The $k\varepsilon 2$ model is shown by Launder *et al.*[26] to significantly improve the calculation of weak shear flows, although there are still problems encountered in the prediction of flows with high density gradients. It is worth noting that in the Harsha[28] and Peters and Phares[31] models special emphasis was placed on the modeling of flows with large density gradients, through the medium of the R_T function defined in Table 8.3. No such special modeling is incorporated in the models described by Launder[26] and the failure of these models to predict the high-density-gradient cases of Reference 57 may indicate that, at least at the level of modeling considered in kinetic energy approaches, the effects of density gradients cannot be included without additional empiricism. A comparison of both the $k\varepsilon 1$ and $k\varepsilon 2$ models with selected cases from the 1972 Langley Conference[57] and with the predictions of other models will be discussed in the following section.

Jones and Launder[23,61] describe the application of the basic $k\varepsilon$ model to the prediction of low Reynolds number[23] and "laminarizing"[61] boundary layers. Again, modifications to the basic formulation are required, in this case, in order to treat the region of the boundary layer very close to the wall. This region of a boundary layer is often handled by matching a computed velocity profile at some distance from the wall to the empirical law of the wall[62]; extensions of this procedure have been developed to handle compressible flows and flows with large temperature gradients. However, in some circumstances it is necessary to perform computations to the wall surface, either because experimental evidence is lacking or because the law of the wall is known in specific circumstances to be inaccurate. Near the wall the local Reynolds number becomes small, and thus the empirical functions devised to predict high-Reynolds-number flow can be expected to fail.

In order to extend the $k\varepsilon$ model to low-Reynolds-number flows, several modifications were introduced by Jones and Launder.[23,61] Because very close to the wall the turbulent length scale is diminished, molecular transport terms in the equation of motion that are neglected at high Reynolds numbers must be retained. The turbulence model itself must be altered: Some of the "constants" become functions of the local turbulence Reynolds number, taken to be defined by $\rho\nu_T/\mu$, or $\rho e^2/\mu\varepsilon$, and additional terms, which are negligible at high Reynolds number, are introduced into the equations for e and ε.

For low-Reynolds-number flows, Jones and Launder[23,61] write

$$C_\mu = 0.09\exp[-2.5/(1+R_T/50)] \tag{8.47}$$

$$C_{\varepsilon 2} = 2.0[1.0-0.3\exp(-R_T^2)] \tag{8.48}$$

All of the other constants retain the values appropriate to the $k\varepsilon 1$ model. Very near a wall the turbulent kinetic energy dissipation rate is set equal to

$$\text{dissipation} = 2\mu\left(\frac{\partial e^{1/2}}{\partial y}\right)^2 \tag{8.49}$$

while the term ε is set equal to zero at the wall. In the dissipation rate equation the term

$$-2\mu\nu_T\left(\frac{\partial^2 U}{\partial y^2}\right)^2$$

is added to improve the k profile in the region very near the wall. Equation (8.49) is introduced primarily to simplify the boundary conditions on the turbulent kinetic energy dissipation rate, since it has been found that the solution of the ε equation is very much simpler if the wall boundary condition is homogeneous. This term is added to allow the use of homogeneous boundary conditions for the energy dissipation rate equation while retaining the physical fact that $\varepsilon \neq 0$ at the wall in the turbulent kinetic energy equation.

With these assumptions, the turbulent kinetic energy and energy dissipation rate equations become

$$\rho U\frac{\partial e}{\partial x}+\rho V\frac{\partial e}{\partial y}=\frac{\partial}{\partial y}\left[\rho(\nu+\nu_T)\frac{\partial e}{\partial y}\right]+\rho\nu_T\left(\frac{\partial U}{\partial y}\right)^2-\rho\varepsilon-2\rho\nu\left[\frac{\partial(e^{1/2})}{\partial y}\right]^2 \tag{8.50}$$

$$\rho U\frac{\partial \varepsilon}{\partial x}+\rho V\frac{\partial \varepsilon}{\partial y}=\frac{\partial}{\partial y}\left[\rho(\nu+\nu_T)\frac{\partial \varepsilon}{\partial y}\right]+\frac{C_{\varepsilon 1}\rho\nu_T\varepsilon}{e}\left(\frac{\partial U}{\partial y}\right)^2-C_{\varepsilon 2}\frac{\rho\varepsilon^2}{e}-2\rho\nu\nu_T\left(\frac{\partial^2 U}{\partial y^2}\right)^2 \tag{8.51}$$

This model has been used to compute several low-Reynolds-number flows with good results. In these calculations, the mean velocity profile is matched to a laminar near-wall solution, rather than a form of the turbulent flow law of the wall. Detailed comparisons with experimental data are shown in References 23, 61, and 63.

8.3.6. Three-Equation Model: Hanjalic and Launder[37]

One of the models discussed in the paper by Launder *et al.*[26] represents an attempt at a higher-order turbulent kinetic energy equation closure. This model, also described by Hanjalic and Launder[37] and termed the $\overline{uv}k\varepsilon$

model in Reference 26, involves the simultaneous solution of transport equations for the turbulent kinetic energy, the turbulent shear stress, and the turbulent kinetic energy dissipation rate. An explicit turbulence model in the form of a relation between τ and e is thus avoided; however, the Prandtl–Kolmogorov[20,21] model does appear implicitly in the modeling of the terms in the model transport equations. The set of equations to be solved to obtain the shear stress in the $\overline{uv}k\varepsilon$ model comprises:

1. An equation for the Reynolds shear stress $(\tau/\rho)=\overline{uv}$,

$$U\frac{\partial \overline{uv}}{\partial x}+V\frac{\partial \overline{uv}}{\partial y}=C_s\frac{\partial}{\partial y}\left(\frac{e^2}{\varepsilon}\frac{\partial \overline{uv}}{\partial y}\right)-C_{\phi 1}\left(\frac{\overline{uv}\varepsilon}{e}+C_\mu e\frac{\partial U}{\partial y}\right) \tag{8.52}$$

2. For the turbulent kinetic energy, e

$$U\frac{\partial e}{\partial x}+V\frac{\partial e}{\partial y}=0.9\,C_s\frac{\partial}{\partial y}\left(\frac{e^2}{\varepsilon}\frac{\partial e}{\partial y}\right)-\overline{uv}\frac{\partial U}{\partial y}-\varepsilon \tag{8.53}$$

3. For the energy dissipation rate, ε

$$U\frac{\partial \varepsilon}{\partial x}+V\frac{\partial \varepsilon}{\partial y}=C_\varepsilon\frac{\partial}{\partial y}\left(\frac{e^2}{\varepsilon}\frac{\partial \varepsilon}{\partial y}\right)-C_{\varepsilon 1}\frac{\overline{uv}\varepsilon}{e}\frac{\partial U}{\partial y}-C_{\varepsilon 2}\frac{\varepsilon^2}{e} \tag{8.54}$$

Note that these equations are written for a plane two-dimensional incompressible flow; also, implicit in them is the assumption that the "Prandtl numbers" for $\overline{uv}$, e, and ε are all unity. Launder[26] reports the following constant for the $\overline{uv}k\varepsilon$ model:

C_s	$C_{\phi 1}$	C_μ	C_ε	$C_{\varepsilon 1}$	$C_{\varepsilon 2}$
0.1	2.8	0.09	0.09	1.40	1.95

8.3.7. Comparison of Turbulence-Model Predictions with Free Shear Layer Data

The 1972 NASA–Langley Conference on Free Shear Flows[57] provided an opportunity for the testing of the predictions of a number of models for turbulent flow against a set of reasonably well-documented data. In this section the predictions of those turbulent kinetic energy models described in this article that were presented at that conference are summarized and compared. Only one model (Peters and Phares[31]) was able to adequately predict the variation in two-dimensional shear layer spread rate with Mach number: all models adequately predicted the variation in the spread rate of a two-dimensional incompressible shear layer with velocity ratio. The variation in spread rate with density ratio for a shear layer between streams of different gases as a function of density ratio is at present poorly understood,

Table 8.6. Turbulent Kinetic Energy Models

Author	Equations: Convection	Diffusion	Production	Dissipation	Shear model
Harsha[29]	$\rho\frac{De}{Dt}$	$\frac{1}{y^\alpha}\frac{\partial}{\partial y}\left(\frac{\rho\nu_T y^\alpha}{\sigma_k}\frac{\partial e}{\partial y}\right)$	$\rho\nu_T\left(\frac{\partial U}{\partial y}\right)^2$	$-a_2\frac{\rho e^{3/2}}{l_k}$	ND
Launder *et al.*[26]					
k	$\rho\frac{De}{Dt}$	$\frac{1}{y^\alpha}\frac{\partial}{\partial y}\left(\frac{\rho\nu_T y^\alpha}{\sigma_k}\frac{\partial e}{\partial y}\right)$	$\rho\nu_T\left(\frac{\partial U}{\partial y}\right)^2$	$-\rho\frac{e^{3/2}}{l_k}$	PK
$k\varepsilon 1$	$\rho\frac{De}{Dt}$	$\frac{1}{y^\alpha}\frac{\partial}{\partial y}\left(\frac{\rho\nu_T y^\alpha}{\sigma_k}\frac{\partial e}{\partial y}\right)$	$\rho\nu_T\left(\frac{\partial U}{\partial y}\right)^2$	$-\rho\varepsilon$	PK
	$\rho\frac{D\varepsilon}{Dt}$	$\frac{1}{y^\alpha}\frac{\partial}{\partial y}\left(\frac{\rho\nu_T y^\alpha}{\sigma_\varepsilon}\frac{\partial \varepsilon}{\partial y}\right)$	$C_{\varepsilon 1}\frac{\varepsilon}{e}\rho\nu_T\left(\frac{\partial U}{\partial y}\right)^2$	$-C_{\varepsilon 2}\frac{\rho\varepsilon^2}{e}$	
$k\varepsilon 2$	as $k\varepsilon 1$	as $k\varepsilon$ 1	as $k\varepsilon 1$	as $k\varepsilon 1$	as $k\varepsilon 1$
$\overline{uv}k\varepsilon$	$\frac{De}{Dt}$	$0.9\,C_s\frac{\partial}{\partial y}\left(\frac{e^2}{\varepsilon}\frac{\partial e}{\partial y}\right)$	$-\overline{uv}\frac{\partial U}{\partial y}$	$-\varepsilon$	None
	$\frac{D\overline{uv}}{Dt}$	$C_s\frac{\partial}{\partial y}\left(\frac{e^2}{\varepsilon}\frac{\partial \overline{uv}}{\partial y}\right)$	$-C_\mu C_{\phi 1}e\frac{\partial U}{\partial y}$	$-C_{\phi 1}\frac{\overline{uv}\varepsilon}{e}$	
	$\frac{D\varepsilon}{Dt}$	$C_\varepsilon\frac{\partial}{\partial y}\left(\frac{e^2}{\varepsilon}\frac{\partial \varepsilon}{\partial y}\right)$	$-C_{\varepsilon 1}\frac{\overline{uv}\varepsilon}{e}\frac{\partial U}{\partial y}$	$-C_{\varepsilon 2}\frac{\varepsilon^2}{e}$	
Morel and Torda[15]	$\frac{De}{Dt}$	$-\frac{\partial}{\partial y}\left(G\frac{\tau}{\rho}\left\|\frac{\tau_{max}}{\rho}\right\|^{1/2}\right)$	$2a_1\frac{\tau\partial U}{\partial y}$	$\frac{(\tau/\rho)^{3/2}}{L}$	ND
Peters and Phares[31]	$\int\frac{De}{Dt}y^\alpha\,dy$	None	$\int\tau\frac{\partial U}{\partial y}y^\alpha\,dy$	$-a_2\int\frac{\rho e^{3/2}}{l_k}y^\alpha\,dy$	ND

Table 8.7. Constants in Turbulent Kinetic Energy Models

Author	Variable	Constants: Diffusion	Production	Dissipation	Length scale	Shear model
Harsha[(29)]	e	$\sigma_k = 0.07$	—	Equation (8.28)	Table 8.2	Equation (8.27)
Launder *et al.*[(26)]						
k	e	$\sigma_k = 0.7$	—	—	Table 8.5[a]	$C_\mu = 0.08$
$k\varepsilon 1$	e	$\sigma_k = 1.0$	—	—	—	$C_\mu = 0.09$ Plane flows
	ε	$\sigma_\varepsilon = 1.3$	$C_{\varepsilon 1} = 1.43$	$C_{\varepsilon 2} = 1.92$[b]	—	Equation (8.41) Axisymmetric
$k\varepsilon 2$	e	$\sigma_k = 1.0$	—	—	—	$C_\mu = 0.09\, g(\overline{G/\varepsilon})$ Figure 8.13
	ε	$\sigma_G = 1.3$	$C_{\varepsilon 1} = 1.40$	$C_{\varepsilon 2} = 1.94$[c]	—	Equation (8.45) Axisymmetric
$\overline{uv}k\varepsilon$	e	$C_s = 0.09$	—	—	—	—
	$\overline{uv}$	$C_s = 0.10$	$C_\mu C_{\phi 1} = 0.252$	$C_{\phi 1} = 2.5$	—	—
	ε	$C_\varepsilon = 0.07$	$C_{\varepsilon 1} = 1.40$	$C_{\varepsilon 2} = 1.95$	—	—
	e	$g(y/\delta)$	—	—	$L(y/\delta)$	See Reference 14
Peters and Phares[(31)]	e	—	—	Equation (8.28)	Table 8.2[d]	Equation (8.27)

[a] $l_k = 0.625\, Y_g$ (axisymmetric flows), $l_k = 0.875\, Y_g$ (plane flows); Y_g from Table 8.5.
[b] For plane flows. Equation (8.42) for axisymmetric flows.
[c] For plane flows. Equation (8.46) for axisymmetric flows.
[d] Fully developed flows only.

either experimentally or theoretically. Thus the predictions to be summarized in this section will be concerned primarily with two-dimensional and axisymmetric jets and wakes.

The features of the turbulent flow models as applied to the prediction of free shear flows in Reference 57 are summarized in Table 8.6, with the constants used in making the calculation summarized in Table 8.7. Reference should be made to the original papers describing the models for further details regarding the constants used and their limits of applicability.

One of the most readily predicted flows of the test data group used at the Langley conference was the subsonic circular jet. Even so, substantial differences exist in the predictions of three turbulent kinetic energy models, as shown in Figure 8.14. The most accurate prediction is by the one-equation model described by Harsha,(28) while the $k\varepsilon 2$ model of Launder(26) produces a slightly lower prediction and the k model presented by the same investigators produces a velocity decay prediction that is slightly too high. It should be noted in this and subsequent comparison that both the Harsha(26) and the Peters and Phares(31) models (both of which produce substantially the same prediction for this flow, see Figure 8.7) begin at the initial station specified for the computation by the organizers of the Langley conference, without any attempt at optimization of the initial profiles of turbulent kinetic

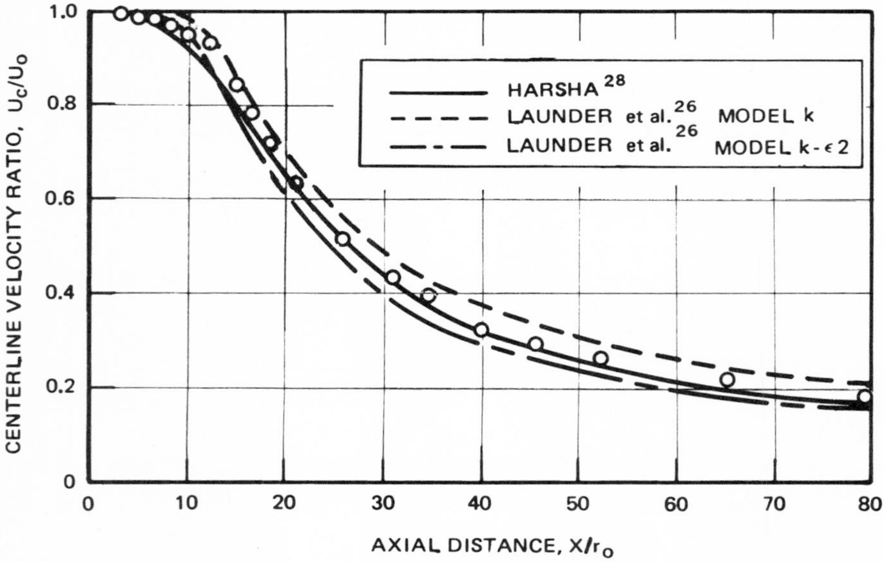

Fig. 8.14. Comparison of one-equation and two-equation TKE models; $M_0 = 0.64$ circular jet, NASA case 6.(53)

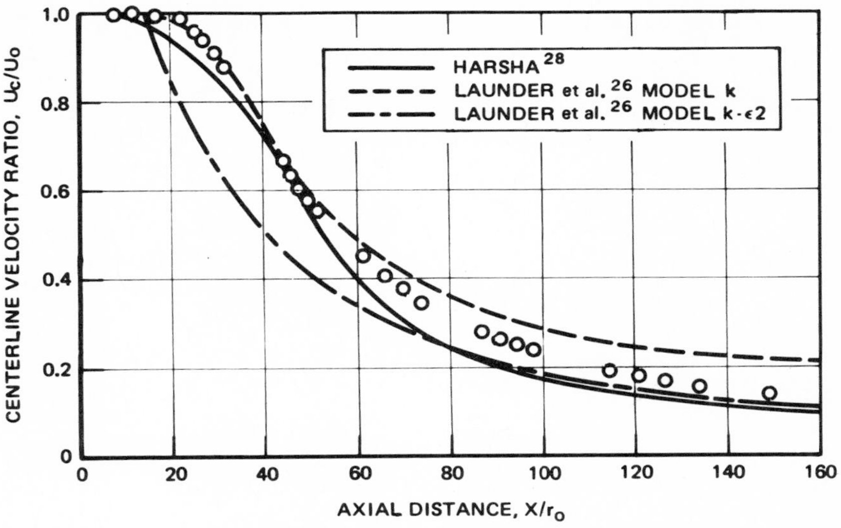

Fig. 8.15. Comparison of one-equation and two-equation TKE models; $M_0 = 2.2$ circular jet, NASA case 7.[53]

energy, which are estimated for each flow by a simple technique.[28,31] The predictions of Launder[26] are obtained after an adjustment of the estimated initial conditions is performed to obtain a potential core (or first regime) length approximately the same as shown by the experiment. To this extent, the calculations of Launder are correlations rather than true predictions (which should be carried out as if the experimental data were unknown).

Predictions of the centerline velocity decay of a supersonic circular jet are shown in Figure 8.15. Except for the Peters and Phares[31] integral technique (Figure 8.8), none of the turbulent kinetic energy model predictions of this flow are highly accurate. The accuracy of the integral model prediction is probably related to its accuracy in predicting the spread rate of a supersonic two-dimensional shear layer, since the effects of compressibility would be expected to be similar in both cases.

Both the simple model described by Harsha[28] and the two-equation $k\varepsilon 2$ model of Launder[26] provide accurate predictions of the incompressible coaxial jet flow, shown in Figure 8.16. Indeed, the Harsha model and the $k\varepsilon 2$ model prediction are virtually indistinguishable for this flow. Note, however, that the Launder model k prediction (which uses an algebraic specification of the dissipation length scale, l_k) underpredicts the velocity decay rate observed in this flow.

All of the kinetic energy models were able to predict the centerline velocity and jet species concentration decay observed in the supersonic

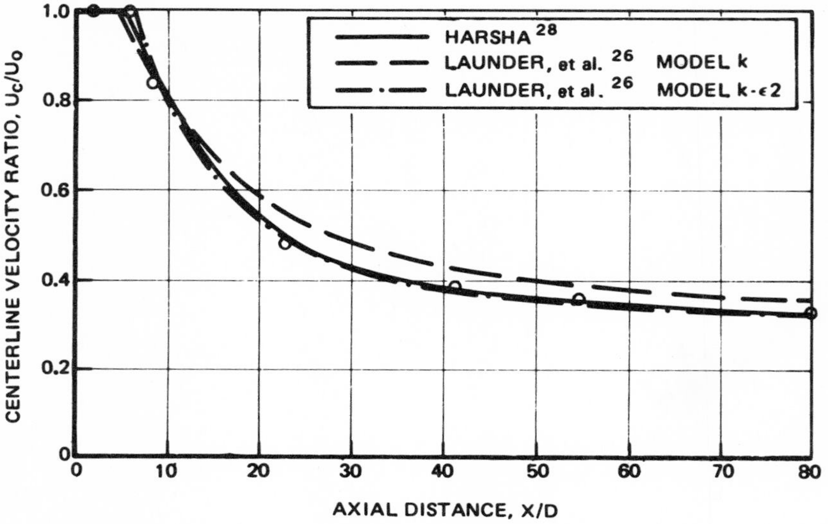

Fig. 8.16. Comparison of one-equation and two-equation TKE models; axisymmetric jet in moving stream, $U_e/U_0 = 0.25$, NASA case 9.[53]

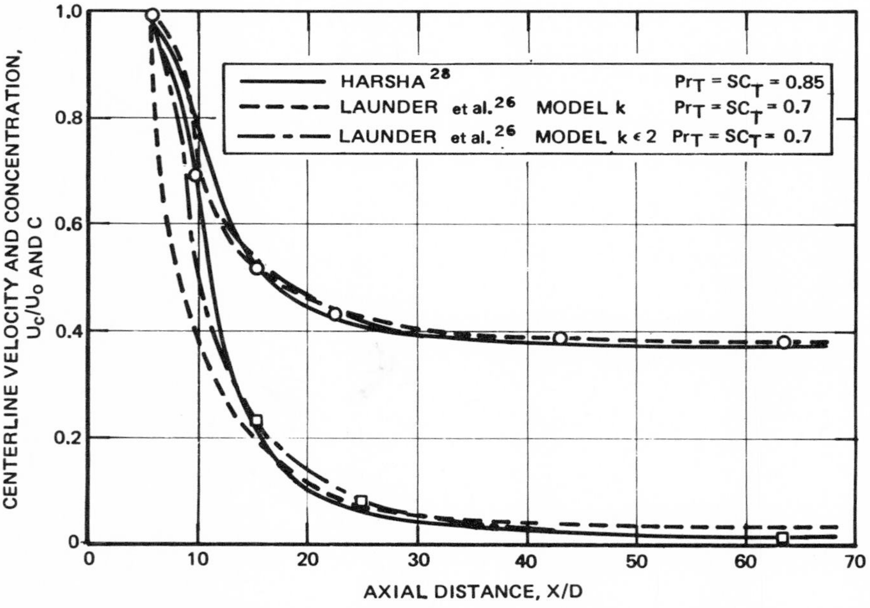

Fig. 8.17. Comparison of one-equation and two-equation TKE models; H_2 jet in moving stream, $M_e = 1.33$, NASA case 12.[53] ○, U_c/U_0; □, c.

hydrogen–air jet experiment used as test case 12, Figure 8.17. This flow had a strongly wakelike character, which accounts for the very rapid decay of jet species, and which resulted in a relatively strong shear field. The strong shear field probably accounts for the accuracy of the prediction of all of the models shown, despite the large density gradient, which would have been expected to cause some problems in the predictions in a flow with weaker shear.

Weak shear flows are generally among the most difficult to predict. The comparison of the one- and two-equation model predictions of the axisymmetric incompressible wake, Figure 8.18, shows that in this weak shear flow, only the Launder $k\varepsilon 2$ model, which has been specifically developed for weak shear flows, provides an accurate prediction. However, the method of plotting this particular figure emphasizes small deviations from the experiment, so that both the Harsha and Peters and Phares models provide a fair prediction of this case (see Figure 8.11) if centerline velocity increase is considered rather than the similarity parameter $1/w^{3/2}$.

The two-dimensional two-stream jet, Figure 8.19, provides another example of a weak shear flow. Here both the Harsha[(28)] model and the Launder[(26)] k model, each of which relies on an algebraic specification of the dissipation length scale, fail to accurately predict the rate of decay of centerline velocity in the early part of the flow. (It might be noted that a later prediction[(29)] using the Harsha model exhibits improved agreement with the experimental data for this case, obtained by a variation in the modeling of the dissipation term. Details of the improved model are given in Reference

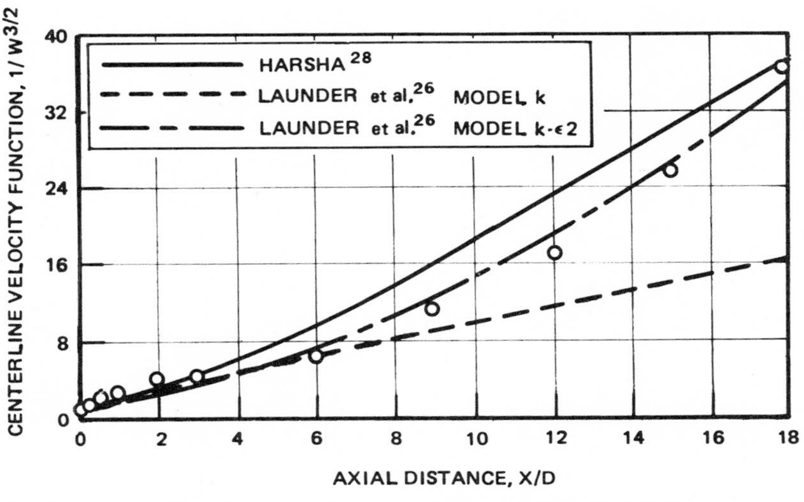

Fig. 8.18. Comparison of one-equation and two-equation models; incompressible axisymmetric wake, NASA case 15.[(53)]

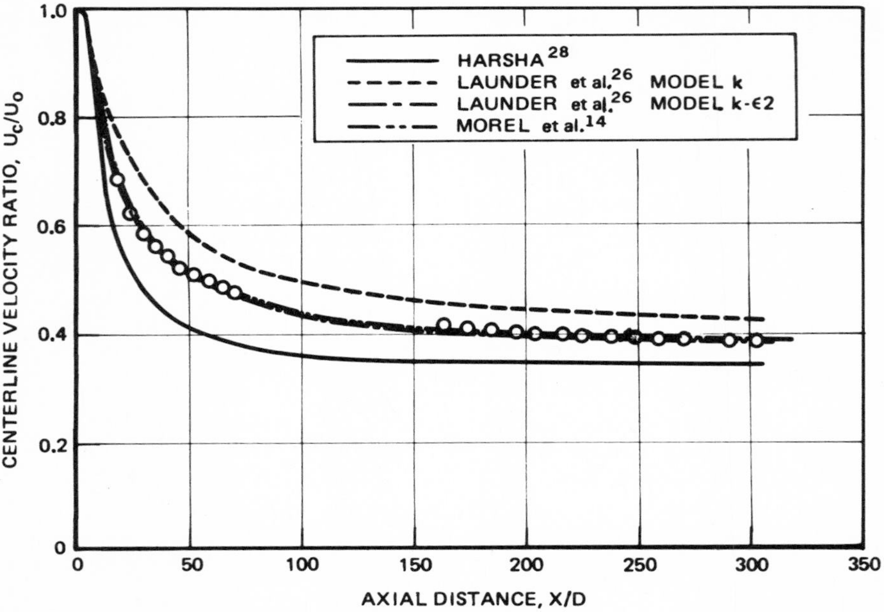

Fig. 8.19. Comparison of one-equation and two-equation TKE models; two-dimensional jet in moving stream, $U_0/U_e = 3.29$, NASA case 13.[53]

29.) Once again, as shown in Figure 8.19, the $k\varepsilon 2$ model provides an excellent prediction of the experimental data, as does the interaction model of Morel *et al.*[14] The Morel model has been optimized for this case, however.

It is clear from Figure 8.20 that the incompressible two-dimensional wake is one of the most difficult flows to properly predict. Indeed, a number of predictors at the 1972 NASA–Langley Conference indicated that their models had difficulty predicting this flow; the difficulties involved in turbulent kinetic energy model predictions of this flow are clearly evident from Figure 8.20. The chief difficulty seems to be in predicting the asymptotic rate of decay of the wake, which again involves the weak-shear problem. Again, the Launder[26] $k\varepsilon 2$ model shows a relatively good prediction of this flow. (Launder notes that the momentum deficit measured in the wake at $X/D \simeq 260$ may be substantially larger than at other stations and thus discounts this point.) It can also be seen that the $\overline{uv}k\varepsilon$ model, which does not include an explicit weak shear flow correction, provides a good prediction of these data. Finally, none of the models that depend on an algebraic length scale formulation seems capable of accurate prediction of this flow.

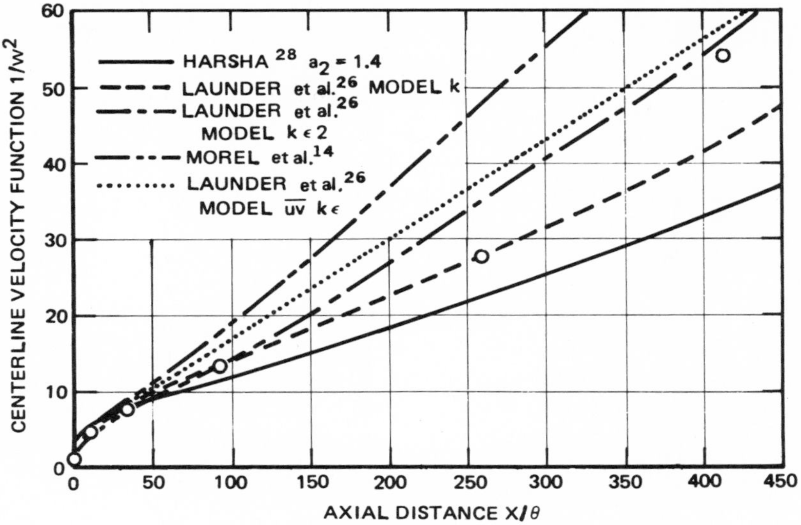

Fig. 8.20. Comparison of one-equation and two-equation models; incompressible plane wake, NASA case 14.[(53)] θ is the initial wake momentum thickness.[(57)]

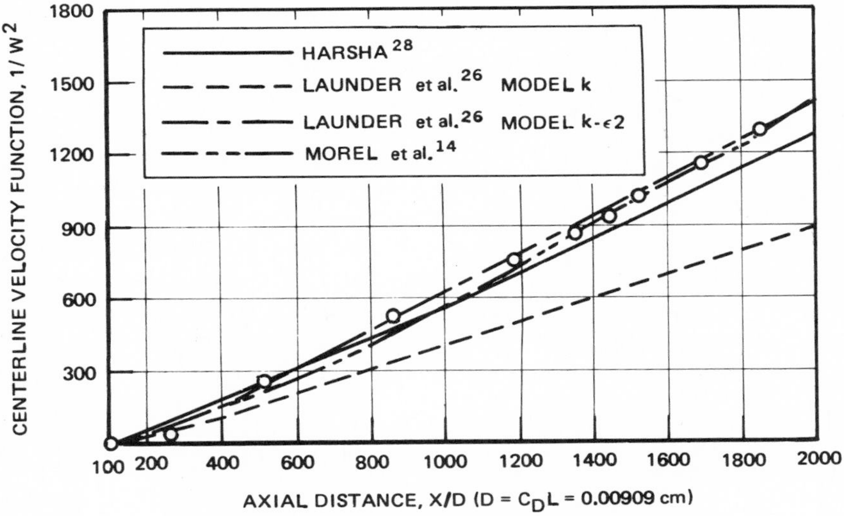

Fig. 8.21. Comparison of one-equation and two-equation models; planar wake, $M = 2.99$, NASA case 16.[(53)]

Interestingly, the situation is not nearly so bad when predictions of the plane compressible wake are considered, Figure 8.21. Besides the Launder[26] $k\varepsilon 2$ model, the algebraic length scale models of both Morel[14] and Harsha[28] do a reasonably good job of predicting this flow. The only poor predictions are those shown in the Launder k model, for which no attempt at generalization of constant was made.

8.4. Summary and Conclusions

It is clear from the number of theoretical models of turbulent flows described in this chapter that a variety of related approaches, all of which involve the use of the turbulent kinetic energy equation, now exist for the computation of turbulent flow phenomena. The results of the 1968 Stanford Conference[41] on turbulent boundary layers and the 1972 NASA–Langley Conference[57] on free turbulent shear layers, taken together, clearly show the considerable increase in generality afforded by turbulent kinetic energy models over the classic locally dependent eddy viscosity approach. A sufficient number of proved turbulent kinetic energy models now exist so that in approaching a given problem the question to answer becomes "what turbulent energy model?" rather than "should a turbulent energy model be chosen?"

It should also be clear from the preceding discussion that to some extent the choice of turbulence energy model should depend on the problem being attacked. The one-equation models presented by Harsha[28,29] and Peters and Phares[31] are capable of providing a good prediction of any strong shear boundary-layer type flow, with a consistent formalism applied to all flows. On the other hand, the one-equation version of the PK model (Launder model k) does not show the same generality as the simple models of Harsha[28,29] and Peters and Phares.[31] This is perhaps related to the shear model used: In the PK model the turbulent length scale appears in the shear-stress definition and is thus explicit in all three terms on the right-hand side of the turbulent kinetic energy equation, while in the ND model the length scale appears explicitly only in the turbulent kinetic energy dissipation term. Methods that use the PK shear stress relation, Equation (8.13), may thus be expected to be more sensitive to length scale specifications than approaches that use the ND model, Equation (8.11).

Of the two-equation models, the Launder[26] $k\varepsilon 2$ model, which incorporates the weak-shear correction embodied in the $\overline{G/\varepsilon}$ ratio concept [Equation (8.44)], has been brought to the highest state of development. This model is particularly useful in non-boundary-layer flows in which length-scale distributions cannot be specified *a priori*. However, in its present form it is not capable of predicting the effects of compressibility or of large density ratio with reliability; further developmental work remains necessary in this area.

Finally, integral versions of turbulent kinetic energy models are often overlooked because the original advantage of integral models, speed of computation, is no longer of very great importance. Indeed, the faster finite-difference computational techniques are often as fast as integral methods. However, integral methods have another advantage. By introducing known empirical data into the solution of a given problem they allow attention to be focused on the unknown part of the problem. Thus, for example, the introduction of empirical turbulent kinetic energy profile in the Peters and Phares[31] model eliminated the turbulent kinetic energy diffusion problem and allowed attention to be directed to the modeling of the turbulent kinetic energy dissipation rate. This in turn allowed the generalization of the model to predict high Mach number and high density ratio phenomena: the Peters and Phares model was the only model at the Langley Conference[57] that was capable of predicting the effects of flow Mach number on the growth rate of a turbulent shear layer. In problems in which some information is reliably known—for example, the law of the wall velocity profiles in wall boundary layers or the cosine- or exponential-form velocity profiles in free jets—inclusion of this information in the model simplifies the problem and provides a known foundation on which to build models of the unknown phenomena.

ACKNOWLEDGMENT

This work was done with partial support of Air Force Office of Scientific Research, Dr. B. T. Wolfson, Project Monitor.

Notation

A, B, C	Constants (may carry subscripts)
D	Decay term
e	Kinetic energy of turbulence $\langle u_k u_k \rangle/2$
f	Function defined in Equation (8.43)
F	Transportable scalar
g	Function defined in Figure 8.13
G	Production term
L	Length scale
l_k	Turbulence length scale
n	$\nu + \nu_T$
p'	Pressure of fluctuation
P	Pressure
r	Radial coordinate
R_T	Function defined in Table 8.3
t	Time
T_{ij}	Viscous stress term
U_i	Mean velocity components
u_i	Fluctuating velocity components
U_∞	Free-stream velocity
ω^*	Mean-square vorticity
x_i	Rectangular coordinates
Y_g	Width of viscous region
$\langle \, \rangle$	Mean value operation
$\alpha, \beta, \gamma, \sigma$	Constants (may take subscripts)
δ_{ij}	Kronecker delta symbol

ε	Kinetic energy dissipation rate
ν	Kinematic viscosity
ν_T	Turbulent eddy viscosity
ε^*	Kinetic energy of the motion due to ω^*
$\phi_{F,j}$	Flux of F due to diffusion
ρ	Density
τ	Shear stress
θ	Boundary-layer momentum thickness
$(\,')$	Denotes turbulence intensities $u' = \sqrt{\langle u^2 \rangle}$
$(\bar{\ })$	Time average or Reynolds stress average

References

1. Brown, G., and Roshko, A., The effect of density difference on the turbulent mixing layer, in: *Turbulent Shear Flows*, North Atlantic Treaty Organization, Advisory Group for Aerospace Research and Development, report No. CP-93 (1972).
2. Kovasznay, L. S. G., Turbulent shear flow, presented at Convegno Sulla Teoria Della Turbulenza, Roma, Italia, April 26–29, 1970.
3. Liu, J. T. C., Developing large scale wavelike eddies and the near jet noise field, *J. Fluid Mech.* **62**, 437–464 (1974).
4. Mellor, G. L., and Herring, H. J., A survey of the mean turbulent field closure models, *AIAA J.* **11**, 590–599 (1973).
5. Prandtl, L., Bericht über Untersuchungen zue ausgebildeten Turbulenz, *Z. Angew. Math. Mech.* **5**, 136–139 (1925).
6. Prandtl, L., Bemerkungen zur Theorie der freien Turbulenz, *Z. Angew. Math. Mech.* **22**, 241–243 (1942).
7. Schetz, J., Turbulent mixing of a jet in a coflowing stream, *AIAA J.* **6**, 2008–2010 (1968).
8. Tufts, L. W., and Smoot, L. D., A turbulent mixing coefficient correlation for coaxial jets with and without secondary flows, *J. Spacecr. Rockets* **8**, 1183–1190 (1971).
9. Zelazny, S. W., Morgenthaler, J. H., and Herendeen, D. L., Shear stress and turbulence intensity models for coflowing axisymmetric streams, *AIAA J.* **11**, 1165–1173 (1973).
10. Nee, V. W., and Kovasznay, L. S. G., Simple phenomenological theory of turbulent shear flows, *Phys. Fluids* **12**, 473–484 (1969).
11. Saffman, P. G., A model for inhomogeneous turbulent flows, *Proc. R. Soc. London, Series A* **317**, 417–433 (1970).
12. Bradshaw, P., Ferriss, D. H., and Atwell, N. P., Calculation of boundary-layer development using the turbulent energy equation, *J. Fluid Mech.* **28**, 593–616 (1967).
13. Saffman, P. G., and Wilcox, D. C., Turbulence—Model predictions for turbulent boundary layers, *AIAA J.* **12**, 541–546 (1974).
14. Morel, T., Torda, T. P., and Bradshaw, P., Turbulent kinetic energy equation and free mixing, in: *Free Turbulent Shear Flows*, Vol. 1, *Conference Proceedings*, NASA report No. SP-321 (1973), pp. 549–573.
15. Morel, T., and Torda, T. P., Calculation of free turbulent mixing by the interaction approach, *AIAA J.* **12**, 533–540 (1974).
16. Glushko, G. S., Turbulent boundary layer on a flat plate in an incompressible fluid, *Izv. Akad. Nauk SSSR Mekh.* **4**, 13–23 (1965) [English translation available from DDC as report No. AD 638 204 (1966), also NASA report No. TT-F-10,080].
17. Beckwith, I. E., and Bushnell, D. M., Calculation of mean and fluctuating properties of the incompressible turbulent boundary layer, in: *Proceedings of the AFOSR–IFP–Stanford Conference*, Vol. 1 (S. J. Kline, M. V. Morkovin, G. Sovran, and D. J. Cockrell, eds.), Stanford University, Stanford, California (1969), pp. 275–299.

18. Mellor, G. L., and Herring, H. J., Two methods of calculating turbulent boundary layer behavior based on numerical solutions of the equations of motion, in: *Proceedings of the AFOSR-IFP-Stanford Conference*, Vol. 1 (S. J. Kilne, M. V. Morkovin, G. Sovran, and D. J. Cockrell, eds.), Stanford University, Stanford, California (1969), pp. 331–345.
19. Ng, K. H., and Spalding, D. B., Turbulence model for boundary layers near walls, *Phys. Fluids* **15**, 20–30 (1972).
20. Kolmogorov, A. N., The equations of turbulent motion in an incompressible fluid, *Izv. Acad. Nauk SSSR Ser. Fiz.* **6**, 56–58 (1942) [English translation: report No. ON/6 Imperial College Department of Mechanical Engineering (1968)].
21. Prandtl, L., and Wieghardt, K., Über ein neues Formelsystem für die ausgebildete Turbulence, *Nach. Akad. Wiss. Göttingen Math. Phys. Kl.* 6–19 (1945).
22. Rotta, J. C., Statistische Theorie nichthomogener Turbulenz, *Z. Phys.* **129**, 547–572 (1951) see also *ibid.* **131**, 51–77 (1951).
23. Jones, W. P., and Launder, B. E., The calculation of low Reynolds number phenomena with a two-equation model of turbulence, *Int. J. Heat Mass Transfer* **16**, 1119–1130 (1973).
24. Rodi, W., The prediction of free turbulent boundary layers by use of a two-equation model of turbulence, Ph.D. dissertation, University of London (1972).
25. Rodi, W., and Spalding, D. B., A two-parameter model of turbulence and its application to free jets, *Wärme Stoffübertrag.* **3**, 85–95 (1970).
26. Launder, B. E., Morse, A., Rodi, W., and Spalding, D. B., Prediction of free shear flows—A comparison of the performance of six turbulence models, in: *Free Turbulent Shear Flows*, Vol. I, *Conference Proceedings*, NASA report No. SP-321 (1973), pp. 361–422.
27. Lee, S. C., and Harsha, P. T., Use of turbulent kinetic energy in free mixing studies, *AIAA J.* **8**, 1026–1032 (1970).
28. Harsha, P. T., Prediction of free turbulent mixing using a turbulent kinetic energy method, in: *Free Turbulent Shear Flows*, Vol. I, *Conference Proceedings*, NASA report No. SP-321 (1973), pp. 463–519.
29. Harsha, P. T., A general analysis of free turbulent mixing, AEDC report No. TR-73-177 (1974).
30. Patel, V. C., and Head, M. R., A simplified version of Bradshaw's method for calculating two-dimensional turbulent boundary layers, *Aeronaut. Quart.* **21**, 243–262 (1970).
31. Peters, C. E., and Phares, W. J., An integral turbulent kinetic energy analysis of free shear flows, in: *Free Turbulent Shear Flows*, Vol. I, *Conference Proceedings*, NASA report No. SP-321 (1973), pp. 577–624.
32. Chou, P. Y., On velocity correlations and the solution of the equations of turbulent fluctuation, *Quart. J. Appl. Math.* **3**, 38–54 (1945).
33. Rotta, J. C., Recent attempts to develop a generally applicable calculation method for turbulent shear flow layers, North Atlantic Treaty Organization, Advisory Group for Aerospace Research and Development, report No. CP-93 (1972).
34. Daly, B. J., and Harlow, F. H., Transport equations in turbulence, *Phys. Fluids* **13**, 2634–2649 (1970).
35. Donaldson, C. du P., A progress report on an attempt to construct an invariant model of turbulent shear flows, North Atlantic Treaty Organization, Advisory Group for Aerospace Research and Development, report No. CP-93 (1972).
36. Lewellen, W. S., Teske, M., and Donaldson, C. duP., Application of turbulence model equations to axisymmetric wakes, *AIAA J.* **12**, 620–625 (1974).
37. Hanjalic, K., and Launder, B. E., Fully developed asymmetric flow in a plane channel, *J. Fluid Mech.* **51**, 301–335 (1972).

38. Reynolds, W. C., Computation of turbulent flow, AIAA Paper No. 74–556, AIAA 7th Fluid and Plasmadynamics Conference, June 17–19, 1974.
39. Yen, J. T., Kinetic theory of turbulent flow, *Phys. Fluids* **15**, 1728–1734 (1972).
40. Nee, V. W., and Kovasznay, L. S. G., The calculation of the incompressible turbulent boundary layers by a simple theory, in: *Proceedings of the AFOSR–IFP–Stanford Conference*, Vol. 1 (S. J. Kline, M. V. Morkovin, G. Sovran, and D. J. Cockrell, eds.), Stanford University, Stanford, California (1969), pp. 300–320.
41. Kline, S. J., Morkovin, M. V., Sovran, G., and Cockrell, D. J., eds., *Proceedings of the AFOSR–IFP–Stanford Conference on Computation of Turbulent Boundary Layers*, Thermosciences Div., Mechanical Engineering Department, Stanford University, Stanford, California (1969).
42. Nevzgljadov, V., A phenomenological theory of turbulence, *J. Phys.* USSR **9**, 235–243 (1945).
43. Dryden, H. L., Recent advances in the mechanics of boundary layer flow, in: *Advances in Applied Mechanics*, Vol. 1 (R. Von Mises and T. von Kármán, eds.), Academic Press, New York (1948), pp. 1–40.
44. Lee, S. C., Harsha, P. T., Auiler, J. E., and Lin, C. L., Heat mass and momentum transfer in free turbulent mixing, in: *Proceedings of the 1972 Heat Transfer and Fluid Mechanics Institute* (R. B. Landis and G. J. Hardeman, eds.), Stanford University Press, Stanford, California (1972), pp. 215–230.
45. Mikatarian, R. R., and Benefield, J. W., Turbulence in chemical lasers, AIAA Paper No. 74–148, AIAA 12th Aerospace Sciences Meeting, January 30–February 1, 1974.
46. Rhodes, R. P., Harsha, P. T., and Peters, C. E., Turbulent kinetic energy analyses of hydrogen–air diffusion flames, *Acta Astronaut.* **1**, 443–470 (1974).
47. Runchal, A. K., and Spalding, D. B., Steady turbulent flow and heat transfer downstream of a sudden enlargement in a pipe of circular cross section, *Wärme Stoffübertrag.* **5**, 31–38 (1972).
48. Bradshaw, P., and Ferriss, D. H., Applications of a general method of calculating turbulent shear layers, *J. Basic Eng. Trans. ASME* **94**, 345–352 (1972).
49. Townsend, A. A., *The Structure of Turbulent Shear Flow*, Cambridge University Press, Cambridge, England (1956).
50. Bradshaw, P. and Ferriss, D. H., Calculation of boundary layer development using the turbulent energy equation: Compressible flow on adiabatic walls, *J. Fluid Mech.* **46**, 83–110 (1971).
51. Bradshaw, P., and Ferriss, D. H., Calculation of boundary layer development using the turbulent energy equation: IV. Heat transfer with small temperature differences, report No. Aero 1271, National Physical Laboratory, Teddington (1968).
52. Bradshaw, P., Calculation of three-dimensional turbulent boundary layers, *J. Fluid Mech.* **46**, 417–445 (1971).
53. Bradshaw, P., Dean, R. B., and McEligot, D. M., Calculation of interacting turbulent shear layers—Duct flow, *J. Fluids Eng. Trans. ASME* **95**, 214–220 (1973).
54. Thompson, B. G. J., A new two parameter family of mean velocity profiles for incompressible turbulent boundary layers on smooth walls, R & M report No. 3463, Aeronautical Research Council (1965).
55. Bradshaw, P., The turbulence structure of equilibrium boundary layers, *J. Fluid Mech.* **29**, 625–645 (1967).
56. Harsha, P. T., and Lee, S. C., Correlation between turbulent shear stress and turbulent kinetic energy, *AIAA J.* **8**, 1508–1510 (1970).
57. *Free Turbulent Shear Flows*, Vol. I, *Conference Proceedings*, Vol. II, *Summary of Data*, NASA, Langley Research Center, Hampton, Virginia, NASA report No. SP 321 (1973).

58. Wygnanski, I., and Fiedler, H., Some measurements in the self-preserving jet, *J. Fluid Mech.* **38**, 577–612 (1969).
59. Albertson, M. L., Dai, Y. B., Jensen, R. A., and Rouse, H., Diffusion of submerged jets, paper No. 2409, *Trans. ASCE* **115**, 639–697 (1950).
60. Launder, B. E., and Spalding, D. B., *Lectures in Mathematical Models of Turbulence*, Academic Press, New York (1972).
61. Jones, W. P., and Launder, B. E., The prediction of laminarization with a two-equation model of turbulence, *Int. J. Heat Mass Transfer* **15**, 301–314 (1972).
62. Lin, C. C., ed., *Turbulent Flows and Heat Transfer*, Princeton University Press, Princeton, New Jersey (1959), p. 132.
63. Launder, B. E., and Spalding, D. B., The numerical computations of turbulent flows, report No. HTS/73/2, Imperial College, Department of Mechanical Engineering (1973).

CHAPTER 9

Use of Invariant Modeling

W. S. LEWELLEN

9.1. Introduction

Previous chapters (7 and 8) have contained several different methods for calculating turbulent flow with varying degrees of complexity and probable computational success. Here I will describe the development and some applications of a method that has been pursued by Donaldson and his colleagues at A.R.A.P. (Aeronautical Research Associates of Princeton) for several years. The name of the model, an invariant model, can be interpreted in two ways. It refers to the constraints imposed on the choice of model terms required for closure. That is, any model term must exhibit the same tensor symmetry and dimensionality as the term it replaces. But the goal of the approach can also be described as an invariant model in the sense that our goal is a model that, although semiempirical, has no varying constants that must be determined for each new flow.

Our starting point is the exact equation for the Reynolds stress. In choosing to attempt closure at the level of the equations for the second-order correlations, we are assuming that this will yield a more general model than first-order closure. Certainly the second-order equations contain a great deal of physical information regarding the dynamics of the turbulent

W. S. LEWELLEN • Aeronautical Research Associates of Princeton, Inc., Princeton, New Jersey fluctuations. Provided this information is not lost by inappropriate modeling of the terms required for closure, this approach should be more general than first-order closure. In fact, it will be seen in a later section that a relatively rational, first-order closure model is obtained as a subset of our invariant model.

No attempt will be made to give a complete review of the literature surrounding turbulence transport modeling. Some surveys have been given by Bradshaw,[1] Mellor and Herring,[2] and Reynolds.[3] A convenient collection of some of the most important literature has been made by Harlow.[4] I will attempt to relate the model described here to those under development by other investigators. A limited number of comparisons between the numerical results of various models will be made.

9.2. Model Development

9.2.1. Closure Requirements

When the flow variables are decomposed into a mean and a fluctuating quantity, the mean flow equations may be written as

$$\frac{\partial U_i}{\partial t}+U_j\frac{\partial U_i}{\partial x_j}=-\frac{\partial \overline{u_iu_j}}{\partial x_j}+\frac{\partial}{\partial x_j}\nu\frac{\partial U_i}{\partial x_j}-\frac{1}{\rho}\frac{\partial P}{\partial x_i}+\frac{g_i(\Theta-\Theta_0)}{\Theta}-2\varepsilon_{ijk}\Omega_jU_k \quad (9.1)$$

$$\frac{\partial U_i}{\partial x_i}=0 \quad (9.2)$$

$$\frac{\partial \Theta}{\partial t}+U_j\frac{\partial \Theta}{\partial x_j}=-\frac{\partial \overline{u_j\theta}}{\partial x_j}+\frac{\partial}{\partial x_j}\left(k\frac{\partial \Theta}{\partial x_j}\right) \quad (9.3)$$

The flow has been assumed to be incompressible but with small variations in density due to changes in temperature. Following the Boussinesq approximation,[5] the only effect of the density variation is in the gravitational body force term in the momentum equation (9.1). The last term in Equation (9.1) is that due to Coriolis forces present in a coordinate system rotating with angular velocity Ω. It is included here in anticipation of atmospheric applications. Equation (9.3) is a diffusion equation for the temperature perturbation.

This system of equations is not complete owing to the presence of the Reynolds stress terms. Exact equations for these correlations of the fluctuating variables may be derived as outlined in the introductory chapters or by Donaldson[6]:

$$\frac{\partial \overline{u_i u_j}}{\partial t} + U_k \frac{\partial \overline{u_i u_j}}{\partial x_k} = -\overline{u_i u_k} \frac{\partial U_j}{\partial x_k} - \overline{u_j u_k} \frac{\partial U_i}{\partial x_k} + \frac{g_i \overline{u_j \theta}}{\Theta} + \frac{g_j \overline{u_i \theta}}{\Theta}$$

$$-2\varepsilon_{ikl}\Omega_k \overline{u_l u_j} - 2\varepsilon_{jlk}\Omega_l \overline{u_k u_i} - \frac{\partial}{\partial x_k}(\overline{u_k u_i u_j})$$

$$-\overline{\frac{u_i}{\rho}\frac{\partial p}{\partial x_j}} - \overline{\frac{u_j}{\rho}\frac{\partial p}{\partial x_i}} + \nu \frac{\partial^2 \overline{u_i u_j}}{\partial x_k^2} - 2\nu \overline{\frac{\partial u_i}{\partial x_k}\frac{\partial u_j}{\partial x_k}} \tag{9.4}$$

$$\frac{\partial \overline{u_i \theta}}{\partial t} + U_j \frac{\partial \overline{u_i \theta}}{\partial x_j} = -\overline{u_i u_j} \frac{\partial \Theta}{\partial x_j} - \overline{u_j \theta} \frac{\partial U_i}{\partial x_j} + \frac{g_i \overline{\theta^2}}{\Theta} - 2\varepsilon_{ijk}\Omega_j \overline{u_k \theta}$$

$$-\frac{\partial \overline{u_i u_j \theta}}{\partial x_j} - \overline{\frac{\theta}{\rho}} \overline{\frac{\partial p}{\partial x_i}} - \nu \theta \frac{\partial^2 u_i}{\partial x_j^2} - k u_i \frac{\partial^2 \theta}{\partial x_j^2} \tag{9.5}$$

$$\frac{\partial \overline{\theta^2}}{\partial t} + U_j \frac{\partial \overline{\theta^2}}{\partial x_j} = -2\overline{u_j \theta}\frac{\partial \Theta}{\partial x_j} - \frac{\partial}{\partial x_j}\overline{u_j \theta^2} + k \frac{\partial^2 \overline{\theta^2}}{\partial x_j^2} - 2k \overline{\frac{\partial \theta}{\partial x_j}\frac{\partial \theta}{\partial x_j}} \tag{9.6}$$

This still does not permit the system to close. In fact, for each new equation added, several more unknown flow variables are added. Clearly, continuing the process by deriving exact expressions for these correlations will add further new correlations in ever-increasing numbers. This is the closure problem.

Closure of the system of equations at the level of Equations (9.4)–(9.6) is called second-order closure. In Equation (9.4) the three terms

$$\frac{\partial}{\partial x_k}(\overline{u_k u_i u_j}), \qquad \left(\overline{\frac{u_i}{\rho}\frac{\partial p}{\partial x_j} + \frac{u_j}{\rho}\frac{\partial p}{\partial x_i}}\right), \qquad 2\nu \overline{\frac{\partial u_i}{\partial x_k}\frac{\partial u_j}{\partial x_k}} \tag{9.7}$$

are to be either neglected or modeled in terms of the other variables. Let us consider these three terms in some detail. The analogous terms in Equations (9.5) and (9.6) can then be modeled in a similar manner.

9.2.2. Dissipation Terms

The last term in Equation (9.7) measures the effect of viscous decay on the structure of the Reynolds stresses. Even in high-Reynolds-number flow, we expect viscous dissipation to be the major loss mechanism for turbulent kinetic energy. Owing to the nonlinear terms in the Navier–Stokes equation, a reduction in viscosity is compensated by a reduction in scale of the smallest eddies in the flow. The dissipation eddies are much smaller than the eddies that receive their energy directly from the mean flow when the Reynolds number is large, and as pointed out in Chapter 4, one may assume that they are statistically independent of the mean-flow geometry.

Just as the end of the cascade contains no information on the scale of the large eddies, the breakup rate of the large eddies should be independent of ν. Therefore, for high Reynolds number, it appears dimensionally correct to have

$$\varepsilon = \nu \overline{\frac{\partial u_i}{\partial x_k} \frac{\partial u_j}{\partial x_k}} \sim \frac{q^3}{\Lambda} \tag{9.8}$$

where q is the root-mean-square value of the total velocity fluctuation and Λ is a macroscale of the eddies. Since the viscous dissipation process is an isotropic process, most investigators model the dissipation terms as an isotropic term, i.e.,

$$\nu \overline{\frac{\partial u_i}{\partial x_k} \frac{\partial u_j}{\partial x_k}} = C_1 \delta_{ij} \frac{q^3}{\Lambda} \tag{9.9}$$

Some [7–9] have modeled it as an anisotropic term. A form that goes to Equation (9.9) in the limit of $\mathrm{Re} \to \infty$, but reflects the probability that at low Re dissipation will be anisotropic, is

$$\nu \overline{\frac{\partial u_i}{\partial x_k} \frac{\partial u_j}{\partial x_k}} = C_1 \delta_{ij} \frac{q^3}{\Lambda} + \nu \frac{C_2}{\Lambda^2} \overline{u_i u_j} \tag{9.10}$$

The first term of the form adopted in Equation (9.10) corresponds to taking the Taylor microscale λ proportional to $\Lambda/(a + bq\Lambda/\nu)^{1/2}$, where a and b are constants, as suggested by Rotta.[10,11]

Several investigators prefer to calculate ε from a dynamic equation obtained by modeling its governing equation. This will be discussed in Section 9.2.6 since, through Equation (9.10), it is equivalent to an equation for Λ.

9.2.3. Pressure Correlations

The correlations involving pressure fluctuations are more difficult to model than the dissipation terms. These terms redistribute the turbulent energy produced by the leading term on the right-hand side of Equation

(9.4). This can perhaps be more readily seen by rewriting the pressure correlations as

$$\pi_{ij} = \frac{\overline{u_i}}{\rho}\frac{\partial p}{\partial x_j} + \frac{\overline{u_j}}{\rho}\frac{\partial p}{\partial x_i} = \frac{\partial \overline{u_i p}}{\partial x_j} + \frac{\partial \overline{u_j p}}{\partial x_i} - \overline{\frac{p}{\rho}\left(\frac{\partial u_i}{\partial x_j} + \frac{\partial u_j}{\partial x_i}\right)} \tag{9.11}$$

Since the flow is incompressible, there will be no contribution from the last term in Equation (9.11) to the kinetic energy of the turbulence, $q^2/2 = \overline{u_i u_i}/2$. It contributes only to a redistribution of energy between the Reynolds stresses. A volume integral of the first two terms on the right-hand side of Equation (9.11) over any finite region of turbulence bounded by laminar flow will yield zero. Thus these two terms must only contribute to a spatial redistribution, i.e., are diffusion terms.

All investigators appear to follow Rotta[(10)] in modeling at least one contribution of the pressure correlation as a return-to-isotropy term proportional to the extent to which the flow is anisotropic, i.e.,

$$\pi_{ij} = \frac{C_3 q}{\Lambda}\left(\overline{u_i u_j} - \delta_{ij}\frac{q^2}{3}\right) \tag{9.12}$$

The term in parentheses, the departure from isotropy, provides the correct tensor symmetry and q/Λ provides the correct dimensionality. Donaldson[(6,7)] adds to this a spatial diffusion term with proper symmetry and dimensionality to represent the first two terms on the right-hand side of Equation (9.11):

$$\pi_{ij} = \frac{C_3 q}{\Lambda}\left(\overline{u_i u_j} - \delta_{ij}\frac{q^2}{3}\right) + C_4\left[\frac{\partial}{\partial x_i}\left(q\Lambda\frac{\partial \overline{u_k u_j}}{\partial x_k}\right) + \frac{\partial}{\partial x_j}\left(q\Lambda\frac{\partial \overline{u_k u_i}}{\partial x_k}\right)\right] \tag{9.13}$$

It is possible to show that the pressure correlation should probably depend on the mean flow strain. By taking the divergence of the fluctuating Navier–Stokes equations, a Poisson equation for p may be obtained:

$$\frac{1}{\rho}\nabla^2 p = -2\frac{\partial U_i}{\partial x_j}\frac{\partial u_j}{\partial x_i} - \frac{\partial^2}{\partial x_i \partial x_j}(u_i u_j - \overline{u_i u_j}) + \frac{g_i}{\Theta}\frac{\partial \theta}{\partial x_i} - 2\varepsilon_{ijk}\Omega_j\frac{\partial u_k}{\partial x_i} \tag{9.14}$$

Equation (9.14) can be formally integrated to obtain p, which may then be differentiated and correlated with u_i to form π_{ij}. The integral cannot be solved, so the procedure can only be used to suggest the form of the modeled term. The simplest addition to Equation (9.13) to incorporate an influence of mean strain is a term of the form

$$C_5 q^2\left(\frac{\partial U_i}{\partial x_j} + \frac{\partial U_j}{\partial x_i}\right) \tag{9.15}$$

In fact, a numerical value for C_5 ($=0.2$) has been obtained by Rotta[(11)] and by Crow[12] for the limiting case of isotropic turbulence in a weak, homogeneous mean strain. However, such a large value appears to make the term too overpowering for any practical calculations. A term such as Equation (9.15) has been included by several modelers[13–15] with C_5 empirically determined.

A form for the mean strain term with a good deal of appeal is that suggested by Launder[16]:

$$C_5(P_{ij} - \delta_{ij}P_{ii}/3) \tag{9.16}$$

where P_{ij} represents the production terms for the Reynolds stress in Equation (9.4). A similar term is included in the more involved model of Naot *et al.*[17] Adding Equation (9.16) to Equation (9.13) would be equivalent to assuming that the pressure correlations not only drive the turbulence towards isotropy with a term linearly proportional to the departure from isotropy, but also redistribute the production terms with a term linearly proportional to the anisotropy of the production terms.

Neither Equation (9.15) nor Equation (9.16) is sufficient to account for the observation that the mean strain effect is often not symmetric between u_2 and u_3 when the flow is one-dimensional in the x_1 direction. More complicated model variations for the pressure correlation terms may be found in the work of Lumley and Khajeh-Nouri,[18] Hanjalic and Launder,[19] Wolfshtein *et al.*,[9] Zeman and Tennekes,[20] and Varma.[21]

9.2.4. Third-Order Velocity Correlations

This term represents a process by which the Reynolds stress is transferred from one part of the flow to another without any net production or loss. Again this can be demonstrated, as it was for the pressure diffusion term of Equation (9.11), by integration over a finite volume. When the volume is bounded by either laminar or by homogeneous isotropic flow, the integral will vanish. The most popular modeling of this transport term is as gradient diffusion, although Bradshaw[1] has suggested that it could well be algebraic.

A number of different gradient diffusion forms have been used as a model for this term. Donaldson[7] proposed a form that satisfies the tensor symmetry of $\overline{u_k u_i u_j}$ with a scalar diffusion coefficient

$$D_{ij} = \frac{\partial}{\partial x_k}(\overline{u_k u_i u_j}) = C_6 \frac{\partial}{\partial x_k}\left[q\Lambda\left(\frac{\partial \overline{u_i u_j}}{\partial x_k} + \frac{\partial \overline{u_k u_i}}{\partial x_j} + \frac{\partial \overline{u_k u_j}}{\partial x_i}\right)\right] \tag{9.17}$$

Hanjalic and Launder[19] used a form with tensorial diffusivity

$$D_{ij} = C_6 \frac{\partial}{\partial x_k}\frac{\Lambda}{q}\left(\overline{u_k u_l}\frac{\partial \overline{u_i u_j}}{\partial x_l} + \overline{u_i u_l}\frac{\partial \overline{u_j u_k}}{\partial x_l} + \overline{u_j u_l}\frac{\partial \overline{u_i u_k}}{\partial x_l}\right) \tag{9.18}$$

They obtained this form by a "firm pruning" of the exact equation for $\overline{u_k u_i u_j}$. The simplest possible form is given by

$$D_{ij} = C_6 \frac{\partial}{\partial x_k}\left(q\Lambda \frac{\partial \overline{u_i u_j}}{\partial x_k}\right) \tag{9.19}$$

Although this form satisfies the tensor symmetry of D_{ij}, it does not satisfy the symmetry of $\overline{u_k u_i u_j}$ itself.

Wolfshtein *et al.*[9] and Lumley and Khajeh-Nouri[18] propose forms that have considerably more terms and involve a number of coefficients. These currently appear to have too many coefficients for manageable computations.

9.2.5. Modeled Equations

Our philosophy is to first attempt calculations with the simplest possible second-order closure models. We choose to use Equation (9.10) for the dissipation term, Equation (9.12) for the pressure–velocity interaction, and the pressure contribution to diffusion incorporated with the velocity diffusion for one diffusion term like Equation (9.19):

$$\nu \overline{\frac{\partial u_i}{\partial x_k}\frac{\partial u_j}{\partial x_k}} = \frac{b\delta_{ij}q^3}{3\Lambda} + \frac{a\nu\overline{u_i u_j}}{\Lambda^2} \tag{9.20}$$

$$\overline{\frac{p}{\rho}\left(\frac{\partial u_i}{\partial x_j} + \frac{\partial u_j}{\partial x_i}\right)} = -\frac{q}{\Lambda}\left(\overline{u_i u_j} - \delta_{ij}\frac{q^2}{3}\right) \tag{9.21}$$

$$\frac{\partial}{\partial x_k}(\overline{u_k u_i u_j}) + \frac{\partial \overline{u_i p}}{\partial x_j} + \frac{\partial \overline{u_j p}}{\partial x_i} = v_c \frac{\partial q\Lambda}{\partial x_k}\frac{\partial \overline{u_i u_j}}{\partial x_k} \tag{9.22}$$

In this restatement of the modeled terms, C_3 has been set equal to unity by using Equation (9.12) to define the macroscale Λ. The other coefficients have been assigned the symbols used in the most recent A.R.A.P. publications.

Equations (9.20)–(9.22) provide the minimum requirements of any second-order closure. The combination of terms in Equations (9.20) and (9.21) provides a destruction term for each of the Reynolds stress components which allows an equilibrium value to be reached in a time long compared with Λ/q. The term in Equation (9.22) prevents any excessively sharp gradients in Reynolds stresses from occurring.

When similar terms are used in the heat-flux and temperature-variance equations [Equations (9.5) and (9.6)], the modeled set of second-order correlation equations may be written as

$$\frac{\partial \overline{u_i u_j}}{\partial t} + U_k \frac{\partial \overline{u_i u_j}}{\partial x_k} = -\overline{u_i u_k}\frac{\partial U_j}{\partial x_k} - \overline{u_j u_k}\frac{\partial U_i}{\partial x_k} + \frac{g_i \overline{u_j \theta}}{\Theta} + \frac{g_j \overline{u_i \theta}}{\Theta}$$

$$-2\varepsilon_{ikl}\Omega_k \overline{u_l u_j} - 2\varepsilon_{jlk}\Omega_l \overline{u_k u_i} + v_c \frac{\partial}{\partial x_k}\left(q\Lambda \frac{\partial \overline{u_i u_j}}{\partial x_k}\right)$$

$$-\frac{q}{\Lambda}\left(\overline{u_i u_j} - \delta_{ij}\frac{q^2}{3}\right) + \nu \frac{\partial^2 \overline{u_i u_j}}{\partial x_k^2} - \delta_{ij}\frac{2bq^3}{3\Lambda} - \frac{2a\nu \overline{u_i u_j}}{\Lambda^2} \tag{9.23}$$

$$\frac{\partial \overline{u_i \theta}}{\partial t} + U_j \frac{\partial \overline{u_i \theta}}{\partial x_j} = -\overline{u_i u_j}\frac{\partial \Theta}{\partial x_j} - \overline{u_j \theta}\frac{\partial U_i}{\partial x_j} + \frac{g_i \overline{\theta^2}}{\Theta} - 2\varepsilon_{ijk}\Omega_j \overline{u_k \theta}$$

$$+ v_c \frac{\partial}{\partial x_j}\left(q\Lambda \frac{\partial \overline{u_i \theta}}{\partial x_j}\right) - \frac{Aq}{\Lambda}\overline{u_i \theta} + k\frac{\partial^2 \overline{u_i \theta}}{\partial x_j^2} + \frac{ak\overline{u_i \theta}}{\Lambda^2} \tag{9.24}$$

$$\frac{\partial \overline{\theta^2}}{\partial t} + U_j \frac{\partial \overline{\theta^2}}{\partial x_j} = -2\overline{u_j \theta}\frac{\partial \Theta}{\partial x_j} + v_c \frac{\partial}{\partial x_j}\left(q\Lambda \frac{\partial \overline{\theta^2}}{\partial x_j}\right) + k\frac{\partial^2 \overline{\theta^2}}{\partial x_j^2} - \frac{2bsq\overline{\theta^2}}{\Lambda} \tag{9.25}$$

This is the set we choose to make calculations with here. Rather than use models with a large number of coefficients that are finely adjusted to fit a few particular flows, we choose to work with a relatively minimum number. Only as experience proves that more terms are required will other terms and coefficients be added. Two additional coefficients were included in Equations (9.24) and (9.25), A and s. Both of these were first set equal to unity,[(6)] but as will be seen in the next section, it was later found desirable to add flexibility.

9.2.6. Scale Determination

To complete closure, it is necessary to provide some means for determining the turbulent scale Λ involved in the modeled terms. This is approached in different ways by various investigators. It may be specified empirically based on the gross features of the particular flow geometry; or it may be predicted from a semiempirically modeled, dynamic differential equation. Each of these approaches has some advantages and disadvantages.

The macroscale Λ is defined by Equations (9.20)–(9.22). It is expected to be related to the integral scale discussed in earlier chapters but, owing to our choice of normalization, not equal to it. It is also related to the mixing length discussed in Chapter 7. As such, there is empirical information that can be used in our determination of this model parameter. It appears fairly clear that the scale cannot exceed some fraction of the total spread of the region of turbulence and that, in some neighborhood of the wall, it should be proportional to the distance from the wall. These two simple ideas, together with empirical information used to determine the two implied constants of proportionality, are sufficient to permit the system to close with relatively good numerical results for many problems.[6,22,23] Others[24,25] have specified a completely empirically determined distribution of Λ across the region of interest in the same manner as is done for first-order mixing length approaches.

In an attempt to remove some of the arbitrariness of the specification of Λ for different flows, a number of investigators have resorted to a modeled dynamic equation for Λ or its equivalent. The starting points for such attempts have varied widely. A two-point-velocity-correlation equation forms the basis for the work of Rotta,[11] Naot *et al.*,[17] Donaldson,[6] and Rodi.[14] By forming an equation for the two-point velocity correlation $\overline{u_i(x)u_i(x+r)}$ and integrating to form an integral scale, it appears appropriate to take

$$q^2\Lambda = c\iiint_v \overline{u_i(x)u_i(x+r)}\frac{dv}{r^2} \tag{9.26}$$

For a flow without body forces, this leads to, following Wolfshtein *et al.*,[9]

$$\begin{aligned}\frac{D(q^2\Lambda)}{Dt} = \iiint_v \Bigg\{ & [U_k(x)-U_k(x+r)]\frac{\partial \overline{u_i(x)u_i(x+r)}}{\partial r_k} \\ & -\overline{u_i(x)u_k(x+r)}\frac{\partial U_i(x+r)}{\partial x_k} \\ & +\overline{u_k(x)u_i(x+r)}\frac{\partial U_i(x)}{\partial x_k} - \frac{1}{c}\frac{\partial}{\partial x_k}\overline{u_k(x)u_i(x)u_i(x+r)} \\ & -\frac{1}{c}\frac{\partial}{\partial r_k}[\overline{u_i(x)u_k(x+r)u_i(x+r)} - \overline{u_i(x)u_k(x)u_i(x+r)}] \\ & -\frac{1}{c\rho}\frac{\partial \overline{p(x)u_i(x+r)}}{\partial x_i} - \frac{1}{c\rho}\frac{\partial}{\partial r_i}[\overline{u_i(x)p(x+r)} - \overline{p(x)u_i(x+r)}] \\ & -\nu\frac{\partial^2\overline{u_i(x)u_i(x+r)}}{\partial x_k^2} - 2\nu\frac{\partial^2\overline{u_i(x)u_i(x+r)}}{\partial x_k\,\partial r_k} + 2\nu\frac{\partial^2\overline{u_i(x)u_i(x+r)}}{\partial r_k^2}\Bigg\}\frac{dv}{r^2}\end{aligned} \tag{9.27}$$

The difficulty with this approach is readily apparent. None of the terms on the right-hand side of Equation (9.27) can be integrated exactly. All of the terms must be modeled. This is also true if one starts with the equation for the dissipation ε, as is favored by many (e.g., Lumley and Khajeh-Nouri,[18] Harlow and Nakayama,[26] and Hanjalic and Launder[19]) or with the equation for the vorticity fluctuations as suggested by Daly and Harlow[8] or Wilcox and Alber.[27] With the model chosen in Section 9.2.5, $\varepsilon = bq^3/\Lambda$. Also the vorticity fluctuation can be taken as proportional to q/Λ. Thus, with the aid of the energy equation, any of these approaches may be reduced to an equation for Λ. As Bradshaw[1] and Mellor and Herring[2] have pointed out, all of the resulting Λ equations have the same form:

$$\frac{D\Lambda}{Dt} = -s_1 \frac{\Lambda}{q^2} \overline{u_i u_j} \frac{\partial U_i}{\partial x_j} - s_2 bq + \text{diffusion terms} \tag{9.28}$$

The difference in the various expressions lies in the construction of the turbulent diffusion terms. Unfortunately, these turn out to be more important in the scale equation than in the Reynolds stress equation. The scale equation we have experimented with at A.R.A.P.[28,29] starts with a rather general diffusion term of the form

$$\text{diffusion terms} = s_0 \frac{\partial}{\partial x_i}\left(v_c q \Lambda \frac{\partial \Lambda}{\partial x_i}\right) + s_3 \frac{\Lambda}{q^2} \frac{\partial}{\partial x_i}\left(q \Lambda \frac{\partial q^2}{\partial x_i}\right)$$
$$+ s_4 q \left(\frac{\partial \Lambda}{\partial x_i}\right)^2 + s_6 \frac{\Lambda}{q^2}\left(\frac{\partial q}{\partial x_i}\right)^2 + s_7 \frac{\partial q}{\partial x_i}\frac{\partial \Lambda}{\partial x_i} \tag{9.29}$$

As pointed out by Mellor and Herring[2] this covers the possibility of starting with equations for the quantity $q^m \Lambda^n$ and in combination with the energy equation reducing it to Equation (9.28).

As in the formulations of References 6 and 8, we will add a term proportional to $(\Lambda/q^2) g_i \overline{u_i \theta}/\Theta$ to permit the direct effect of stratification. This will require introducing another constant s_5.

It is immediately obvious that the scale equation contains much more arbitrariness than the Reynolds stress equations, where many of the terms were determined precisely without recourse to modeling or coefficients. With such a large number of coefficients in the scale equation, a correspondingly large number of different experiments must be matched concurrently if the resulting coefficients are to have any invariant validity.

There is also the question as to what extent it is really appropriate to have the integral quantity Λ determined by point values of the other variables. It seems that spatial boundary conditions may be expected to play a much more important role in the determination of Λ than they do for the Reynolds stress.

9.3. Evaluation of Model Coefficients

With the exception of the model scale variable Λ, the simple model chosen in Section 9.2.5 contains five coefficients: a, b, v_c, A, and s. Ideally these coefficients should be constants, but conceptually the approach is still valid if they turn out to be unique functions of the dimensionless parameters. For the flow specified in Equations (9.1)–(9.3), there are three such dimensionless parameters: a Reynolds number, a Richardson number, and a Rossby number. The coefficients might also be a function of some dimensionless number characterizing the state of the turbulence, such as that proposed by Lumley and Khajeh-Nouri[18]:

$$(\overline{u_i u_j} - \delta_{ij} q^2/3)(\overline{u_j u_i} - \delta_{ji} q^2/3)/q^4$$

We will initially, optimistically, assume that the coefficients are constant. If the coefficients have to be functions of the dimensionless variables, then the second-order approach will lose some of its advantage over first-order closure. Insofar as is possible, each coefficient should be determined from a critical flow experiment that involves only that coefficient.

9.3.1. Dissipation Coefficient b

For high-Reynolds-number turbulence, the most sensitive model coefficient in Equation (9.23) is the coefficient b. The value of this coefficient can be isolated by considering steady, homogeneous turbulence with no body forces. When a one-dimensional mean flow in the x_1 direction is considered with a gradient in only one direction, x_3, Equation (9.23) reduces to

$$0 = -\overline{u_i u_3}\frac{\partial U_1}{\partial x_3}\delta_{1j} - \overline{u_j u_3}\frac{\partial U_1}{\partial x_3}\delta_{1i} - \frac{q}{\Lambda}\left(\overline{u_i u_j} - \delta_{ij}\frac{q^2}{3}\right) - \delta_{ij}\frac{2bq^3}{3\Lambda} \qquad (9.30)$$

From these separate component equations, one can obtain several different expressions for b independent of Λ:

$$b = \frac{(\overline{u_1 u_3})^2}{\overline{u_3 u_3}\, q^2} = \frac{3}{2}\left(\frac{1}{3} - \frac{\overline{u_3 u_3}}{q^2}\right) = \frac{3}{2}\left(\frac{1}{3} - \frac{\overline{u_2 u_2}}{q^2}\right) \qquad (9.31)$$

The data of Champagne *et al.*[30] for such a homogeneous shear flow in a wind tunnel yield values of 0.12, 0.14, and 0.08, respectively, for the three expressions in Equation (9.31). With the exception of the last value, these are all close to the value of 1/8 chosen by Donaldson[7] by comparison with other data. A closer fit to all the data will require additional model terms as discussed in Section 9.2.3.

The computational results to be described here have all been made with $b = 0.125$.

9.3.2. Diffusion Coefficient v_c

The diffusion coefficient v_c is difficult to decouple from the determination of the scale Λ. We have chosen a value of 0.3 based on computer studies of jets and wakes.[22] However, owing to the relative insensitivity of our results to this coefficient, it is considered known only to ± 0.05. This is the basic reason we have chosen to use the simplest form of diffusion, as given in Equation (9.22). Critical measurements may show one of the more complicated forms to be preferable. The difficulty of measuring the pressure correlation contribution to diffusion makes the diffusion terms particularly difficult to isolate empirically. The best chance for finer detailed modeling of those terms probably lies in the possibility of detailed comparisons with ensemble averages from the type of simulated turbulence calculation described in Chapter 10.

9.3.3. Scale Determination

Section 9.2.6 discussed two approaches to the determination of Λ. Let us first determine the two constants needed to fix the bounds on Λ. Close to a wall, we will assume $\Lambda = \alpha z$ with z the distance normal to the wall. The coefficient α is related directly to von Kármán's constant κ. In the "constant flux" surface layer region of a flat-plate boundary layer, Equation (9.30) holds, with the subscript 1 denoting the direction of the free stream flow and $x_3 = z$. When the definition of κ

$$\kappa = \frac{u^*}{z}\left(\frac{\partial U}{\partial z}\right)^{-1} \tag{9.32}$$

is used to eliminate the mean-flow gradient, Equation (9.30) can be arranged to yield

$$\alpha = \kappa b\left[\frac{b(1-2b)}{3}\right]^{3/4} \tag{9.33}$$

With our previously assigned value of $b = 0.125$, Equation (9.33) yields

$$\alpha = 1.68\kappa \tag{9.34}$$

The generally accepted value of κ from laboratory flows is 0.40, but rather extensive measurements in the atmospheric surface layer at much higher Reynolds numbers[31] give $\kappa = 0.36$. The corresponding values of α are 0.67 and 0.60. We have used 0.65 in most of the calculations to follow, although some have been performed with $\alpha = 0.60$.

The other bound of Λ is that tied to the spread of the region. This we have picked by computer optimization. For an axisymmetric wake or jet, it is 0.20 times the radius at which the turbulent energy falls to one-fourth its maximum value. For a two-dimensional boundary layer or wake, it is 0.3 times the similar measure of the turbulent spread. For a two-dimensional shear layer, it might be expected to be 0.6. It is clear that the more complicated the flow geometry is, the more difficult it becomes to specify Λ.

The second approach is an attempt to circumvent this difficulty by using a dynamic Λ equation with coefficients that are independent of flow geometry. Let us attempt to determine the coefficients in Equations (9.28) and (9.29).

The coefficient s_2 can be estimated from the decay of homogeneous grid turbulence. If homogeneous turbulence is assumed to decay as

$$q^2 \sim x^{-n} \tag{9.35}$$

then from the contraction of Equation (9.30)

$$\Lambda = bq^3 \Big/ \left(-\frac{1}{2} U \frac{\partial q^2}{\partial x} \right) \tag{9.36}$$

Equation (9.28) then leads to

$$s_2 = -\frac{U}{bq}\frac{\partial \Lambda}{\partial x} = -\frac{(2-n)}{n} \tag{9.37}$$

A recent review of grid turbulence experiments by Gad-el-Hak and Corrsin[32] shows values of n predominantly between 1 and 1.3 with more of the values lying near 1.25. This value of n gives a value of $s_2 = -0.6$.

To reduce the number of diffusion coefficients, it is desirable to keep the coefficient of the direct diffusion term s_0 the same as in the Reynolds stress equation. This calls for $s_0 = 1$. It is compatible with this assumption to take $s_3 = 0$, since with $s_0 = 1$ any combination of m and n for an expression of the form

$$\frac{Dq^m \Lambda^n}{Dt} - v_c \frac{\partial}{\partial x_i}\left(q\Lambda \frac{\partial q^m \Lambda^n}{\partial x_i} \right) \tag{9.38}$$

would lead to an equation for $D\Lambda/Dt$ with $s_3 = 0$.

A simple relationship between s_0, s_1, s_2, and s_4 may be obtained from the reduced form of the scale equation in the steady, neutral, constant shear-stress layer near a solid boundary where $\Lambda = \alpha z$. In this region, the scale equation reduces to

$$0 = s_1 \frac{\alpha z u^{*2}}{q^2} \frac{\partial U}{\partial z} - s_2 bq + v_c q\alpha^2 + s_4 q\alpha^2 \tag{9.39}$$

which, with the aid of the energy equation, may be reduced to

$$-s_4 = v_c + (b/\alpha^2)(s_1 - s_2) \tag{9.40}$$

This leaves s_1, s_6, and s_7 to be determined by computer optimization. Prior to doing this, we must determine boundary conditions for the scale.

The boundary conditions on the scale equation are not as straightforward as those on the Reynolds stress equations, since the scale need not go to zero at the free boundary of a region of turbulence. In fact, the eddies extending the farthest are expected to be the largest eddies present in the region. At a free boundary, we therefore set Λ equal to twice the bound established by the previous approach to Λ.

The edge boundary condition appears simpler if one chooses to use the dissipation equation, since $\varepsilon = bq^3/\Lambda$ clearly approaches zero. However, this leaves Λ free to approach any value from zero to infinity as $q \to 0$, and no independent information is gained.

Since the production term is small for a momentumless wake,[21] this provides a flow for estimating s_6 and s_7 to lie between 0.3 and 1. To reduce the uncertainty in these coefficients, we have rather arbitrarily set $s_4 = s_6 = s_7/2$. The last coefficient, s_1, may now be determined by computer fit with the free jet result to be ≈ -0.35. This is in the middle of the range of values used by Rodi,[14] who found it necessary to vary his analogous coefficient from 1.2 to 2 (corresponding to varying s_1 in our case from -1 to 0.6).

With these choices for the coefficients, our proposed scale equation for no body forces becomes

$$\frac{D\Lambda}{Dt} = 0.35\frac{\Lambda}{q^2}\overline{u_i u_j}\frac{\partial U_i}{\partial x_j} + 0.6\frac{\nu}{\lambda^2}\Lambda + 0.3\frac{\partial}{\partial x_i}q\Lambda\frac{\partial \Lambda}{\partial x_i} - \frac{0.375}{q}\left(\frac{\partial q\Lambda}{\partial x_i}\right)^2 \tag{9.41}$$

As noted earlier, we cannot assign as high a confidence level to this equation as to our modeled Reynolds stress equations. In fact, it apppears that unless the turbulence is far out of equilibrium, the approach of simply limiting the Λ to be equal to the lower of the two bounds previously described is adequate for a simple model.

9.3.4. Low-Reynolds-Number Dependence

The coefficient a in Equation (9.23) is only important at low Reynolds numbers, i.e., when $bq\Lambda/\alpha\nu \leq 0(1)$. The best measured low-Reynolds turbulent flow is the transition region in a boundary layer just outside the so-called laminar sublayer. In this region, Equation (9.1) reduces to

$$-\overline{u_1 u_3} + \nu\frac{\partial U_1}{\partial x_3} = \text{const} = u^{*2} \tag{9.42}$$

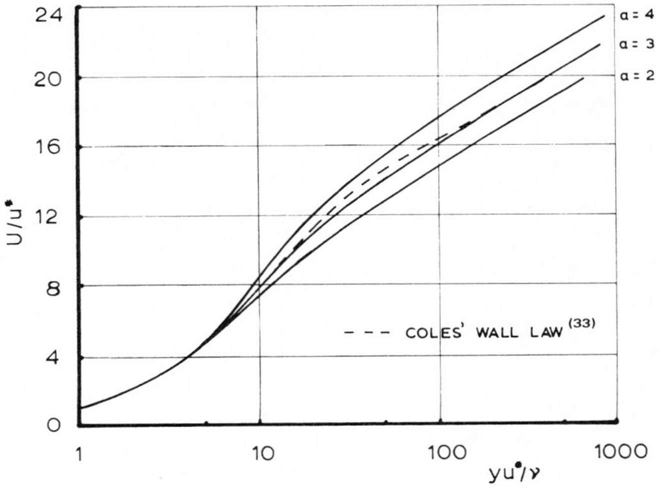

Fig. 9.1. Influence of a on model predictions of law of the wall.

Equation (9.42) and the four nonzero Reynolds stress component equations from Equation (9.23) may be solved in this region and compared to Coles' empirical law of the wall.[33] This is done in Figure 9.1 for three values of a for the previously chosen values of α, b, and v_c. The best choice appears to be $a = 3$.

9.3.5. Additional Coefficients Required to Compute Temperature Fluctuations A, s, and s_5

The coefficient A in Equation (9.24) can be determined by again considering the constant flux region of a high-Reynolds-number boundary layer with no body forces, i.e., the limit of small temperature fluctuations. The normal component of Equation (9.24) then reduces to

$$-\overline{u_3u_3}\frac{\partial\Theta}{\partial x_3}-\frac{Aq}{\Lambda}\overline{u_3\theta}=0 \qquad (9.43)$$

while the shear stress equation from Equation (9.30) for the same conditions reduces to

$$-\overline{u_3u_3}\frac{\partial U_1}{\partial x_3}-\frac{q}{\Lambda}\overline{u_1u_3}=0 \qquad (9.44)$$

Equations (9.43) and (9.44) may be used to form the ratio of eddy diffusivity for momentum to that for heat. This is sometimes called the turbulent

Prandtl number

$$\mathrm{Pr}_t = \frac{\overline{u_1 u_3} \partial \Theta / \partial x_3}{\overline{u_3 \theta} \partial U_1 / \partial x_3} = A \tag{9.45}$$

Based on the measurements of Businger *et al.*[31] in the neutral atmospheric surface layer, we chose a value of 0.75 for this coefficient.

The coefficient of the temperature dissipation term s can be evaluated by comparing the frequency spectrum of the temperature fluctuations with the spectrum of the turbulent kinetic energy. In the inertial subrange, it is experimentally observed that the three-dimensional energy spectrum in scalar wave-number space is given by (see Chapter 4)

$$E(k) = \alpha \varepsilon^{2/3} k^{-5/3} \tag{9.46}$$

with α a constant.

The temperature-variance spectrum is similarly given by

$$E_\theta(k) = \beta N \varepsilon^{-1/3} k^{-5/3} \tag{9.47}$$

with β a constant and N the dissipation of temperature variance. By integrating Equation (9.46) to form $q^2/2$ and Equation (9.47) to form $\overline{\theta^2}/2$, it is possible to show that the coefficient s in Equation (9.25) should be equated to α/β. With the values of $\alpha = 1.5$ and $\beta = 0.5$ given by Tennekes and Lumley[34] this would yield $s = 3.0$. But some values of the ratio are quoted as low as 0.6.[35] We have chosen a value of 1.8.[36] Mellor[15] chose a value of 15/8, while Wyngaard and co-workers[37,38] chose a value of 2.8.

An estimate for the coefficient of the stratification term in the scale equation can be obtained from the stable, constant flux layer, for Richardson number equal to its critical value ≈ 0.21.[31] Assuming Λ and q constant, and picking the value of Λ that leads to $\mathrm{Ri}_c = 0.21$,[36] we find a coefficient of 0.8 on the buoyancy term. Thus the form of the scale equation used here is

$$\frac{D\Lambda}{Dt} = 0.35 \frac{\Lambda}{q^2} \overline{u_i u_j} \frac{\partial U_i}{\partial x_j} + 0.6 \frac{\nu \Lambda}{\lambda^2} + 0.3 \frac{\partial}{\partial x_i}\left(q\Lambda \frac{\partial \Lambda}{\partial x_i} \right) - \frac{0.375}{q}\left(\frac{\partial q \Lambda}{\partial x_i} \right)^2 + \frac{0.8\Lambda}{q^2} \frac{g_i \overline{u_i \theta}}{\Theta} \tag{9.48}$$

9.4. Model Verification

In viewing the comparisons between model predictions and experiments in this section, the reader should remember our goal of making reasonable predictions for a wide class of flows rather than very accurately predicting any one flow. Flows for which significant discrepancies occur between model predictions and observations for some of the variables will be noted. If one is interested in a relatively narrow class of flows, these discrepancies could be reduced by changes in the model coefficients. A better general model would require more sophisticated modeling of the individual terms, as discussed in Section 9.2.

9.4.1. Axisymmetric Free Jet

Comparisons between model predictions and the experimental data of Wygnanski and Fiedler[39] for the self-similar region of an axisymmetric free jet are given in Figure 9.2. The computations have been made[22,29] by

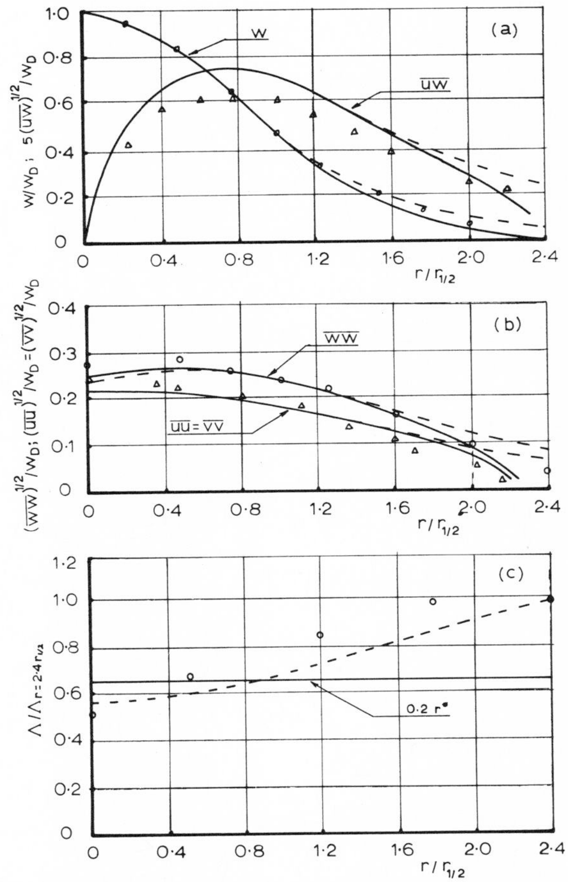

Fig. 9.2. Similarity profiles for axisymmetric free jet with constant Λ (——) and dynamic Λ (– – –). Data from Wynganski and Fiedler.[39]

starting with some arbitrary velocity and turbulence distributions and letting the jet develop until self-similar distributions are obtained. The model predictions have been made using both of the approaches to the scale determination discussed in Sections 9.2.6 and 9.3.3. The scale variations for the two approaches are compared with the measured longitudinal integral scale in Figure 9.2(c). The variation predicted by Equation (9.41) is closer to the observations as expected, but the improvement, if any, in the predicted flow variables is not significant. The fact that the predicted energy is low while the predicted shear stress is high is the type of discrepancy one might expect to overcome by incorporating additional terms in the model of the pressure correlations discussed in Section 9.2.3.

9.4.2. Free Shear Layer

Results for a two-dimensional, self-similar shear layer are compared with observations(40,41) in Figure 9.3. The agreement with the mean-flow profile is good, but there exists considerable scatter in the data for the shear-stress distribution. The components of the normal stress predicted by the model were compared with Wygnanski and Fiedler's data by Donaldson.(6) Because of some uncertainty surrounding the data,(41) these comparisons are not shown here.

9.4.3. Two-Dimensional Wake

Predictions are compared with Townsend's(42) self-similar wake measurements in Figure 9.4. Lumley and Khajeh-Nouri's(18) model predictions

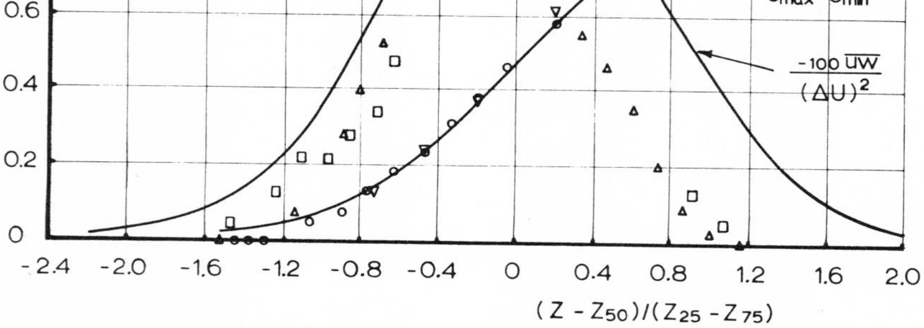

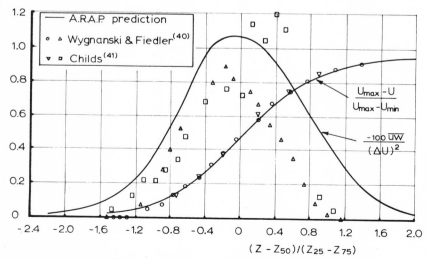

Fig. 9.3. Similarity profiles for free shear layer.

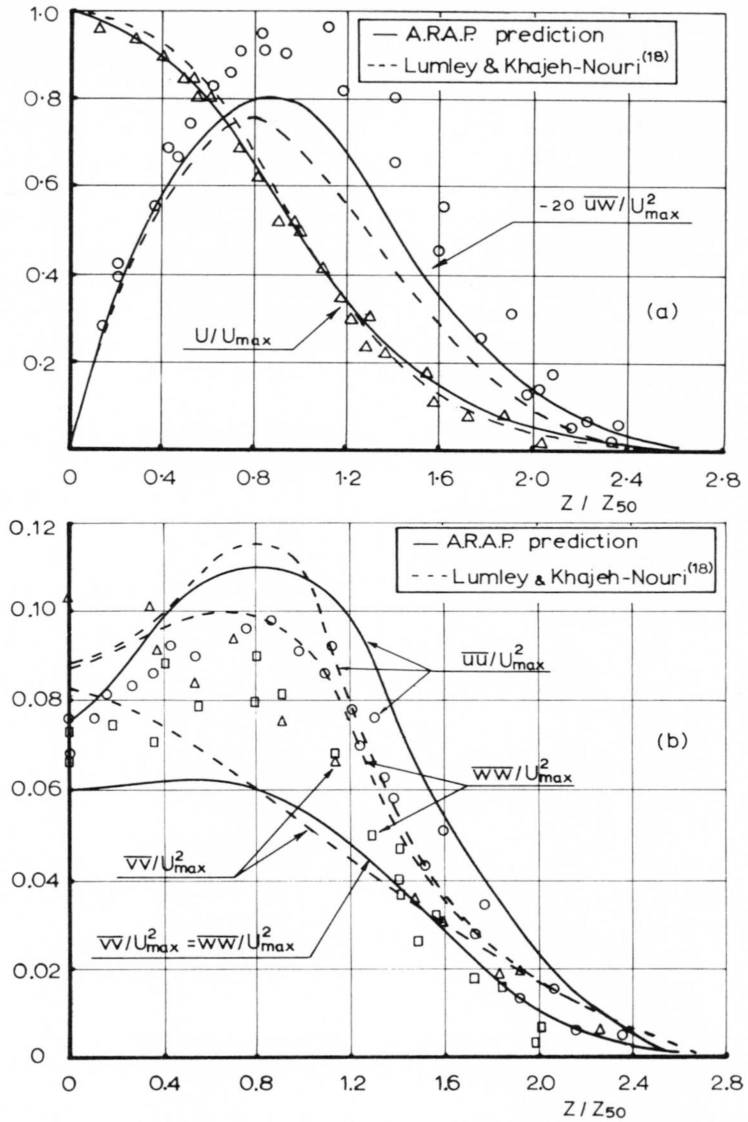

Fig. 9.4. Similarity profiles for two-dimensional wake. Data from Townsend.[42]

are also included for comparison. Their model, which contains many more terms and coefficients, makes a marginally better prediction for the normal components of the Reynolds stress for this flow. The lateral and transverse velocity fluctuations are equal in the present model predictions while the observations are significantly different. Again, there is some uncertainty in the data since Thomas[43] reports measurements of the longitudinal root-mean-square turbulence intensity that are about 33% higher than Townsend's reported values.

9.4.4. Axisymmetric Wake

The capability of the model to predict the development of a wake was tested by matching the velocity and Reynolds stress distributions as observed by Chevray[44] at one axial station and then comparing the predicted and observed development.[22] The decay of the maximum velocity defect w_D, the maximum shear stress $\overline{uw}_{max}$, and the maximum axial

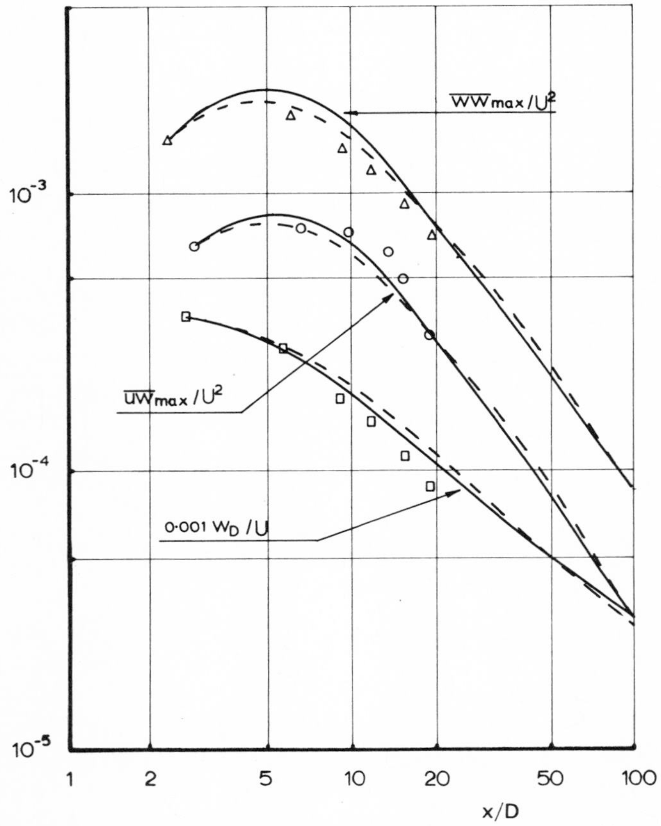

Fig. 9.5. Downstream decay of axisymmetric wake with constant Λ (——) and dynamic Λ (– – –). Data from Chevray.[44]

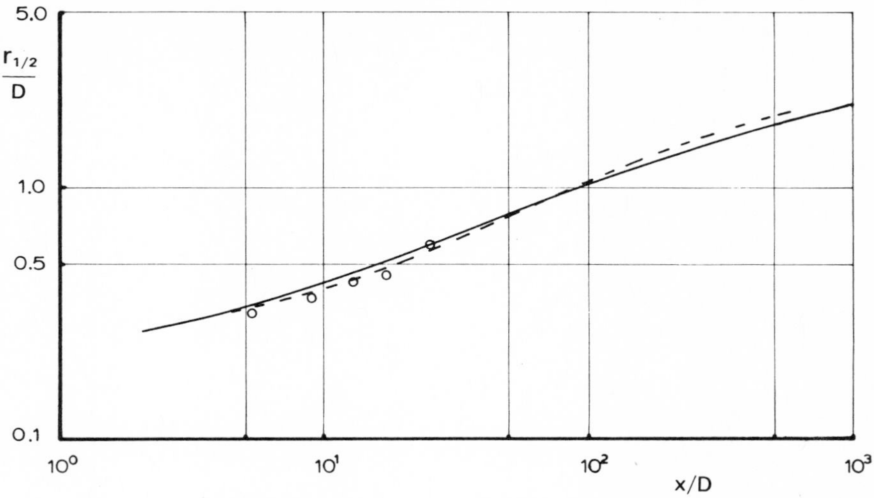

Fig. 9.6. Wake radius growth for conditions of Figure 9.5.

velocity fluctuation $\overline{ww}_{max}$ are given in Figure 9.5. The largest discrepancy is in the decay of the axial velocity fluctuations. The model predicts an increase of 35% over the initial value at $z/D = 3$ which the observations do not show.

The wake spread as measured by the radius for which the defect in the mean velocity is equal to half of its maximum value is given in Figure 9.6. The agreement is good. Although Chevray fits a $\frac{1}{3}$-power law growth rate to his data 18 diameters behind the body, the model indicates that ≈ 100 diameters is required for the wake to reach self-similarity. The self-similar profiles obtained by the model are compared in Figures 9.7(a) and (b).

Comparisons between predictions and the momentumless wake measurements of Naudascher[45] have also been made[22] with results similar to the comparisons shown here with Chevray's measurements. In this case, no true self-similarity exists.

9.4.5. Flat-Plate Boundary Layer

Predicted velocity distributions in the neighborhood of the wall were compared with Coles' law of the wall[33] in Figure 9.1 in order to choose a. Comparisons with Klebanoff's[46] velocity and Reynolds stress measurements are given in Figure 9.8 for a Reynolds number of 5×10^6. The agreement with U and $\overline{uw}$ is good but the sharp peaks observed in $\overline{uu}$ and $\overline{vv}$ very near the wall are not predicted. There also appears to be a little too much diffusion in the outer region.

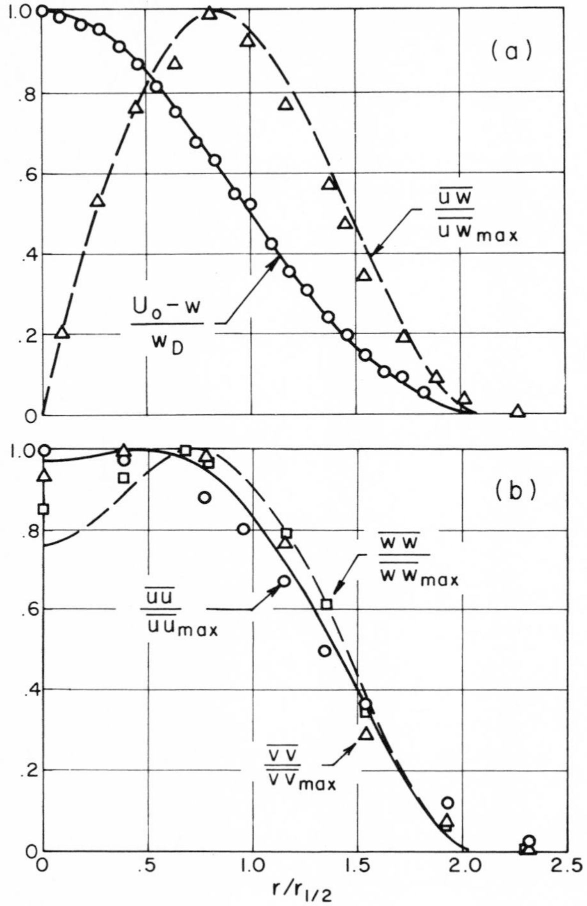

Fig. 9.7. Similarity profiles for conditions of Figure 9.5.

A plot of surface shear stress versus Reynolds number ($Re_x = Ux/\nu$) is given in Figure 9.9 along with the laminar, Blasius value, and Coles[33] compilation for fully turbulent flow. For this comparison, the calculation is started with the laminar velocity profile at a relatively small value of Re_x. All turbulent correlations are initially zero except for a small spot of turbulent energy. The model does a creditable job at predicting transition.

9.4.6. Flow over an Abrupt Change in Surface Roughness

A calculation was made in simulation of the atmospheric surface data of Bradley,[47] who recorded velocities and surface shear stress at a little-used airfield in Australia. Transitions are shown in Figure 9.10 for rough-to-smooth and smooth-to-rough, with the roughness measured in terms of the aerodynamic roughness z_0. The calculations[28] begin with the velocity and turbulence components in equilibrium for one value of z_0. The surface boundary condition is changed to that appropriate for the other value of z_0 at $x = 0$. When both runs are allowed to continue far downstream, the turbulence comes into equilibrium with the new z_0 value. Figure 9.10 also includes the predictions of Rao *et al.*,[48] who used the second-order closure model developed in Reference 38.

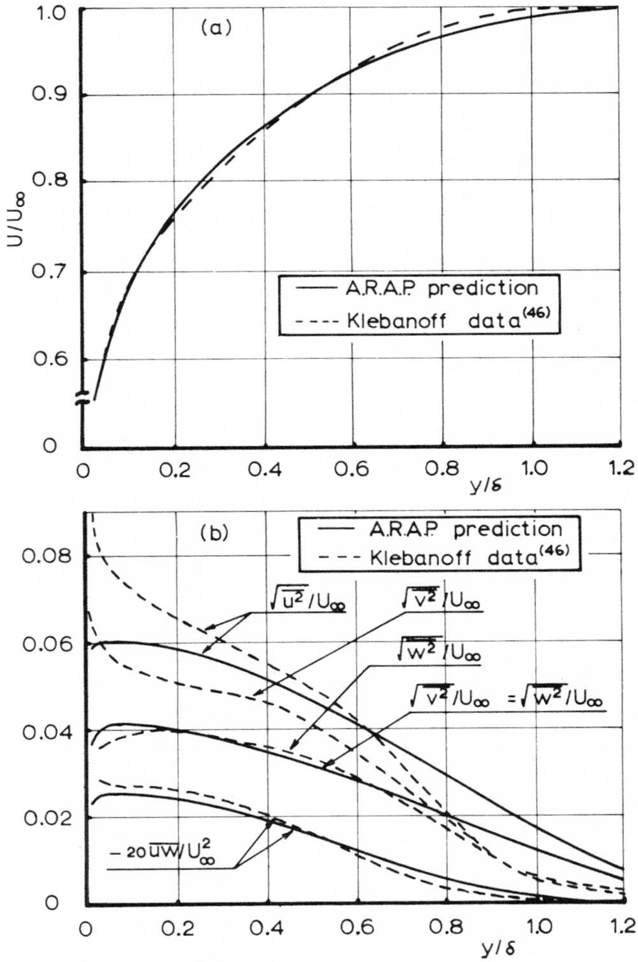

Fig. 9.8. Similarity profiles for flat-plate boundary layer.

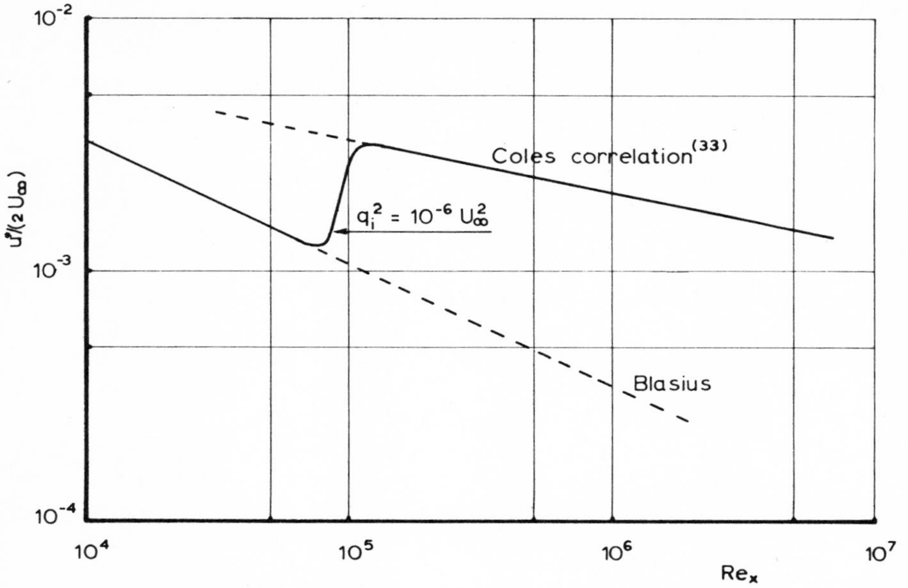

Fig. 9.9. Surface shear stress on a flat plate.

9.4.7. Temperature Fluctuations in the Plane Turbulent Wake

Measurements behind a slightly heated cylinder have been made by Freymuth and Uberoi[49] and LaRue and Libby.[50] Figure 9.11 shows a comparison between the measured distributions far downstream of the cylinder and the self-similar profiles predicted by the model. The normalizing length z_{50} is fixed for the mean velocity profile given in Figure 9.4. The agreement is quite good, particularly in the concurrence of the amplitude and position of the maximum temperature fluctuation.

9.4.8. Stability Influence in the Atmospheric Surface Layer

The atmospheric surface layer extending a few tens of meters above the earth's surface is the most extensively studied example of a turbulent shear layer incorporating the influence of stratification on the dynamics of turbulence. Following Monin and Obukhov,[51] considerable success has been achieved in experimentally determining similarity functions that describe the dependence of the mean turbulence characteristics on height, shear stress, and heat flux. In this layer, the shear stress and heat flux may be considered constant. From these turbulent transport constants, a characteristic length may be found:

$$L = -\frac{\Theta u^{*3}}{kg\overline{w\theta}}$$

This is the Monin–Obukhov length. At altitudes below which any cross flows appear owing to the Coriolis force caused by the earth's rotation and above which the direct viscous contributions to momentum transport are important, z/L is the only variable determining the normalized turbulence quantities for a steady homogeneous layer. Equations (9.23)–(9.25) have been integrated for this case by Lewellen and Teske.[36] Results for the normalized wind and temperature gradients are compared with atmospheric data in Figure 9.12.

In these computations, the scale was set equal to $0.6z$ with an upper bound of $0.20L$ in stable flow. This was chosen to yield the observed critical Richardson number of 0.21. That is, no matter how stable the flow becomes

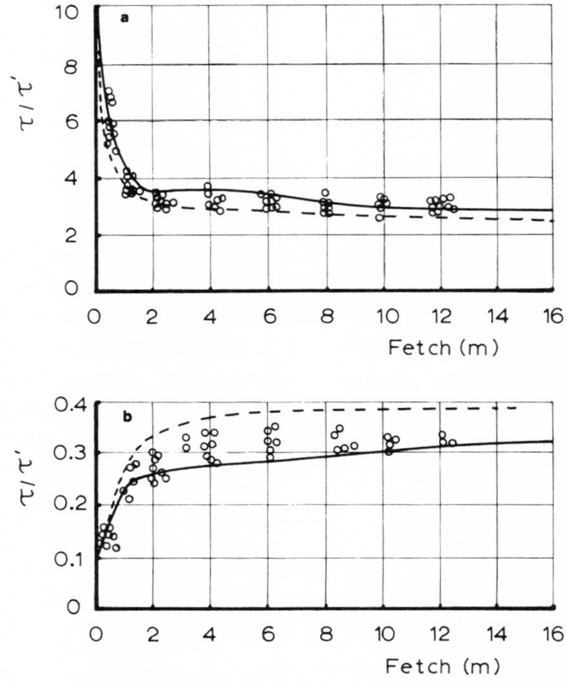

Fig. 9.10. Step change in surface roughness. A.R.A.P. prediction (——); prediction by Rao *et al.*[48] (- - -). Data from Bradley.[47] (a) Smooth to rough $z_0' = 0.002$ cm, $z_0 = 0.25$ cm; (b) rough to smooth $z_0' = 0.25$ cm, $z_0 = 0.002$ cm.

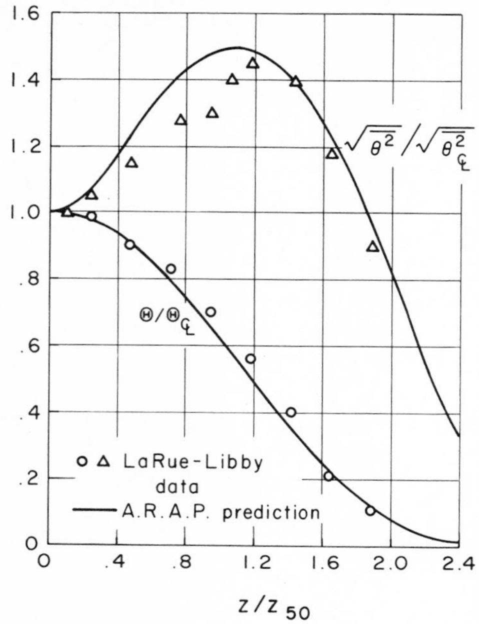

Fig. 9.11. Temperature similarity profiles in a two-dimensional wake. Data from LaRue and Lilley.[50]

in terms of large z/L, Ri never exceeds 0.21. This bound was also used in determining the coefficient of the buoyant term in the scale equation in Section 9.3.3.

The vertical velocity and temperature fluctuations are compared with the data of Wyngaard *et al.*[52] in Figures 9.13 and 9.14. The agreement for the vertical velocity is very good but that for the temperature fluctuations leaves a bit to be desired. As pointed out by Wyngaard *et al.*[52] there is considerable uncertainty in $(\overline{\theta^2})^{1/2}u^*/\overline{w\theta}$ at $z/L = 0$ because both variables go to zero.

Mellor[15] made a somewhat similar calculation of these surface-layer similarity functions. The major difference is that Mellor eliminated the diffusion terms so that the differential equations reduce to an algebraic set. In this case the length scale may be normalized out of the problem, eliminating any need for, or any possibility of, incorporating an influence of stratification on the scale. His model coefficients must then be chosen so that they will match the critical Richardson number.

9.4.9. Shear Layer Entrainment in a Stratified Fluid

Kato and Phillips[53] measured turbulent entrainment in an annular tank of density-stratified water with a shear stress applied at the surface by a

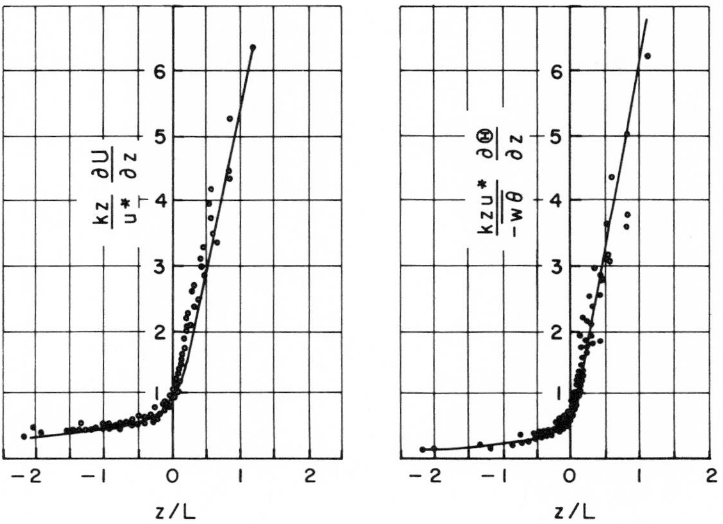

Fig. 9.12. Normalized atmospheric surface layer gradients as a function of stability. Data from Businger *et al.*[31]

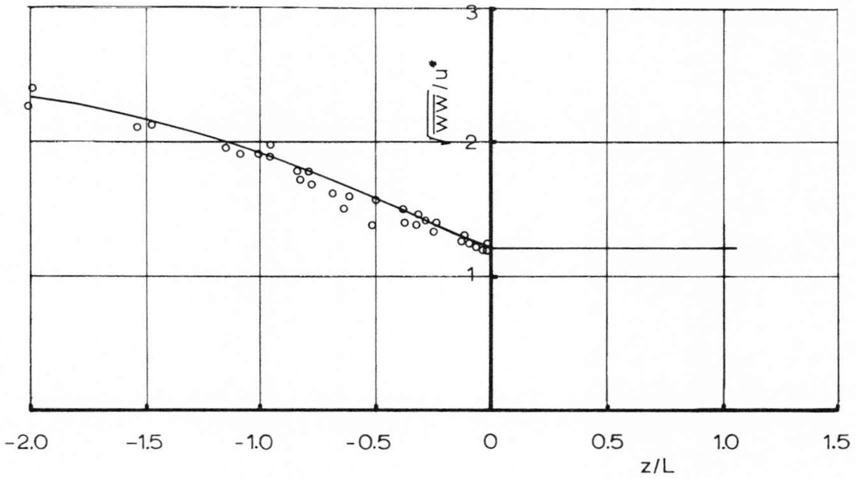

Fig. 9.13. Vertical velocity fluctuation in the atmospheric surface layer. Data from Wyngaard *et al.*[52]

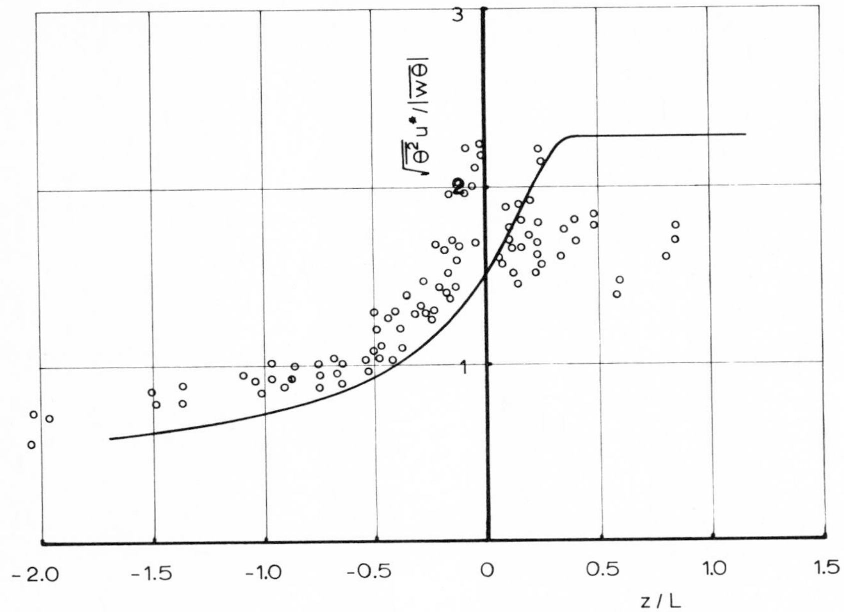

Fig. 9.14. Temperature fluctuation in the atmospheric surface layer. Data from Wyngaard *et al.*[52]

rectangular mesh screen. The rate of spreading of the turbulent fluid into the linear density field was measured visually by injected dye. Their experimental results of entrainment rate versus flow Richardson number are given in Figure 9.15. The two model prediction curves[29] in the figure represent calculations using the two different approaches to the scale discussed in Section 9.2.6. The dynamic scale rounds the apparently straight-line prediction of the gross scale model. Neither model predicts quite as steep a decrease of entrainment with Richardson number as the data indicate.

9.4.10. Free Convection

Willis and Deardorff[54] have measured the velocity and temperature fluctuations in an unstable layer bounded above by a stable density gradient and below by a surface with a positive heat flux. Their experiment called for establishing an initial stable temperature gradient between two surfaces and then applying a heat flux to the region through the lower surface. With mean velocities restricted to a minimum, free convection occurs in a mixed layer above the lower surface. The thickness of this mixed layer increases with time, but the fluctuating velocity distributions exhibit self-similarity when normalized by the characteristic velocity

$$w^* = \left(\frac{g\overline{w\theta_0}z_i}{\Theta}\right)^{1/3}$$

where $\overline{w\theta_0}$ is the surface heat flux and z_i is the depth of the mixed layer.

Figure 9.16 shows our model predictions for the normalized vertical velocity fluctuations and the experimental observations. For the model predictions, we assumed no mean velocity to simulate the laboratory experiment and a constant surface heat flux. The agreement between predictions and observations is heartening, particularly since no model constant adjustments were involved.

The horizontal velocity fluctuations are given in Figure 9.17. Here the comparison with experiment is not as close since the observations show a peak near the lower surface that is not evidenced in the model results. The distributions of temperature fluctuations are shown in Figure 9.18. The agreement between model and experiment is very good except at the top of

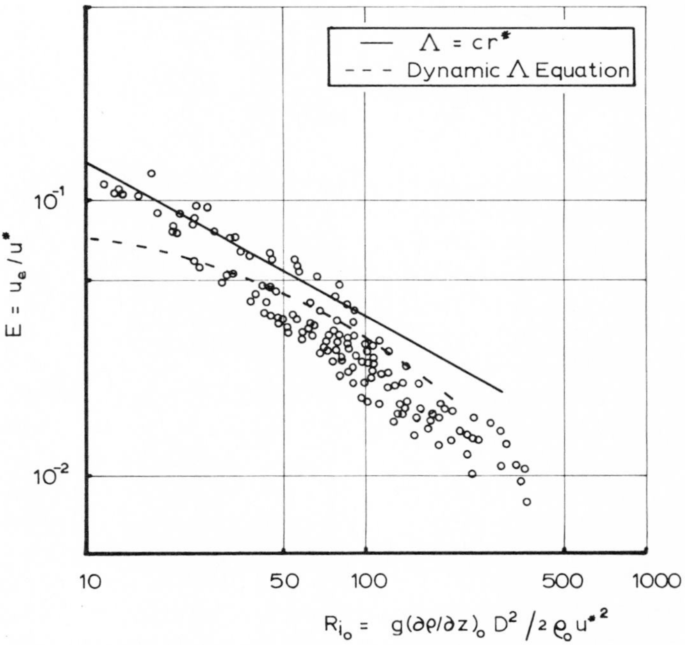

Fig. 9.15. Shear layer entrainment into a stratified fluid. Data from Kato and Phillips[53]; r^* is a measure of the spread of turbulence.

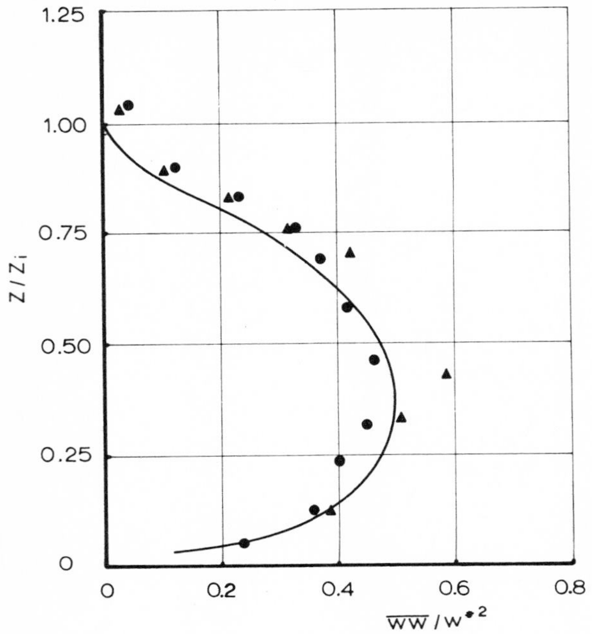

Fig. 9.16. Similarity profile of vertical velocity fluctuation in an unstable mixed layer. Data from Willis and Deardorff.[54]

the mixed layer where the observations show a much stronger local maximum than is predicted. This may be due to the fact that inertial oscillations exist in this region. Overall, the comparison between model predictions and experimental observations is quite favorable.

9.4.11. Planetary Boundary Layer for Neutral Steady State

Height variation of the mean velocities and Reynolds stress components is given in Figure 9.19 as predicted by our model for the atmospheric Ekman layer. The distributions are compared with Deardorff's[55] results of ensemble averages of integrations of the unsteady equations of motion for all the turbulent fluctuations above a prescribed grid size. The mean velocity distributions are in good agreement. The value of the surface shear stress ($u^*/U_g = 0.036$) is within the scatter of field data for the same value of Rossby number. At the surface q is 14% lower than Deardorff's result for the same value of u^*. A large part of this discrepancy occurs in the lateral velocity fluctuations. This tends to reinforce the need for some refinement of our pressure correlation modeling, as previously discussed. Near the top of

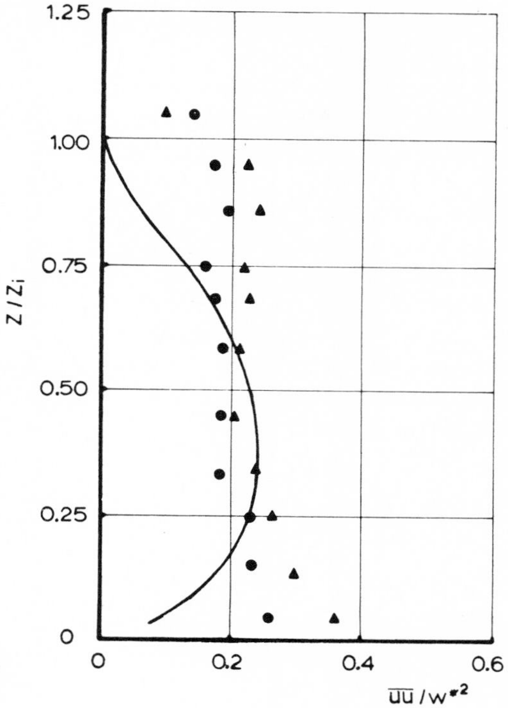

Fig. 9.17. Horizontal velocity fluctuations for the conditions of Figure 9.16.

the boundary layer, it appears that the accuracy of Deardorff's results is affected by his choice of a low value for his fixed boundary-layer height.

9.5. Local Equilibrium Approximations

If the second-order correlations are assumed to be in local equilibrium so that there is no time evolution or spatial diffusion of the correlations, Equations (9.23)–(9.25) may be reduced to

$$0 = -\overline{u_i u_k}\frac{\partial U_j}{\partial x_k} - \overline{u_j u_k}\frac{\partial U_i}{\partial x_k} + \frac{g_i \overline{u_j \theta}}{\Theta} + \frac{g_j \overline{u_i \theta}}{\Theta} - 2\varepsilon_{ikl}\Omega_k \overline{u_l u_j} - 2\varepsilon_{jlk}\Omega_l \overline{u_k u_i}$$

$$-\frac{q}{\Lambda}\left(\overline{u_i u_j} - \delta_{ij}\frac{q^2}{3}\right) - \delta_{ij}\frac{2bq^3}{3\Lambda} \qquad (9.49)$$

$$0 = -\overline{u_i u_j}\frac{\partial \Theta}{\partial x_j} - \overline{u_j \theta}\frac{\partial U_i}{\partial x_j} + \frac{g_i \overline{\theta^2}}{\Theta} - 2\varepsilon_{ijk}\Omega_j \overline{u_k \theta} - \frac{Aq\overline{u_i \theta}}{\Lambda} \qquad (9.50)$$

$$0 = -2\overline{u_j\theta}\frac{\partial\Theta}{\partial x_j} - \frac{2bsq\overline{\theta^2}}{\Lambda} \tag{9.51}$$

The laminar terms have also been neglected since low Reynolds number is incompatible with equilibrium turbulence. Equations (9.49)–(9.51) form a closed set of algebraic relationships between the second-order correlations and the gradients of the mean velocity and temperature. This is what Donaldson[(6)] calls the superequilibrium approximation.

The relationships between the second-order correlations and the mean-flow gradients determined by Equations (9.49)–(9.51) form a first-order closure or K theory for turbulence. This will be a valid approximation, provided (1) any changes in the mean flow are very slow compared with the characteristic time of the turbulence Λ/q and (2) spatial variations in the turbulence are small over the scale length Λ. Note that only rarely are both conditions satisfied because Λ is usually related to spatial variations in the

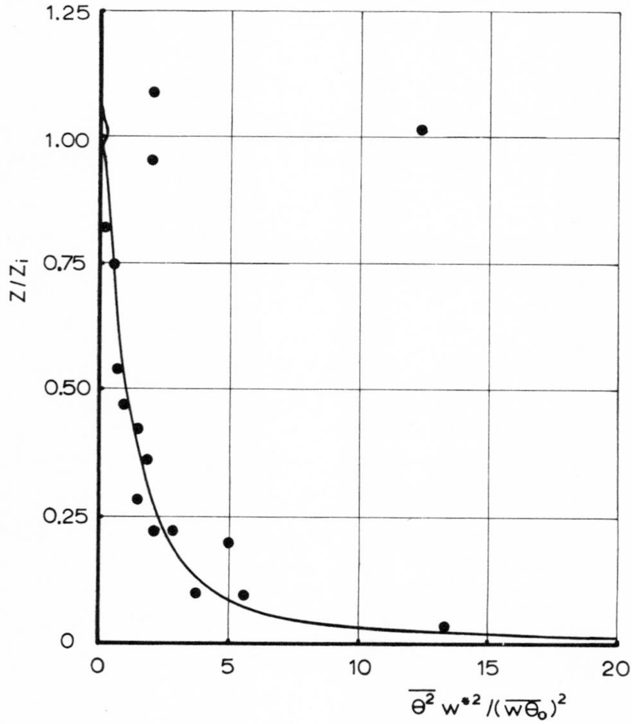

Fig. 9.18. Temperature fluctuations for the conditions of Figure 9.16.

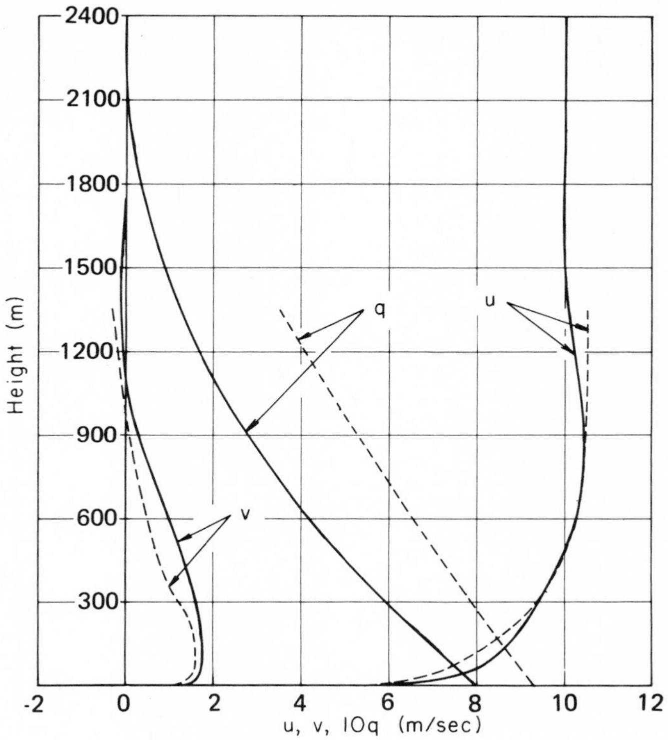

Fig. 9.19. Steady neutral atmospheric boundary layer profiles. A.R.A.P. prediction (——); prediction from Deardorff[55] (– – –).

flow. One particular region where both conditions are satisfied is in the constant flux region of the boundary layer when Λ/q is approaching zero.

The functional dependence of the correlations on the mean-flow gradients obtained from Equations (9.49)–(9.51) for the case where $U_i = (U(z), 0, 0)$, $g_i = (0, 0, -g)$, and $\Theta = \Theta(z)$ is shown in Figure 9.20. The only independent variable is the Richardson number of the mean flow:

$$\mathrm{Ri} = \frac{g}{\Theta}\frac{\partial\Theta}{\partial z}\bigg/\left(\frac{\partial U}{\partial z}\right)^2$$

A critical Ri is reached, above which no turbulence can exist. This critical value of 0.56 given in Figure 9.20 is higher than the value 0.21 observed in the atmospheric surface layer by Businger *et al.*[31] as reported in Section 9.4. However, in this surface layer, the turbulent correlations vary with

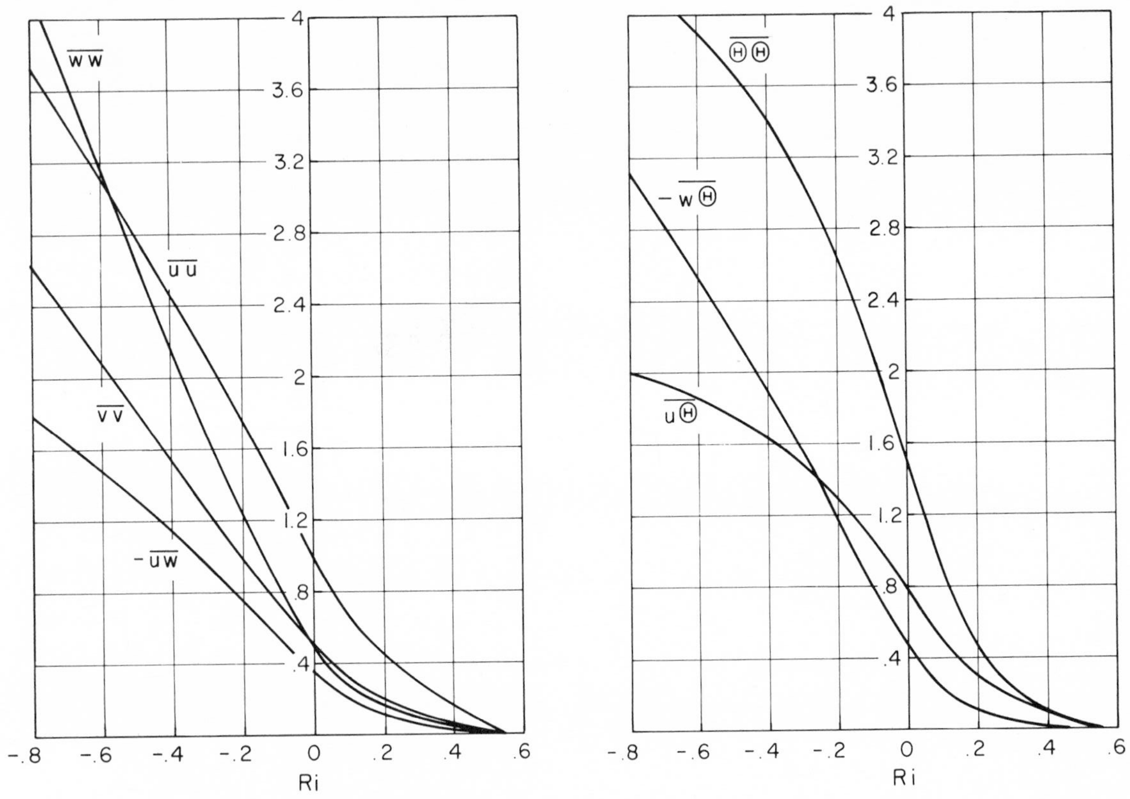

Fig. 9.20. (a) Superequilibrium Reynolds stress profiles normalized by $\Lambda^2(\partial U/\partial z)^2$. (b) Superequilibrium temperature correlation profiles normalized by $\Lambda^2(\partial\Theta/\partial z)(\partial U/\partial z)$, etc.

height as stability varies with height, so one should expect the diffusion terms to have some influence.

Superequilibrium relationships have also been computed for constant-density swirling flows.[55–58] In this case, there is both a lower bound and an upper bound on the range of values the swirl parameter may have for turbulence to exist.

Other approximations to Equations (9.23)–(9.25) may be tried short of reducing them to Equations (9.49)–(9.51). Mellor and Yamada[25] have experimented with four levels of approximation. We have had some success with a type of energy equation approach we call quasiequilibrium.[23,29] This consists of carrying the energy equation formed by the contraction of Equation (9.23) in full but using Equations (9.49)–(9.51) to relate the individual turbulent correlations to q^2. This should be considerably more accurate than superequilibrium because overall levels of time evolution and spatial diffusion are included through the turbulent energy equation. It cannot accurately predict flows when the transport of a quantity is counter to its gradient.

9.6. Applications

We believe the verifications of the model predictions made in Section 9.4 are sufficiently accurate over a wide variety of conditions to justify detailed calculations of flow situations for which no detailed measurements are available. A.R.A.P. is currently applying the model to a number of such areas. A few interesting results of these applications are presented here. Progress is also being made on applications to other areas, including compressible boundary layers,[59] compressible shear layers,[60,61] three-dimensional isolated vortex,[62] chemically reacting wakes[63] and jets,[64] swirling flow in an annulus,[65] and moisture in the atmospheric boundary layer.

9.6.1. Diurnal Variations in the Planetary Boundary Layer

Over homogeneous terrain, the atmospheric boundary layer may be considered a function of time and altitude only. The results of calculations for a fixed geostrophic wind and upper level temperature lapse rate with a cyclic surface heat flux approximating conditions over the Midwestern plains during summer[66] are presented in Reference 23. These results were obtained using the quasiequilibrium approach outlined in Section 9.5.2 and the gross feature approach to the scale. The results presented here were obtained using the full equations. There is relatively little difference in the numerical results.

The predicted contours of constant turbulent fluctuations are presented in Figure 9.21 as a function of time and altitude. The boundary-layer thickness grows during the day and shrinks during the evening and morning hours. The turbulent kinetic energy reaches a maximum of 4.8% of the geostrophic mean kinetic energy in the midafternoon at an altitude of approximately 500 m. As the sun sets, the turbulence begins to decay until in the early morning hours the maximum kinetic energy is ≈0.25% of the mean kinetic energy. The biggest difference between the results presented here and those in Reference 23 occurs during the early morning hours. The full equations with a dynamic scale predict a slower decay of q^2 and consequently considerably larger q^2 in the altitude range from 100 m to 1 km in the morning hours. But even this larger value is still quite small compared with afternoon values. Both model representations predict such features as the temperature inversion and local wind jet observed to occur nocturnally at low levels.

Surface shear stress is plotted as a function of the stability variable $\kappa u^*/fL$ in Figure 9.22. Data points for $\mathrm{Ro} \approx 10^7$, as taken from Tennekes' summary,[67] are included on the plot. The model predictions show a hysteresis loop with the surface shear stress significantly larger when the surface heat flux is decreasing rather than increasing. The data show

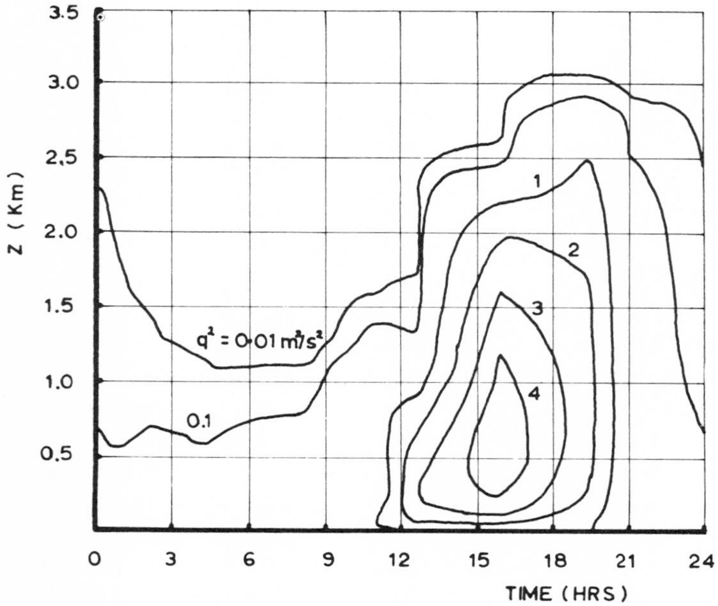

Fig. 9.21. Isopleths of turbulent energy in diurnal planetary boundary layer.

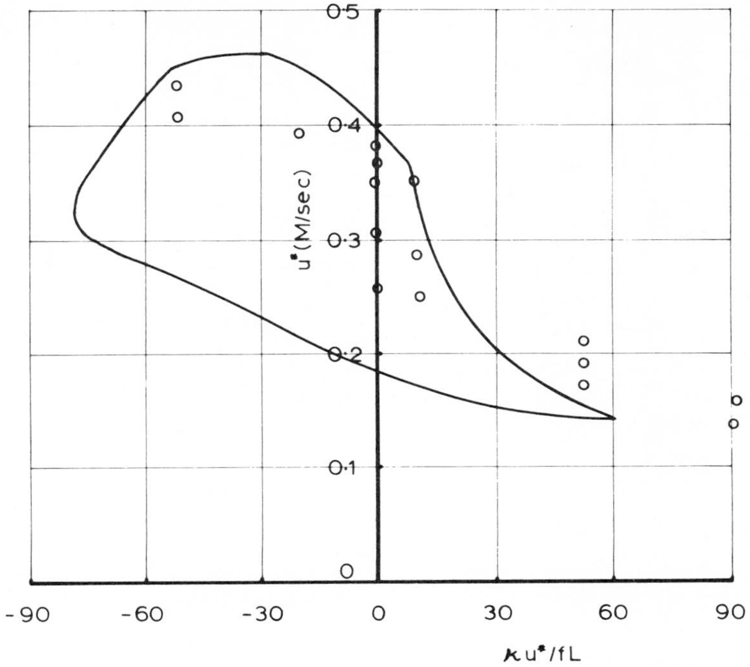

Fig. 9.22. Surface shear velocity vs. stability for diurnal planetary boundary layers. Data from Tennekes.[67]

considerable scatter but tend to be biased toward the upper bound. The factor of 2 difference in u^* at neutral conditions demonstrates the influence of unsteady dynamics on the atmospheric boundary layer.

9.6.2. Stratified Wake

The passage of a body through a stratified fluid generates a turbulent wake containing kinetic and potential energy. The buildup of potential energy is caused by the comixing of heavier-density fluid above the lighter-density fluid. In the initial stages of wake development in a weakly stratified medium, the turbulent kinetic energy dominates the potential energy, and the wake spreads. This spreading in turn increases the potential energy until, at a time comparable to the natural oscillating period of the fluid, inertial waves are generated. Computations for this flow using the quasiequilibrium version of the present model have been reported in References 29 and 68.

Figure 9.23 shows typical contours of the turbulent kinetic energy and secondary flow stream function at two instants of time following wake

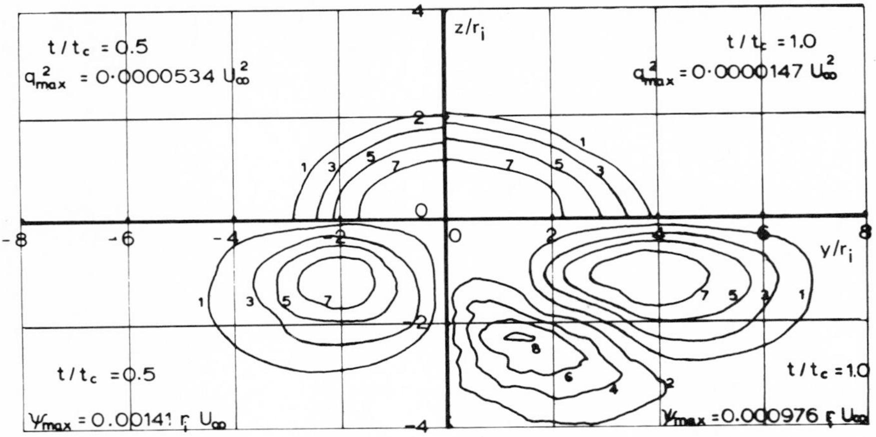

Fig. 9.23. Contours of turbulent energy and secondary flow streamlines for two instances of time after wake initialization for $\mathrm{Ri}_0 = 0.00925$; r_i is a measure of initial turbulence spread. $1 = +10\%$ of maximum value for that quadrant; $2 = -10\%$, $3 = +30\%$; $4 = -30\%$, etc.

initialization. Time is normalized by the characteristic Brunt–Väisälä period of the fluid

$$t_c = 2\pi\left[\left(\frac{g}{\Theta}\right)\frac{\partial\Theta}{\partial z}\right]^{-1/2}$$

The flow is symmetric about both axes so only one quadrant is shown for each variable at each time. At $t/t_c = 0.5$, the secondary flow exhibits a simple collapsing motion, while at $t/t_c = 1$, a second wave mode has been added. At both times the secondary flow extends well outside the region of turbulence.

Figure 9.24 is a summary plot showing the sensitivity of the wake shape to initial Froude number, $F_D = Ut_c/2\pi D$, or initial Richardson number of the turbulence, $\mathrm{Ri}_0 = (2\pi r_i/q_{\max} t_c)^2$, along with a qualitative comparison with laboratory flow visualization results.[69] There is good qualitative agreement between the observations and predictions.

9.6.3. Pollutant Dispersal

By adding equations for species continuity C and the correlations of species fluctuations $\overline{u_i c}$, $\overline{cc}$, and $\overline{c\theta}$, Equations (9.1)–(9.3) and (9.23)–(9.25) may be used to estimate pollutant dispersal in the atmospheric boundary layer. As long as the pollutant is assumed neutrally buoyant and nonreacting, the equations for C, $\overline{u_i c}$, $\overline{cc}$, and $\overline{c\theta}$ should be identical to Equations (9.3), (9.24), and (9.25), respectively.

Results of computations of pollutant dispersal in atmospheric turbulence as represented by our model predictions have been presented in References 6, 28, and 70. Figure 9.25 shows the influence of atmospheric stability on the vertical spread σ_z of a plume released at ground level:

$$\sigma_z^2 = \int_0^\infty (Cz^2)\,dz \Big/ \int_0^\infty C\,dz$$

The Richardson number $\hat{\mathrm{Ri}}$ is based on the difference between the values of temperature and velocity at 10 m altitude and at the surface, specifically,

$$\hat{\mathrm{Ri}} = \frac{g}{\Theta}\,\frac{(\Theta_{10\mathrm{m}} - \Theta_0)}{\hat{u}^2}(1\ \mathrm{m})$$

where $\hat{u} = 0.1U$ at 10 m. The spread rate for neutral conditions is very close to that predicted by Pasquill.[71] But the present model predicts a stronger effect of stability under unstable conditions and a weaker effect on the stable side.

Deardorff and Willis[72] have published results of particle releases into an unstable mixed layer, the same experimental flow used for comparison in Figures 9.16–9.18. Reference 70 shows comparable model predictions for a surface pollutant released into the calculated, unstable mixed layer field of turbulence shown in those figures. The model predicts that the maximum concentration will rise above the surface. As pointed out in Reference 72

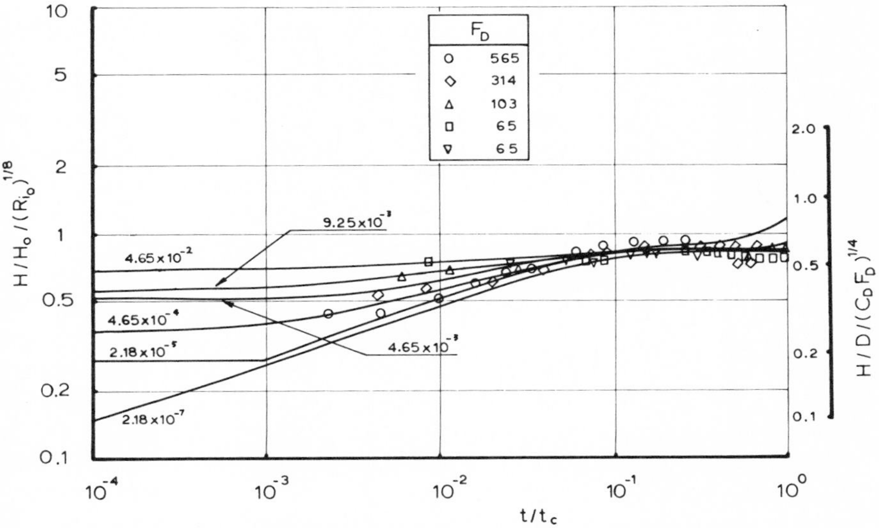

Fig. 9.24. Vertical spread of stratified wakes. Data from Lin and Pao.[69]

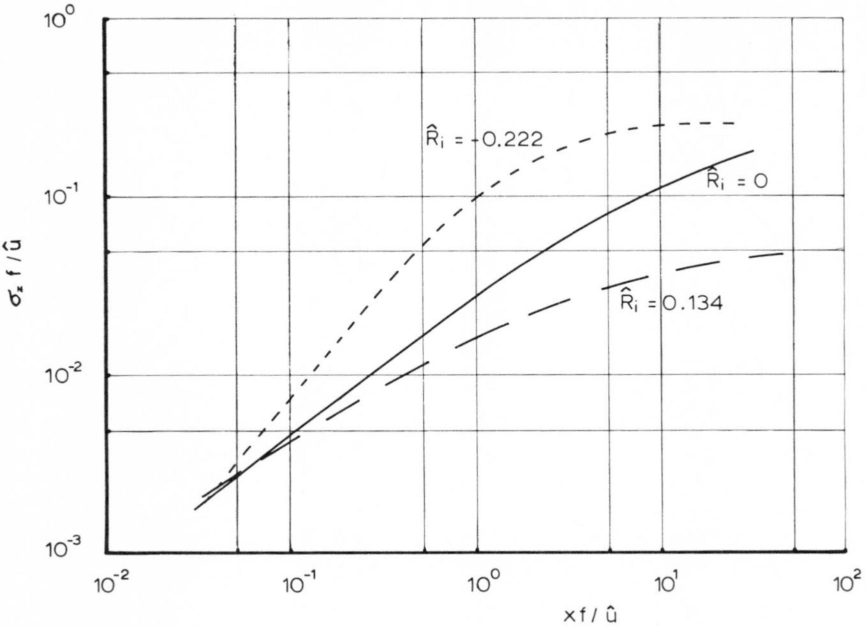

Fig. 9.25. Vertical plume spread from ground release into planetary boundary layers of various stabilities.

this feature could not be predicted by any eddy viscosity model or a Gaussian plume model.

The model predicts several non-Gaussian features for pollutant plumes caused by such mechanisms as wind shear variations in magnitude and direction with altitude, interaction with the inversion layer capping the boundary layer, and regions where turbulent scales are much larger than plume scales.

9.7. Concluding Remarks

Second-order closure has been established as a viable approach to practical turbulent flow computations. There are still about as many models as there are modelers—each model differing from the others in some detail. But this is a healthy situation, since there is still much to be learned. This is particularly true of the scale equation, since in many of the models every term in this equation must be modeled.

The relatively simple model presented here does a commendable job over a wide variety of flows. Thus, practical calculations can proceed concurrently with attempts to improve the model.

ACKNOWLEDGMENTS

This review has been partially funded with Federal funds from the Environmental Protection Agency under contract No. EPA 68-02-1310. The content of this paper does not necessarily reflect the views or policies of the U.S. Environmental Protection Agency, nor does mention of trade names, commercial products, or organizations imply endorsement by the U.S. Government.

Notation

A	Coefficient
a, b, c	Constants (with subscripts)
D_{ij}	Diffusion function [Equation (9.18)]
E	Energy spectrum
E_θ	Temperature-variance spectrum
F_D	Froude number
g_i	Acceleration vector
H	Vertical height of wake
k	Thermal conductivity
L	Characteristic length
m, n	Indices
N	Dissipation of temperature variance
p	Pressure fluctuation
P	Mean pressure
P_{ij}	Reynolds stress production term
Pr_t	Turbulent Prandtl number
q	Root-mean-square velocity fluctuation
Re	Reynolds number
Ri	Richardson number
Ro	Rossby number
s	Coefficients (with subscripts)
t	Time
u^*	Friction velocity
u_i	Fluctuating velocity components
U_i	Mean velocity components
v_c	Diffusion coefficient
W_D	Maximum velocity defect
x_i	Rectangular coordinates
α	1.68κ
β	Constant
δ_{ij}	Kronecker delta
κ	von Kármán constant
λ	Taylor microscale
Λ	Eddy macroscale
θ	Temperature fluctuations
Θ	Mean temperature
σ_z	Vertical spread of plume
ν	Kinematic viscosity
ρ	Density
π_{ij}	Pressure correlation function in Equation (9.12)
Ω	Angular velocity

References

1. Bradshaw, P., The understanding and prediction of turbulent flows, *Aeronaut. J.* **76**, 403–418 (1972).
2. Mellor, G. L., and Herring, H. J., A survey of the mean turbulent field closure models, *AIAA J.* **11**, 590–599 (1973).
3. Reynolds, W. C., Computation of turbulent flows—State of the art, 1970, report No. MD-27, Department of Mechanical Engineering, Stanford University, Stanford, California (1970).
4. Harlow, F. H. (ed.), Turbulence transport modeling, *AIAA Selected Reprints Series*, Vol. XIV, American Institute of Aeronautics and Astronautics, New York (1973).
5. Phillips, O. M., *The Dynamics of the Upper Ocean*, Cambridge University Press, Cambridge, England (1969).
6. Donaldson, C. du P., Atmospheric turbulence and the dispersal of atmospheric pollutants, *AMS Workshop on Micrometeorology* (D. A. Haugen, ed.), Science Press, Boston (1973), pp. 313–390.

7. Donaldson, C. du P., Calculation of turbulent shear flows for atmospheric and vortex motions, *AIAA J.* **10**, 4–12 (1972).
8. Daly, B. J., and Harlow, F. H., Transport equations in turbulence, *Phys. Fluids* **13**, 2634–2649 (1970).
9. Wolfshtein, M., Naot, D., and Lin, A., Models of turbulence, Ben-Gurion University of the Negev report No. ME-746 (N) (1974).
10. Rotta, J. C., Statistische Theorie nichthomogener Tubulenz, *Z. Phys.* **129**, 547–572 (1951).
11. Rotta, J. C., *Turbulence Strömungen*, Teubner Press, Stuttgart, West Germany (1972).
12. Crow, S. C., Viscoelastic properties of fine-grained incompressible turbulence, *J. Fluid Mech.* **33**, 1–20 (1968).
13. Launder, B. E., and Spalding, D. B., *Lectures in Mathematical Models of Turbulence*, Academic Press, New York (1972).
14. Rodi, W., The prediction of free-turbulent boundary layers by use of a two-equation model of turbulence, Ph.D. dissertation, Mechanical Engineering Department, Imperial College, London, December (1972).
15. Mellor, G. L., Analytic prediction of the properties of stratified planetary surface layers, *J. Atmos. Sci.* **30**, 1061–1069 (1973).
16. Launder, B. E., On the effects of gravitational field on the turbulent transport of heat and momentum, *J. Fluid Mech.* **67**, 569–581 (1975).
17. Naot, D., Shavit, A., and Wolfshtein, M., Two-point correlation model and the redistribution of Reynolds stresses, *Phys. Fluids* **16**, 738–743 (1973).
18. Lumley, J. L., and Khajeh-Nouri, B., Computational modeling of turbulent transport, *Advances in Geophysics*, Vol. 18A, Academic Press, New York (1974), pp. 169–192.
19. Hanjalic, K., and Launder, B. E., A Reynolds stress model of turbulence and its application to thin shear flows, *J. Fluid Mech.* **52**, 609–638 (1972).
20. Zeman, D., and Tennekes, H., A self-contained model for the pressure terms in the turbulent stress equations of the atmospheric boundary layer, unpublished manuscript (1975).
21. Varma, A. K., Second-order closure of turbulent reacting shear flows, Aeronautical Research Associates of Princeton report No. 235 (1975).
22. Lewellen, W. S., Teske, M., and Donaldson, C. du P., Application of turbulence model equations to axisymmetric wakes, *AIAA J.* **12**, 620–625 (1974).
23. Lewellen, W. S., Teske, M., and Donaldson, C. du P., Turbulence model of diurnal variations in the planetary boundary layer, *Proceedings 1974 Heat Transfer and Fluid Mechanics Institute* (L. R. Davies and R. E. Wilson, eds.), Stanford University Press, Stanford, California (1974), pp. 301–319.
24. Shir, C. C., A preliminary numerical study of atmospheric turbulent flows in the idealized planetary boundary layer, *J. Atmos. Sci.* **30**, 1327–1339 (1973).
25. Mellor, G. L., and Yamada, T., A hierarchy of turbulence closure models for planetary boundary layers, *J. Atmos. Sci.* **31**, 1791–1806 (1974).
26. Harlow, F. H., and Nakayama, P. I., Turbulence transport equations, *Phys. Fluids* **10**, 2323–2332 (1967).
27. Wilcox, D. C., and Alber, I. E., A turbulence model for high speed flows, *Proceedings* 1972 *Heat Transfer and Fluid Mechanics Institute*, Stanford University Press, Stanford (1972), pp. 231–252.
28. Lewellen, W. S., Teske, M., Contiliano, R. M., Hilst, G. R., and Donaldson, C. du P., Invariant modeling of turbulent diffusion in the planetary boundary layer, Environmental Protection Agency report No. EPA-650/4-74-035 (1974).

29. Lewellen, W. S., Teske, M., and Donaldson, C. du P., Turbulent wakes in a stratified fluid, A.R.A.P. report No. 226 (1974).
30. Champagne, F. H., Harris, V. G., and Corrsin, S., Experiments on nearly homogeneous turbulent shear flow, *J. Fluid Mech.* **41**, 81–139 (1970).
31. Businger, J. A., Wyngaard, J. C., Izumi, Y., and Bradley, E. F., Flux-profile relationships in the atmospheric surface layer, *J. Atmos. Sci.* **28**, 181–189 (1971).
32. Gad-el-Hak, M., and Corrsin, S., Measurements of the nearly isotropic turbulence behind a uniform jet grid, *J. Fluid Mech.* **62**, 115–143 (1974).
33. Coles, D., Measurements in the boundary layer on a smooth flat plate in supersonic flow. 1: The problem of the turbulent boundary layer, JPL report No. 20–69 (1953).
34. Tennekes, H., and Lumley, J. L., *A First Course in Turbulence*, MIT Press, Cambridge, Massachusetts (1972).
35. Gibson, C. H., Stegen, G. R., and Williams, R. B., Statistics of the fine structure of turbulence velocity and temperature fields measured at high Reynolds number, *J. Fluid Mech.* **41**, 153–167 (1970).
36. Lewellen, W. S., and Teske, M. E., Prediction of the Monin–Obukhov similarity functions from an invariant model of turbulence, *J. Atmos. Sci.* **30**, 1340–1345 (1973).
37. Wyngaard, J. C., Coté, O. R., and Rao, K. S., Modeling the atmospheric boundary layer, *Advances in Geophysics*, Vol. 18A, Academic Press, New York (1974), pp. 193–212.
38. Wyngaard, J. C., and Coté, O. R., The evolution of a convective planetary boundary layer—A higher order–closure model study, *Boundary Layer Meteorol.* **7**, 289–308 (1974).
39. Wygnanski, I., and Fielder, H., Some measurements in the self-preserving jet, *J. Fluid Mech.* **38**, 577–612 (1969).
40. Wygnanski, I., and Fiedler, H., The two-dimensional mixing region, *J. Fluid Mech.* **41**, 327–361 (1970).
41. Birch, S. F., and Eggers, J. M., A critical review of the experimental data for developed free turbulent shear layers, in: *Free Turbulent Shear Flows*, Volume I: *Conference Proceedings*, NASA report No. SP-321, NASA Langley Research Center, Hampton, Virginia (1973).
42. Townsend, A. A., *The Structure of Turbulent Shear Flow*, Cambridge University Press, Cambridge, England (1956).
43. Thomas, R. M., Conditional sampling and other measurements in a plane turbulent wake, *J. Fluid Mech.* **57**, 549–582 (1973).
44. Chevray, R., The turbulent wake of a body of revolution, *J. Basic Eng.* **90**, 275–284 (1968).
45. Naudascher, E., Flow in the wake of self-propelled bodies and related sources of turbulence, *J. Fluid Mech.* **22**, 625–656 (1965).
46. Klebanoff, P. S., Characteristics of turbulence in a boundary-layer with zero pressure gradient, NACA report No. 1247 (1955).
47. Bradley, E. F., A micrometerological study of velocity profiles and surface drag in the region modified by a change in surface roughness, *Quart. J. R. Meteorol. Soc.* **94**, 361–379 (1968).
48. Rao, K. S., Wyngaard, J. C., and Coté, O. R., The structure of the two-dimensional internal boundary layer over a sudden change of surface roughness, *J. Atmos. Sci.* **31**, 738–746 (1974).
49. Freymuth, P., and Uberoi, M. S., Structure of temperature fluctuations in the turbulent wake behind a heated cylinder, *Phys. Fluids* **14**, 2574–2580 (1971).
50. LaRue, J. C., and Libbey, P. A., Temperature fluctuations in the plane turbulent wake, *Phys. Fluids* **17**, 1956–1967 (1974).
51. Monin, A. S., and Obukhov, A. M., Dimensionless characteristics of turbulence in the atmospheric surface layer, *Dokl. Akad. Nauk. SSSR* **93**, 223–226 (1953).
52. Wyngaard, J. C., Coté, O. R., and Izumi, Y., Local free convection, similarity and the budgets of shear stress and heat flux, *J. Atmos. Sci.* **28**, 1171–1182 (1971).

53. Kato, H., and Phillips, O. M., On the penetration of a turbulent layer into stratified fluid, *J. Fluid Mech.* **37**, 643–656 (1969).
54. Willis, G. E., and Deardorff, J. W., A laboratory model of the unstable planetary boundary layer, *J. Atmos. Sci.* **31**, 1297–1307 (1974).
55. Deardorff, J. W., Numerical investigation of neutral and unstable planetary boundary layers, *J. Atmos. Sci.* **29**, 91–115 (1972).
56. Donaldson, C. du P., The relationship between eddy transport and second-order closure models for stratified media and for vortices, in: *Free-Turbulent Shear Flows*, Volume I; *Conference Proceedings*, NASA report No. SP-321, NASA Langley Research Center, Hampton, Virginia (1973), pp. 233–255.
57. Donaldson, C. du P., and Bilanin, A. J., Vortex wakes of conventional aircraft, AGARDograph, No. 204 (1975).
58. Mellor, G. L., A comparative study of curved flow and density stratified flow, unpublished (1974).
59. Donaldson, C. du P., and Sullivan, R. D., An invariant second order closure model of the compressible turbulent boundary layer on a flat plate, Aeronautical Research Associates of Princeton report No. 178 (1972).
60. Varma, A. K., Beddini, R. A., Sullivan, R. D., and Donaldson, C. du P., Application of an invariant second-order closure model to compressible turbulent shear layers, AIAA paper No. 74-592 (1974).
61. Varma, A. K., Beddini, R. A., Sullivan, R. D., and Fishburne, E. S., Turbulent shear flows in laser nozzles and cavities, Air Force Office of Scientific Research report No. AFOSR-TR-74-1786 (1974).
62. Sullivan, R. D., A program to compute the behavior of a three-dimensional turbulent vortex, Aerospace Research Laboratories report No. ARL-TR-74-0009 (1973).
63. Hilst, G. R., The chemistry and diffusion of aircraft exhausts in the lower stratosphere during the first few hours after flyby, Aeronautical Research Associates of Princeton report No. 216 (1974).
64. Donaldson, C. du P., and Varma, A. K., Remarks on the construction of a second order closure description of turbulent reacting flows, in: *Combustion Science and Technology*, Vol. 13 (1976), pp. 55–78.
65. Bilanin, A. J., Snedeker, R. S., Sullivan, R. D., and Donaldson, C. du P., Final report on an experimental and theoretical study of aircraft vortices, Aeronautical Research Associates of Princeton report No. 238 (1975).
66. Wyngaard, J. C., Notes on surface layer turbulence, in: *AMS Workshop on Micrometeorology* (D. A. Haugen, ed.), Science Press, Boston (1973), pp. 101–149.
67. Tennekes, H., Similarity laws and scale relations in planetary boundary layers, in: *AMS Workshop in Micrometeorology* (D. A. Haugen, ed.), Science Press, Boston (1973), pp. 177–216.
68. Lewellen, W. S., Teske, M., and Donaldson, C. du P., Second-order turbulence modeling applied to momentumless wakes in stratified fluids, Aeronautical Research Associates of Princeton report No. 206 (1973).
69. Lin, J. T., and Pao, Y. H., Turbulent wake of a self-propelled slender body in stratified and nonstratified fluids: Analysis and flow visualizations, Flow Research Corp. report No. 11 (1973).
70. Lewellen, W. S., and Teske, M., Second-order closure modeling of diffusion in the atmospheric boundary layer, *Boundary Layer Meteorol.* **10**, 69–90 (1976).
71. Pasquill, F., *Atmospheric Diffusion*, Second Edition, Halstead Press, John Wiley and Sons, Inc., New York (1974).
72. Deardorff, J. W., and Willis, G. E., Physical Modeling of Diffusion in the Mixed Layer, in: *Proceedings of the Symposium on Atmospheric Diffusion and Air Pollution*, Santa Barbara (1974), American Meteorological Society, Boston, pp. 387–391.

CHAPTER 10

Numerical Simulation of Turbulent Flows

S. A. ORSZAG

10.1. Introduction

Digital computers capable of 1–10 MIPS (millions of instructions per second) are now readily available (e.g., CDC 6600, 7600, IBM 360–195), while it is hoped that machines currently being developed will be capable of 10–100 MIPS (CDC STAR, Illiac, TI ASC, Cray Research). The great power of these machines has made possible the solution of some of the most challenging fluid dynamical problems, those of turbulence, by numerical solution of the Navier–Stokes equations.

Early progress towards this goal was reported by Orszag and Patterson.[1] In that work, simulations of three-dimensional homogeneous isotropic turbulence at moderate Reynolds numbers were performed and the results compared with the predictions of turbulence theory. Since then several studies have been made of other turbulent flows using similar methods. In the present paper, we review the work on these calculations. While it has not yet been possible to simulate three-dimensional inertial ranges from first principles, there has been substantial progress in the simulation of inertial ranges in two dimensions and the energy-containing range in three dimensions.

S. A. ORSZAG • Massachusetts Institute of Technology, Cambridge, Massachusetts

10.2. Methods

The principal problems of interest involve incompressible flows governed by the Navier–Stokes equations

$$\frac{\partial \mathbf{v}(\mathbf{x}, t)}{\partial t}+[\mathbf{v}(\mathbf{x}, t) \cdot \boldsymbol{\nabla}]\mathbf{v}(\mathbf{x}, t)=-\boldsymbol{\nabla} p(\mathbf{x}, t)+\nu\nabla^2\mathbf{v}(\mathbf{x}, t) \tag{10.1}$$

$$\boldsymbol{\nabla} \cdot \mathbf{v}(\mathbf{x}, t)=0 \tag{10.2}$$

where $\mathbf{v}(\mathbf{x}, t)$ is the velocity field, $p(\mathbf{x}, t)$ is the pressure, and ν is the kinematic viscosity. Equations (10.1) and (10.2) isolate the basic nonlinear mechanisms of turbulent flow; other effects may be important including compressibility, buoyancy, chemical reactions, multiphase flows, etc. It is appropriate to recall that it follows from Equations (10.1) and (10.2) that

$$\nabla^2 p(\mathbf{x}, t)=-\boldsymbol{\nabla} \cdot [\mathbf{v}(\mathbf{x}, t) \cdot \boldsymbol{\nabla}]\mathbf{v}(\mathbf{x}, t) \tag{10.3}$$

This Poisson equation determines the pressure in order that the incompressibility constraint of Equation (10.2) be satisfied. The pressure is governed by the diagnostic equation [Equation (10.3)] rather than a prognostic equation for $\partial p/\partial t$, as would be the case in a compressible flow with finite sound speed. [In fact, the pressure in Equation (10.1) is quite analogous to a Lagrange multiplier for maintenance of the kinematical constraint of Equation (10.2).]

Homogeneous turbulence can be simulated by imposition of periodic boundary conditions on Equation (10.1), i.e.,

$$\mathbf{v}(\mathbf{x}+2\pi\mathbf{n}, t)=\mathbf{v}(\mathbf{x}, t) \tag{10.4}$$

where $\mathbf{n}$ is a vector with integer components. With these boundary conditions, an attractive method for numerical solution of the Navier–Stokes equations is to seek an approximate solution of the form of a truncated Fourier series

$$\mathbf{v}(\mathbf{x}, t)=\sum_{|\mathbf{k}|<K} \mathbf{u}(\mathbf{k}, t)\, e^{i\mathbf{k}\cdot\mathbf{x}} \tag{10.5}$$

where the wave vectors $\mathbf{k}$ must have integer components if the periodicity interval is 2π as in Equation (10.4). When Equation (10.5) is used, the Navier–Stokes equations become

$$\left(\frac{\partial}{\partial t}+\nu k^2\right)u_\alpha(\mathbf{k}, t)=-ik_\beta\left(\delta_{\alpha\gamma}-\frac{k_\alpha k_\gamma}{k^2}\right)\sum_{\substack{|\mathbf{p}|<K\\ |\mathbf{k}-\mathbf{p}|<K}} u_\beta(\mathbf{p}, t)u_\gamma(\mathbf{k}-\mathbf{p}, t) \tag{10.6}$$

where summation over repeated Greek indices is implied and $\delta_{\alpha\gamma}$ is the Kronecker delta; the pressure has been eliminated from Equation (10.6) by use of Equation (10.3). Efficient techniques to solve the coupled system of ordinary differential equations in t are discussed by Orszag.[2] This method of solving the Navier–Stokes equations is called a spectral method, in contrast to conventional finite-difference methods in which Equations (10.1) and (10.2) are discretized on a finite grid of points.

More generally, spectral methods involve the representation of the velocity field by a truncated series in terms of smooth functions:

$$\mathbf{v}(\mathbf{x}, t) = \sum_{|m|<M} \sum_{|n|<N} \sum_{|p|<P} \mathbf{a}_{mnp}(t)\psi_m(x_1)\phi_n(x_2)\chi_p(x_3) \tag{10.7}$$

The proper choice of expansion functions is crucial to the success of the method: The criteria are that there be rapid convergence to the exact solution as the cutoffs $M, N, P \to \infty$, and that there be efficient methods to solve the coupled system of differential equations for $\mathbf{a}_{mnp}(t)$ (cf. Orszag and Israeli[3]). Some appropriate choices of expansion functions are given in Table 10.1.

Table 10.1. Choice of Expansion Functions

Type of boundary condition	Function	Comment
1. Periodic	$\exp(ikx)$	
2. Free-slip (stress-free)	$\sin(kx)$, $\cos(kx)$	Same as (1) with symmetry constraint
3. Rigid (no-slip)		
a. Cartesian coordinates	$T_n(x)$	Chebyshev polynomials—series similar to cosine series since $T_n(\cos x) = \cos nx$
	$P_n(x)$	Legendre polynomials—less efficient than $T_n(x)$ but more natural for maintenance of conservation laws because orthogonality weight factor is 1 instead of $1/(1-x^2)^{1/2}$
b. Cylindrical (polar coordinates)	$r^s T_n(r)$	See Orszag[28]
4. Spherical polars	$Y_n^m(\theta, \phi)$ $\sin^s\theta \cos n\theta \exp(im\phi)$	Surface harmonics—generalized Fourier series, see Orszag[28]
5. Semi-infinite and infinite geometry	$T_n(x)$	With mapping or truncation to finite domain
	$L_n(x)$	Laguerre polynomial $0 \le x < \infty$
	$H_n(x)$	Hermite polynomial $-\infty < x < \infty$, see Gottlieb and Orszag[29]

When these spectral methods are properly designed and implemented, they offer a number of advantages over finite-difference methods. Five general areas of comparison are as follows (cf. Orszag and Israeli[(3)]).

a. Rate of Convergence. If $\mathbf{v}(\mathbf{x}, t)$ is smooth (infinitely differentiable) then the error incurred by use of the spectral representation of Equation (10.7) goes to zero faster than any finite power of the cutoffs. In contrast, finite-difference methods yield finite-order rates of convergence (most usually second-order in the resolution). The important consequence is that high accuracy is achieved with little or no extra effort in spectral methods once moderate accuracy is achieved.

b. Efficiency. The development of fast transform methods has allowed spectral codes to be developed that are competitively as fast as finite-difference codes with the same number of independent degrees of freedom. However, the spectral codes require a factor of 2–5 fewer degrees of freedom in each of the (three) space directions to resolve the flow with order 5% error. Hence, an order of magnitude or better improvement in utilization of computer resources is achieved.

c. Boundary Conditions. There are three aspects to boundary conditions of importance for spectral methods. First, if Chebyshev or Legendre polynomials are used to represent the direction normal to a boundary layer of normalized thickness ε, only about $1/\varepsilon^{1/2}$ polynomials need be retained to achieve high accuracy. On the other hand, if a uniform grid were used then $1/\varepsilon$ grid points would be required. Second, if coordinate transformations are employed to assist in the resolution of a boundary or internal layer of thickness ε, the spectral errors are decreased faster than any finite power of ε as $\varepsilon \to 0$. On the other hand, finite-difference results are improved by a finite power of ε. Third, with spectral methods high-order accuracy is achieved at inflow and outflow boundaries without the need for special methods to impose the boundary conditions. On the other hand, high-order difference methods break down near boundaries because grid points outside the physical domain must be invoked, and the necessary modifications to maintain accuracy near the boundary can get quite complicated (Kreiss and Oliger[(4)]).

d. Discontinuities. Surprisingly, spectral methods do a better job of localizing errors than difference schemes and hence require considerably less local dissipation to smooth discontinuities.

e. "Bootstrap" Estimation of Accuracy. It has been shown by Herring *et al.*[(5)] that spectrum shape provides a built-in measure of accuracy of spectral calculations. If the spectral amplitudes approach zero regularly, it is possible to estimate the accuracy of the calculation from the results of the calculation itself. On the other hand, no such *internal* measure of accuracy of difference methods has yet been found; to determine the accuracy of a finite-difference calculation, it is necessary to compare calculations of varying resolution.

Turbulence computations are performed by a numerical solution of Equations (10.1) and (10.2) together with suitable initial and boundary conditions. With homogeneous turbulence, periodic boundary conditions [Equation (10.4)] are applied while random initial conditions are set up as follows. The spectral representation [Equation (10.5)] is used with the choice

$$u_\alpha(\mathbf{k}, 0) = \left(\delta_{\alpha\gamma} - \frac{k_\alpha k_\gamma}{k^2}\right) r_\gamma(\mathbf{k}) \tag{10.8}$$

where $\mathbf{r}(\mathbf{k})$ are independent, Gaussianly distributed random variables with variance proportional to a specified (nonrandom) energy spectrum $E(\mathbf{k})$. The choice contained in Equation (10.3) guarantees the incompressibility constraint $k_\alpha u_\alpha(\mathbf{k}, 0) = 0$. When Equation (10.8) is used in Equation (10.5), there results a random initial flow field with Gaussian distribution and energy spectrum $E(\mathbf{k})$.

The establishment of random initial conditions for inhomogeneous turbulent flows like wakes and jets is somewhat more complicated. It has been found (cf. Orszag and Pao[6]) that the most convenient way to maintain the incompressibility constraint is to express the velocity in terms of a vector potential

$$\mathbf{v}(\mathbf{x}, 0) = \nabla \times \mathbf{A}(\mathbf{x}) \tag{10.9}$$

where $\mathbf{A}(\mathbf{x})$ is chosen to model the inhomogeneous statistics.

10.3. Problems

The fundamental difficulty with numerical simulation of turbulence can be illustrated by considering the possibility for calculating a flow similar to that observed by Grant *et al.*[7] in their study of the inertial-range spectrum of the flow in Discovery Passage, a tidal channel in British Columbia. The parameters of this flow are roughly

$\bar{v}$ = root-mean-square turbulent velocity ~ 150 cm/sec

L = large eddy size $\sim 10^4$ cm

l = small eddy size ~ 1 cm

Here l is the scale on which energy dissipation by viscosity occurs; in water the kinematic viscosity $\nu \simeq 0.01\ \text{cm}^2/\text{sec}$. The Reynolds number is $R = \bar{v}L/\nu \simeq 1.5 \times 10^8$. The inertial range spectrum of this flow is plotted in Figure 10.1, where $\phi_1(k)$ is the one-dimensional spectrum (Batchelor[8]),

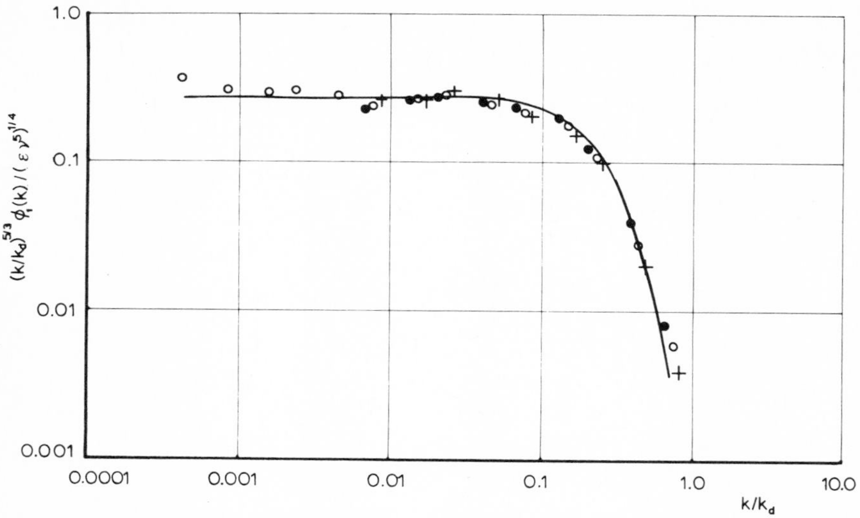

Fig. 10.1. Nondimensional one-dimensional spectrum function $k^{5/3}\phi_1(k)$ in the inertial and dissipation ranges as computed by the Lagrangian-history direct-interaction approximation (solid curve: Kraichnan[(25)]) and compared with the October 1959 data of Grant *et al.*[(7)] Here $k_d = (\varepsilon/\nu^3)^{1/4}$ and $\phi_1(k)$ is related to the isotropic energy spectrum $E(k)$ by

$$E(k) = k^3\left[\frac{d}{dk}\frac{1}{k}\frac{d}{dk}\phi_1(k)\right]$$

(from Kraichnan[(25)]).

ε is the rate of energy dissipation, and $k_d = (\varepsilon/\nu^3)^{1/4}$ is the Kolmogorov dissipation wave number. Notice that the inertial range spectrum $\phi_1(k) \propto k^{-5/3}$ extends over two decades of wave number.

In order to simulate this flow accurately over all dynamically important scales, it is necessary to resolve the range of scales 10^4–1, so that at least 10^4 degrees of freedom are required in each of three space directions. Consequently, about $(10^4)^3 = 10^{12}$ degrees of freedom (velocity, pressure values, etc.) must be retained. To be useful, the calculation of Equation (10.1) should proceed for some dynamically significant time, typically the large-eddy circulation time, which is about $L/\bar{v} \simeq 60$ sec. However, numerical solution of Equation (10.1) must proceed in small time steps, to maintain both accuracy and numerical stability, typically of the size $l/\bar{v} \simeq 6\times 10^{-3}$ sec. Consequently, about 10^4 time steps are required to calculate one eddy circulation time. A typical numerical solution of Equation (10.1) requires about 100 computer operations (instructions) per time step per retained degree of freedom. Therefore, with 10^4 time steps and 10^{12} degrees of freedom, about 10^{18} operations are required to complete the simulation.

Even with the next generation of machines, which may operate at 1 nsec ($=10^{-9}$ sec) per operation at a cost of about \$1/sec, about 10^9 sec $\simeq 30$ yr and \$$10^9$ are required to complete a single computation!

The estimates just given may be formalized as follows: The large-scale L of a turbulent flow is the size of the energy-containing eddy motions and is typically related to the size of the body or scale of the forces generating the motion. The turbulent motions on smaller scales give rise to enhanced rates of transport in the flow over corresponding molecular transfer rates, including enhanced energy dissipation, momentum transfer, heat transfer, and particle diffusion. These enhanced transport rates are often modeled by eddy transfer coefficients which turn out to be many orders of magnitude larger than molecular coefficients when the Reynolds number R is large.

It has been conjectured [with some experimental and theoretical support (cf. Batchelor[8] and Orszag[9])] that the energy dissipation rate ε remains $O(1)$ as $R \to \infty$ in a turbulent flow, in contrast to the elementary estimate $\varepsilon = O(1/R)$ in a laminar flow. This property that $\varepsilon = O(1)$ as $R \to \infty$ is intimately related to the idea that the small scales in a turbulent flow adjust themselves to provide the required enhanced transport as $R \to \infty$. In the case of energy dissipation, this behavior requires that the dissipation scale l, which is the smallest dynamically important scale in the flow, must be of order

$$l = (\nu^3/\varepsilon)^{1/4} \simeq L/R^{3/4} \qquad \text{as } R \to \infty \tag{10.10}$$

Equation (10.10) follows from dimensional analysis: $[l] = \text{cm}$, $[\varepsilon] = \text{cm}^2\,\text{sec}^{-3}$, $[\nu] = \text{cm}^2\,\text{sec}^{-1}$, while l can depend only on ε and ν since (a) in homogeneous turbulence, l adjusts itself to maintain $\varepsilon = O(1)$ because ε provides the only dynamical coupling of small scales to the large energy-containing eddies; (b) the precise scale of dissipation is governed by the molecular viscosity ν, which must do the actual dissipation of energy. A similar argument implies that, in the inertial range, which consists of small eddies that are sufficiently large that molecular viscosity ν has no direct effect on them, the energy spectrum $E(k)$ must satisfy Kolmogorov's law

$$E(k) = C\varepsilon^{2/3} k^{-5/3} \tag{10.11}$$

Equation (10.11) also follows by dimensional analysis since $[E(k)] = \text{cm}^3\,\text{sec}^{-2}$ and $[k] = \text{cm}^{-1}$, while $E(k)$ can depend only on ε and k, the wave number, in the inertial range. The qualitative spectrum of turbulence predicted by this dimensional theory is plotted in Figure 10.2.

It follows from Equation (10.10) that the range of spatial scales that must be accurately resolved in a calculation at Reynolds number R is of order

$$L/l \simeq R^{3/4}$$

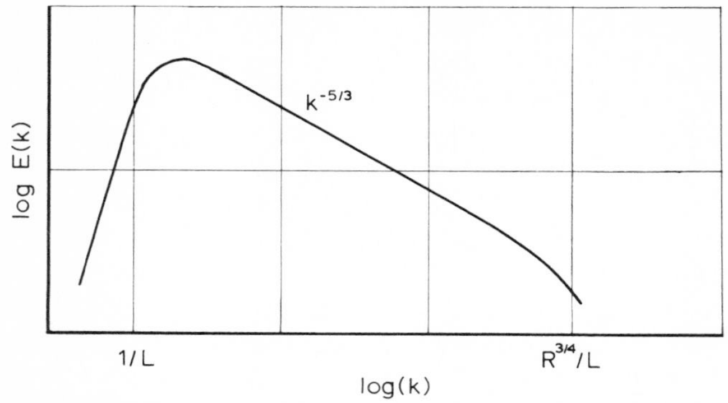

Fig. 10.2. Qualitative spectrum of turbulence in the inertial and dissipation ranges on the basis of the Kolmogorov theory.

so that the total number of spatial degrees of freedom is of order $(R^{3/4})^3 = R^{9/4}$. The calculation must proceed for a time of order $L/\bar{v}$, while time steps must be restricted to order $l/\bar{v}$, so that order $L/l \simeq R^{3/4}$ time steps must be taken per simulation run. Consequently, the total number of operations scales as $R^{9/4}R^{3/4} = R^3$.

The discouraging result of this argument is that every increase of computer power by a factor of 10 allows only an increase by a factor of $10^{1/3} \simeq 2.15$ in the Reynolds number.

Estimates like those just given have seemed to spell doom for all reasonable attempts at numerical solution of the turbulence problem. However, it is unreasonable to be so pessimistic. If a calculation at $R = 10^8$ requires 10^{21} operations (which is a better estimate than the 10^{18} obtained above) then the R^3 dependence implies that a calculation at $R = 10^4$ requires only 10^9 operations or 15 min on a 1-MIPS machine. There are many interesting turbulent flows at Reynolds numbers $R \sim 10^4$–10^5. In fact, most laboratory turbulence experiments fall within this range—only geophysical environments provide tractable sources of data at much higher Reynolds numbers.

The numerical simulations of turbulence reviewed in Section 10.4 have been performed at Reynolds numbers in the range 10^3–10^5. At these Reynolds numbers, the computer experiments can complement and supplement laboratory experimental data. The question of the relevance of these moderate-Reynolds-number turbulence experiments to huge-Reynolds-number geophysical flows is explored further in Section 10.5.

There are two important practical difficulties with numerical simulations of turbulence in addition to the resolution problem outlined above for homogeneous turbulence; these additional difficulties also relate to inadequate resolution but are of a somewhat different nature than the difficulty discussed above. First, there is the difficulty with obtaining adequate statistics to compute accurate statistical averages, and, second, there are a variety of difficulties relating to the imposition of boundary conditions on the flow.

The difficulty with statistics lies in the fact that the arithmetical mean of N independent values of a random variable with average A and standard deviation D fluctuates about A with amplitude of order $D/N^{1/2}$. Consequently, increasing the sample size by a factor of 100 only decreases errors by a factor of 10. Large samples are necessary for high accuracy.

Large samples are readily available for homogeneous turbulence, since three-dimensional spatial averages may be used. In the isotropic turbulence simulations discussed in Section 10.4, averages were obtained as arithmetical means over bands in wave space, i.e., as averages over all wave vectors satisfying $k-\frac{1}{2}\Delta k<|\mathbf{k}|<k+\frac{1}{2}\Delta k$, where the bandwidth was chosen sufficiently large that most spectra so obtained were accurate to better than 10%. On the other hand, in shear flows, sample size is a serious problem. In two-dimensional shear flows, averages are functions of x (the downstream direction) and z (the cross-stream direction), but not of y (the spanwise direction). In this case, spatial averages may be obtained as averages over the y direction; if the flow is longitudinally homogeneous (statistically homogeneous in x) then the x direction may also be used for averaging. Similarly, in a statistically axisymmetric shear flow, the azimuthal direction and perhaps the longitudinal direction may be used for spatial averages. However, realistic shear flows have at least one direction of inhomogeneity and this means a significant degradation of the accuracy of spatial statistics compared with corresponding statistically homogeneous flows.

In a general turbulent flow, the situation with regard to statistics is very serious. If there is no direction of spatial homogeneity, then averages may be computed in but two ways. First, time averages may be used *if* the flow is statistically stationary. However, it is not sufficient to use the flow values at N successive time steps to get N independent measurements, because the flow is strongly correlated from one time step to the next. In order to get N independent samplings of an energy-containing scale of motion, it is necessary to calculate through order N large-eddy circulation times $L/\bar{v}$. This increases our previous operation estimates by a factor of N.

Second, averages of an arbitrary (statistically inhomogeneous, nonstationary) flow can always be computed as an ensemble average, in which the average is taken over N distinct flows with the same statistics. However, computing an ensemble average increases the amount of work by a factor of N over that necessary to compute a single flow.

The final difficulty with numerical simulation of turbulence that we wish to discuss concerns boundary conditions. In Figure 10.3, we give a schematic

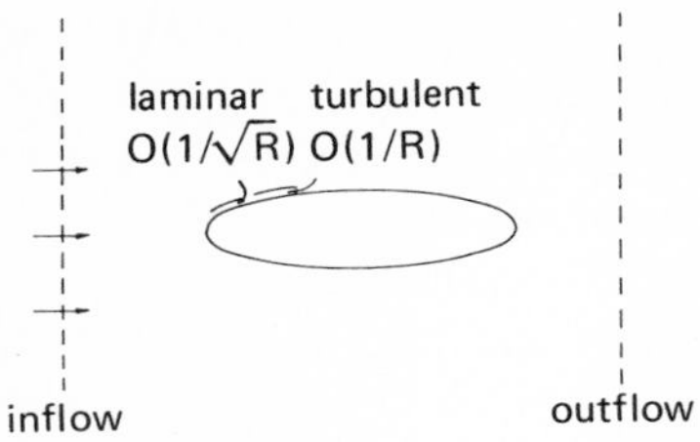

Fig. 10.3. Schematic representation of computational domain for simulation of turbulent flow past a body, including inflow and outflow boundaries and the boundary layers on the body.

representation of the computational region for turbulent flow past a body. There are three general regions where boundary conditions must be imposed: at the inflow boundary, at the outflow boundary, and on the body.

a. Inflow Boundaries. It is not possible to deal directly with infinite spatial regions because of the finite capacity of computers. There are two possible procedures for dealing with this problem. One way is to transform the infinite region into a finite region by use of a map, like $z = x/(x+1)$, which maps $0 \leq x < \infty$ into $0 \leq z < 1$. As explained by Grosch *et al.*[10] the use of such maps requires considerable care; mapping is successful only if the asymptotic behavior of the solution being sought is sufficiently simple. In the case of flow past a body, the downstream outflow is not at all simple (since it comprises the turbulent wake region), so mapping does not seem to hold promise there. However, the flow at the upstream boundary may be "simple" and mapping may be appropriate, though to the author's knowledge it has not yet been tried in this situation. The other method is to truncate the computational domain at some finite distance, as shown in Figure 10.3. In this case, the inflow velocity field must be completely specified. The arbitrariness of this specification influences the rest of the computation. It may seem that the truncation method involves more arbitrariness than mappings do. However, this is not so, since it may be shown (Grosch *et al.*[10]) that both procedures have similar properties. In summary, the difficulty with upstream (inflow) boundaries concerns the lack of knowledge of the required upstream flow. A similar arbitrariness appears in the specification of initial conditions in turbulence simulations. Presumably this lack of knowledge does not affect the final results strongly, in the sense that while individual realizations of turbulence are highly sensitive and unstable, average properties are strongly stable to perturbations.

b. Outflow Boundaries. The situation concerning outflow boundary conditions is potentially very serious. Here it seems physically unrealistic to permit specification of the complete flow field. Yet, *viscous* flow theory shows that the complete velocity field should be specified at an outflow boundary; *inviscid* flow theory suggests that specification of either the normal component of the outflow velocity or the pressure is sufficient to set the problem. However, all these flow features are *a priori* unknown; in fact, it is certain that a large obstacle, say, placed some distance downstream of the outflow boundary *will* affect the flow within the computational domain. In a turbulent flow, it seems that the only hope is that specific details of downstream boundary conditions have little effect upstream. There is some evidence for this behavior in simulations of boundary-layer transition but it remains to be verified. The real danger with downstream boundaries is that by overspecification the flow may be modified in a very important way. For example, if both the normal component of velocity and the pressure are given at outflow, then the drag coefficient of the body is completely specified, most likely in error compared with the true drag coefficient. Drag should be a result of the computation, not something unwittingly imposed during problem formulation.

c. Boundary Conditions on the Body. The difficulty here shows itself in several ways. If there is a *laminar* boundary layer on the body, its thickness is of order $R^{-1/2}$ compared with the body size. At large R, it seems that high resolution is required to resolve the boundary layer. However this is not difficult to achieve, even at high Reynolds number, if a mapping is used to stretch the coordinate normal to the body. This mapping solves the problem *if* the length scale of the turbulence exterior to the boundary layer is much larger than the boundary-layer thickness. *Because the boundary layer is laminar, only low resolution is required streamwise along the boundary layer.* On the other hand, if the boundary layer is turbulent, the viscous sublayer thickness is of order $1/R$. Again the direction normal to the boundary can be handled by transformation (if turbulent length scales outside the boundary layer are much larger than the sublayer thickness). However, the principal difficulty with a turbulent boundary layer or a laminar one undergoing transition to turbulence is that the streamwise direction must also be resolved. Since typical streamwise length scales are of the same order as the boundary-layer thickness (as for a turbulent boundary layer) or at most an order of magnitude larger (as for Tollmien–Schlicting waves in a boundary layer undergoing transition), it follows that *streamwise resolution must be of order $1/R$ in the boundary layer.* This restriction is exceedingly severe and will hinder any attempt to do a good job of simulating turbulent flow about realistic body shapes.

A possible way around the difficulty just raised is to model the turbulent boundary layer by invoking conditions on wall stress or other similar quantities. There is no general theory concerning this kind of turbulence modeling, though some recent attempts have been moderately successful (e.g., Deardorff[(11)]). We do not enter into a discussion of these questions here because it falls within the general realm of turbulence modeling (Section 10.5), not direct solution of the Navier–Stokes equations.

10.4. Survey of Applications

Operational computer codes for the numerical simulation of turbulence include codes capable of using up to 65,000 degrees of freedom to represent each of the dynamical variables, including velocity, temperature, pressure, etc. Two-dimensional homogeneous turbulence codes are operational with up to 256×256 modes [cutoff $K=128$ in Equations (10.5) and (10.6)], while three-dimensional shear and homogeneous turbulence codes are running with $32\times 32\times 32$ modes ($K=16$), $64\times 8\times 128$ modes, etc. For example, a $32\times 32\times 32$ three-dimensional turbulence calculation can be used to simulate wind-tunnel turbulence at a Reynolds number of order 25,000 and requires about 3 sec per time step and several hundred time steps on a CDC 7600.

The most elementary application of numerical methods to gain information of interest about homogeneous turbulence is the Taylor–Green vortex (Taylor and Green[12]). Here the initial velocity field is nonrandom:

$$\begin{aligned} v_1(\mathbf{x}, 0) &= \cos x_1 \sin x_2 \cos x_3 \\ v_2(\mathbf{x}, 0) &= -\sin x_1 \cos x_2 \cos x_3 \\ v_3(\mathbf{x}, 0) &= 0 \end{aligned} \tag{10.12}$$

for which the vortex lines are twisted. The flow field does not remain two-dimensional for $t>0$ and the vortex lines are stretched by the self-induced shear. Enhancement of vorticity by stretching of vortex lines in a local shear flow is the fundamental mechanism involved in the enhanced energy dissipation of turbulence relative to laminar flow that is the basis of the Kolmogorov theory of turbulence (Section 10.3). In fact,

$$\varepsilon = \nu\overline{\omega^2} \tag{10.13}$$

where $\boldsymbol{\omega}$ is the vorticity; Equation (10.13) shows directly how vortex stretching by convection can enhance energy dissipation.

Numerical solution of the Navier–Stokes equations with the initial conditions of Equation (10.12) gives the results shown in Figure 10.4 for the dissipation rate $\varepsilon(t)$. Here the Reynolds number $R=1/\nu$ and the calculations were done using a cutoff $K=16$ spectral code (Orszag[13]).

The increase of $\varepsilon(t)$ for $t\lesssim 5$ is due to the vortex stretching mechanism. As remarked in Section 10.3, the basic premise of the Kolmogorov and related theories of turbulence is that $\varepsilon=O(1)$ as $R\to\infty$. According to Equation (10.13), this behavior is possible only if the limiting behavior of $\overline{\omega^2}$ as $\nu\to 0$ ($R\to\infty$) is as follows: $\overline{\omega^2}$ is finite as $\nu\to 0$ for $t<t_*$, while $\overline{\omega^2}\to\infty$ as $\nu\to 0$ for $t>t_*$ in such a way that $\varepsilon=O(1)$. This hypothetical behavior is plotted as the curve $R=\infty$ in Figure 10.4 (Orszag[13]). Admittedly, the

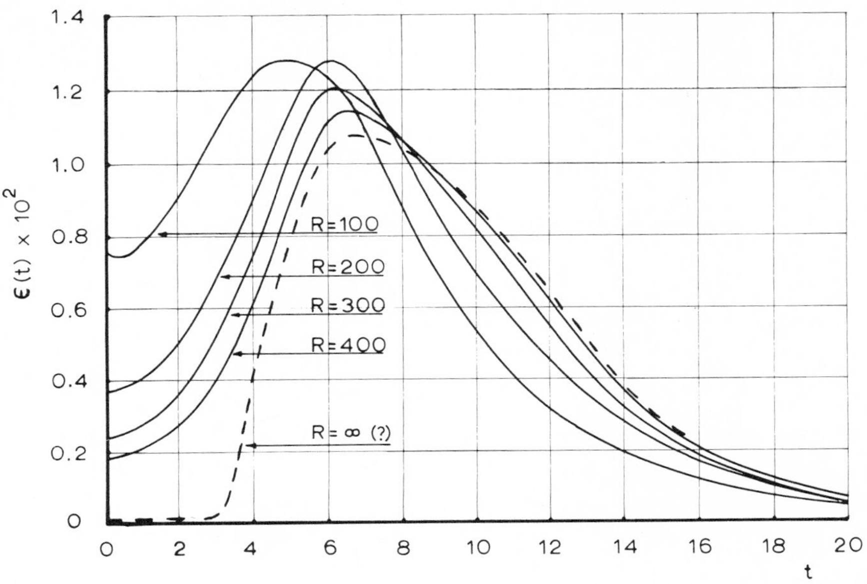

Fig. 10.4. Rate of energy dissipation ε versus t for the Taylor–Green vortex at $R = 100, 200, 300, 400$. Curve labeled $R = \infty$ is schematic of hypothetical extrapolated behavior of $\varepsilon(t)$.

Taylor–Green results shown in this figure are far from conclusive evidence that $\varepsilon = O(1)$ as $\nu \to 0$, but there is scant additional theoretical information on which to test the hypothesis, first inferred experimentally.

Some results of numerical simulations of two-dimensional turbulence are given in Figures 10.5–10.8. In Figure 10.5 we plot four enstrophy (mean-square vorticity) dissipation spectra $k^4E(k)$ vs. k (cf. Herring *et al.*[5]). In two dimensions, vortex lines cannot be stretched, so $\overline{\omega^2(t)} \leq \overline{\omega^2(0)}$ and energy dissipation $\varepsilon(t) \leq \varepsilon(0) = O(\nu)$ as $\nu \to 0$ (cf. Orszag[9]). Therefore, Kolmogorov's dimensional reasoning invoked in Section 10.3 must be reconsidered. In fact, while $\varepsilon(t)$ decreases with t in two dimensions, the rate of enstrophy dissipation $\eta(t)$ *may* be increased by shear. Here

$$\eta(t) = -\frac{d}{dt}\overline{\omega^2} = 2\nu \int_0^\infty k^4E(k)\,dk \tag{10.14}$$

so $2\nu k^4E(k)$ is the spectrum of enstrophy dissipation. It turns out that this new quantity $\eta(t)$ plays a dynamical role in two dimensions similar to that played by $\varepsilon(t)$ in three dimensions. In fact, accurate resolution of the enstrophy dissipation spectrum guarantees an accurate numerical simulation of the dynamically important scales of two-dimensional turbulence.

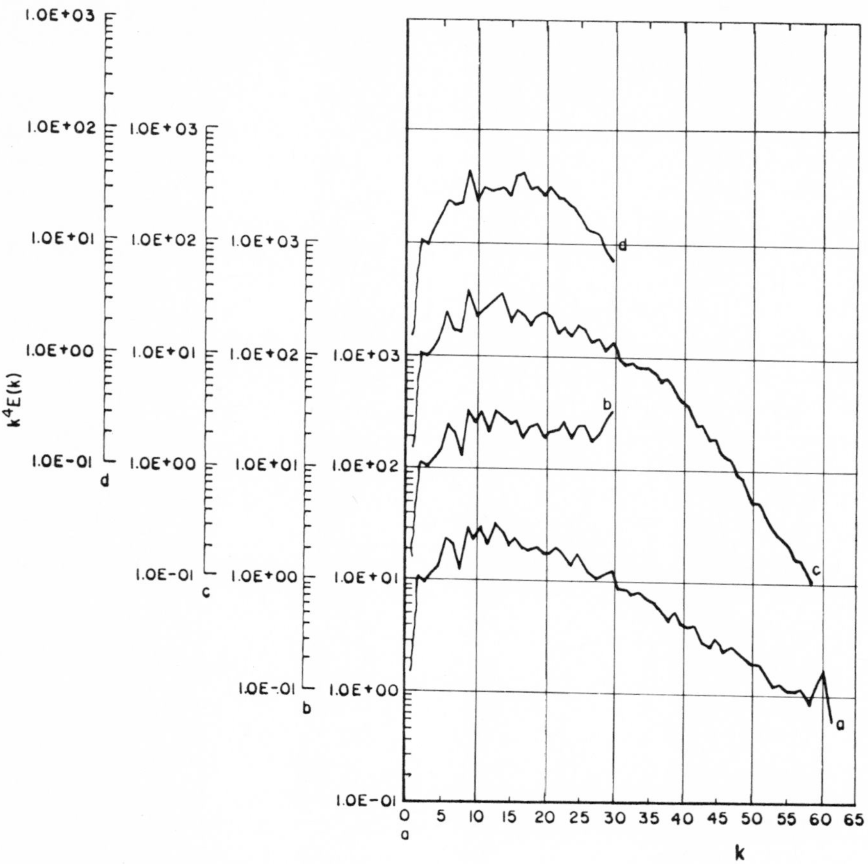

Fig. 10.5. Plot of $k^4E(k)$ versus K at $t = 2$. Curve (a): Spectral method with $(128)^2$ independent wave numbers (cutoff $K = 64$). Curve (b): Spectral method with $(64)^2$ wave numbers. Curve (c): Arakawa[27] finite-difference scheme with $(64)^2$ grid points. Curve (d): Arakawa finite-difference scheme with $(64)^2$ grid points. Initial energy spectrum is $E(k) = \frac{2}{3}ke^{-3k/2}$ and Reynolds number $R_L = 349$.

The enstrophy dissipation spectra in Figure 10.5 are obtained by solution of the Navier–Stokes equations from the same initial conditions out to $t = 2$ (a significant time of evolution) using four different numerical methods: (a) 128×128 (cutoff $K = 64$) spectral method; (b) 64×64 spectral method; (c) 128×128 finite-difference method; (d) 64×64 finite-difference method. The Reynolds number of these simulations is roughly 350 based on integral scale and root-mean-square velocity. Figure 10.5 illustrates several attractive features of the spectral calculations already discussed in Section

10.2. First, comparison of Figure 10.5(a) with Figure 10.5(b) shows that increasing the spectral resolution from 64×64 to 128×128 does not affect wave numbers $k \lesssim 15$, while similar comparison of Figures 10.5(c) and 10.5(d) suggests that increasing resolution of the finite-difference codes does affect the results for $k \gtrsim 3$. On the assumption that the 128×128 spectral results are accurate at all scales (and they may be shown to be very nearly so), it is apparent that the 64×64 spectral results are comparably as accurate as the 128×128 finite-difference results, even at the crude accuracy level of comparing graphs.

Second, Figure 10.5 illustrates the bootstrap capability of the spectral calculations, for the 64×64 spectral results show that $\eta(t)$ is only marginally resolved so that the 64×64 spectral results are *not* accurate at all scales. The 128×128 spectral results show that $\eta(t)$ is adequately resolved and that the results are accurate at all scales. In contrast, the finite-difference results shown in Figures 10.5(c) and 10.5(d) seem to show that $\eta(t)$ is adequately resolved since the spectrum decreases rapidly to large k; yet neither calculation is accurate at all scales.

Figures 10.6 and 10.7 present some further results of the calculations used to construct Figure 10.5. In Figure 10.6, there is plotted the mean-

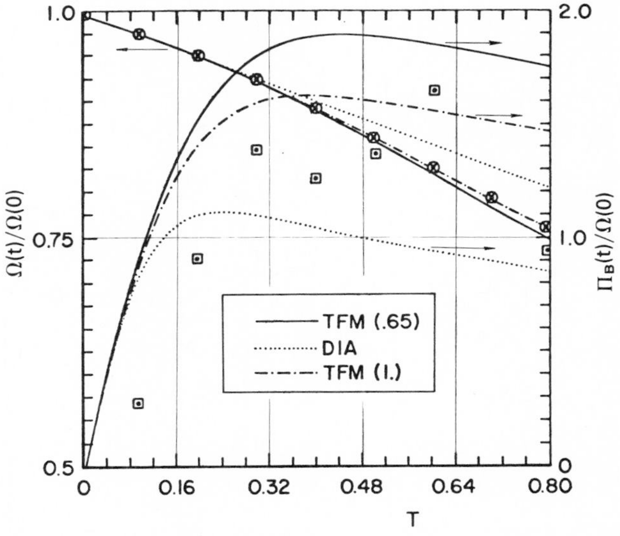

Fig. 10.6. Normalized enstrophy $\Omega(t)$ and normalized total energy back-transfer $\Pi_B(t)$ versus t for the run shown in Figure 10.5. Curves show three theoretical results: Test-field model (TFM, Kraichnan[16]) with $\lambda = 0.65$ and $\lambda = 1$ and the direct-interaction approximation (DIA, Kraichnan[17]). Encircled points give computer experimental values of $\Omega(t)$ averaged over two realizations, while the points in squares give similarly averaged values of $\Pi_B(t)$.

square vorticity $\Omega(t)=\overline{\omega^2(t)}$ vs. t for the 128×128 spectral calculation (circled dots) and for several analytical theories of turbulence (curves). The curves labeled $\Pi_B(t)$ will not be discussed here (cf. Herring *et al.*[5]). In Figure 10.7, a similar comparison is made between the 128×128 spectral results (points) for $k^4E(k)$ and the theories (curves). The important result of Figures 10.6 and 10.7 for the present discussion is the illustration that it is possible to obtain comparison between theory and numerical experiment; these comparisons have proved to be extremely useful in understanding the limitations of turbulence theories and in generating ideas for their improvement.

Another basic question concerns the possibility for verifying theories of the inertial range by direct numerical simulation. In two dimensions, dimensional analysis similar to Kolmogorov's (Kraichnan[14]) implies that there is a two-dimensional range with

$$E(k)=C'\eta^{2/3}k^{-3} \tag{10.15}$$

with C' a constant. It does not seem possible to give a definitive test of Equation (10.15) with only 128×128 spectral resolutions (cf. Herring *et al.*[5]); it is hoped that new 512×512 spectral calculations will be more useful in this regard. Some results from high-Reynolds-number ($R\simeq10^3$)

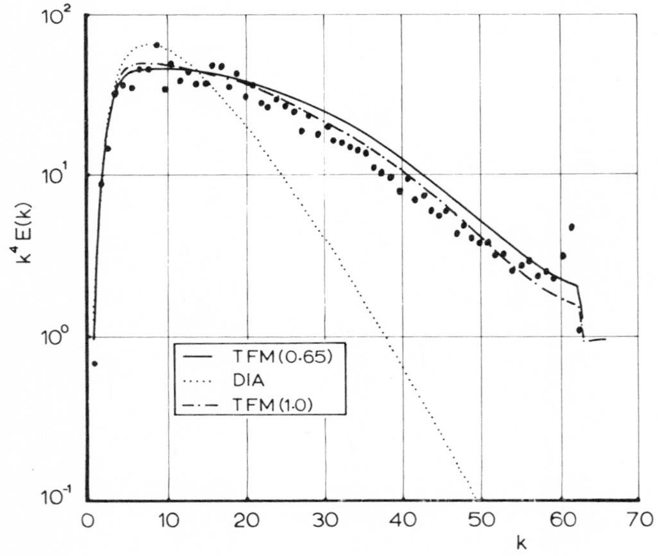

Fig. 10.7. Comparison of computer experimental enstrophy dissipation spectrum $k^4E(k)$ (points) with theories TFM ($\lambda=0.65$), TFM ($\lambda=1$), and DIA at $t=0.8$ for the run described in the caption to Figure 10.5.

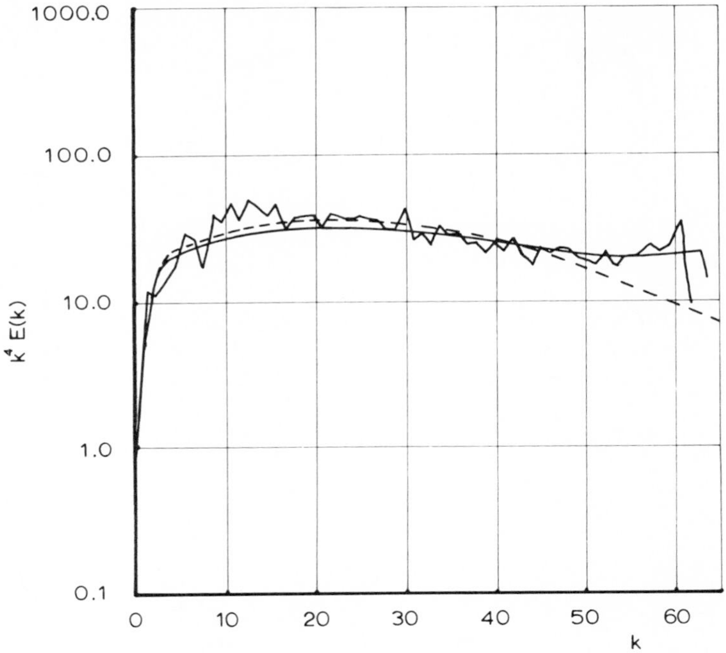

Fig. 10.8. Comparison of $k^4E(k)$ versus k for a run with the same initial conditions as described in the caption to Figure 10.5 but at the higher Reynolds number $R = 1100$. Jagged solid line: Computer experimental results. Smooth solid line: TFM with cutoff $k_{max} = 64$ [identical to the cutoff of the $(128)^2$ spectral computations]. Dashed line: TFM with cutoff $k_{max} = 128$.

128×128 computations are plotted in Figure 10.8. The heavy solid curve is the result of the Navier–Stokes calculations for $k^4E(k)$; rather than Equation (10.15) it seems that there is a broad spectral range over which $E(k) \alpha k^{-4}$, as predicted for the inertial range by Saffman.[15] However, this conclusion is premature, as shown by the other two curves in Figure 10.8. The dashed line is the result of numerical solution of the integro-differential equations of the test-field model, an analytical turbulence theory (cf. Kraichnan[16]) using a wave-number cutoff $K = 128$; the light solid curve is the result of a similar calculation with $K = 64$, the same cutoff used to obtain the heavy solid curve. The conclusion from these comparisons is that the apparent k^{-4} spectral range is due to the spectral cutoff, not necessarily the basic physics of the turbulence problem. In fact, it is known analytically that the test-field model, which is in good agreement with the numerical simulations of two-dimensional turbulence as shown by Figures 10.6 to 10.8, yields a k^{-3} inertial range behavior.

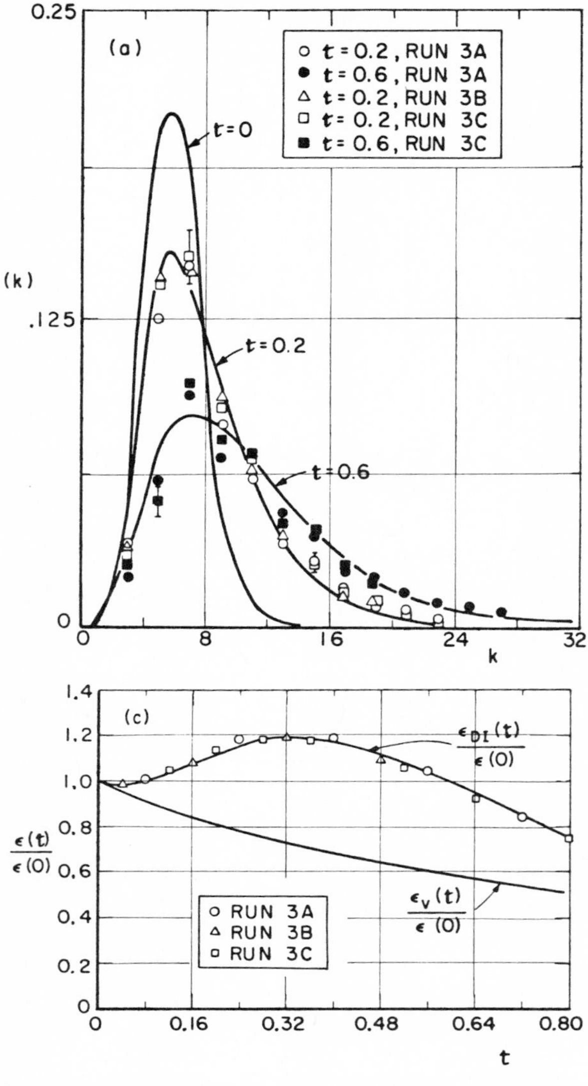

Fig. 10.9. Comparisons between numerical simulations (data points) and direct-interaction (DI) theory (solid curves). Run 3 is a numerical simulation performed using the initial energy spectrum $\varepsilon(k, 0) \propto k^4 \exp[-2(k/k_{\max})^2]$ with $v_{\mathrm{rms}}(0) = 1$, $k_{\max} = 4.75683$, $\nu = 0.01189$,

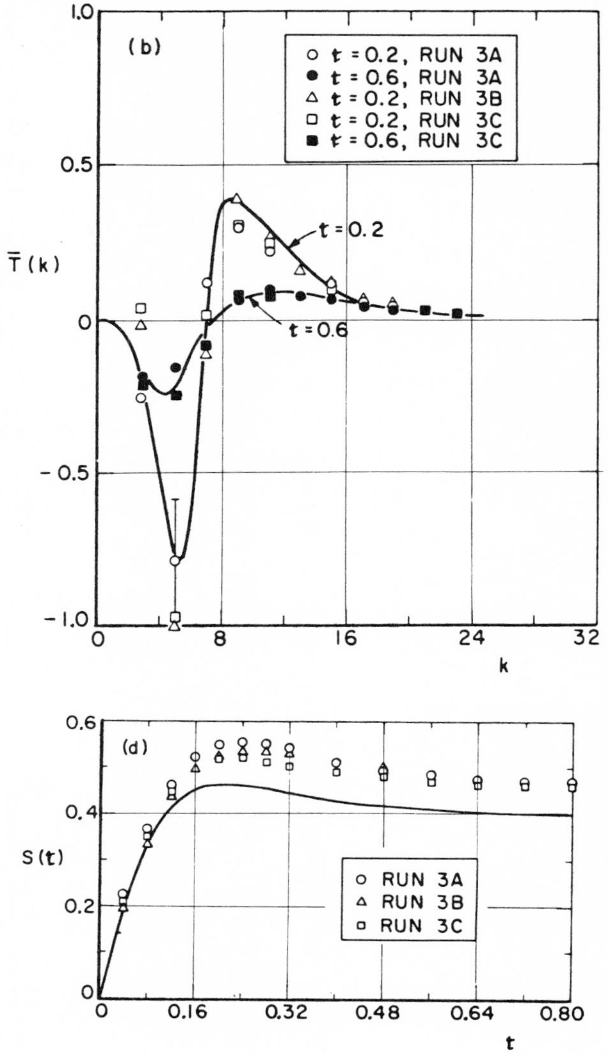

$R_\lambda(0) = 35.4$. Other details are given by Orszag and Patterson.[1] (a) Dissipation spectrum $D(k) = 2k^2E(k)$; (b) transfer spectrum $T(k)$; (c) energy dissipation rate. The curve labeled ε_v is for pure viscous decay. (d) Skewness of the longitudinal velocity derivative.

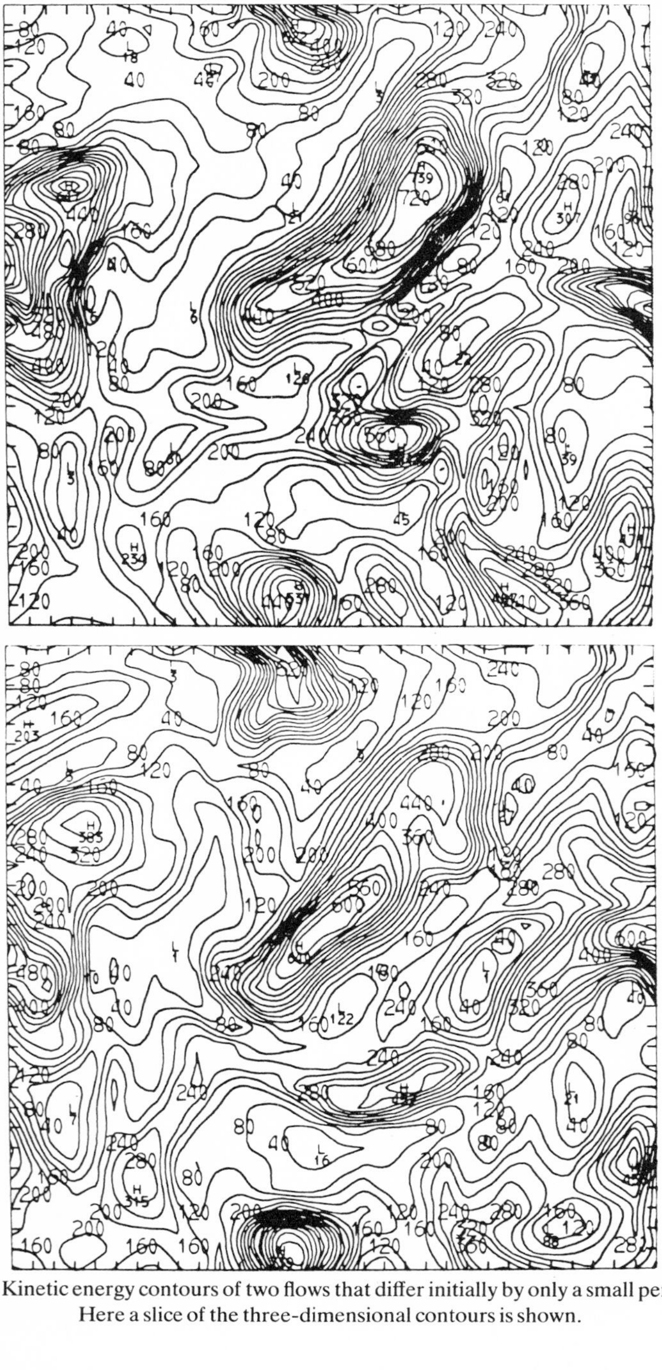

Fig. 10.10. Kinetic energy contours of two flows that differ initially by only a small perturbation. Here a slice of the three-dimensional contours is shown.

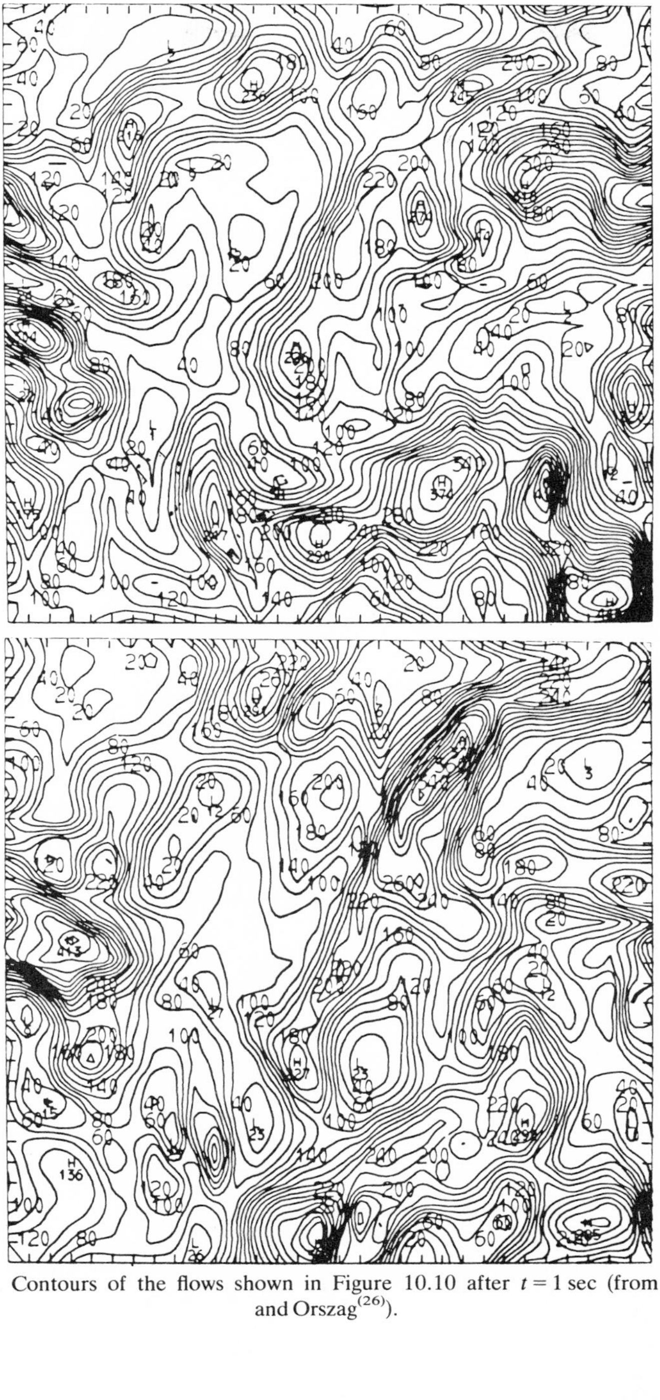

Fig. 10.11. Contours of the flows shown in Figure 10.10 after $t = 1$ sec (from Patterson and Orszag[26]).

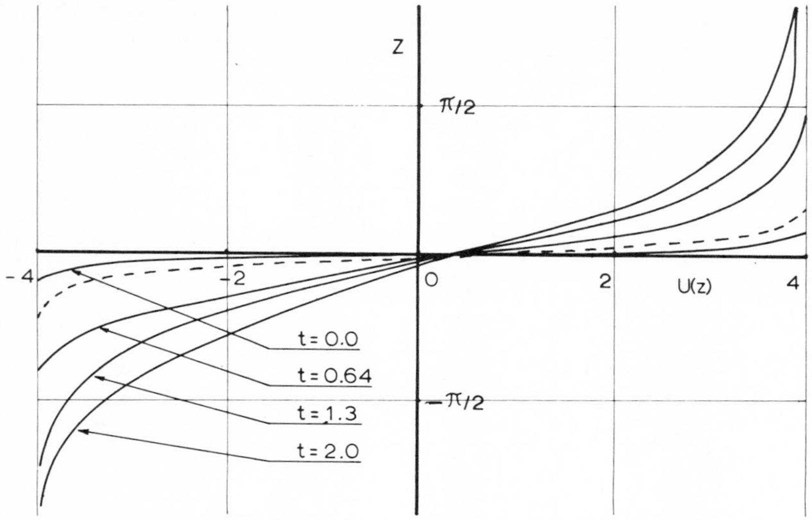

Fig. 10.12. Mean velocity profiles $\bar{U}$ versus z from numerical simulation of a free-shear layer at several values of t. Dashed curve: Pure viscous diffusion at $t = 2$. Solid curves: Direct simulation at $t = 0.0, 0.64, 1.3$, and 2.0.

Some results of numerical simulations of three-dimensional homogeneous turbulence are shown in Figure 10.9 (Orszag and Patterson[1]). Again, the results of the simulations are compared with those of another analytical turbulence theory, the direct-interaction approximation (cf. Kraichnan[17]). These simulations were performed at wind-tunnel-like Reynolds numbers (35, based on the Taylor microscale; roughly 15,000, based on mesh separation and air speed in the corresponding wind tunnel).

Additional results of three-dimensional turbulence simulations are shown in Figures 10.10 and 10.11. Here the evolution of two velocity fields that differ from each other initially by only a small perturbation is studied, a problem of interest with regard to the instability of turbulence with important applications to the predictability of atmospheric motions (cf. Leith[18]). In Figures 10.10(a) and 10.10(b), the kinetic energy contours of the two initial velocity fields are plotted. In Figures 10.11(a) and 10.11(b), the corresponding contours after evolution for about one large-eddy circulation time are plotted; a $32 \times 32 \times 32$ spectral code was used to obtain these results. It is striking how the relatively small initial perturbation grows into a large perturbation during time evolution.

Numerical simulations of turbulent shear flows by direct solution of the Navier–Stokes equations have been pursued only very recently. Orszag and Pao[6] report the results of simulations of the wake of a self-propelled body.

Because of the general difficulties with inhomogeneous turbulence simulations (especially statistics), these calculations were performed by isolating a slab of the wake and following its downstream evolution by Taylor's hypothesis.

Some results of a simulation of a free turbulent shear layer are shown in Figures 10.12 and 10.13. In this flow, all average quantities are functions of z alone, while the mean velocity $\bar{U}$ is in the x direction; of course, fluctuating (turbulent) velocities exist in all three space directions. The mean velocity $\bar{U}$ is plotted as a function of z in Figure 10.12, for several values of the time t. Initially ($t = 0$), the mean velocity undergoes an abrupt jump near $z = 0$. The dashed curve would be the result for $\bar{U}(z)$ at $t = 2$ if (molecular) viscosity were totally responsible for the broadening of the shear layer. Instead, the actual rate of spreading is significantly faster, as shown by the results for

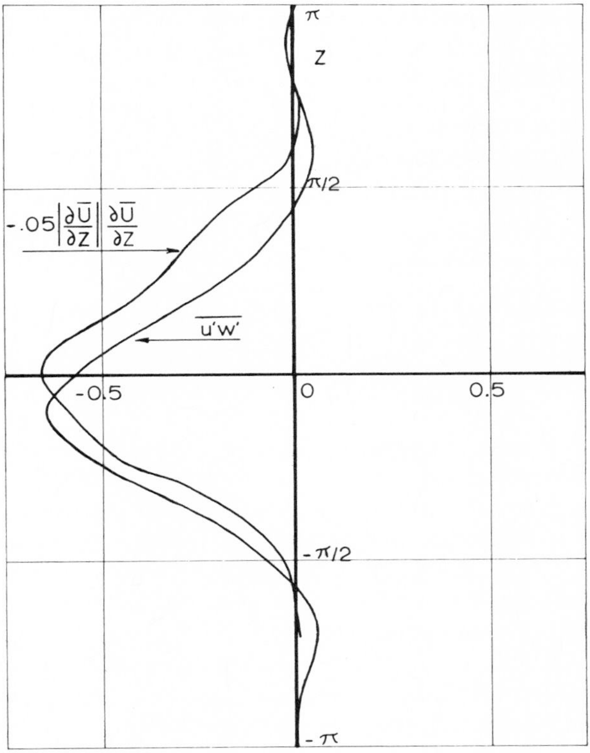

Fig. 10.13. Comparisons of Reynolds stresses $\overline{uw}$ at $t = 2$ for free shear layer (cf. Figure 10.12) with eddy viscosity (10.16).

$t = 0.64$, 1.3, 2, which is due to the effect of the turbulent velocities in diffusing the discontinuity.

This enhanced rate of diffusion of the turbulent flow over molecular diffusion is often characterized by an eddy viscosity coefficient. The validity of one particular type of eddy viscosity formulation, due to Prandtl,[19] is tested by the results shown in Figure 10.13. Prandtl asserted that the Reynolds stress $\overline{uw}$ and the mean velocity $\bar{U}(z)$ are related by the mixing length hypothesis

$$\overline{uw} = -L^2 \frac{\partial \bar{U}}{\partial z} \left| \frac{\partial \bar{U}}{\partial z} \right| \tag{10.16}$$

where L is the so-called mixing length, typically related to the large eddy size in the flow. The results plotted in Figure 10.13 show reasonably good agreement with the simple transfer expression, Equation (10.16), for the choice $L \simeq 0.22$.

More sophisticated transfer models of turbulence may be similarly tested by numerical experiment. The results of these comparisons will be reported elsewhere.

10.5. Comparison with Other Methods

There have been at least four parallel assaults on the turbulence problem, which may be classified as:

(i) analytical turbulence theories
(ii) turbulence transport models
(iii) subgrid closure models
(iv) direct numerical simulations

We have been discussing method (iv) up to this point.

Method (i) includes all attempts at a fundamental theory of turbulence. The prototype examples are the direct-interaction approximation (Kraichnan[17]) and the test-field model (Kraichnan[16]); these theories are discussed critically by Orszag.[9] They are characterized by their attempt to isolate and understand the fundamental difficulties of turbulence, i.e., the closure problem, nonlinear interaction, strong interaction, etc., usually without any free or adjustable parameters. These theories are nearly universally statistical in nature, the fundamental dynamical quantities being averages of the turbulent flow variables, like moments, average response functions, etc. These theories are very complicated and they have been applied only to homogeneous turbulence, one notable exception being the work of Herring[20] on large-Prandtl-number thermal convection. The extensions of these theories to general turbulent shear flows is quite difficult, although algorithms have been recently developed that should permit the economical application of these theories to a variety of shear flows.

Turbulence transport models are reviewed by Launder and Spalding[21] and Harlow.[22] These theories seek simplified models of turbulent flows, hopefully based on good physical insight, that are capable of application to complicated flows encountered in engineering practice. Several adjustable parameters typically appear that must be chosen by comparison with experimental data. In common with the analytical theories, only averaged turbulent fields appear as dynamical quantities in these transport models. These smoothed fields exhibit symmetries (like axisymmetry for a circular jet flow in unstratified fluid) that the detailed, unaveraged, turbulent flow fields do not themselves exhibit. Thus, instead of having to perform a three-dimensional time-dependent simulation, it may be possible to reduce the problem to a steady state problem in fewer than three space dimensions with consequent enormous savings of computational effort (assuming numerical solution of the equations of the theory is ultimately necessary).

In contrast to the analytical theories, the transport models deal directly with only the gross properties of the turbulent flow, with little or no attempt at consideration of the fundamental dynamics of interaction between the various turbulent scales of motion. The advantage of this approach over the analytical theories is clear: Even for complicated inhomogeneous turbulent flows, the dynamical quantities appearing in the transport models are relatively simple. The disadvantage is equally clear: Potentially important dynamical information is forever thrown away by the cavalier disregard of detailed dynamics.

There is a very wide variety of transport models now being touted in the literature. All authors have in common the use of the Reynolds equation for the mean velocity field obtained by averaging the Navier–Stokes equations. This equation relates the evolution of the mean velocity to the Reynolds stresses. Second-order closure models develop further equations of motion for the Reynolds stresses themselves. However, these additional equations do not contain enough information to predict evolution of the turbulence without additional hypotheses because this further evolution depends, in an essential way, on the details of the flow that were wiped out by the averaging. In the end, several ad hoc assumptions relating dynamical quantities must be made to close the equations of these models; all differences between these models enter from the precise closure conditions that are chosen.

In the subgrid closure models [method (iii)], turbulence transport approximations are made only on those scales of motion not explicitly resolved by a numerical approximation to the Navier–Stokes equations. The chief architect of these methods has been Deardorff,[11,23] who has applied both simplified eddy viscosity closures and sophisticated second-order transport closures to the subgrid component of the flow. The large scales (those explicitly resolved by the numerical approximation) are treated as they are in direct numerical solution of the Navier–Stokes equations. The very small scales are treated by statistical approximation, while the large scales are treated in detail. The effect of unresolved small scales on larger resolved scales is represented by an eddy viscosity K replacing molecular viscosity. Here K is chosen (cf. Smagorinsky *et al.*[24]) as

$$K = (c\,\Delta x)^2 \left| \left(\frac{\partial v_i}{\partial x_j} + \frac{\partial v_j}{\partial x_i} \right)^2 \right|^{1/2}$$

in three dimensions, and

$$K = (c'\Delta x)^3 \left| \frac{\partial}{\partial x_i} (\nabla \times \mathbf{v}) \right|$$

in two dimensions; Δx is the spatial resolution and c, c' are constants ($\sim$0.1–0.2). These expressions for K are appropriate only if first- or second-order space differencing is used; higher-order schemes require different eddy viscosities.

This approach has very much to commend it, including the important fact that there is no direct Reynolds number limitation on the simulations, since the effect of the subgrid turbulence is taken into account. Ultimately, it is the author's opinion that the subgrid approach must be involved in solution of all turbulence problems, since one *is* interested in details of the large scales and not just their statistical properties. However, the current methods for inclusion of the subgrid effect have a number of disadvantages, including the use of ad hoc turbulence transport models for the subgrid component and the neglect of any stochastic effect of fluctuations in subgrid flow variables on the generation of fluctuations in large scales.

It appears that the principal advantage of the subgrid approach over direct simulation is the absence of any Reynolds number limitations in the former. However, this is not so. The accuracy of the subgrid closures requires that the separation of the flow into subgrid scale and large-scale components does not have a significant effect on the evolution of large scales.

In the same way, consider the possibility that direct numerical simulation [method (iv)] of a huge-Reynolds-number flow may be accomplished by artificially decreasing the Reynolds number to the point where the flow can be accurately simulated on existing machines. Of course, it is not possible that all scales of motion of the reduced-Reynolds-number flow be unchanged, but it is possible that sufficiently large scales be unchanged by the Reynolds number change. In fact, large-scale features of turbulent flows do not seem to change with Reynolds number, *if* boundary and initial conditions are fixed independently of Reynolds number (cf. Herring *et al.*[5]). Some evidence for this behavior is shown in Figure 10.14. In Figure 10.14, we plot the evolved vorticity contours of the initial vorticity field described in the caption to Figure 10.5 for three different Reynolds numbers. The initial flow field is identical for all three Reynolds numbers. The similarity in large-scale structure is apparent.

As a consequence, it does not seem necessary to simulate huge-Reynolds-number turbulence to gain information on large and moderate scales of motion. It is only necessary to simulate flows at Reynolds numbers at which the desired scales of motion have achieved Reynolds number independence. This behavior is illustrated pictorially in Figure 10.15. As the Reynolds number increases, the small scale of three-dimensional turbulence adjusts to maintain the energy dissipation rate (cf. Section 10.3), giving a mean-square vorticity spectrum $k^4E(k)$ that extends to higher and higher k. The peak in $k^2E(k)$ occurs near $k \sim 1/l$, where l is given by Equation (10.10). For $k \lesssim 1/l$, the flow is nearly Reynolds number independent, not just statistically but apparently in detail as well.

Some results of Deardorff's[23] simulation of the planetary boundary layer are plotted in Figure 10.16. Here K_m is the coefficient of subgrid scale eddy viscosity (suitably normalized);

$$K_{MX} = K_m - \overline{uw}/(\partial\bar{u}/\partial z) \qquad \text{and} \qquad K_{MY} = K_m - \overline{vw}/(\partial\bar{v}/\partial z)$$

measure the effective eddy viscosity due to the Reynolds stresses of resolved small scales acting on resolved large scales; K_θ is a similar measure of the eddy conductivity of heat due to resolved scale motions. It is apparent from Figure 10.16 that the Reynolds stresses due to resolved motions are much larger than those due to the imposed subgrid eddy viscosity. Consequently, most of the eddy transport is being accomplished by the resolved scales with the subgrid component providing only the necessary small-scale excitation to ensure the Reynolds number independence of energy dissipation.

In summary, it appears that the large scales of turbulent flows have a strong tendency to adjust themselves to be independent of the details of the dissipation mechanism; subgrid eddy motions or molecular viscous action are both effective dissipators provided that the viscosity is small enough that the large scales have achieved Reynolds number independence. It seems that either method (iii) or (iv) can be used to achieve simulations of turbulent flows with limited resolution; the outstanding question concerns the efficiency with which invariance to the scale of separation into subgrid and resolved motions [for method (iii)] or Reynolds number independence [for method (iv)] is achieved.

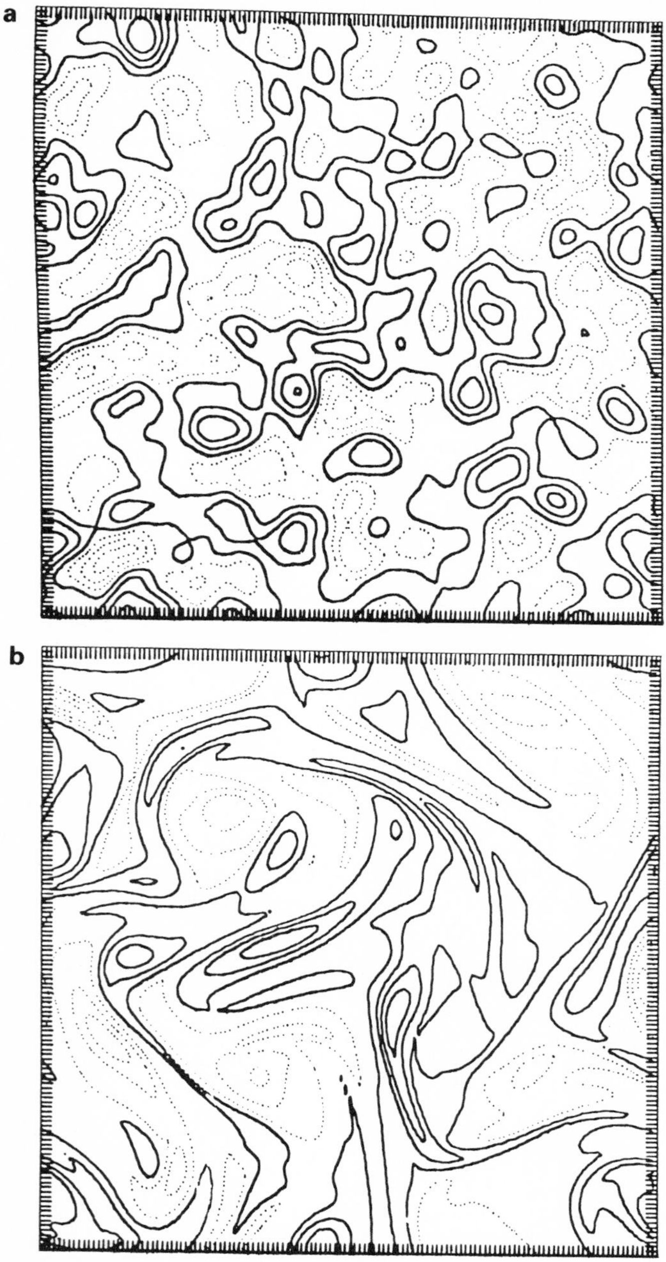

Fig. 10.14. Contours of the vorticity field for $(128)^2$ spectral calculations of two-dimensional turbulence. (a) Contours of initial vorticity field (identical to that used in the calculations

c

d

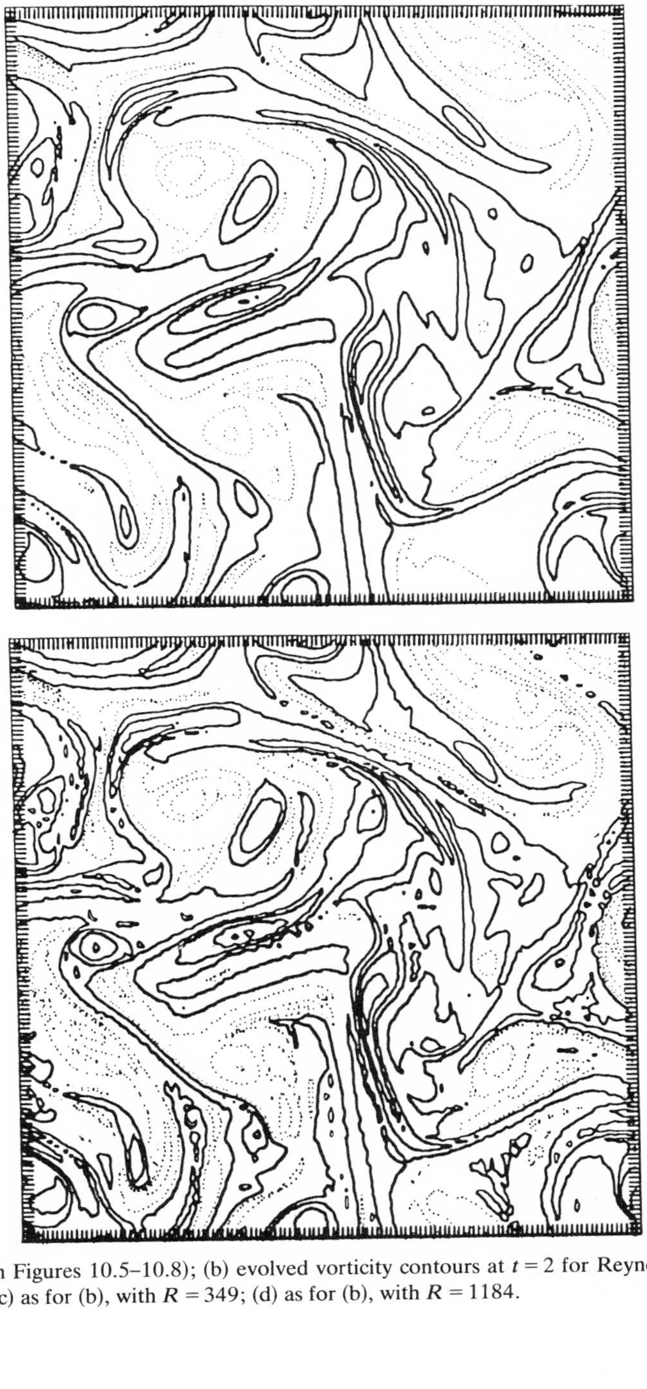

reported in Figures 10.5–10.8); (b) evolved vorticity contours at $t = 2$ for Reynolds number $R = 138$; (c) as for (b), with $R = 349$; (d) as for (b), with $R = 1184$.

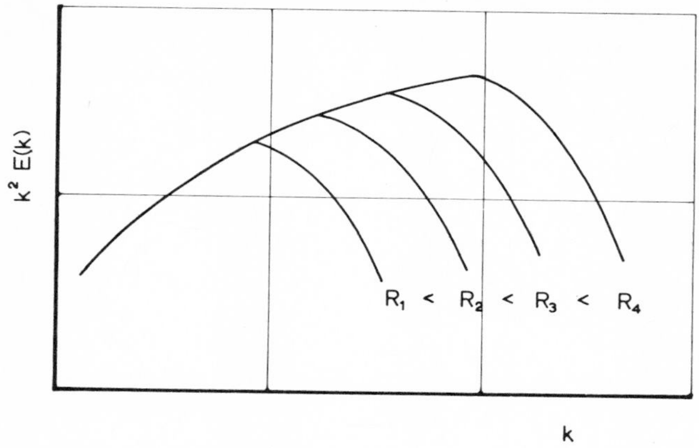

Fig. 10.15. Schematic representation of the effect of increasing Reynolds number on the energy dissipation spectrum. The hypothesis of Reynolds number independence asserts that the scales of motion with wave numbers less than that of the peak in the $k^2E(k)$ spectrum are approximately Reynolds number independent.

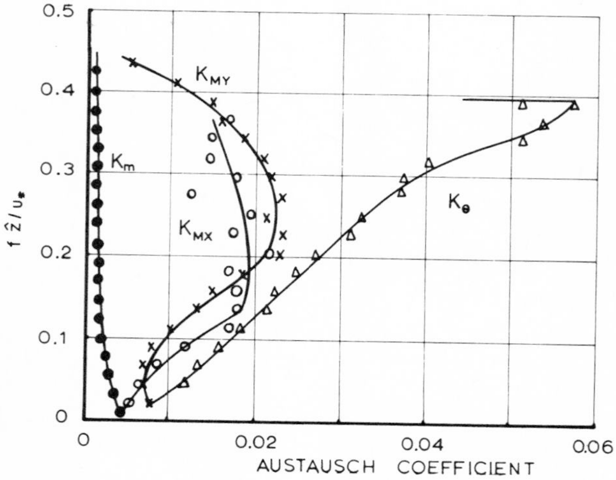

Fig. 10.16. Eddy transport coefficients in Deardorff's[23] simulation of the planetary boundary layer. K_m: Subgrid scale eddy viscosity. K_{MX}, K_{MY}: Effective eddy viscosity of resolved small scales acting on resolved large scales in the x, y directions, respectively. K_θ: Eddy heat conductivity due to resolved small scales on large scales.

10.6. Prospects

Computers that will be available by 1980 are not likely to be more than an order of magnitude or so faster than those available today. Reynolds number restrictions similar to those imposed on current simulations will still be present. However, the order of magnitude improvement in speed can be used to give an order of magnitude improvement in statistics, a prospect that is especially important for inhomogeneous turbulence simulations.

It seems that the best hope for achieving simulations of huge-Reynolds-number flows is to investigate further effects like Reynolds number independence. These effects should be used to give both moderate-Reynolds-number simulation models and subgrid closure models having the property that they give essentially the same evolution for suitably large-scale eddies as the required huge-Reynolds-number flow.

Techniques must also be investigated for simulation of flows in the presence of solid boundaries when the available resolution is not sufficient for an exceedingly thin turbulent boundary layer. The latter problem has not received as much attention as the (interior) turbulence modeling problem, though it is equally important. One possible approach is the consideration of small sections of the flow in detail, then parametrizing their properties, and finally the use of the parametrization to provide boundary conditions in the large-scale simulation (as done by Deardorff(23) in his parametrization of the planetary boundary layer for use in large-scale general circulation models of the atmosphere).

In conclusion, there has been some modest progress towards the solution of the problem of numerical simulation of turbulence. However, many interesting and important problems remain to be solved, hopefully in the not too distant future.

ACKNOWLEDGMENT

This work was supported by the Office of Naval Research under Contract No. N00014-72-C-0355.

Notation

D	Standard deviation
E	Energy spectrum
$\mathbf{k}$	Wave vector
k_d	Kolmogorov dissipation wave number
l, L	Eddy sizes
N	Number of random events
$\mathbf{n}$	Integer vector
P	Pressure
$\mathbf{r}$	Random variables (Gaussian)
R	Reynolds number
t	Time
$\mathbf{u}$	Fourier transform of $\mathbf{v}$
$\bar{U}$	Mean velocity
$\mathbf{v}$	Velocity vector
$\bar{v}$	Root-mean-square velocity

$\mathbf{x}$	Space vector	ϕ_1	Inertial range spectrum
$\boldsymbol{\nabla}$	Gradient operator	η	Enstrophy dissipation rate
∇^2	Laplacian operator	ν	kinematic viscosity
δ_{ij}	Kronecker delta	$\boldsymbol{\omega}$	Vorticity vector
ε	Energy dissipation rate		

References

1. Orszag, S. A., and Patterson, G. S., Numerical simulation of three-dimensional homogeneous isotropic turbulence, *Phys. Rev. Lett.* **28**, 76–79 (1972).
2. Orszag, S. A., On the resolution requirements of finite difference schemes, *Stud. Appl. Math.* **50**, 395–397 (1971).
3. Orszag, S. A., and Israeli, M., Numerical simulation of viscous incompressible flows, in: *Annual Review of Fluid Mechanics*, Vol. 6 (M. Van Dyke, W. G. Vincenti, and J. V. Wehausen, eds.), Annual Reviews, Inc., Palo Alto, California (1974), pp. 281–318.
4. Kreiss, H. O., and Oliger, J., *Methods for the Approximate Solution of Time Dependent Problems*, Monograph Number 10, GARP Publ. Service, World Meteorology Organization (1973).
5. Herring, J. R., Orszag, S. A., Kraichnan, R. H., and Fox, D. G., Decay of two-dimensional homogeneous turbulence, *J. Fluid Mech.* **66**, 417–444 (1974).
6. Orszag, S. A., and Pao, Y-H., Numerical computation of turbulent shear flows, in: *Advances in Geophysics*, Vol. 18A (F. N. Frenkiel and R. E. Munn, eds.), Academic Press, New York (1974), pp. 225–236.
7. Grant, H. L., Stewart, R. W., and Moilliet, A., Turbulence spectra from a tidal channel, *J. Fluid Mech.* **12**, 241–268 (1962).
8. Batchelor, G. K., *The Theory of Homogeneous Turbulence*, Cambridge University Press, Cambridge (1953).
9. Orszag, S. A., Lectures on the statistical theory of turbulence, in: *Fluid Dynamics* (R. Balian and J.-L. Peube, eds.), Gordon and Breach, New York (1977).
10. Grosch, C. E., and Orszag, S. A., Numerical solution of problems in unbounded regions: Coordinate transforms, *J. Comp. Phys.*, to appear (1977).
11. Deardorff, J. W., A numerical study of three-dimensional turbulent channel flow at large Reynolds number, *J. Fluid Mech.* **41**, 453–480 (1970).
12. Taylor, G. I., and Green, A. E., Mechanism of the production of small eddies from large ones, *Proc. R. Soc. London, A*, **158**, 499–521 (1937).
13. Orszag, S. A., Numerical simulation of the Taylor–Green vortex, *Computing Methods in Applied Science and Engineering Proceedings of the International Symposium, Pt. 2*, Versailles, France, Springer, Berlin (1974), pp. 50–64.
14. Kraichnan, R. H., Inertial ranges in two-dimensional turbulence, *Phys. Fluids* **10**, 1417–1423 (1967).
15. Saffman, P. G., On the spectrum and decay of random two-dimensional vorticity distributions at large Reynolds number, *Stud. Appl. Math.* **50**, 377–383 (1971).
16. Kraichnan, R. H., An almost Markovian Galilean invariant turbulence model, *J. Fluid Mech.* **47**, 513–524 (1971).
17. Kraichnan, R. H., The structure of isotropic turbulence at very high Reynolds numbers, *J. Fluid Mech.* **5**, 497–543 (1959).
18. Leith, C. E., Atmospheric predictability and two-dimensional turbulence, *J. Atmos. Sci.* **28**, 145–161 (1971).

19. Prandtl, L., Bericht über Untersuchungen zur ausgebildeten Turbulenz, *Z. Angew. Math. Mech.* **5**, 136–139 (1925).
20. Herring, J. R., Statistical theory of thermal convection at large Prandtl numbers, *Phys. Fluids* **12**, 39–52 (1969).
21. Launder, B., and Spalding, D. B., *Lectures in Mathematical Models of Turbulence*, Academic Press, New York (1972).
22. Harlow, F. H. (ed.), *Turbulence Transport Modeling*, American Institute of Aeronautics and Astronautics, New York (1973).
23. Deardorff, J. W., A three-dimensional numerical investigation of the idealized planetary boundary layer, *Geophys. Fluid Dyn.* **1**, 377–410 (1970).
24. Smagorinsky, J., Manabe, S., and Holloway, J. L., Numerical results from a nine-level general circulation model of the atmosphere, *Mon. Weather Rev.* **93**, 727–768 (1965).
25. Kraichnan, R. H., Isotropic turbulence and inertial range structure, *Phys. Fluids* **9**, 1728–1752 (1966).
26. Patterson, G. S., and Orszag, S. A., Numerical simulation of turbulence, *Atmos. Technol.* **3**, 71–78 (1973).
27. Arakawa, A., Computational design for long-term numerical integration of the equations of fluid motion: Two-dimensional incompressible flow Part 1, *J. Comput. Phys.* **1**, 119–143 (1966).
28. Orszag, S. A., Fourier series on spheres, *Mon. Weather Rev.* **102**, 56–75 (1974).
29. Gottlieb, D., and Orszag, S. A., *Numerical Analysis of Spectral Methods*, SIAM Monograph, Philadelphia (1977).

CHAPTER 11

Laboratory Instrumentation in Turbulence Measurements

V. A. SANDBORN

11.1. Introduction

Measurements of turbulence can arise in a great number of fluid flow problems. The possible range of turbulent fluctuations can vary over many decades of time, amplitude, and/or space scales. Some ideas of the vast time scales that are associated with "turbulent" motion can be gained by considering Figure 11.1.

Within the laboratory we need only consider frequencies greater than approximately 10^2/h. Obviously the large-scale, lower frequencies are associated with atmospheric motion. On the linear energy scale of Figure 11.1 the high-frequency regions associated with "local isotropy" and "viscous dissipation" are shown as too large in magnitude. These high-frequency regions, while several orders of magnitude smaller than the lower-frequency regions in energy, are of major interest in the overall structure of turbulence. Frequency, as used in Figure 11.1, is somewhat

V. A. SANDBORN • College of Engineering, Colorado State University, Fort Collins, Colorado

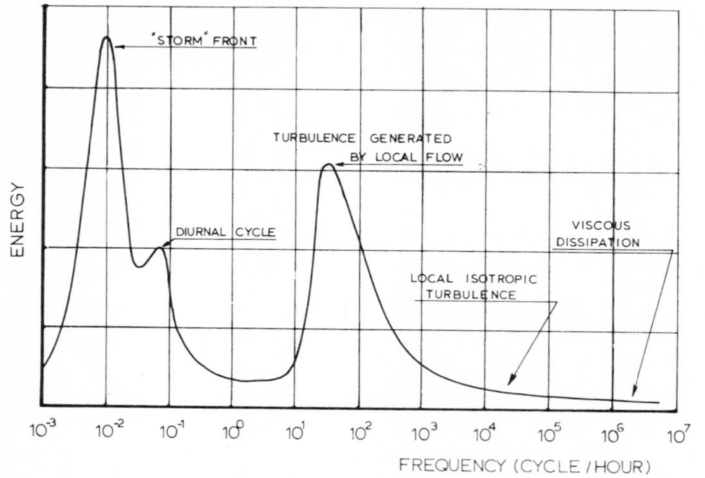

Fig. 11.1. Spectrum of turbulence energies (atmospheric spectrum after Reference 1).

misleading, since it will depend on the local mean flow velocity. The frequencies shown in Figure 11.1 correspond to those of low-speed atmospheric flow. If the local mean velocities were *supersonic* the frequencies would be extended by several orders of magnitude. Thus, the time scales of interest in the laboratory study of turbulent motion range from approximately 10^2 to 10^8 or 10^9 cycles/hr.

The physical size of the turbulent motion depends on the size of the flow facility. For the atmosphere it is possible to have scales of the order of kilometers. For the laboratory the scale sizes are at most of the order of a meter or less. Figure 11.2 shows a set of turbulent scales evaluated for a "large"-size boundary layer (thickness: 82.6 cm, equivalent length Reynolds number: 3×10^8) made at Colorado State University. The "macroscale" of the turbulence [$L_x \equiv \int_0^\infty R_x\, dx = UF(0)/4$, where $F(0)$ is the spectral energy at "zero frequency," U is the local mean velocity, and R_x is the longitudinal turbulent velocity correlation in the x direction] is seen to be only a few centimeters in length. The macroscale shown in Figure 11.2 is probably the largest scale associated with the turbulence in this particular flow. The smaller scales shown in Figure 11.2 are the "microscale" of the turbulence

$$1/\lambda_x^2 \equiv \lim_{x \to 0} \left(\frac{1 - R_x}{x^2} \right) \simeq \frac{2\pi^2}{U^2} \int_0^\infty f^2 F(f)\, df$$

which is associated with the local isotropic region of the spectrum; and the

"Kolmogorov scale" [$\eta \equiv (\nu^3/\varepsilon)^{1/4}$, where ν is the kinematic viscosity and ε is the turbulent dissipation], which is associated with the viscous dissipation region of the spectrum. The "turbulent mixing length" calculated from measurements of the turbulent shear stress $\overline{uv}$ and the local mean velocity gradient

$$l \equiv \left[\frac{(\overline{uv})}{|dU/dy|(dU/dy)}\right]^{1/2}$$

are also shown in Figure 11.2. There are a great many different types of turbulent flows that can be generated in the laboratory, so the range of values of the macroscale may vary by an order of magnitude. At the lower limit the Kolmogorov scale depends mainly on the viscous properties of the fluid, so it would not be expected to change greatly with flow conditions.

While a general concept of the magnitude of the frequency and size of turbulent fluctuations can be specified for laboratory measurements, it is more difficult to specify the amplitudes of the fluctuations. The spectral curve of Figure 11.1 shows a large amount of energy, and thus large

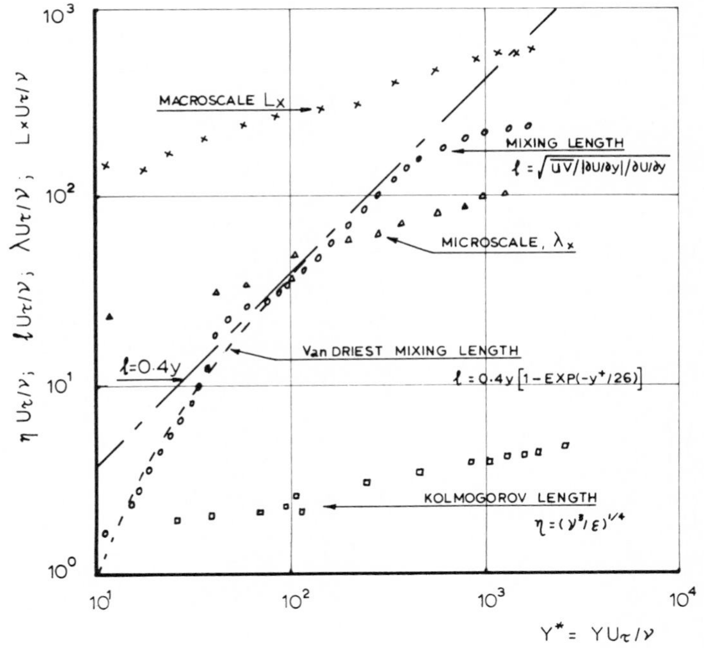

Fig. 11.2. Turbulent scale lengths near the wall in a boundary layer.

amplitudes, in the region where the turbulence was generated by local flow conditions. The actual magnitude of the turbulence will depend on the turbulent generator. Rough surfaces, and in particular flexible roughness, such as plastic strips used to model crops like corn, can generate very large velocity fluctuations.

Turbulence-measuring instruments are available to evaluate most of the turbulent flows encountered in the laboratory. Even the very high frequencies required for supersonic flows can be measured by existing instrumentation. Likewise, the size of the measuring instruments can be made small enough to reach the Kolmogorov scales. The most difficult requirements will be found in the area of amplitude evaluation. For smooth-surface, flat-plate, subsonic turbulent boundary layers the velocity fluctuations can be evaluated by several techniques. However, there are many flows where the velocity fluctuations are as large as or larger than the mean flow velocity, which makes evaluation difficult or impossible. For cases where the turbulent velocity actually produced reversal of the total velocity vector reliable measurements are at best difficult to make.

Turbulence measurements have in the past been focused on the evaluation of velocity fluctuations. However, turbulent temperature, density, pressure, concentration, and surface shear stress fluctuations also require evaluation. For incompressible flows, temperature measurements pose no major instrumentation problem. For compressible flows it is at present nearly impossible to sort out temperature, density, and pressure fluctuations. Most supersonic flow studies simply assume that pressure fluctuations can be neglected. In fact very little information exists for density and pressure fluctuations. Concentration fluctuation measurements have been made only in recent years.

Turbulent Quantities. The evaluation of turbulent motion is one of the most complex measurements required in fluid mechanics. Since no adequate theory exists for turbulent flow, the measurements are the main information available. As noted, the measuring transducer must have a very fast response and be quite small. Details of velocity, temperature, density, pressure, concentration, and shear stress fluctuations are required.

Fluctuating Velocities. The statistical theory of turbulence requires that the total velocity can be separated into a mean and a fluctuating part. The fluctuations are highly three-dimensional, so that three orthogonal fluctuating velocity components are defined. If the concept of a mean plus a fluctuating velocity is introduced into the Navier–Stokes equations of motion of a fluid, a set of equations, termed the Reynold's equations, is obtained:

$$\rho\frac{DU_i}{Dt}=\frac{\partial}{\partial x_j}\left(\mu\frac{\partial U_i}{\partial x_j}-\rho\overline{u_iu_j}-\delta_{ij}p\right) \tag{11.1}$$

where U is the mean-flow velocity and u is the turbulent component. In this equation the turbulence term $\rho\overline{u_i u_j}$ is similar to a stress term (although it evolved directly from the inertia terms). Thus, the statistical turbulent velocity terms of interest have come to be called the "turbulent stress" terms.

Denoting the three turbulent velocity components by u, v, and w in the three orthogonal directions, a turbulent shear stress tensor can be constructed:

$$T_T = \rho \begin{Vmatrix} \overline{u^2} & \overline{uv} & \overline{uw} \\ \overline{vu} & \overline{v^2} & \overline{vw} \\ \overline{wu} & \overline{wv} & \overline{w^2} \end{Vmatrix} \tag{11.2}$$

Not only the three components $\overline{u^2}$, $\overline{v^2}$, and $\overline{w^2}$ are obtained, but also the cross-product velocities $\overline{uv}$, $\overline{vu}$, $\overline{uw}$, $\overline{wu}$, $\overline{vw}$, and $\overline{wv}$ appear. It is possible to simplify the tensor by assuming $\overline{uv} = \overline{vu}$, $\overline{uw} = \overline{wu}$, and $\overline{vw} = \overline{wv}$. For most of the turbulent flows studied a quasisymmetry, either plane or cylindrical, is obtained, so that $\overline{vw}$ and $\overline{uw}$ can be neglected. Measurement of the tensor for turbulent boundary layers was reported by Sandborn and Slogar.[(2)] These measurements were made with hot-wire anemometers.

For most shear flows it was found (Reference 2 for boundary-layer flows) that the cross product, $\overline{uv}$, was the only turbulent velocity term of major importance in Equation (11.1). The three velocity components $\overline{u^2}$, $\overline{v^2}$, and $\overline{w^2}$ are of basic importance in diffusion studies.

In order to study the turbulent velocities, it is necessary to consider the higher-order "turbulent kinetic energy equation," Reference 3, p. 326. The higher-order equations require not only the terms of Equation (11.2) but also triple correlations $\overline{vu^2}$, $\overline{v^3}$, $\overline{vw^2}$. The turbulent energy dissipation term

$$\varepsilon \equiv \nu\left[2\overline{\left(\frac{\partial u}{\partial x}\right)^2} + 2\overline{\left(\frac{\partial v}{\partial y}\right)^2} + 2\overline{\left(\frac{\partial w}{\partial z}\right)^2} + \overline{\left(\frac{\partial u}{\partial y} + \frac{\partial v}{\partial x}\right)^2} + \overline{\left(\frac{\partial u}{\partial z} + \frac{\partial w}{\partial x}\right)^2} + \overline{\left(\frac{\partial v}{\partial z} + \frac{\partial w}{\partial y}\right)^2}\right] \tag{11.3}$$

and a pressure–velocity correlation term, $\overline{vp'}$, also appear in the energy equation. Direct measurement of all the terms in the turbulent energy dissipation has not been reported. Only for flows where some approximation, such as local isotropy, can be made, have values of the dissipation been evaluated. The Kolmogorov scales shown in Figure 11.2 were calculated assuming ε could be estimated from local isotropic relations. The assumption was questionable for the smaller values of yU_τ/ν.

Associated with the evaluation of the turbulent velocities in the laboratory are a number of statistical quantities, such as probability distributions, spectral distributions, intermittency, and auto-and-space correlations.

Details of the electronic evaluation of turbulent signals are given in Chapter VIII of Reference 3.

Fluctuating Temperatures. A great number of flow conditions exist where fluctuating temperatures as well as fluctuating velocities are present. Examination of the energy equation indicates that the cross product of the vertical velocity, v, and the temperature, T', namely, $\overline{vT'}$, is the major turbulent transporter of energy. For low-speed flows considerable information has been obtained on the temperature fluctuations. In compressible flows evaluation of the temperature fluctuations are made difficult by the fact that most transducers sense total temperature and not static temperature. For compressible flows the problem is further compounded in that temperature fluctuations must also be associated with density and pressure fluctuations by the equation of state.

Fluctuating Density and Pressure. As noted above, both density and pressure can fluctuate in a particular flow. For liquids it is of course not likely that density fluctuations need be considered. For gases the equation of state required that temperature, density, and pressure be coupled. In general, it has not been possible in the past to measure directly density or pressure fluctuations within a flow field. A great deal of information has been recorded on pressure fluctuations at surfaces of bodies. Techniques of measuring pressure fluctuations in the flow field are currently being developed and should be of value in future studies.

Detailed evaluation of density fluctuations in compressible flows is needed. For supersonic flow the major fluctuation is one of mass flow. The heat-transfer sensors (hot-wire and hot-film anemometers) are directly sensitive to mass flow fluctuations. The main difficulty arises in that one needs to "uncouple" the density fluctuations, so the velocity fluctuations can be evaluated. For a few special cases, such as the freestream of a supersonic wind tunnel, it is possible to show that the fluctuations are of a specific nature. In the future, optical techniques should become available to greatly improve the evaluation of density fluctuations.

Concentration Fluctuations. Measurements of concentration fluctuations can be made in certain areas. In the past, most interest has been in the measurement of mean concentrations, so not a great deal of data on the fluctuations have been reported. The advent of laser light sources will greatly improve the ability to measure such properties as aerosol concentration fluctuations. As with temperature, the terms of interest are velocity–concentration correlations.

The diffusion equation may be written as follows:

$$\frac{\partial c}{\partial t}+u_i\frac{\partial c}{\partial x_i}=-\frac{\partial}{\partial x_i}\overline{(u_i c')} \qquad (11.4)$$

The cross product of the velocity and concentration fluctuation is the turbulent term of interest. Direct measure of this term has not normally been made, although recent techniques have been developed that make this measurement feasible.

Surface Shear Stress Fluctuations. In recent years a number of experiments have been reported on the evaluation of the fluctuating surface shear stress. Although early boundary-layer analysis suggested an ideal steady flow in the wall region, it is now obvious that large fluctuations in the local wall shear stress can occur. Studies of sediment transport suggest that much of the movement is due to the large shear "pulses" observed at the wall. The time variation of wall shear appears to show a skewed type of signal, which contains periods of high shear.

11.2. Measurement of Velocity Fluctuations

A number of techniques exist for the evaluation of turbulent velocity fluctuations. The hot-wire and hot-film anemometers have been extensively used to measure turbulent velocity fluctuations. The recent developments in laser velocimeters have greatly improved the older techniques of using tracers to evaluate turbulent velocity fluctuations. A third technique employing electrochemical phenomena is of value in liquids. Techniques employing lift and drag on bodies have also been developed to evaluate turbulent velocities. A number of techniques that respond mainly to large-scale turbulent fluctuations, such as acoustical anemometers, propellers, etc., may also be of value in specific laboratory applications.

11.2.1. Heat-Transfer Techniques

Hot-wire anemometers have been used to measure turbulence for the past thirty years. The early development of the anemometer is reviewed by Schubauer and Burgers.(4) In recent years the hot-film anemometer has been developed to measure turbulence in liquids. Detailed evaluation of all aspects of these resistance-temperature transducers is covered in the book by Sandborn.(3) The relation between heat loss and flow velocity for circular cylinders was demonstrated in Reference 3.

Sensor Operation. The hot-wire sensor can be operated in either a constant-current or a constant-temperature mode. For the constant-current mode, which might also be called constant mean resistance, the operator varies the current, as the flow changes, to maintain the bridge in balance and thus keeps the mean resistance of the sensor constant. For the constant-temperature mode, which might also be called instantaneous resistance, a feedback amplifier is employed to keep the bridge in balance. The hot-film sensor is normally operated in the constant-temperature mode, because of its frequency response characteristics.

If a hot wire is placed in a turbulent air flow the heat loss will fluctuate with the turbulence. As discussed in Reference 3, the hot wire follows a transient change roughly as a first-order system. The frequency response of a 0.0002-in.-diam tungsten wire is nearly perfect to approximately 70 Hz. For frequencies greater than 70 Hz, the response drops off approximately as a first-order system. The drop-off in frequency for a hot-wire and a hot-film sensor are compared in Figure 11.3.

The hot-film sensor responds as a higher-order system, References 3 and 6. In order to measure the high-frequency turbulence the frequency response of the hot wire must be improved. There are, of course, fluid flows, such as atmospheric and liquid flows, where a response of the order of 70 Hz is adequate to evaluate the root-mean-square turbulence level.

The frequency response of a hot-wire or hot-film sensor can be improved by electronic means. As noted in Reference 3, the hot-wire storage of energy term, $c_p(\partial T_w/\partial t)$, leads to the reduction in response of the sensors. If the wire temperature T_w can be kept constant, independent of time, the storage term will be zero. An electronic system can be employed to keep the wire temperature constant. The hot wire or film is operated in a bridge circuit, such as is shown in Figure 11.4. Unbalance of the bridge occurs when the wire resistance changes owing to changes in heat transfer. A dc, feedback, amplifier circuit is employed to sense the unbalance of the bridge. The amplifier senses the unbalance and in turn varies the current to

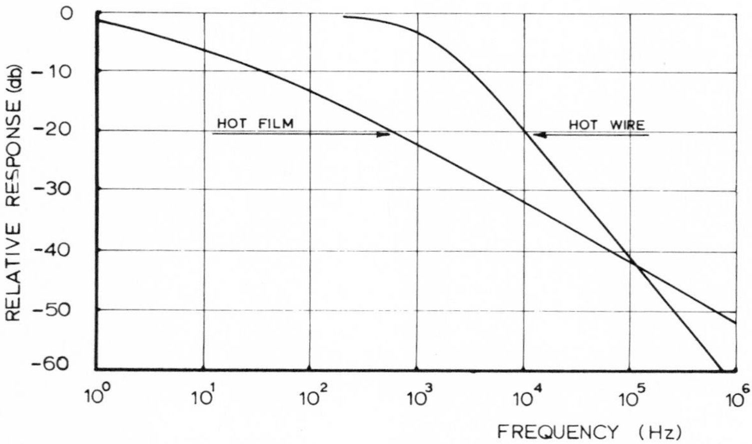

Fig. 11.3. Frequency response of a hot-wire and a hot-film sensor.[(5)]

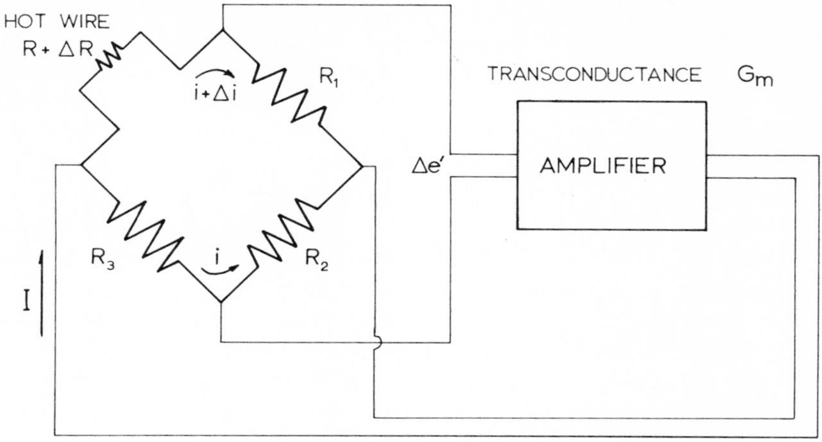

Fig. 11.4. Constant-temperature anemometer circuit.

the bridge, so that the wire is maintained at a constant resistance. Keeping the wire at constant resistance is equivalent to maintaining the wire at constant temperature. The frequency response of this type of system, which is called the constant-temperature anemometer, is strictly a function of how fast the electronics can rebalance the bridge. Commercial constant-temperature systems are available with frequency response greater than 50,000 Hz. For detailed information on the design of constant-temperature anemometer systems the reader should consult Reference 3.

Hot-Wire Anemometers. The frequency loss of the hot wire may also be corrected by either analog or digital techniques. For the constant-current mode of operation analog electronic circuits may be used to increase the amplitude of the hot-wire signal at high frequencies. The hot-wire response decreases approximately as a first-order system, so electrical circuits that have an increase in gain with frequency like a first-order system (Reference 7) will compensate for the wire drop-off. Note that it should be possible to employ more complex analog electronic circuits to account for the higher-order response errors discussed in Reference 3. Operation of analog compensation systems is covered in Reference 3.

It is shown by Sandborn[(3)] that the sensitivity of the hot wire for constant-current operation (where the wire mean resistance is held constant by varying the current, so that it is independent of the mean flow) compared to the sensitivity for constant temperature operation is given as

$$\frac{de\ \text{(constant current)}}{de\ \text{(constant temperature)}} = 2\left(\frac{R - R_a}{R_a}\right) \tag{11.5}$$

Where R is the hot resistance of the wire and R_a is the cold resistance at ambient temperatures. For most wire materials this ratio of resistances will not exceed 0.5. Tungsten wire begins to oxidize at temperatures on the order of 600°F, which correspond to $R \simeq 1.6R_a$. For platinum alloys, the wire will glow red at about $R = 1.25R_a$ and will fail before values of $R = 1.5R_a$ are reached. Thus, there is no major sensitivity advantage in selecting either the constant-current or constant-temperature mode of operation.

If a hot wire or hot film (Figure 11.5) is placed in a three-dimensional turbulent flow with its axis normal to the mean flow, it is found to be sensitive mainly to the turbulent component in the direction of the mean flow.[3] The major restriction is that the turbulent velocity components are much smaller than the mean velocity. For cases where the turbulent components are of the same order as the mean velocity it is not possible to interpret the output in a simple way. If the cylinder of Figure 11.5 is yawed to the mean flow then the output is proportional to both the velocity component in the direction of the mean flow and the component in the plane of the yaw.[3]

A hot wire or hot film operated by a constant-temperature system will have a calibration curve, such as shown in Figure 11.6. The wire voltage is the output from the anemometer required to keep the wire at a constant temperature as the velocity changes. The present analysis assumes that only the velocity sensitivity of the hot wire need be considered. Temperatures and density sensitivities will be considered in later sections.

Assuming the output of the hot wire is proportional to the longitudinal component of the turbulence, the mean voltage–velocity calibration curve

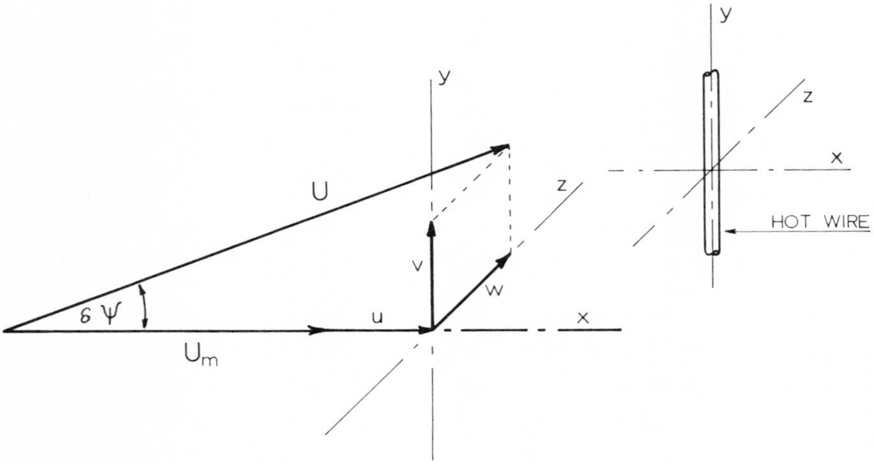

Fig. 11.5. Total velocity in a three-dimensional turbulent field.

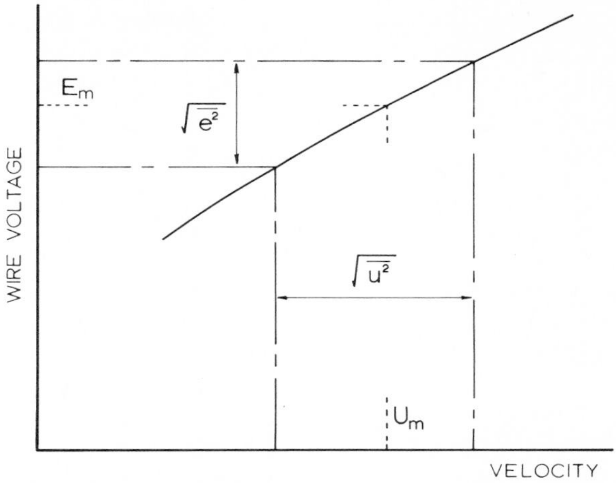

Fig. 11.6. Evaluation of velocity fluctuations from a hot-wire anemometer output.

can be used to compute $(\overline{u^2})^{1/2}$. The evaluation of the root-mean-square or mean-square velocity fluctuation requires two electrical measurements: the mean dc voltage, and the root-mean-square of the ac voltage. For a hot-wire calibration curve, such as shown in Figure 11.6, the measurement of mean voltage, E_m, determines the working point on the calibration curve. The root-mean-square voltage $(\overline{e^2})^{1/2}$ is assumed to be symmetrically distributed about the mean voltage. The assumption of symmetry may lead to some error if the actual fluctuations are skewed about the mean. From the location of the $(\overline{e^2})^{1/2}$ points on the calibration curve the value of $(\overline{u^2})^{1/2}$ can be determined from the velocity scale. The mean velocity is determined as the midpoint of the $(\overline{u^2})^{1/2}$ points. If the calibration curve is linear, no error due to nonsymmetry would occur.

Empirical rather than graphic techniques are normally employed to evaluate the output of hot-wire anemometers. The general approach is to "locally" linearize the calibration curve and relate the voltage fluctuation to the velocity fluctuation by the relation

$$(\overline{e^2})^{1/2} = \left(\frac{\partial E}{\partial U}\right)_{E_m} (\overline{u^2})^{1/2} \tag{11.6}$$

where $(\partial E/\partial U)_{E_m}$ is the slope of the calibration curve at the point E_m. This relation points out the fact that the hot-wire calibration curve must be sufficiently accurate so that its local first derivative can be determined. This requirement is equally true, whether or not the calibration curve is electronically linearized, since the derivative is equivalent to the constant of a linearized calibration curve.

Equation (11.6) is a general relation for nearly every type of measurement of a time-varying physical quantity. If a resistance thermometer were employed, rather than a hot-wire anemometer, the velocities, U and u, in Equation (11.6) would be changed to temperatures, T and T'. Modern analog applications employ linearizing circuits in order to simplify the calibrations. The linearized calibration curve has a constant value of $(\partial E/\partial U)$ that is independent of E_m. The setup of the linearizer circuit for hot wires will require the selection of the constants in an empirical relation from calibration information about each particular wire. The most accurate approach to selecting the constants is a curve fit of the first derivative of the calibration curve, since the derivative is the important term of Equation (11.6).

Corresponding to Equation (11.6) for the normal wire, the fluctuating voltage for a yawed wire may be written as

$$\overline{e^2} = \left(\frac{\partial E}{\partial U}\right)^2 \overline{u^2} + 2\left(\frac{\partial E}{\partial U}\right)\left(\frac{1}{U}\frac{\partial E}{\partial \psi}\right)\overline{uv} + \left(\frac{\partial E}{\partial \psi}\right)^2 \frac{\overline{v^2}}{U^2} \tag{11.7}$$

The angle, $\delta\psi$, as shown in Figure 11.5, may be expressed as (neglecting the w component)

$$\delta\psi = \tan^{-1}\frac{v}{U_m + u} \simeq \frac{v}{U_m + u} \simeq \frac{v}{U_m} \tag{11.8}$$

where u can be neglected compared to U_m. These assumptions are equivalent to the linearization made in connection with Equation (11.6). Equation (11.7) may be written as

$$e = \frac{\partial E}{\partial U}u + \frac{\partial E}{\partial \psi}\frac{v}{U_m} \tag{11.9}$$

Equation (11.9) can be evaluated directly from the experimental calibration curves by measuring the differentials. Figure 11.7 shows typical values of $\partial E/\partial U$ and $\partial E/\partial \psi$ evaluated from a calibration curve. A comparison of the terms shows that $(1/U)(\partial E/\partial \psi)$ increases with angle much faster than $\partial E/\partial U$ decreases. As a result, the hot-wire output voltage will increase as the wire approaches the parallel condition. Figure 11.8 shows the actual measured output of a hot wire as it is yawed to the flow.

A check of the yawed-wire analysis has been made by computing the $\overline{v^2}$ component as a function of angle for data similar to those shown in Figure 11.8. These calculations, shown in Figure 11.9, were taken at the center of a fully developed pipe flow (thus $\overline{uv}$ is equal to zero). The fact that $\overline{v^2}$ is not independent of angle is difficult to understand. A great number of measurements have all indicated that minimum values of $\overline{v^2}$ are always found around an angle of 40°. The deviations below 35° may reflect the failure of the inequality approximations at small angles.

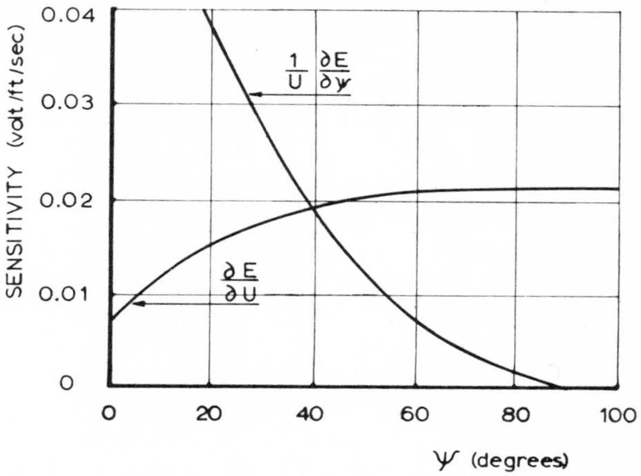

Fig. 11.7. Hot-wire sensitivity to velocity and yaw.

The evaluation of the turbulent velocity components from yawed wires normally requires measurements at three yaw angles. Rewriting Equation (11.7) in terms of the measurable voltage $\overline{e^2}$ gives

$$\overline{e^2} = S_u^2\overline{u^2} + 2S_uS_v\overline{uv} + S_v^2\overline{v^2} \tag{11.10}$$

where $S_u = \partial E/\partial U$, $S_v = (1/U)(\partial E/\partial \psi)$, and $\overline{uv}$ is the correlation between the two velocity components. There are three turbulent quantities to be evaluated. For the data of Figure 11.9 it was possible to set $\overline{uv} = 0$. The $\overline{u^2}$ component of turbulent velocity can be calculated from the wire normal to

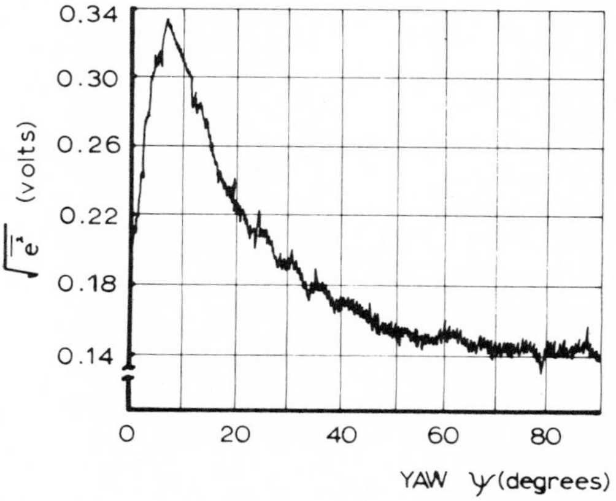

Fig. 11.8. Root-mean-square output of a hot wire as it is yawed to the flow.

the flow ($\partial E/\partial\psi = 0$ at 90°). For flows where $\overline{uv}$ is not zero the wire might be yawed to +40°, −40°, and 90°. If the wire has no support or mounting interference effect it is found that $S_{u_{+40}} = S_{u_{-40}}$ and $S_{v_{+40}} = S_{v_{-40}}$. The three voltage readings for the three angles make it possible to solve three equations for the velocity components. This technique of evaluating the turbulence requires a calibration of the hot wire for velocity at the three angle settings, and angle calibration at two angles.

Evaluation of the turbulent components normal to the mean flow and the turbulent shear stress $\overline{uv}$ or $\overline{uw}$ are made with either a single yawed wire or an X probe. Figure 11.10 shows typical hot-wire probe configurations. The X probes are employed in areas where rotation of the wire is inconvenient. The two wires of the X probe are set at fixed angles $+\psi$ and $-\psi$. The wires operate identically to the single yawed wire, so that Equation (11.10) applies for their output also. Three equations are required from the X probe output in order to obtain the three unknown velocities $\overline{u^2}$, $\overline{v^2}$ or $\overline{w^2}$ and $\overline{uv}$ or $\overline{uw}$. The output of each wire gives two independent equations, and the third is obtained by taking the sum or difference of the two output signals. Details of analyzing the output of X wires are covered in Reference 3. The X probe was originally employed by Schubauer and Klebanoff(7) to obtain instantaneous values of the normal turbulent velocity component normal to the flow. If the two wires of the X probe are identical in sensitivity to velocity, then the electronic difference of the signals is directly proportional to v or w. In practice it has proved difficult to ideally match the two wires. Schubauer and Klebanoff(7) give an approximate technique for matching the wires. The problems and errors encountered in X-probe measurements are discussed in Reference 3.

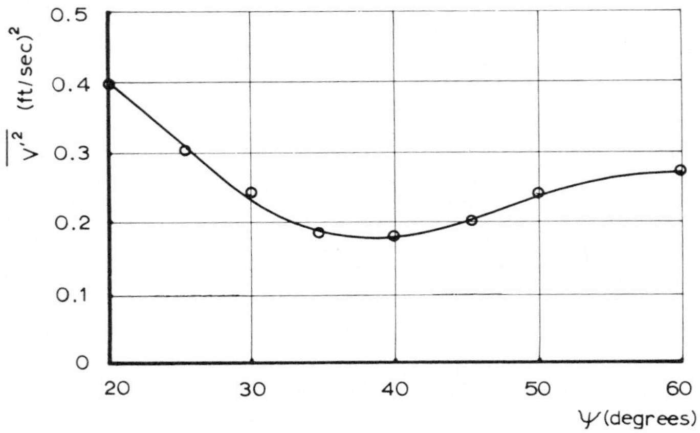

Fig. 11.9. Values of $\overline{v^2}$ as a function of a wire yaw angle.

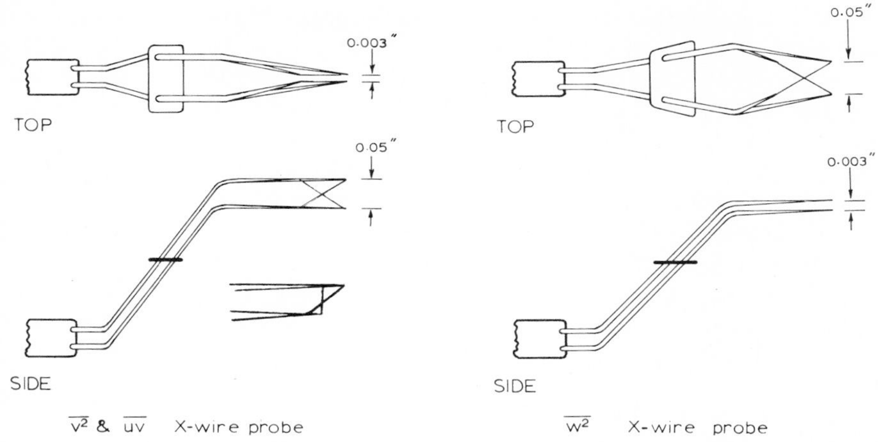

Fig. 11.10. Typical hot-wire probes used to measure the turbulent velocity components.

Two wire probes may also be employed in shapes other than the specific "*X*" shape. An alternate form is noted in Figure 11.10. One of the wires is mounted normal to the direction of the mean flow and the second wire is mounted at an angle ψ. The output of the normal wire, as given by Equation (11.6), is related only to the longitudinal velocity component. The yawed wire, as given by Equation (11.9), is related to both the longitudinal and normal velocity components. The product of the two wires outputs will be

$$\overline{e_1 e_2} = S_{u1} S_{u2} \overline{u^2} + S_{v1} S_{v2} \overline{uv} \tag{11.11}$$

The above equation is solved directly for $\overline{uv}$, since $\overline{u^2}$ is known from the normal wire output. The value of $\overline{v^2}$ is obtained from the mean-square output of the yawed wire, once $\overline{u^2}$ and $\overline{uv}$ are known. This modification of the *X*-probe technique should offer a somewhat improved accuracy in evaluating $\overline{uv}$ over that of solving three simultaneous equations.

Hot-Film Anemometers. The present discussion of hot-wire anemometer measurements applies equally for either gases or liquids. The hot-film anemometer is employed extensively in the measurement of liquids, since it is more rugged and can be insulated with a molecular layer of quartz. The heat loss from a cylinder-type hot-film probe is much the same as that for a hot wire. A detailed evaluation of the sensitivity of hot-film probes is given by Richardson *et al.*[(8)] (see also Reference 3). A major difficulty encountered in the measurement of turbulence in water was the change in sensor calibration due to contaminates. McQuivey[(9)] has shown that the contamination effect is approximately the same as a reduction in the temperature difference between the sensor and the liquid. Thus, it is possible to account for the change in sensitivity by calibrating the sensor at a number of different overheats.

The film sensor can be made in almost any shape desired. Cone and wedge shaped sensors have been employed as a means of reducing the contamination effects. Unfortunately, difficulties can be encountered if the heat transfer varies around the probe. Bellhouse and Rasmussen[(10)] demonstrate that for uneven heat transfer the probe substrate may drive the film sensor at very low frequencies. It appears desirable to employ the small-diameter cylinder probes to reduce this low-frequency substrate effect.

The operation of hot-film anemometers in non-Newtonian fluids has been reported by Friehe and Schwarz.[(11)] For certain non-Newtonian fluids the heat transfer is greatly reduced. Over specific Reynolds number ranges the data of Friehe and Schwarz show no heat-transfer change with velocity. The heat-transfer characteristics for a yawed sensor are quite unusual. The Nusselt number is found to increase with angle of yaw, rather than decrease, as is found in Newtonian fluids. Theoretically, it should be possible to employ the hot-film or hot-wire sensor in any liquid, as long as the sensitivities can be determined.

A novel new type of film probe has become available in recent years: a "split film," where the upper half of a 0.006-in.-diam sensor is a film independent of the lower half of the film. This type of probe is extremely sensitive to the flow angle. The sum of the heat transfer from the two films is mainly proportional to the local velocity. The difference between the heat transfer of the two films was found to be proportional to the flow direction.[(12)] Also the ratio of the heat transfers is proportional to the flow direction. Difficulties have been reported in operating the split film near surfaces in boundary layers. No doubt a need exists to develop operating techniques for this new instrument.

11.2.2. Tracer Techniques

A number of "tracer" methods have been employed to evaluate the turbulent motion. The tracer technique depends on the convection of a tracer by the turbulent motion. The recent development of laser velocimeters represents the most advanced method of employing tracers. Tracers are of great value not only for velocity evaluations, but also in flow visualization.

One of the earliest attempts to measure the turbulent motion with tracers was reported by Fage and Townend.[(13)] They employed an "ultramicroscope" to observe the motion of minute particles in water flow. The ultramicroscope employed an intense light and a dark background to make visable ultrasmall particles that are normally present in water. The microscope used a magnification of 50 or greater to view the details of water flow in a pipe. The particles appear as streaks of light at the focal point of the microscope. To measure speed the objective lens of the microscope was moved in the same direction as the particles. The objective lens was mounted on an eccentric wheel, such that it rotated into line with the microscope axis once every revolution. When the particle appeared to be stationary, the speed of the particle could be obtained from the rotational speed of the objective lens. For turbulent flow, it was possible to determine the maximum and minimum velocities at a point. The direction of the particles could also be obtained with the microscope. These measurements provided the first insight into much of the character of the turbulent motion.

The tracer particle is normally a great deal larger than the molecules of the fluid, so it will not follow exactly the ultra-small-scale motion of the flow. It is generally assumed that the motion of a particle responds as a first-order system:

$$\frac{dU_p}{dt} = \frac{1}{\tau_p}(U - U_p) \tag{11.12}$$

where U_p is the particle velocity, U is the fluid velocity, and τ_p is a "time constant" for the particle motion. The solution of this first-order equation is

$$U_p = U + (U_{p,0} - U)\,e^{-t/\tau_p} \tag{11.13}$$

where $U_{p,0}$ is the initial velocity of the particle. If the particles are spheres then the drag of the particle can be computed from a Stokes-type analysis to determine the time constant. For viscous Stokes flow the time constant would be $m/3\pi\nu d$, where d is the diameter of the sphere, m is the mass, and ν is the fluid viscosity. Bourot[(14)] measured a value of $m/6.6\nu d$ for aluminum flakes that were about 4 μ square and 1 μ thick. He obtained a time constant of the order of 67 μsec for the aluminum flakes in air. The flakes would respond to a sine wave variation of 2500 Hz.

The Laser Anemometer. The most recent advances in transient velocity measurements have been in the area of light-scattering optical techniques. These advances were made practical by the development of the "Light Amplification by Stimulated Emission of Radiation" (LASER) light source. The laser is in principle equivalent to the electronic sine wave oscillator that employs amplifiers and feedback, except that the laser operates in the range of 10^{14}–10^{15} Hz (present discussion taken from Reference 15). The basic properties of the laser are high intensity, high monochromaticity, and directivity of its output beam. Electromagnetic radiation is emitted from a material when atoms of the material drop from one energy level to a lower energy level. The laser action is that of "pumping" or exciting a collection of atoms in the material such that a selected transition will have more atoms in its upper energy level than in its lower energy level. When the excited atoms emit, they are forced to contribute to the electromagnetic wave that already exists in the neighborhood. Thus, all atoms emit in phase with the existing wave, and a high degree of time coherence is achieved. The structure of the laser is that of a resonator, so that the frequencies generated are confined to specific modes of the resonator. This confining results in a high degree of space coherence. The wavelength of radiation is determined by the energy levels of the active material, which may be either a gas or a crystal. A broad range of frequencies can be obtained by considering all possible materials.

The laser anemometer measures the velocities of microscopic particles suspended in a gas or liquid flow. The principle of the laser anemometer depends on a Doppler shift of the frequency of light scattered from the moving particles. It would be desirable to employ the molecules of the fluid as the light scatters; however, the light intensity is generally not available in continuous form. Secondly, the magnitude of the light intensity required to measure scattering from molecules would also cause measurable changes in the molecular structure of the fluid. The present laser measurements employ particles that are large compared to the molecules. Major problems of size and dispersion of particles have been encountered. The particles must be able to follow the fluid velocity accurately. One may question whether any particle that is large compared to the molecules can follow accurately the turbulent motion.

Simply, the laser velocimeter may be viewed as the forming of a light interference fringe pattern at the measuring point in the flow. As a tracer passes through the pattern it produces an alternating light output. The frequency of the "scattered" light will be proportional to the particle velocity perpendicular to the fringes. Details of the Doppler evaluation of the light frequency were given by Goldstein and Kreid.[16]

The optical system employed by Goldstein and Kreid is shown in Figure 11.11. The laser light is incident upon the splitter plate, where there is a division of amplitude of the beam due to partial reflection into a main beam and a reference beam. The main beam is focused by a lens into the flowing fluid. Some of the incident illumination is scattered in the direction of the photomultiplier with a Doppler shift.

The reference beam is reflected by a mirror and is focused by another lens onto the same region of the fluid as the main beam. The reference beam is directed toward the photomultiplier and the two apertures, which aid in limiting the volume element (and angular direction) that the photomultiplier sees. The two beams (scattered and reference) incident upon the photocathode produce a beat signal with the frequency of the Doppler shift. For convenience, the incident beam and the reference beam make equal angles with the plane normal to the tube axis. The Doppler shift is therefore proportional to the velocity component parallel to the axis. The

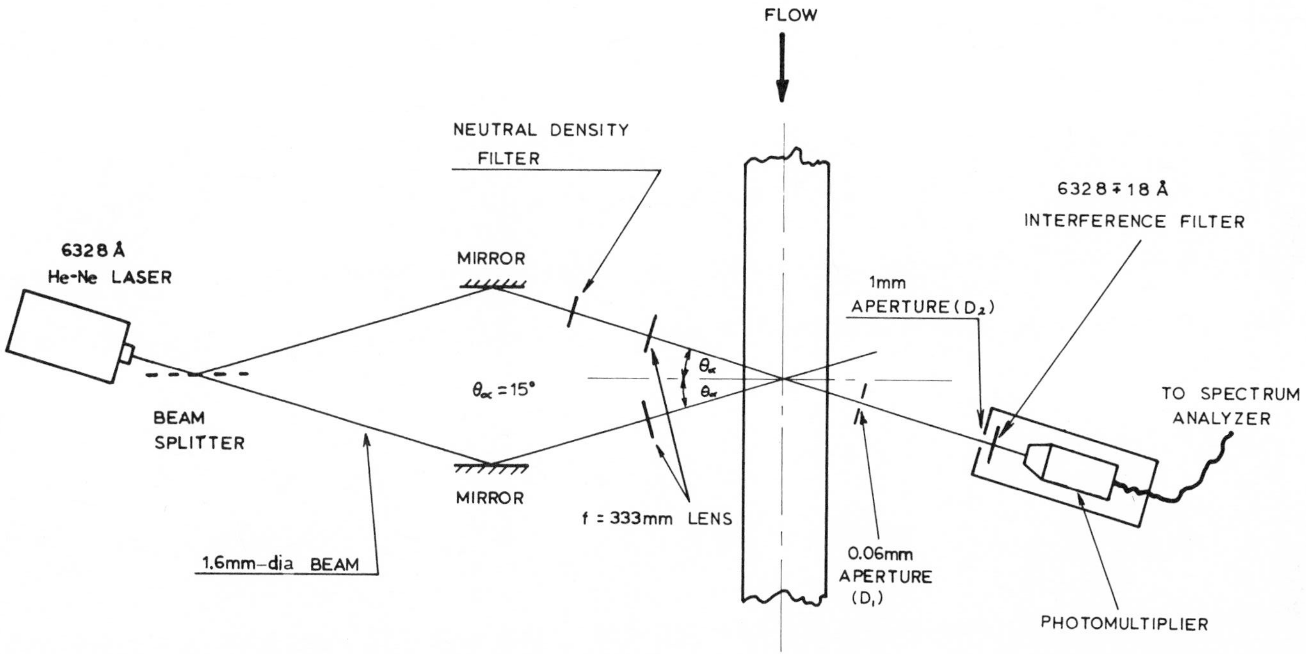

Fig. 11.11. Typical laser Doppler anemometer. Test fluid is water with 1 : 50,000 concentration of 0.557-μ polystyrene particles.

photomultiplier output is scanned with a spectrum analyzer. The Doppler shift corresponding to the mean velocity is usually related to the peak output frequency.

The Doppler technique can be applied to any scattered electromagnetic wave and is not limited to just the laser light source. Microwave Doppler systems are employed as velocity-measuring instruments in the atmosphere. The acoustic Doppler anemometer appears to be of major interest in the measurement of blood flow. Additional details of these techniques and of the laser anemometer are given in Chapter 13.

Heat Tracers. In 1935 Schubauer(17) employed the diffusion of heat from a line source to indicate the lateral turbulent velocity. The theory of diffusion by continuous movements developed by Taylor(18) showed that the spread of a heated wake behind a line source could be characterized by the lateral turbulence. The angle of spread, α, between the line source and positions in the wake where the temperature rise due to the source is one-half that at the center of the wake, as shown in Figure 11.12, will be equal to

$$\alpha^2 = \alpha_m^2 + \alpha_\varepsilon^2 \tag{11.14}$$

where α_m is the spreading of heat due to molecular diffusion and α_ε is the spreading due to the lateral turbulent velocity. The molecular spreading angle in degrees was computed from the relation

$$\alpha_m = 190.8\left(\frac{k}{\rho c_p U_m x}\right)^{1/2} \tag{11.15}$$

where ρ is the density, c_p is the specific heat, k is the thermal conductivity, U_m is the mean air speed, and x is the distance downstream where the wake

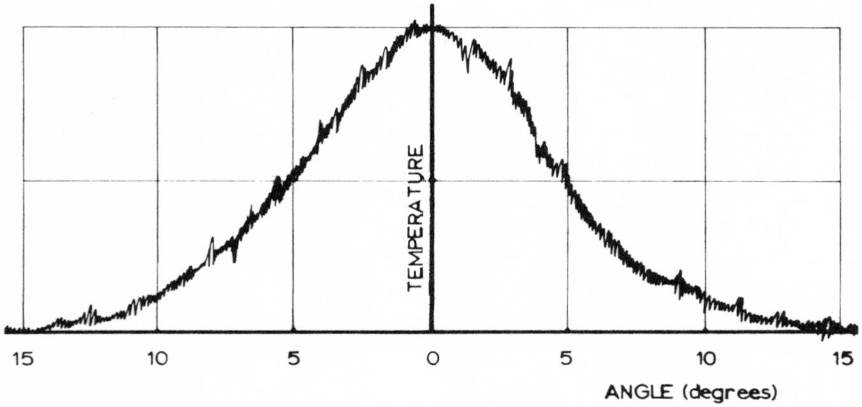

Fig. 11.12. Spread of heat from a line source.

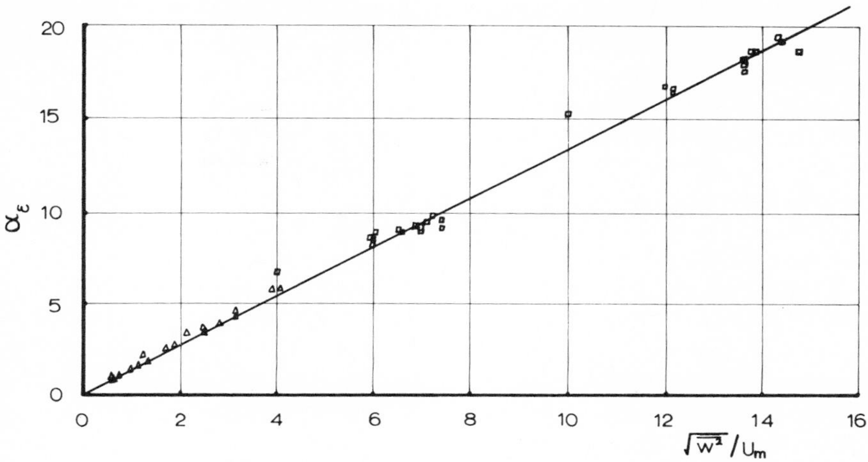

Fig. 11.13. Variation of α_ε with the lateral turbulent intensity.

is measured. Taylor assumed a Gaussian distribution for the wake, which produced a turbulent spreading angle of (in degrees)

$$\alpha_\varepsilon = 134.7(\overline{w^2}/U_m)^{1/2} \tag{11.16}$$

Schubauer compared measured spreading angles with the longitudinal turbulent component and found a value of 151.5 for the constant. The difference in constants was thought to be due to the use of $(\overline{u^2})^{1/2}$ rather than $(\overline{w^2})^{1/2}$. Hagist[19] has repeated Schubauer's measurements in a boundary layer. Figure 11.3 shows Hagist's measurements in a turbulent boundary layer. The flow turbulence was increased by placing blocks upstream of the probe. The turbulent intensity was evaluated using a yawed hot wire. The line drawn through the "center of gravity" of the measured points is that given by Equation (11.16).

The heat-diffusion technique may also serve as a flow direction indicator. The maximum of the temperature wake will be aligned with the mean flow. Both Schubauer and Hagist found some nonsymmetry of the measured temperature wake. The instruments employed thermocouples to sense the small change in temperature. Further development might include more sensitive temperature sensors, such as the thermistor, to increase the accuracy of the wake measurement.

Townend[20] employed periodic electric sparks to tag a small mass of heated air. The heated air can be observed with a schlieren optical system and evaluated in terms of the turbulent diffusion of the hot mass of air. Townend was able to analyze a series of several hundred photographs to obtain the mean velocity and also the root-mean-square values of the lateral and axial turbulent velocity components in a 3-in.-square pipe. These thermal tracer methods were for the most part replaced by the hot wire as a turbulence-measuring technique, the main objection to the diffusion techniques being that they are not a point measurement. The measurement is an integral average over the distance behind the sample point.

Related Tracer Techniques. A number of particle tracer techniques have been advanced for evaluation of turbulent flow fields. While the laser velocimeter greatly improves previous techniques of evaluating turbulent velocities, a great deal of information is still lacking on the turbulent flow field. Tracers can be employed to indicate the local instantaneous flow field, as well as point measurements. If the tracers are observed by a fluctuating light source or a shutter arrangement, a photograph can be obtained with a series of dashes representing the local velocity and vector of the particle. Use of stereoscopic observations makes it possible to observe the complete three-dimensional flow field. Cady[(21)] describes a rotating-mirror technique (due to Zwicky and Sutton) used to measure the speed of tracers. The rotation of the mirror causes the image of the particles to move diagonally across the photographic plate. The inclination of the particle path is proportional to its velocity. The measurement of the inclination angle is simplified by also photographing on the same plate the particle image not affected by the rotating mirror.

The hydrogen-bubble technique has been developed in recent years as a quantitative measuring tool in water-flow studies. Use of combined-time-streak-markers produces simultaneous visual images of the flow structure and quantitative measurement of velocity over a finite region. The hydrogen-bubble technique is credited to Geller, and was mainly developed in its present form by Clutter *et al.*[(22)] at Douglas Aircraft Co. and by Schraub *et al.*[(23)] at Stanford University. The technique consists of using a fine wire as one electrode of a dc circuit to electrolyze water. A very small hydrogen bubble is produced at the wire. These bubbles are small enough to form good markers, without major difficulty due to their rise rate.

A fine wire of the order of 0.0005–0.002 in. in diameter is employed as the negative electrode of a dc circuit in water flow. The second electrode is any convenient part of the water-flow facility. Detailed circuits for operating the bubble wires are reported in References 22 and 23. Care is required to obtain uniform bubble production and sharp on–off characteristics if time lines are used. Platinum, copper, and stainless-steel wire have all been used successfully in the production of bubbles. Large currents are needed to obtain sufficient optical density of the bubbles for photographic requirements. The bulk of the bubbles formed have diameters on the order of one-half to one wire diameter.

Fig. 11.14. Streak squares in a contraction.[23] Flow from left to right.

The hydrogen bubbles allow simultaneous velocity measurements over a whole plane at any instant in time. A short pulse of bubbles marks a local group of fluid particles. Figure 11.14 shows a typical marking of the flow in a contraction, obtained by Schraub *et al.*[23]

The hydrogen-bubble data may be evaluated by the techniques employed for other types of particles. The x component of velocity can be evaluated from the relation

$$\tilde{u} = (\Delta x/S)f \tag{11.17}$$

where $\tilde{u}$ is the estimate of the x component of the actual local Eulerian field velocity $u(x, y, z, t)$ at some point (x_0, y_0, z_0, t_0); Δx is the x component of displacement of a bubble during a known time interval $\Delta T = 1/f$ measured on the film; S is a scale factor for the film measurement; and f is the bubble-wire pulse rate or camera framing speed. It is shown[23] that the technique can measure velocities within approximately 4% at a confidence level of 95%.

11.2.3. Electrochemical Techniques

In liquids it is possible to employ chemical reactions as a measure of the turbulence. An electrochemical technique has been developed at the University of Illinois, under the direction of Hanratty,[24,25] which is employed to measure both velocity and concentration fluctuations. A chemical reaction is carried out at the surface of an electrode. The voltage applied to the electrode is sufficient to reduce the concentration of the reacting species to zero at the electrode surface. The electrode is analogous to a constant-temperature, hot-wire anemometer, in that the surface concentration is kept constant, and the current flowing in the circuit is related to the mass-transfer rate. The mass-transfer measurement is related to the local velocity gradient at the electrode. The problems involved in defining the relation between current and mass-transfer rate are similar to those encountered in relating hot-wire anemometer measurements to the fluctuating velocity field.

The actual technique employed is to use the electrode as part of an electrochemical cell, which is composed of an aqueous solution of potassium ferri- and ferrocyanide and nickel electrodes. Thus, the measurements can only be made in one particular fluid. The electrolyte was ferrocyanide, and 2 molar sodium hydroxide. The sodium hydroxide acts as a low-resistance vehicle for current flow throughout all of the cell except very near the electrodes. The measuring electrode is the cathode of the cell, and the anode is a much larger area of the flow facility. The larger anode allows the current flow at the cathode to control the probe output.

The following reactions occur at the surface of the electrodes:

$$\begin{array}{ll} \text{cathode (test electrode)} & Fe(CN)_6^{3-} + e \rightleftarrows Fe(CN)_6^{4-} \\ \text{anode (reference electrode)} & Fe(CN)_6^{4-} \rightleftarrows Fe(CN)_6^{3-} + e \end{array} \tag{11.18}$$

The current flow, I, is related to the rate of reaction of ferricyanide per unit area of the test electrode, N, as given by the equation

$$N = (I/AF)(1 - T) \tag{11.19}$$

where A is the area of the test electrode, F is Faraday's constant, and T is the transfer number. The large concentration of sodium hydroxide relative to the ferricyanide leads to a very small value for the transfer number (approximately 0.001), so it can be neglected. The experiments are done in an atmosphere of nitrogen to prevent cathodic oxygen reactions at the electrode.

A mass-transfer coefficient K is defined as

$$K = N/(C_b - C_w) \tag{11.20}$$

where C_b is the bulk concentration of the ferricyanide ion and C_w is the concentration at the surface of the electrode. At sufficiently large voltages across the electrodes the reaction rate is fast enough that C_w approaches zero at the test electrode and the current is controlled by the magnitude of the mass-transfer coefficient. The operation of the probe consists of controlling the voltage at a value such that $C_w \simeq 0$, and measuring the current. Figure 11.15 shows typical polarization curves for a 0.064-in.-diam electrode in the wall of a 1-in.-diam pipe. The limiting current is a function of the flow rate in the pipe. The measurements of Reiss,[26] Mitchell,[27] and Son[28] show that the electrochemical technique gives accurate values of the wall velocity gradient over a Reynolds number range from 300 to 70,000.

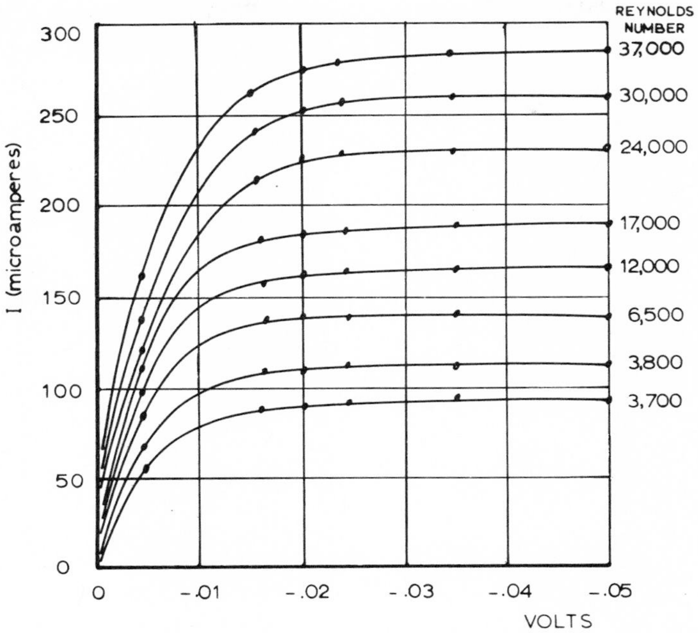

Fig. 11.15. Limiting current curves for the reduction of ferricyanide.[24] Electrode is 0.064 in. in diameter.

The technique of evaluating fluctuating velocity gradients is similar to that employed for the hot wire. The mass-transfer coefficient is assumed to have a steady as well as an unsteady component:

$$K = \bar{K} + k \tag{11.21}$$

Sirkar and Hanratty[25] give the following linear relations between the mass-transfer coefficients and the local velocity gradients:

$$\text{(mean)} \qquad \frac{\bar{K}L}{D} = 0.807\left(\frac{\bar{S}_x L^2}{D}\right)^{1/3} \tag{11.22}$$

$$\text{(fluctuating)} \qquad \frac{k}{\bar{K}} = \frac{1}{3}\left(\frac{s_x}{\bar{S}_x}\right) \tag{11.23}$$

where $\bar{S}_x$ is the local mean velocity gradient and s_x is the fluctuating velocity gradient in the direction of the mean flow. An electrode yawed to the mean flow can be employed to measure the velocity gradient fluctuations in directions normal to the mean flow. A V electrode is employed to measure the transverse gradients, much the same as an X-wire hot-wire probe.

The paper by Mitchell and Hanratty[24] gives a detailed set of fluctuation measurements made along the wall of a fully developed pipe flow. Velocity intensities, mass-transfer intensities, spectra, and correlations are included. Probability density functions show both positive and negative deviations from the mean that are of the same magnitude as the mean.

An "electrokinetic" technique of indicating turbulence in liquid, mainly water, has been reported by Baldwin and Cermak.[29] Voltage fluctuations are observed to exist when an electrode arrangement, similar to those discussed for the electrochemical experiments, is in a turbulent liquid. The voltage fluctuations are of the order of microvolts. The fluctuations both in the distribution of root-mean-square values and for the frequency spectrum can be related to the corresponding velocity fluctuations. The measurements are of a relative nature, since techniques of direct calibration were not developed.

11.2.4. Sonic Anemometer

The propagation of a sound wave can be employed to measure the velocity and temperature of a gas flow. This type of measuring instrument has found extended application in atmospheric measurements (see Chapters 12 and 13). Recent developments in sonic-Doppler-shift- and sounder-type instruments have also been made. The present discussion is directed mainly toward the simple sound-wave propagation type of instrument. The basic concept of the sonic anemometer is demonstrated in Figure 11.16. A source

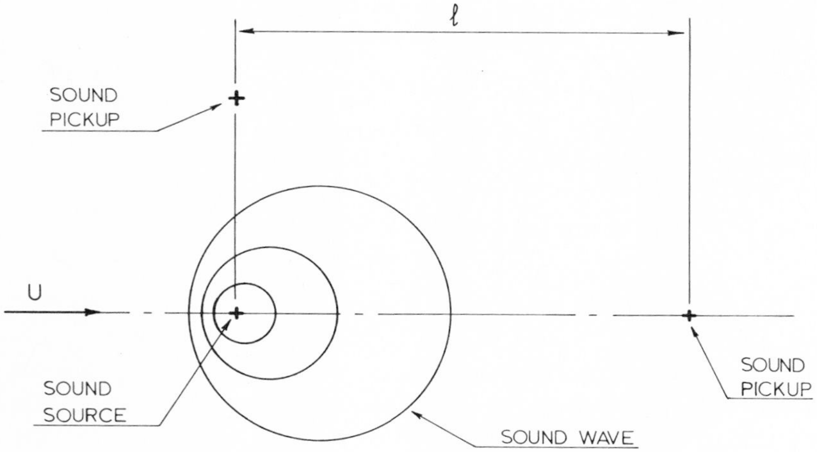

Fig. 11.16. Measurement of velocity with a sound wave.

of sound is produced at a point in the flow. A sound pickup downstream measures the time between the pulse of sound and the arrival of the wave. The simple velocity of propagation is

$$V = a + U = l/t \tag{11.24}$$

where a is the speed of sound, which for an ideal gas is a function of the absolute temperature. By using two equally spaced sound pickups, one at right angles to the flow, both U and a are measured directly. The difference in time between the two pickups (which are both at a distance l from the source) is directly proportional to the local velocity.

For atmospheric measurements the direction of the mean wind varies. Thus, a three-dimensional sonic system is employed (see Reference 30). Mitsuta gives the following general relation for the sonic anemometer with a sound source at the origin of a rectangular coordinate system:

$$(X - V_x t)^2 + (Y - V_y t)^2 + (Z - V_z t)^2 = at^2 \tag{11.25}$$

where (V_x, V_y, Y_z) are the rectangular velocity components of the wind. The time required for the sound wave to reach a point $(l, 0, 0)$ is

$$t_1 = \frac{l[(a^2 - V_n^2)^{1/2} - V_x]}{a^2 - V^2} \tag{11.26}$$

where V_n is the velocity component normal to the x axis $(V_n^2 = V_y^2 + V_z^2)$ and V is the total velocity $(V^2 = V_x^2 + V_n^2)$. The time to reach a point $(-l, 0, 0)$ is

$$t_2 = \frac{l[(a^2 - V_n^2)^{1/2} + V]}{a^2 - V^2} \tag{11.27}$$

The difference of the transit times is

$$t_1 - t_2 = \frac{2lV_x}{a^2 - V^2} = -\frac{1}{A}\frac{2lV_x}{a^2} \tag{11.28}$$

The sum of the transit times is

$$t_1 + t_2 = \frac{2l(a^2 - V_n^2)^{1/2}}{a^2 - V^2} = \frac{1}{B}\frac{2l}{a} \tag{11.29}$$

For low velocities (50 m/sec or less) the values of $A = (1 - V^2/a^2)$ and $B = (1 - V^2/a^2)/(1 - V_n^2/a^2)^{1/2}$ are nearly equal to unity. Thus, for atmospheric measurements, sum and difference of the transit times to l and $-l$ can be employed to measure V_x, or the other velocities depending on the pickup locations.

The speed of sound is a function of the air temperature, with a slight effect due to the humidity. Barret and Suomi[(31)] give the following relation

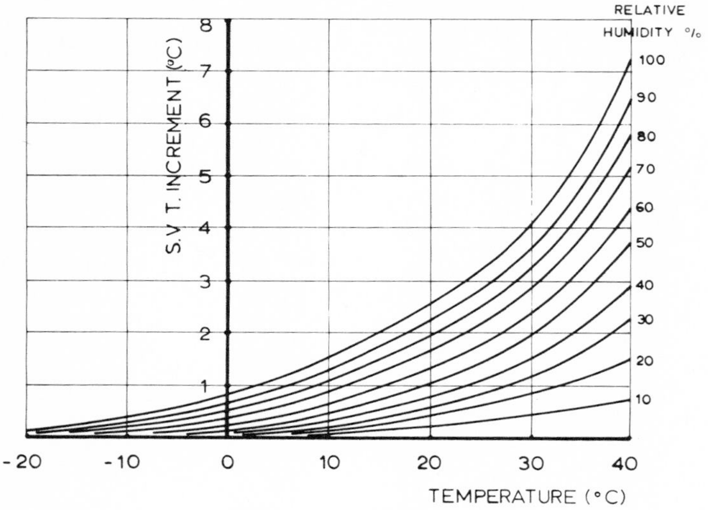

Fig. 11.17. Variation of the sound virtual temperature increment with temperature and relative humidity.[30]

for the speed of sound:

$$a = 20.067\,[T(1+0.32e/p)]^{1/2} \tag{11.30}$$

where T is the absolute temperature, e is the water vapor pressure, and p is the atmospheric pressure. The quantity in the square brackets of Equation (11.30) is called the sound virtual temperature, T_{sv}. Figure 11.17 is a plot of the sound virtual temperature increment $(T_{sv} - T)$ as a function of temperature and relative humidity. The sound virtual temperature is given by

$$T_{sv} = \left[\frac{2l}{20.067}\frac{1}{(t_1+t_2)}\right]^2 \tag{11.31}$$

The approach to measuring fluctuating velocities and temperatures is to emit a pulse of sound and sample the transit times, t_1 and t_2. Since the pulse is a sample at one time, a series of samples are made to obtain statistical values. The pulse technique limits the frequency response of the instrument. The atmospheric sonic anemometers have responses of the order of 100 Hz. The space resolution is limited to eddy sizes that are larger than the sound path length. Typical path lengths are of the order of 50 cm. A comparison of a number of different acoustic instruments is reported in Reference 32.

11.2.5. Lift and Drag Sensors

A great number of techniques based on the variation of lift or drag with velocity have been used as turbulence sensors. These sensors are of particular value in large-scale turbulence, such as the atmosphere. Devices such as cup, vane, and propeller anemometers are all employed to evaluate atmospheric turbulence; however, they are usually too large to be employed in laboratory studies. Small rotary-flow meters[33] have been developed to measure turbulent fluctuations in large water-flow facilities. In recent years there has been an attempt to develop lift and drag sensors for use in the laboratory.

Airfoil Probe. A technique developed by Siddon and Ribner[34] employs the lift of a small airfoil as a measure of the transverse component of turbulence. Consider the simple airfoil shown in Figure 11.18. The airfoil is at an angle of attack α with respect to the total velocity component. In turbulent flow the total velocity, U_{tot}, and the angle of attack, α, will vary in a random way. The quasisteady, linear, airfoil lift relation is

$$L = \tfrac{1}{2}\rho U_{tot}^2 S[dC_L/d\alpha]\alpha \tag{11.32}$$

For most airfoil sections the slope of the life curve is linear for small angles of attack. Thus, employing the linear approximations that $U_{tot} \simeq U_m$, and $\alpha \simeq v/U_m$ (which are equivalent to the assumptions made for the yawed hot-wire linear approximation) the vertical velocity component is a linear function of the lift:

$$\frac{v}{U_m} = \frac{L}{\text{const}} \tag{11.33}$$

where the constant is $\tfrac{1}{2}\rho S(dC_L/d\alpha)$. The linear restrictions imposed are roughly the same as those of the hot wire. Siddon and Ribner[34] suggest the probe is usable to root-mean-square vertical velocities that are less than 30% of the mean velocity.

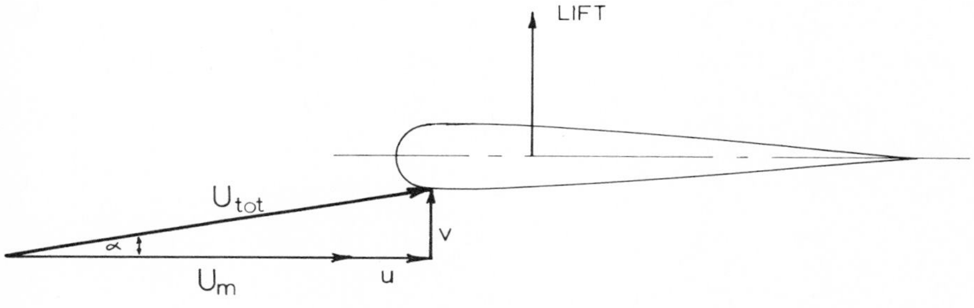

Fig. 11.18. Lift on an airfoil.

The physical construction of the probe consisted of attaching the airfoil to a tapered cantilever beam. A piezoelectric transducer was imbedded in the cantilever beam. The actual airfoil sensing element must be a trade-off between two conflicting requirements. For maximum output signal it is desirable to have the maximum possible lifting surface. For high-frequency response and space resolution the lift sensor must be kept small. For the study of Siddon and Ribner a disk airfoil of 0.070-in. diameter was employed. Their measurements were made in a 4-in. free airjet facility with a mean velocity of 85 ft/sec. Rectangular planforms of comparable area and $AR = 2$ were also employed with equal success. A conical aluminum beam with a short outer section of uniform diameter was employed to support the airfoil. The first critical frequency of the beam was approximately 12.5 kHz. A shroud was employed to ensure that only the sensing element responded to the turbulent impulses. A ceramic piezoelectric transducer deformed in the length expander mode was used to sense the airfoil lift.

Figure 11.19 is a comparison of the v-component frequency spectrum measured with a hot-wire anemometer and the airfoil probe reported by Siddon and Ribner. Some slight disagreement is noted at low frequencies, which was due to the electronic readout of the piezoelectric transducer. The probe tested by Siddon and Ribner agrees with the hot-wire data out to approximately 4000 Hz. The drop in response above 4000 Hz is due to the inertia of the airfoil. A resonant peak is noted at the critical frequency of 12.5 kHz.

A number of devices similar to the airfoil probe have been studied. For atmospheric measurements, bivanes, which respond by aligning with flow directions, are of great value. Gust probes, also employed in atmospheric studies, measure both the lift and drag on either vanes or bodies. Recently, Cheng[35] has reported the development of a small sphere sensor rather than an airfoil to measure all the turbulent velocity components.

11.2.6. Corona-Discharge Anemometer

The electrical discharge between two electrodes in air is a function of pressure, velocity, and humidity. A number of experimental studies have been reported on the use of glow-discharge techniques to measure air flow.[36,37] Small electrodes of the order of 0.001 in. in diameter and spacing of less than 10 mm are employed. The voltage across the gap may vary from 1000 to 5000 V. The experiments of Werner[36] show that the discharge current is markedly sensitive to pressure and less sensitive to velocity. A typical measure of the variation in current with velocity for a discharge probe is shown in Figure 11.20. Werner[36] reports measurements of turbulent velocity fluctuations in supersonic flow with a glow-discharge

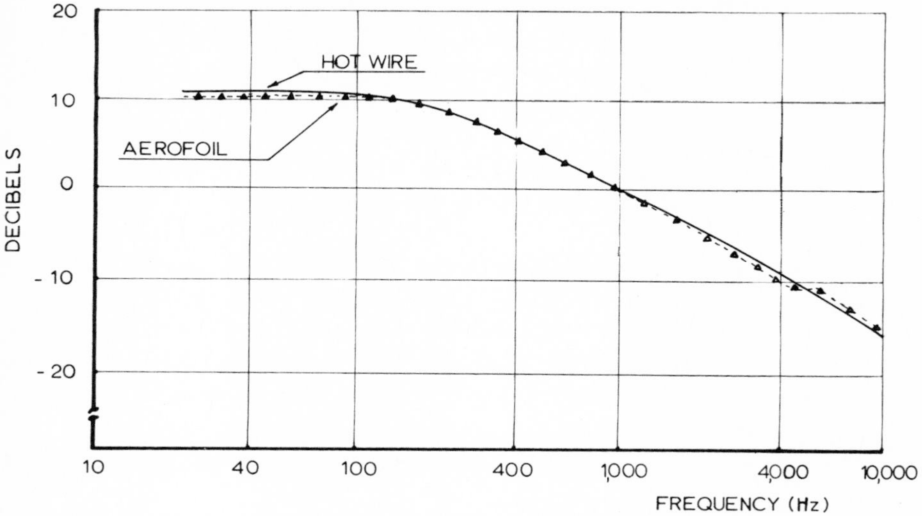

Fig. 11.19. Comparison of the v-component spectrum with a hot wire and an airfoil probe.[34] ($U = 85$ ft/sec; $x/D = 4.5$; $y/r = 1.0$ and $D = 4$ in.)

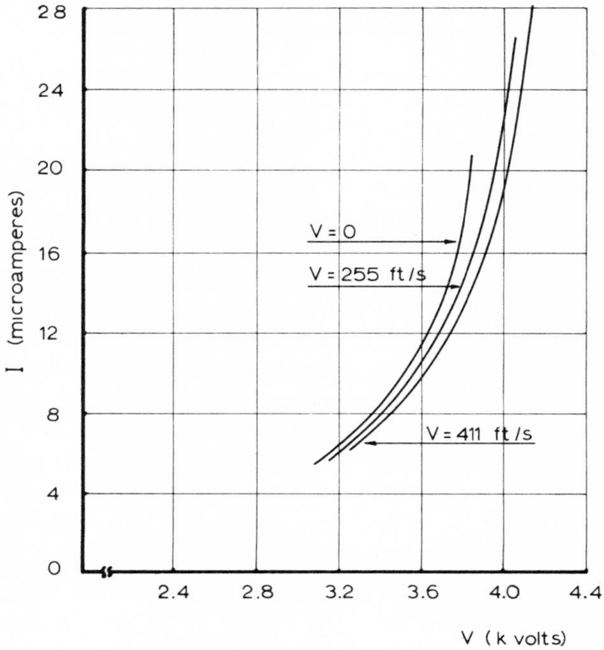

Fig. 11.20. Velocity dependence of a glow-discharge probe.[36]

anemometer. In general, this type of instrument has not been used extensively in transient measurements. Recently, an ion-drift-type instrument has been developed for mean flow measurements. This drift type of application could be employed to measure turbulence much as the acoustic instrument.

11.3. Measurement of Temperature Fluctuations

Measurements of temperature fluctuations in turbulent flow fields have been made mainly with resistance thermometers. While many different mean sensors are available to measure temperature, it is difficult to make the sensor small enough to evaluate the fluctuation. For low-frequency fluctuations a thermocouple may be employed, but it is difficult to make and use very small thermocouples.

11.3.1. Resistance Thermometer

Resistance thermometers for fluctuation measurements are similar to hot-wire anemometers. For very sensitive resistance thermometers it is possible to use wire diameters of the order of 0.00002 in. in diameter.[3] The sensitivity of a resistance thermometer is

$$\frac{\Delta E}{\Delta T} = I\alpha R_0 \tag{11.34}$$

where E is the voltage output, T is the temperature, I is the current used to detect the resistance, α is the thermal coefficient of resistance, and R_0 is the reference resistance of the sensor. The smaller the wire diameter the larger the resistance will be for a given length of sensor. The transient response of the resistance thermometer is also similar to that of the hot-wire anemometer.[3] The very-small-diameter wires can respond directly to fluctuations of the order of several thousand cycles per second. Improvement in the frequency response can be made by use of the analog "constant-current" compensation systems. If the transient response of the sensor is known[3] then an electronic amplifier, which increases in gain, as the sensor decreases in gain with frequency, can be employed to compensate for the loss in transient response. This technique can greatly increase the transient response of the system. It is possible to follow frequencies of the order of tens of thousands of cycles per second. Note that the "constant-temperature" technique of compensating for the frequency loss of the hot wire cannot be employed for the resistance thermometer application.

The output of the resistance thermometer can be evaluated similarly to the hot-wire technique:

$$\overline{(e^2)}^{1/2} = \frac{\partial E}{\partial T}\overline{(T'^2)}^{1/2} \tag{11.35}$$

For most metallic sensors the relation between resistance and temperature is nearly linear, so the sensitivity is nearly constant over large ranges of temperature.

The major difficulty in evaluating the output of the resistance thermometer is that it senses not static, but total temperature. The problem is further complicated by the fact that the wire sensor is not a perfect transducer. The total temperature, T_0, can be written as

$$T_0 = T_s + \frac{1}{2}\frac{U^2}{c_p} \tag{11.36}$$

where T_s is the static temperature, U is the flow velocity, and c_p is the specific heat at constant pressure. The transducer must convert the flow kinetic energy $[U^2/2c_p]$ into a temperature. For low-speed flows the velocity term is not important in Equation (11.36), so the sensor gives a reasonable measure of the static temperature. For velocities greater than roughly 150–200 ft/sec, the velocity term becomes important. Figure 11.21 shows typical measured values of the "recovery temperature ratio" ($\eta \equiv T_w/T_0$) as a function of flow Mach number for wires normal to flowing air. (See Reference 3 for identification of the experimenters listed in Figure 11.21.) The ratio is a measure of the failure of the transducer to convert the velocity

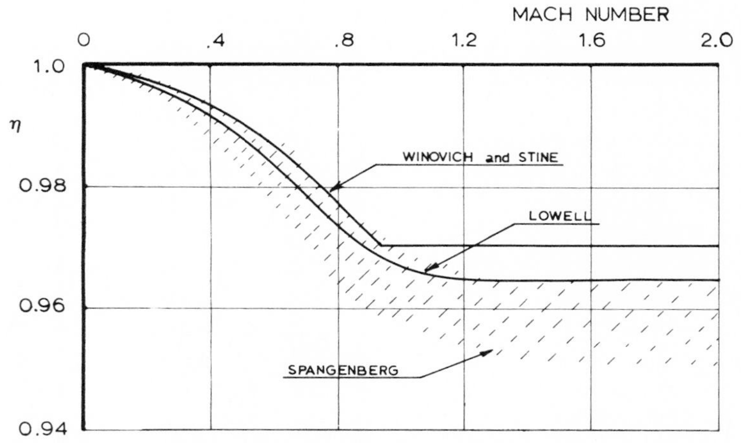

Fig. 11.21. Wire recovery temperatures.

to temperature. The actual deviation from the perfect transducer is never greater than 5%; however, note that the temperature ratio is for absolute temperatures. Thus, the actual deviation may be as much as 20° or 30°, depending on the value of T_0. Figure 11.21 is for the case where the wires are operating in continuum flow. If the flow density is low, as is often the case in supersonic flows, the wire may be operating in slip flow conditions. Slip and free molecular flow further complicate the recovery temperature (see Reference 3, p. 73).

Figure 11.22 shows measurements of the total temperature fluctuations that have been reported for a number of supersonic boundary-layer surveys[38] (see Reference 38 for identification of the experimenters). The fluctuations are nondimensionalized by the difference between the surface recovery temperature, T_r, and the flow total temperature. By independently evaluating the velocity fluctuations (such as with a hot-wire or laser velocimeter) it is possible to separate the total temperature fluctuation into a static temperature and a velocity fluctuation. As demonstrated in Reference 38, this leads to considerably more scatter in the static temperature data than that found for the total temperature. For the simpler incompressible flows the output of the resistance thermometer is related directly to the static temperature.

11.3.2. Measurement of Temperature–Velocity Correlations

For the incompressible flows it is possible to measure the static temperature fluctuations directly with a resistance thermometer. It should be

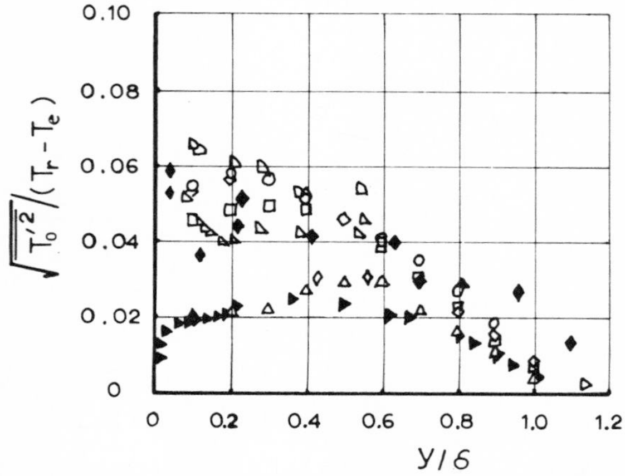

Fig. 11.22. Measured total temperature fluctuations in supersonic boundary layers.[38] (Data as in Reference 38.)

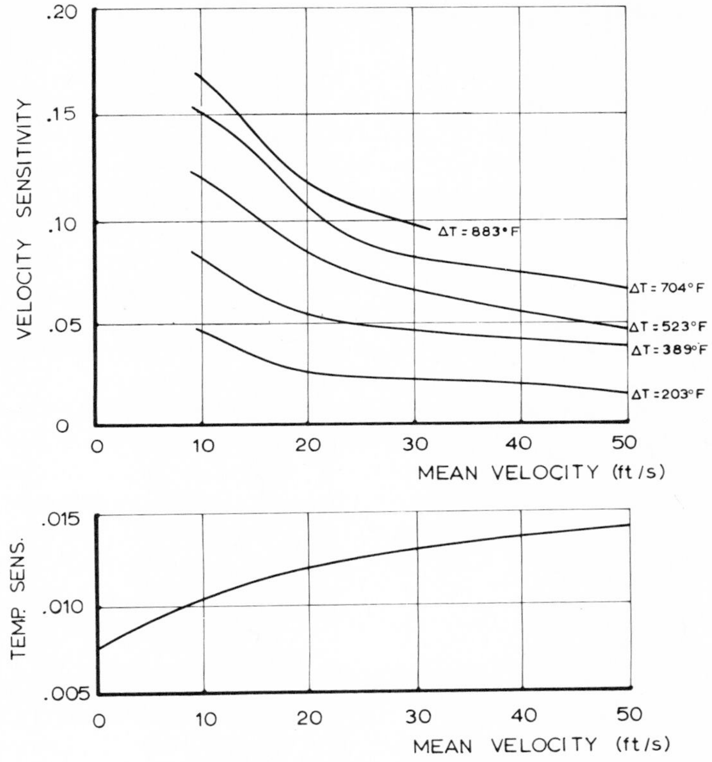

Fig. 11.23. Variation of hot-wire velocity and temperature sensitivities.

obvious that if temperature fluctuations are present it will not be possible to neglect them in the evaluation of the hot-wire anemometer output. In the literature it has been suggested that by operating the hot wire at high temperatures the sensitivity to velocity can be made much larger than the sensitivity to temperature. Figure 11.23 shows a comparison of the magnitudes of the velocity and temperature sensitivities. While the temperature sensitivity was found to vary slightly with velocity (because of end loss effects at the supports and also variations in the wire temperature distribution due to the varying heat transfer), the velocity sensitivity increased considerably with increasing ΔT.

Considering the hot-wire output in the form

$$e = \frac{dE}{dU}u + \frac{dE}{dT}T' \tag{11.37}$$

leads to a relation for the statistically measured quantity

$$\overline{e^2} = S_u^2\overline{u^2} + 2S_uS_T\overline{uT'} + S_T^2\overline{T'^2} \quad (11.38)$$

where $S_u = dE/dU$ and $S_T = dE/dT$. From Equation (11.38) it can be seen that if $S_T = 0.1S_u$ and $\overline{uT'} = 0$, then the hot-wire output is proportional to the velocity. However, it is not obvious that the velocity–temperature correlation can be neglected. In order to evaluate the output of a hot wire in the presence of temperature fluctuations, the wire is operated at several values of temperature, ΔT. As may be seen in Figure 11.23, each value of ΔT produces a new value of S_u, while S_T remains approximately the same. A set of three equations can be constructed from Equation (11.38) and solved for three unknowns. Obviously, the value of $\overline{T'^2}$ can be evaluated directly from the $\Delta T \to 0$ case and only two equations need be solved.

A "fluctuation diagram" technique was introduced by Kovasznay[39] for evaluation of experimentally obtained simultaneous equations. When the equations are obtained experimentally the accuracy to be expected may be quite poor. Thus, a larger sample of measurements (several values of ΔT, instead of just three) can eliminate some of the uncertainty.

11.4. Measurement of Density and Pressure Fluctuations

The direct measurement of both density and pressure in turbulent flows is of great interest. Techniques are currently being studied for direct measurements of these fluctuations. At the present time the measurements are still of an exploratory nature and cannot be viewed as well-established techniques. Recent developments in optical schlieren techniques[40] and interferometry techniques[41] have shown the feasibility of measuring density fluctuations. For low densities "electron beams" can be used to measure density fluctuations.[42] In each of these studies it is difficult to assess the accuracy of the data, since there is no known standard to compare with. Of specific interest is the use of hot-wire anemometers in supersonic flows, where density, temperature, and velocity are all sensed by the transducer.

Supersonic Flow Measurements. At the present time almost all supersonic turbulence measurements have been made using either hot-wire or laser velocimeters. The general heat-loss relation for a circular cylinder will be a function of Mach number, Knudsen (or Reynolds) number, temperature loading, total temperature, and angle of yaw:

$$\mathrm{Nu} = f(M, \mathrm{Kn}, \Delta T, T_w, \psi) \quad (11.39)$$

Sandborn[3] gives detailed information on the evaluation of the general hot-wire relation for both constant-temperature and constant-current operation. The general relation may be reduced to the form

$$e = S_u \frac{\delta U}{U} + S_\rho \frac{\delta \rho}{\rho} + S_{T_t} \frac{\delta T}{T_t} + S_\psi \frac{\delta \psi}{\psi} \tag{11.40}$$

where δU, $\delta \rho$, δT, and $\delta \psi$ represent the fluctuations in velocity, density, temperature, and angle, respectively.

Equation (11.40) indicates that the hot wire, or film, can respond as an anemometer, resistance thermometer, and a "manometer" all at the same time. The contribution of each mode is assumed to be additive. In a supersonic compressible gas flow it is possible that all three modes of fluctuation may be present. It is also possible that velocity fluctuations may correlate with the density and temperature fluctuations. Thus, the squared output of Equation (11.40) could contain six unknown fluctuations. A total of six independent equations are required in order to evaluate all of the turbulent terms.

In general it has been very difficult to evaluate the output of hot wires in supersonic flow. It has been necessary to reduce the number of variables in order to deal with the problems. While Equation (11.40) implies that one might use a technique similar to that used to separate temperature and velocity fluctuations, it is not usable directly in supersonic flow. Experimentally it is found[3] that the velocity and density sensitivities of the hot wire are identical. The hot wire is sensitive to the mass flow rather than the individual components. For subsonic-compressible flows the sensitivities to density and to velocity are different, but this area of flow has received little or no attention. Equation (11.40) can be rewritten for supersonic flow as

$$e = S_{\rho u} \frac{\delta \rho u}{\rho U} + S_{T_t} \frac{\delta T}{T_t} + S_\psi \frac{\delta \psi}{\psi} \tag{11.41}$$

For a hot wire normal to the mean flow the angle sensitivity will be zero. Thus, Equation (11.41) can be evaluated similarly to the velocity–temperature technique of Section 11.3.2. The mass-flow fluctuations and the correlation between mass flow and total temperature can be obtained.

For the free stream of supersonic wind tunnels it appears possible to relate the fluctuations directly to sound waves. For a Mach number range from 1.6 to 5, Laufer[43] found from hot-wire measurements that the correlation coefficient between mass flow and total temperature fluctuations in the free stream had values of approximately −1. Secondly, it was known that neither static temperature nor velocity fluctuations by themselves were sufficient to produce the high turbulence levels indicated by the hot wire. On the basis of the analysis of Kovasznay,[39] it was apparent that the only simple fluctuating field consistent with the measurements was a pure sound field. For a pure sound field the isentropic relations between pressure, density, and temperature (and their fluctuations) are valid. Evidence for sound domination of the supersonic free stream has been reported for a number of facilities. The sound fluctuations are related directly to the mass-flow fluctuations measured with the hot-wire anemometer. The only deviations appear in the hypersonic flow facilities, where inlet flows must be heated to high temperatures. For the hypersonic facilities,[44,45] large temperature, as well as mass-flow fluctuations, are encountered.

Figure 11.24 is a plot of the mass-flow fluctuation intensity versus Mach number reported by a number of experimenters. For Mach numbers below 2, Laufer[46] and Morkovin[47] found that the velocity fluctuations upstream of the sonic throat affect the free-stream turbulence in the supersonic flow. At a Mach number of approximately 2.5 the upstream velocity fluctuations could no longer be related to the test-section turbulence.

The measurements of Laufer,[43] Sandborn and Wisniewski,[48] and Donaldson and Wallace[49] all show that the mass-flow fluctuation intensity decreases with increasing Reynolds number at a fixed Mach number. The decrease is estimated to vary as $\mathrm{Re}^{-0.25}$. It was pointed out by Laufer[43] that the wall boundary layers, which generate the free-stream sound fluctuations, become thinner as the Reynolds number increases. The wall boundary-layer effect was dramatically demonstrated by Laufer[43] by reducing the tunnel Reynolds number to where the wall layers are laminar. The point shown in Figure 11.24 at $M=4.5$ (Laufer, $\mathrm{Re/in.} \sim 2.6\times 10^4$) shows almost an order-of-magnitude reduction in tunnel mass-flow fluctuations. Although the data of Figure 11.24 fail to agree in all cases, there is evidence to suggest that the larger-size supersonic wind tunnels (for a given Mach number) will have the lower turbulence levels ($M>2$). The ratio of free-stream area to boundary-layer perimeter should be an important parameter.

The increase of the mass-flow fluctuation intensity with Mach number is a result of the sound field being proportional to the fourth power of the Mach number.[43] A curve fit of the data of Laufer for $M>2$ shows that the actual mass-flow fluctuations increase as M^2. As an approximate curve fit, the following relation was obtained:

$$\frac{\overline{(\rho u)'}}{\rho U} = A\,(\mathrm{Re/in.})^{-0.25}(M^2-0.5) \qquad (11.42)$$

where the constant A accounts for the variation due to tunnel size (and/or origin of the sound source). For Laufer's data a value $A=0.0095$ was estimated. Curves for $A=0.0095$ and Re/in. = 90,000 and 330,000, corresponding to Laufer's measurements, are plotted in Figure 11.24. A curve for $A=0.017$ and Re/in. = 90,000 is plotted to demonstrate the effect of A.

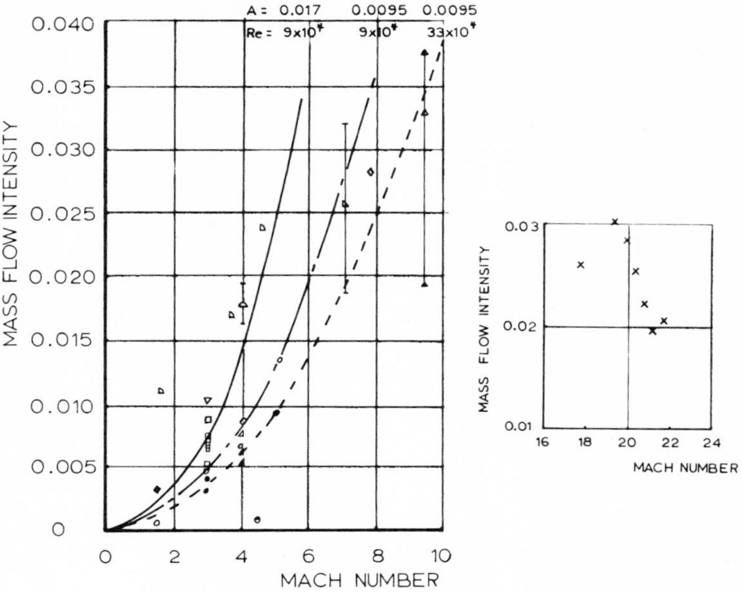

Fig. 11.24. Measured free-stream mass-flow fluctuations. Curves are for the relation $\overline{(\rho u)'}/U = A(\text{Re/in.})^{-0.25}(M^2-0.5)$. (Data as in Reference 38.)

Summary of Data (Reference Numbers Taken from Reference 38)

Symbol	Reference	Reynolds No./in.	Tunnel size
◇	Laderman[63]	$1.4–22\times10^5$	50-in. diameter
⊿	Donaldson and Wallace[18]	$5–24\times10^4$	12×12-in. tunnel
△	Laderman and Demetriades[17]	12.5×10^4	21×21-in. tunnel
◓	Laufer[16]	2.6×10^4	18×20-in. tunnel
○	Laufer[16]	9×10^4	18×20-in. tunnel
●	Laufer[16]	33×10^4	18×20-in. tunnel
□	Sandborn and Wisniewski[13]	$7.4–16.5\times10^4$	6×6-in. tunnel
D	Kistler[4]	$51.5–71.4\times10^4$	6×6-in. tunnel
⊾	Stainback, Wagner, Owen, and Horstman[11]	2769×10^4	40-in. diameter
◆	Wise and Schulz[64]	—	9×3-in. tunnel
⌓	Rose[37]	42×10^4	2-in. diameter
▽	Johnson and Rose[23]	125×10^4	8×8-in. tunnel

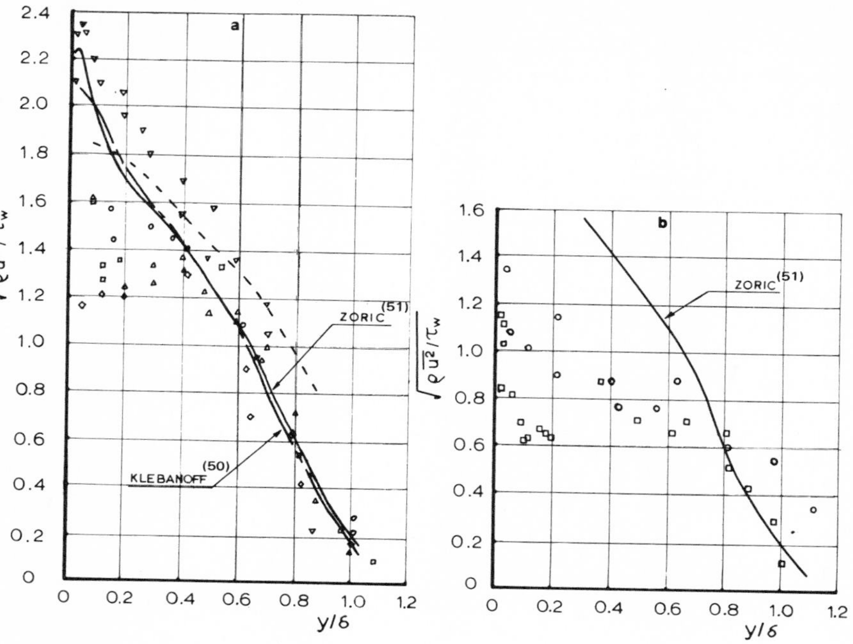

Fig. 11.25. Longitudinal turbulent velocity measurements in zero-pressure gradient boundary layers. (Data as in Reference 38.) (a) Adiabatic; (b) cooled wall.

This latter curve is an approximate fit of the Sandborn–Wisniewski data for a 6×6-in. tunnel. The measurements of Kistler,(67) shown in Figure 11.24, may not represent a true value for the wind tunnel, as they were taken at the edge of the wall boundary layer. Kistler reports that the free-stream levels were below the noise level of his measuring system.

The very high Mach number results, shown as the insert in Figure 11.24, do not vary according to Equation (11.42). The particular facility is such that the Mach number increases as the Reynolds number is increased. Thus, the higher Mach numbers also represent higher Reynolds numbers. Both the variation with Mach number and the variation with Reynolds number are greater than would be predicted by Equation (11.42). Note also that the mass-flow intensities are much lower than the extrapolation of the supersonic curves, shown in Figure 11.24, would predict.

For boundary-layer measurements it is not obvious just how to evaluate the hot-wire output. The general approach has been to neglect fluctuations in pressure and couple directly the temperature and density fluctuations. The recent measurements of Rose and Johnson(50) have compared hot-wire evaluations with laser velocimeter measurements in a supersonic flow. The laser responds to particle velocity, and should be independent of density, temperature, or pressure. Figure 11.25a shows the results of the laser and hot-wire measurements (inverted triangles). Also included are a number of other hot-wire measurements reported by other experimenters. Incompressible measures of $\overline{u^2}$ by Klebanoff[51] at a low Reynolds number and Zoric[52] at a high Reynolds number are noted in Figure 11.25. The measurements of Rose and Johnson appear to be in reasonable agreement with the incompressible results. These results would appear to justify the neglect of pressure fluctuations in the flow measured by Rose and Johnson. The drop-off of the turbulent fluctuations near the surface for the other measurements shown may suggest measurement difficulties. Figure 11.25b shows measurements reported for hypersonic flow facilities, where it may be unrealistic to neglect the pressure fluctuations.

Figure 11.26 shows the measurements of the vertical turbulent velocity component reported by Rose and Johnson. The hot-wire measurement (open points) are somewhat higher than might be expected, but this may represent the uncertainty in evaluation of experimental simultaneous equations. The evaluation of the turbulent shear stress, $\rho\overline{uv}$, is shown in Figure 11.27. The shaded area is a "best estimate" of the expected values.[38] The drop-off of the points near the wall appears to be a major question in the

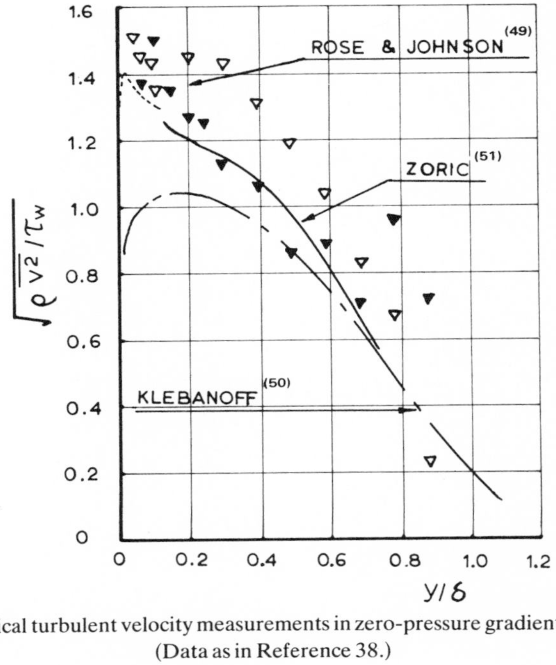

Fig. 11.26. Vertical turbulent velocity measurements in zero-pressure gradient boundary layer. (Data as in Reference 38.)

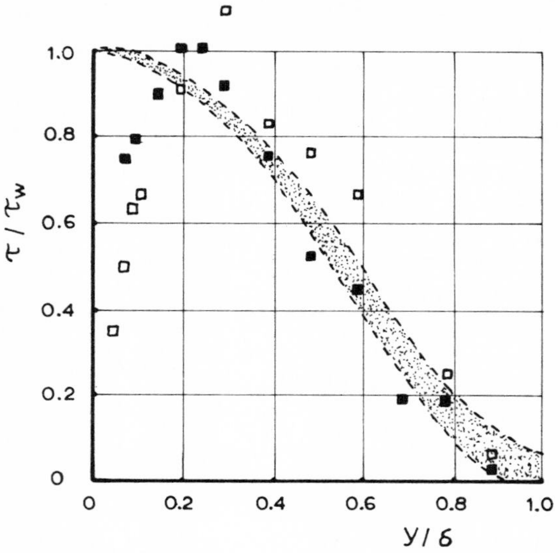

Fig. 11.27. Measured turbulent shear stress in a zero-pressure gradient boundary layer.

accuracy of the measuring techniques. A major effort is presently underway to improve the techniques of measuring the turbulent shear stress in supersonic flow.

Little information is available on measurements in subsonic compressible flows. This flow regime offers some new problems. The hot wire is no longer equally sensitive to density and velocity, so it is possible in theory to separate them. There are major difficulties associated with probe interference effects at the high subsonic velocities. No doubt the laser anemometers will offer the best techniques for measurements in this flow regime.

Pressure Measurements. The measurement of pressure fluctuations has mainly been reported only for surfaces. Microphones and pressure transducers are well developed for surface measurements. Techniques of measuring surface pressure fluctuations have been reported by Corcos.[(53)] Measurements of static pressure fluctuations in the flow field are at best very difficult to make. Recently Jones and Planchon[(54)] have developed a static pressure probe, which appears to measure static pressure fluctuations up to 10,000 cycles per second. The basic probe configuration is shown in Figure 11.28. The probe is designed so that a constant gas pressure in the probe produces a flow out into the stream through the static holes. The velocity of the gas flow through the orifice will depend on the static pressure surrounding the tube.

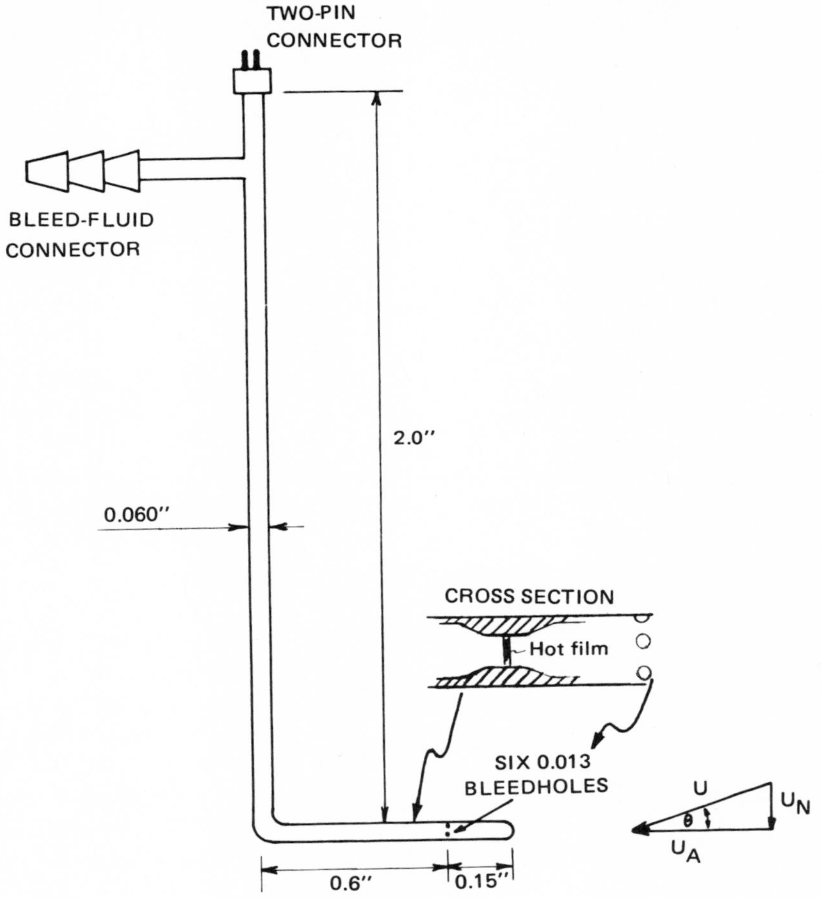

Fig. 11.28. Static pressure probe for fluctuating pressure measurements.

The hot film element will sense the variation in velocity due to the variation in the static pressure in the stream. Figure 11.29 shows a calibration of the sensor.

Questions arise as to the effect of normal turbulent velocities on the static-type pressure probes. It appears that as long as the fluctuation scales of the velocity are larger than the probe diameter they are not a major factor in the static pressure reading. There is a need to develop this type of instrumentation for use in the compressible flow areas.

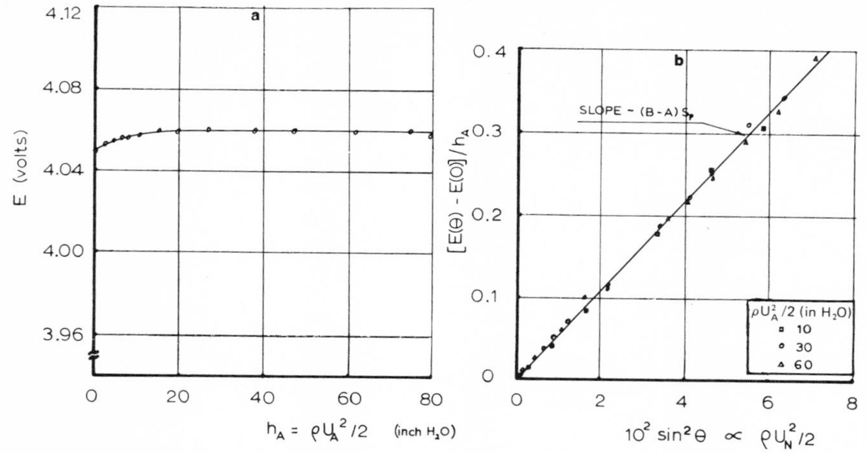

Fig. 11.29. Calibration of the static pressure sensor shown in Figure 11.28. (Reference pressure $h_B = 15$ cm water). (a) Axial velocity ($\theta = 0°$); (b) transverse velocity.

11.5. Measurement of Concentration Fluctuations

The measurement of concentration fluctuations has only been attempted in a limited number of cases. Way and Libby[55] employed a hot film and a hot wire to evaluate the concentration–velocity correlation fluctuations in a flow of helium and air. It appears that recent techniques of measuring aerosol fluctuations with laser probes can be coupled with a hot wire to evaluate the correlation (see Yang and Meroney[56]). Concentration fluctuation measurements by themselves have been reported using a number of different techniques, such as the electrochemical method noted in Section 11.2.3.

11.5.1. Heat-Transfer Techniques

Heat-transfer elements, such as small wires, foils, or coils, have been employed for many years to measure concentrations of gases in the area of gas chromatography. The heat transfer from a surface will depend directly on the thermal conductivity, k_a, of the gas according to the Fourier law of heat transfer:

$$Q = k_a A \frac{\Delta T}{\delta} \tag{11.43}$$

where Q is the heat flux, A is the area of the surface normal to the direction of heat flow, ΔT is the temperature difference, and δ is the thickness over which the heat flows. The use of a heated sensor to analyze an unknown gas sample is termed a katharometer. Such a device was patented in 1915, with the name meaning "purity meter."

In recent years, hot-wire techniques for evaluating concentration fluctuations have been developed by Way and Libby[55] and also by Brown and Rebollo.[57]

Hot-Wire–Hot-Film Sensor. Way and Libby[55] have employed the hot-wire–hot-film sensor, shown in Figure 11.30, to measure the velocity and concentration fluctuation. The sensor consists of a hot wire, operated at relatively cool temperature, orthogonal to, and slightly upstream of, a hot-film sensor. Both the wire and film are set orthogonal to the mean flow. The hot wire is within the "thermal field" of the film.

The wire was of platinum, 0.0001 in. in diameter, approximately 0.015 in. long. The hot film was 0.001 in. in diameter. It was a quartz-coated platinum film 0.010 in. long. The two sensors were approximately 0.001 in. apart. The hot wire was operated at 125°C and the film at 300°C above the flow temperature.

Figure 11.31 is a calibration "map" for the probe operating in helium–air mixtures. The hot-wire output voltage is E_w and the film voltage is E_f. Lines of constant velocity, *U*, and concentration, *C*, are noted in Figure 11.31. The sensor fails to give useful signals at low velocities and high concentrations.

Way and Libby[58] give a detailed account of the digital analysis of the sensor output. They demonstrate the measurement of fluctuations of concentration and of streamwise velocity in turbulent flows of helium and air.

In theory one could employ a single hot wire, much the same as outlined for the separation of velocity and temperature (Section 11.3.2), to evaluate

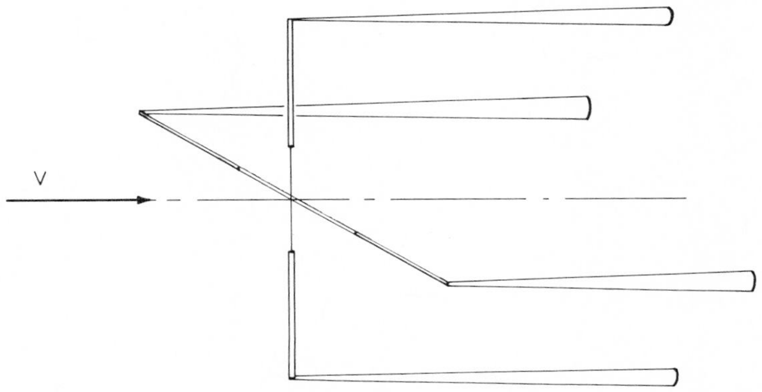

Fig. 11.30. Schematic representation of a velocity–concentration sensor.

velocity and thermal conductivity fluctuations. If the temperature of the mixture remains constant the fluctuation in thermal conductivity is directly related to the concentration. The hot-wire sensitivity to thermal conductivity should vary approximately as the wire temperature. The temperature of the wire will cause some variation in the local gas temperature, which would alter the thermal conductivity slightly. The velocity sensitivity should be similar to that shown in Figure 11.23. Corrsin[(59)] originally discussed this approach to measuring concentration fluctuations.

Katharometer Technique. Brown and Rebollo[(57)] have developed a katharometer-type probe for fast-response measurements of gas concentration. The probe, shown in Figure 11.32, is constructed from a 2-mm glass tube drawn to a point and polished to expose a fine hole. The hole is of the order of 0.001 in. in diameter, so that the flow through the probe will not be a function of the local velocity (choked flow). Figure 11.33 shows a typical calibration of the probe. Brown and Rebollo were able to construct a smaller probe than shown in Figure 11.32, which had a time response of about 200 μsec. The response appears to be very close to a first-order exponential.

Atomic Concentration Sensor. The atomic concentration sensor operates on the principle that heat is released when two or more atoms combine into a molecule. The transducer sensor acts as a "catalyzer" to bring about the recombination of the atoms, and then as a calorimeter to measure the heat produced owing to recombination. An analysis of the hot wire as a catalytic sensor of atom concentration was given by Wray.[(60)] The theory of

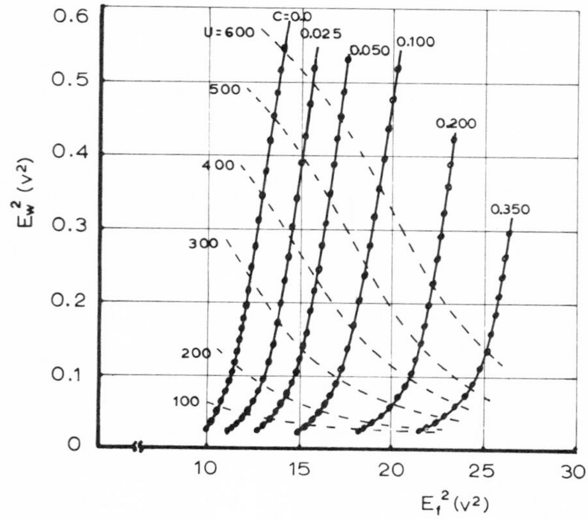

Fig. 11.31. Calibration "map" for the sensor shown in Figure 11.30.

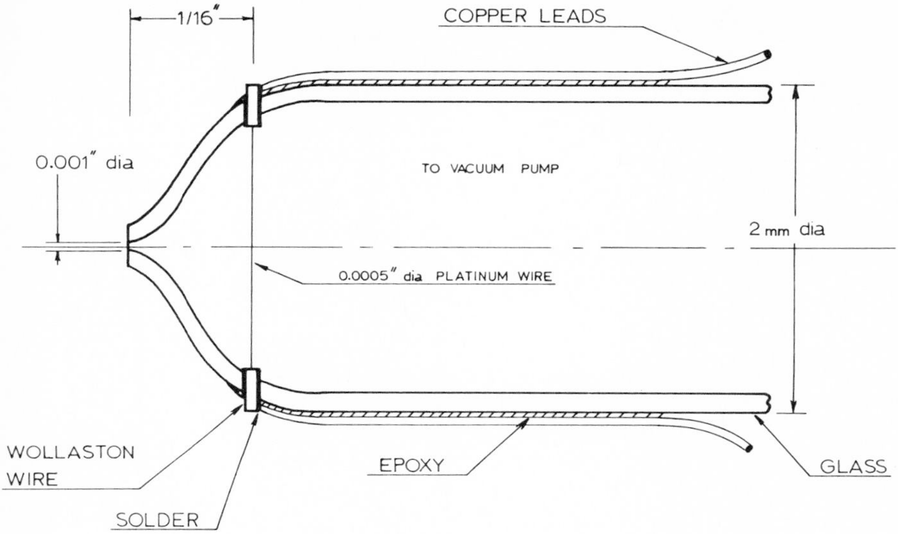

Fig. 11.32. Katharometer-type concentration measuring probe.

the probe in supersonic flow was outlined by Rosner.(61) The hypersonic applications of the probe were evaluated by Hartunian.(62)

The general method of measuring atom concentration employs a differential technique. Two sensors are employed: One sensor has a catalytic surface, which the second does not. Thus, by measuring the difference in heat transfer between the two surfaces, only the heat transfer due to recombination is measured. The catalytic surface is normally a platinum film or wire. The noncatalytic surface is a platinum film coated with a thin film of silicon monoxide (about 5–10 μ thick). These gauges can have a time response of the order of 10 μsec. This type of concentration sensor has mainly been used in the nonequilibrium gas flows encountered in shock tube tests. Recombination-type atoms are not normally present in most gas flows at room temperature.

11.5.2. Light Scattering

Light-scattering techniques can be applied to measure aerosol concentrations. The development of light-scattering techniques has been greatly improved with the laser light source. It is possible to employ the scattering both as a concentration-measuring device and as a velocimeter at the same time, although this has not as yet been reported. Figure 11.34 shows a light-scattering probe, developed by Yang and Meroney.(56) The output of

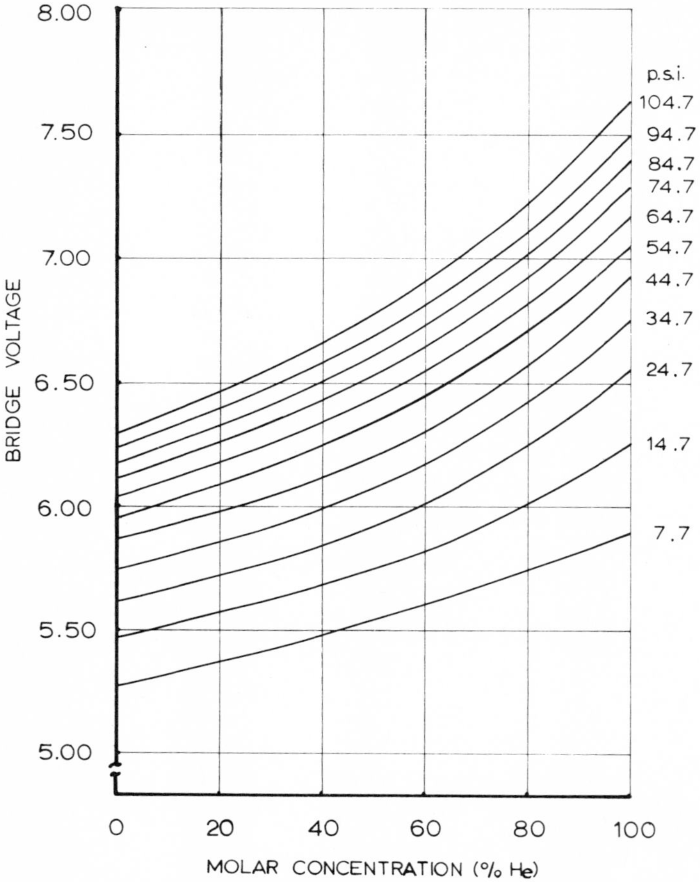

Fig. 11.33. Calibration of the katharometer probe of Figure 11.32.

the sensor is led to a photomultiplier tube. The response of the probe appears to be as great as 100 μsec, although the aerosol particles may not be able to fluctuate at such a fast response. The output of the photomultiplier was found to be a linear function of the concentration.

11.6. Measurement of Surface Shear Fluctuations

The measurement of fluctuations in the surface shear stress can be made both with surface heat-transfer gauges, and with the electrochemical techniques discussed in Section 11.2.3.

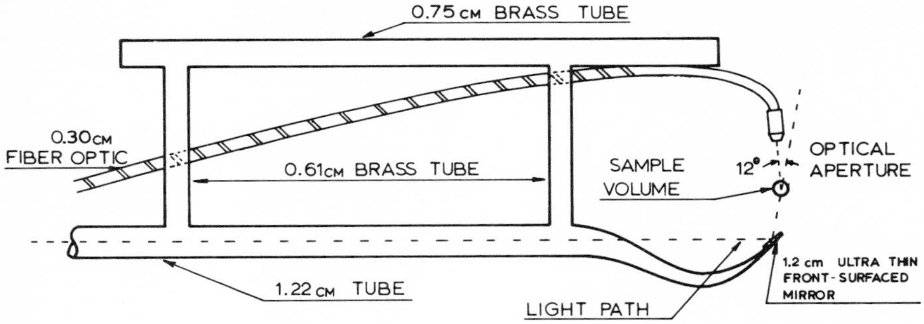

Fig. 11.34. A laser light-scattering probe.

Heat-Transfer Technique. This method makes use of the analogy between the diffusion of heat and momentum. The concept of an analogy between heat transfer and shear stress was first applied by Fage and Falkner[(63)] for laminar flow and by Ludwieg[(64)] for turbulent boundary layers. In both cases it is found that the heat transfer is proportional to the one-third power of the wall shear stress. Figure 11.35 shows typical calibration curves for thin-film platinum gauges obtained by Owen and Bellhouse.[(65)] By inclusion of the value of the density at the surface it is possible to account for the effect of compressibility. The calibration is also found to be valid for both laminar and turbulent flow. Similar results may also be obtained for quartz-coated film gauges in water and other liquids.

The film gauge can be operated with a constant-temperature (feedback) type of electronic system, identical to that employed for hot-wire anemometers. Thus, the frequency response of the gauges can be many thousands of cycles per second. Details of the measurement of fluctuations in the skin friction with thin-film heated elements were first given by Bellhouse and Schultz.[(66)]

The thin-film gauges have dimensions of roughly 0.1 in. long by 0.025–0.05 in. wide. Hot wires mounted on the surface have also been used. The temperature rise at the gauge is kept small to ensure that the heat transfer is directly related to the wall region. A question may arise as to whether the fluctuation in heat transfer is a true measure of the fluctuation in wall shear stress or related to fluctuations directly above the wall region. Details of the operation of the wall shear stress gauges are also covered by Sandborn.[(3)]

In turbulent boundary layers the fluctuating surface shear stress is found to be very large compared to the mean value.[(68)] The probability distribution of the surface shear is highly "skewed," with positive fluctuations greater than five times the mean shear stress observed for flat-plate flow. Major problems are encountered in the evaluation of these large

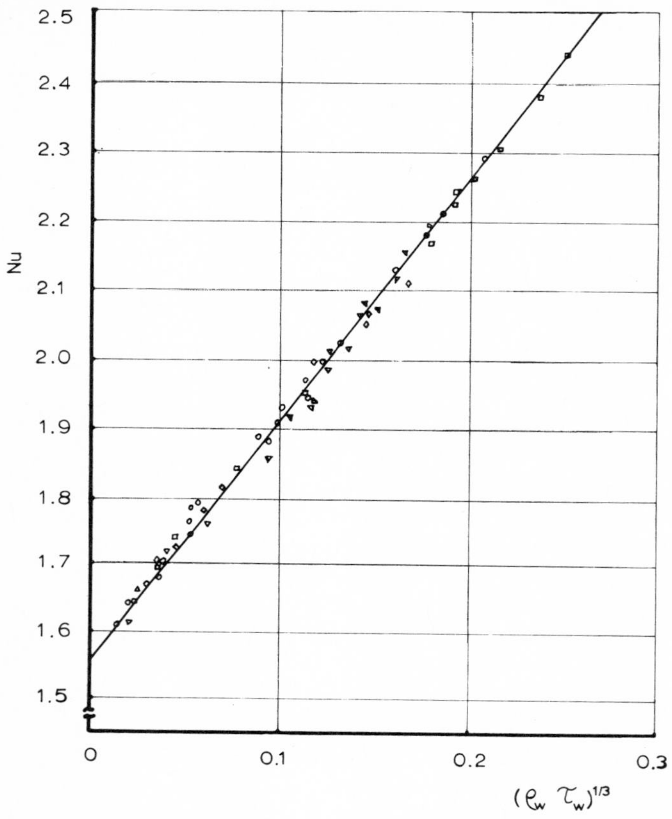

Fig. 11.35. Calibration of thin-film heat-transfer–shear-stress gauges. Data are for both subsonic and supersonic flows under laminar, transitional, and turbulent conditions.

fluctuations, owing to "nonlinear" averaging effects. Difficulties arise in that the heat-transfer sensor must be calibrated in a turbulent flow (in order to obtain the required magnitude of the surface shear), thus the nonlinear errors are directly present in the calibration curve. An approximate technique of correcting for the effect, employing the probability distribution, is outlined in Reference 68.

Notation

A	Area	C_L	Lift coefficient
a	Sound speed	de	Voltage increment
C_p	Specific heat	e	Voltage fluctuations
c	Concentration	E	Mean voltage

f	Pulse rate
F	Faraday constant
$F(0)$	Spectral energy at "zero frequency"
I	Current
K	Heat-transfer coefficient
K_n	Knudsen number
k	Thermal conductivity
L	Lift
L_x	Macroscale of turbulence
l	Distance and mixing length
M	Mach number
Nu	Nusselt number
N	Reaction rate per unit area
P	Pressure
Q	Heat flux
R	Wire resistance
Re	Reynolds number
R_x	Longitudinal velocity correlation
S	Surface area
S_u, S_v	$\partial E/\partial U$; $(1/U)\partial E/\partial \psi$
T_T	Turbulent shear stress tensor
T	Temperature
t	Time
T_w	Hot-wire temperature
T	Transfer number
u, v, w or u_i	Fluctuating velocity components
U, V, W or U_i	Mean velocity components
U_τ	Friction velocity
U_p	Particle velocity
x, y, z or x_i	Rectangular coordinates
α	Heat spread rate or incidence
δ_{ij}	Kronecker delta
ε	Turbulent dissipation
η	Kolmogorov scale
ρ	Density
ν	Kinematic viscosity
μ	Viscosity coefficient
λ_x	Microscale of turbulence
ψ, θ	Angles
τ_p	Time constant

References

1. Lumley, J. L., and Panofsky, H. A., *The Structure of Atmospheric Turbulence*, Interscience Publishers, New York (1964).
2. Sandborn, V. A., and Slogar, R. J., Study of the momentum distribution of turbulent boundary layers in adverse pressure gradients, NACA Report No. TN 3264 (1955).
3. Sandborn, V. A., *Resistance Temperature Transducers*, Metrology Press, Fort Collins, Colorado (1972).
4. Schubauer, G. B., and Burgers, J. M., in: *Advances in Hot Wire Anemometry* (W. L. Melnik and J. R. Weske, eds.), Proceedings, International Symposium on Hot Wire Anemometry, University of Maryland (1967).
5. Ling, S. C., and Hubbard, P. G., The hot-film anemometer—A new device for fluid mechanics research, *J. Aerosp. Sci.* **23**, 890–891 (1956).
6. Lowell, H. H., and Patton, N., Response of homogenous and two-material laminated cylinders to sinusoidal environmental temperature change, with applications to hot-wire anemometry and thermocouple pyrometry, NACA report No. TN-3514 (1955).
7. Schubauer, G. B., and Klebanoff, P. S., Theory and application of hot wire instruments in the investigation of turbulent boundary layers, NACA report No. WR W-86 (1946).
8. Richardson, E. V., McQuivey, R. S., Sandborn, V. A., and Jog, P. M., Comparison between hot film and hot wire measurements of turbulence, in: *Proceedings of the Tenth Midwestern Mechanics Conference, Developments in Mechanics*, Vol. 4 (J. E. Cermak and J. R. Goodman, eds.), Johnson Publishing Company, Boulder, Colorado (1968), pp. 1213–1223.

9. McQuivey, R. S., Turbulence in a hydrodynamically rough and smooth open channel flow, Ph.D. dissertation, Colórado State University (1967). See also Richardson, E. V., and McQuivey, R. S., Measurements of turbulence in water, *ASCE J. Hydraulics* **95**, 411–430 (1968).
10. Bellhouse, B. J., and Rasmussen, C. G., Low frequency characteristics of hot-film anemometers, DISA Information No. 6 (1968).
11. Friehe, C. A., and Schwarz, W. H., The use of Pitot static tubes and hot film anemometers in dilute polymer solutions, in: *Viscous Drag Reduction* (C. S. Wells, ed.), Plenum Press, New York (1969), pp. 281–296.
12. Spencer, B. W., and Jones, B. G., Turbulence measurements with the split film anemometer probe, in: *Proceedings of Symposium on Turbulence in Liquids*, University of Missouri at Rolla (1971).
13. Fage, A., and Townend, H. C. H., An examination of turbulent flow with an ultramicroscope, *Proc. R. Soc. London* **A135**, 656–677 (1932).
14. Bourot, J. M., Chromophotography des champs aerodynamiques, Publications Scientifiques et Techniques, No. 226, Ministere Air, Paris, France (1949).
15. Gee, T. H., *An Introduction to the Laser*, von Karman Institute for Fluid Dynamics Lecture Series No. 39: Laser Technology in Aerodynamic Measurements (1971).
16. Goldstein, R. J., and Kreid, D. R., Measurement of laminar flow development in a square duct using a laser-Doppler flow meter, *J. Appl. Mech.* **34**, 813–818 (1967).
17. Schubauer, G. B., A Turbulence indicator utilizing the diffusion of heat, NACA report No. TR-524 (1935).
18. Taylor, G. I., Statistical theory of turbulence—IV, Diffusion in a turbulent air stream, *Proc. R. Soc. London* **A151**, 465–478 (1935).
19. Hagist, W. H., Measurements of turbulent intensity by a diffusion method (unpublished work done at Colorado State University) (1968).
20. Townend, H. C. H., Statistical measurements of turbulence in the flow of air through a pipe, *Proc. R. Soc. London* **A145**, 180–211 (1934).
21. Cady, W. M., Velocity measurements by illuminated or luminous particles, in: *Physical Measurements in Gas Dynamics and Combustion* (R. W. Ladenburg, B. Lewis, R. N. Pease, and H. S. Taylor, eds), *High Speed Aerodynamics and Jet Propulsion*, Vol. 9, Princeton University Press, Princeton, New Jersey (1954).
22. Clutter, E. W., Smith, A. M. O., and Braxier, J. G., Techniques of flow visualization using water as the working medium, Douglas Aircraft Company report No. ES 29075 (1959). Also Clutter, D. W., and Smith, A. M. O., Flow visualization by electrolysis of water, *Aerosp. Eng.* **20**, 24–27 (1961).
23. Schraub, F. A., Kline, S. J., Henry, J., Runstadler, P. W., and Littell, A., Use of hydrogen bubbles for quantitative determination of time dependent velocity fields in low speed water flow, Department of Mechanical Engineering report No. MD-10, Stanford University, Stanford, California (1964).
24. Mitchell, J. E., and Hanratty, R. J., A study of turbulence at a wall using an electrochemical wall shear stress meter, *J. Fluid Mech.* **26**, 199–221 (1966). See also Reference 25.
25. Sirkar, K. K., and Hanratty, R. J., The use of electrochemical techniques to study turbulence close to a wall, in: *Symposium on Turbulence Measurements in Liquids*, University of Missouri at Rolla (1971).
26. Reiss, L. P., Investigations of turbulence near a pipe wall using a diffusion controlled electrode, Ph.D. dissertation, Chemical Engineering Department, University of Illinois (1962).
27. Mitchell, J. E., Investigation of wall turbulence using a diffusion controlled electrode, Ph.D. dissertation, Chemical Engineering Department, University of Illinois (1965).

28. Son, J. S., and Hanratty, T. J., Limiting relation for the eddy diffusivity close to a wall, *AIChE J.* **13**, 689–696 (1967).
29. Baldwin, L. V., and Cermak, J. C., Fluid mechanics, paper No. 2, Colorado State University (1964).
30. Mitsuta, Y., Sonic Anemometer—Thermometer for general use, *J. Meteorol. Soc. Japan* **44**, 12–24 (1966).
31. Barrett, E. W., and Suomi, V. E., Preliminary report on temperature measurement by sonic means, *J. Meteorol.* **6**, 273 (1949).
32. Miyake, M., Stewart, R. W., Burling, R. W., Tsuang, L. R., Koprov, B. M., and Kuxnetzov, O. A., Comparison of acoustic instruments in an atmospheric turbulent flow over water, *Boundary Layer Meteorol.* **2**, 228–245 (1971).
33. Plate, E. J., and Bennett, J. P., Rotary flow meter as turbulence transducer, *J. Eng. Mech. Div. ASCE Proc.* **95** (No. EM6, paper No. 6955), 1307–29 (1969).
34. Siddon, T. E., and Ribner, H. S., An aerofoil probe for measuring the transverse component of turbulence, *AIAA J.* **3**, 747–749 (1965).
35. Cheng, D. Y., Introduction of the viscous force sensing fluctuating probe technique with measurement in the mixing zone of a circular jet, AIAA paper No. 73-1004 (1973).
36. Werner, F. D., An investigation of the possible use of a glow discharge as a means of measuring air flow characteristics, *Rev. Sci. Instrum.* **21**, 61–68 (1950).
37. Fuchs, W., Investigations of the operating properties of the leakage current anemometer, NACA report No. TM-1178 (1947).
38. Sandborn, V. A., A review of turbulence measurements in compressible flow, NASA report No. TM X 62-337 (1974).
39. Kovasznay, L. S. G., Turbulence in supersonic flow, *J. Aeronaut. Sci.* **20**, 657–674, 682 (1953).
40. Wilson, L. N., and Domkevala, R. J., Statistical properties of turbulent density fluctuations, *J. Fluid Mech.* **43**, 291–303 (1970).
41. Wehrmann, O. H., Velocity and density measurements in a free jet, in: *Turbulent Shear Flows*, AGARD CP-93 (1970), paper No. 15.
42. Wallace, J. E., Hypersonic turbulent boundary layer measurements using an electron beam, *AIAA J.* **7**, 757–759 (1969).
43. Laufer, J., Aerodynamic noise in supersonic wind tunnels, *J. Aeronaut. Sci.* **28**, 685–692 (1961).
44. Stainback, P. C., Wagner, R. D., Owen, F. K., and Horstman, C. C., Experimental studies of hypersonic boundary layer transition and effects of wind tunnel disturbances, NASA report No. TN D-7453 (1974).
45. Laderman, A. J., and Demetriades, A., Measurements of the mean and turbulent flow in a cooled-wall boundary layer at Mach 9.37, AIAA paper No. 72-73, San Diego, California (1972).
46. Laufer, J., Factors affecting transition Reynolds numbers on models in supersonic wind tunnels, *J. Aeronaut. Sci.* **21**, 497–498 (1954).
47. Morkovin, M. V., On supersonic wind tunnels with low free-stream disturbances, Air Force Office of Scientific Research report No. TN 56-540 (1956).
48. Sandborn, V. A., and Wisniewski, R. J., Hot wire exploration of transition on cones in supersonic flow, *Proceedings of the 1960 Heat Transfer and Fluid Mechanics Institute* (D. M. Mason, W. C. Reynolds, and W. G. Vincenti, eds.), Stanford University Press, Stanford, California (1960).
49. Donaldson, J. C., and Wallace, J. P., Flow fluctuation measurements at Mach number 4 in the test section of the 12 inch supersonic tunnel (D), AEDC report No. TR-71-143 (1971).

50. Rose, W. C., and Johnson, D. A., Turbulence in a shock-wave boundary layer interaction, *AIAA J.* **13**, 884–889 (1975).
51. Klebanoff, P. S., Characteristics of turbulence in a boundary layer with zero pressure gradient, NACA report No. TR-1247 (1955).
52. Zoric, D. L., Approach of turbulent boundary layers to similarity, Ph.D. dissertation, Colorado State University, Fort Collins, Colorado (1968).
53. Corcos, G. M., Pressure measurements in unsteady flows, *ASME Symposium on Measurements in Unsteady flow*, Worcester, Massachusetts (1962), pp. 15–21.
54. Jones, B. G., and Planchon, H. P., A study of the local pressure field in turbulent shear flow and its relation to aerodynamic noise generation, NASA report No. CR 134493 (1972).
55. Way, J., and Libby, P. A., Hot wire probes for measuring velocity and concentration in helium–air mixtures, *AIAA J.* **8**, 976–977 (1970).
56. Yang, B. T., and Meroney, R. N., On diffusion from an instantaneous point source in a naturally stratified turbulent boundary layer with a laser light scattering probe, Colorado State University Technical Report No. 20 (CER 70-73 BTY-RNM-17), Fort Collins, Colorado (1972).
57. Brown, G. L., and Rebollo, M. R., A small, fast response probe to measure composition of a binary gas mixture, *AIAA J.* **10**, 649–652 (1972).
58. Way, J., and Libby, P. A., Application of hot wire anemometry and digital techniques to measurements in a turbulent helium jet, *AIAA J.* **9**, 1567–1573 (1971).
59. Corrsin, S., Extended application of the hot wire anemometer, NACA report No. TN 1864 (1949).
60. Wray, K. L., A quantitative rapid response atom detector, AVCO Research Report No. 46 (1959).
61. Rosner, D. E., Catalytic probes for the determination of atom concentrations in high speed gas streams, *ARS J.* **32**, 1065–1073 (1962).
62. Hartunian, R. A., Local atom concentrations in hypersonic dissociated flows at low densities, *Phys. Fluids* **6**, 343–348 (1963).
63. Fage, A., and Falkner, V. M., An experimental determination of the intensity of friction on the surface of an aerofoil, *Proc. R. Soc. London* **A129**, 378–410 (1930).
64. Ludwieg, H., Instrument for measuring the wall shearing stress of turbulent boundary layers, NACA report No. TM-1284 (1950).
65. Owen, F. K., and Bellhouse, B. J., Skin-friction measurement at supersonic speeds, *AIAA J.* **8**, 1358–1360 (1970).
66. Bellhouse, B. J., and Schultz, D. L., Determination of mean and dynamic skin friction, separation and transition in low-speed flow with a thin film heated element, *J. Fluid Mech.* **24**, 379–400 (1966).
67. Kistler, A. L., Fluctuation measurements in a supersonic turbulent boundary layer, *Phys. Fluids* **2**, 290–296 (1959).
68. Sandborn, V. A., and Pyle, W. L., Evaluation of the surface shear stress along rearward facing ramps, Research Memorandum No. 25, Colorado State University (1975).

CHAPTER 12

Techniques for Measuring Atmospheric Turbulence

J. R. CONNELL

12.1. Introduction

In many respects measurement of atmospheric turbulence is similar to measurement of turbulence in a wind tunnel. Differences between tunnel and the atmosphere permit and often require unique instrumentation and techniques. A few of the factors unique for atmospheric turbulence are listed in Table 12.1.

A useful diagram of scales of atmospheric phenomena and measuring techniques is given by Lilly and Lenschow[1] and is repeated here as Figure 12.1.

Consider, briefly, the somewhat simpler problem of measurement of airflow properties in wind tunnels. Wind-tunnel techniques of measurement are often successful because the conditions in the tunnel are quite controllable. Thus, hot wires and fine cold wires do not as often experience impacts due to particulates in the air. Temperature and air density and chemical-species-concentration variations may be held to relatively small magnitudes so that hot-wire anemometer and laser concentration signals contain only small portions of ambiguous information. The ratio of turbulent component of velocity to the mean component is usually small so that, for example,

J. R. CONNELL • The University of Tennessee Space Institute, Tullahoma, Tennessee 37388

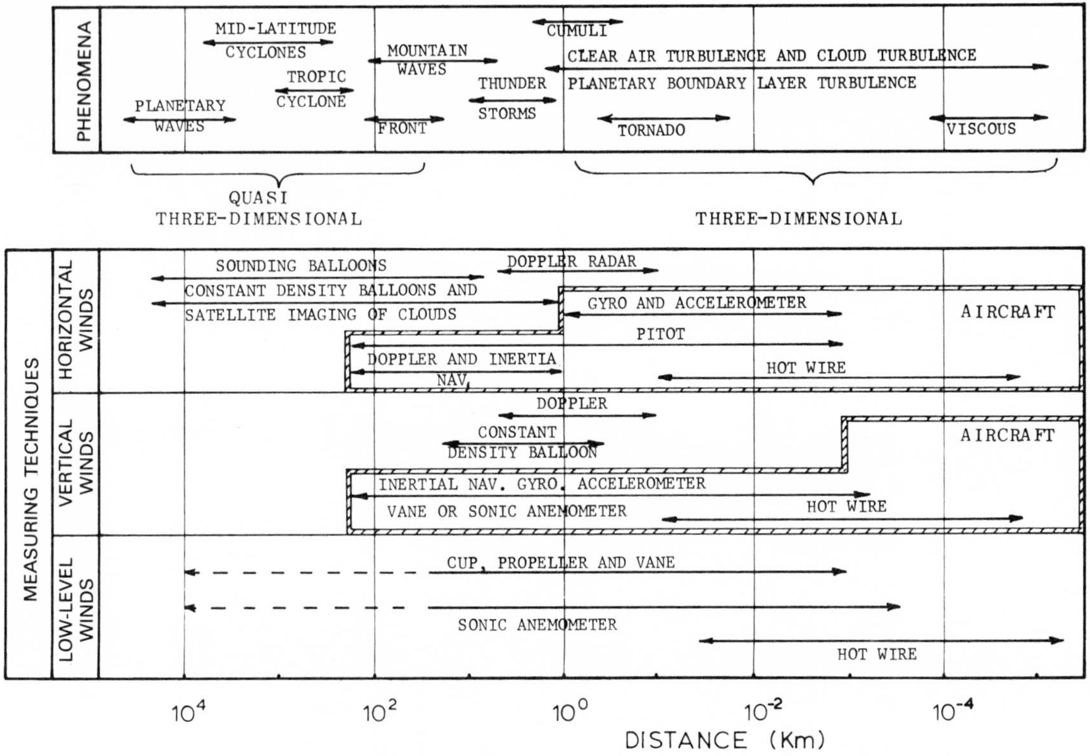

Fig. 12.1. Horizontal scales of various atmospheric phenomena and scale limitations of techniques for measuring atmospheric motions (Lilly and Lenschow[1]).

Table 12.1. Some Factors Affecting the Choice of Methods for Measurement of Atmospheric Turbulence

1. Large scales and great ranges of scales.
2. Moving phenomena (wave motions, mean winds and propagating cellular motion).
3. Long periods and trends.
4. Temperature, pressure, water vapor, and atmospheric particulate variations occur concurrently with velocity fluctuations.
5. Both wave and turbulence, or mixing fluctuations, may occur.
6. Mechanical, thermodynamic, and physiological stresses encountered by instruments and observers.
7. Great spatial separation between observer and phenomena.

directional ambiguity in hot-wire anemometry is reduced. Finally, scattering aerosols may be introduced and ranges kept small so that laser velocimeters are useful. All of the above methods can be and are used in the atmosphere for carefully restricted problems or sometimes with more uncertainty in the result. A great variety of additional methods are utilized for atmospheric measurements over larger scales and time intervals, lower frequencies, and more remote regions of the atmosphere. Table 12.2 lists more of the general methods of turbulence measurement.

Table 12.2. Methods of Measurement of Atmospheric Turbulence

I. Remote sensing (sensing at a distance from the instrument), wave propagation methods using scattering intensity and Doppler shift.
    - Radio wave (RADAR): developed, advancing slowly in public domain
    - Light wave (LIDAR): rapidly developing
    - Acoustic wave (ACDAR): economical, developing

II. Local sensing. Towers and poles.
    - Cup and propeller anemometers and vanes
    - Wind-pressure anemometers
    - Sonic anemometers
    - Hot-film anemometers
    - Thermometers
    - Barometers
    - Humidity meters

III. Sensing local to the instrument on platforms at a distance from the normal location using automatic data recording and/or telemetering.
    - Balloons
    - Kites
    - Aircraft with or without inertial platforms
    - Remote land station
    - Meteorological buoys

Ground- and tower-based direct sensing mostly utilizes old concepts for which reliability and technique are steadily improving. Wave propagation methods of sensing and tracking and instrumented aircraft carrying inertial navigation equipment for instrument position and velocity reference have widely expanded capabilities for measurement of turbulence in the atmosphere.

It is worth noting that turbulence or turbulencelike phenomena of importance in the atmosphere vary in length scale from about one centimeter to a few thousand kilometers. In some instances it is very difficult to view the motion as anything but random (e.g., friction-driven turbulence); in others it is a greater conceptual problem not to view the motion as mostly deterministic (thunderstorm and cyclone waves). One discussion of the latter category is the book on *Negative Viscosity* by Victor Starr.[(2)] Thus measurement of airflow by tracking a set of constant-density balloons floating in the atmosphere would at first glance seem to be a mean-flow measurement with only a turbulence perturbation upon it. But the "mean flow" which they describe over many days, is a fluctuation flow that transports heat, water vapor, and momentum and contributes to the wind and precipitation patterns over a much larger time average. Finally, there are phenomena that do not always mix atmosphere but are called turbulent because of the bumpiness they create for aircraft.

This chapter will concentrate on small mesoscale (10 km) and lesser scales of turbulence in the atmosphere. After a brief excursion into a variety of methods of measurement with emphasis on aircraft platforms, instruments, techniques, and applications, examples of measurement in the planetary boundary layer will be shown. Each involves an obstruction at the lower surface of the atmosphere: a mountain and a building.

12.2. Measurements: Background, Instruments, Platforms, and Techniques

12.2.1. Instrument Response

Terminology is not uniform even within the atmospheric science community. For instrument-response definitions see Gill and Hexter[(3)] and Mazzarella.[(4)]

An instrument is selected in consideration of the best available information on the characteristics of the time series of the variable to be measured, the environment in which the property is to be measured, and the response characteristics of the instrument. The next problem is to check its response to the expected forcing function and further to understand what influence the platform that supports the instrument has upon the measure of and the actual value of the variable. What is known about these influences may permit one to alter the techniques of measurement or to filter and otherwise correct the data.

12.2.2. Tower-Based Cup Anemometers

To be specific, consider the measurement of mean and turbulent properties of airflow in the first 20 m of the atmosphere over, say, a clearing in a forest using rotating-cup anemometers and wind-direction vanes. The available instruments of this type are of the order of 0.5 m in longest dimension and have a moving mass of about 0.1 kg. As a consequence of these factors and others, such as area of cups and vanes and turning friction, the cups and vanes do not respond accurately to variations of wind speed and direction with frequencies greater than 3 Hz at a wind speed of, say 3 m sec^{-1}. The characteristic of the response is often stated as the distance constant, here about, 1 m. The time constant, phase lag, and damping ratios of a number of types of anemometers and vanes are shown in Mazzarella.[(4)] A good definition of terms is given by MacCready.[(5)] Figure 12.2 shows response curves for a typical anemometer taken from Carter.[(7)] It is clear that turbulent energy levels cannot be well measured above a low frequency. The uncompensated phase lags result in incorrect cross-correlation estimates of eddy fluxes. It has been shown that slightly inaccurate leveling of cup and propeller anemometers can also give significant cross-correlation errors. Calibration or phase and frequency-response curves in this "imperfect" or falloff region are not always available for each anemometer.

Cup anemometers overestimate mean wind and underestimate the fluctuation component and generally must be corrected.

12.2.2.1. Exposure. In addition to the frequency response of the cups or propellers careful positioning and exposure of the instruments is of considerable importance. Wind-tunnel studies of wind defect at typical anemometer placement distances from open support towers show extrema of −40% and +12% error in mean wind speed (Gill *et al.*,[(6)] Cermak and Horn,[(8)] and Carter[(7)]). See Figure 12.3. Only a narrow lateral angle of placement about the exact upwind location results in less than 1% error, and the distance from the tower must be somewhat greater than 1 width of a side of the triangular tower. Vertical differences in placement relative to cross-support members of the tower structure cause variations in maximum error.

12.2.2.2. Alignment. If Reynolds stress is to be computed by cross correlation of measured components of velocity the instruments must be well aligned with respect to true vertical. An error of 1° in alignment can result in a 5–10% error in Reynolds stress (Deacon[(9)] and Pond[(10)]) and possibly as much as 100% error (Kraus[(11,12)]). This leveling error effect may be reduced by filtering the frequencies not operative in the stress from the

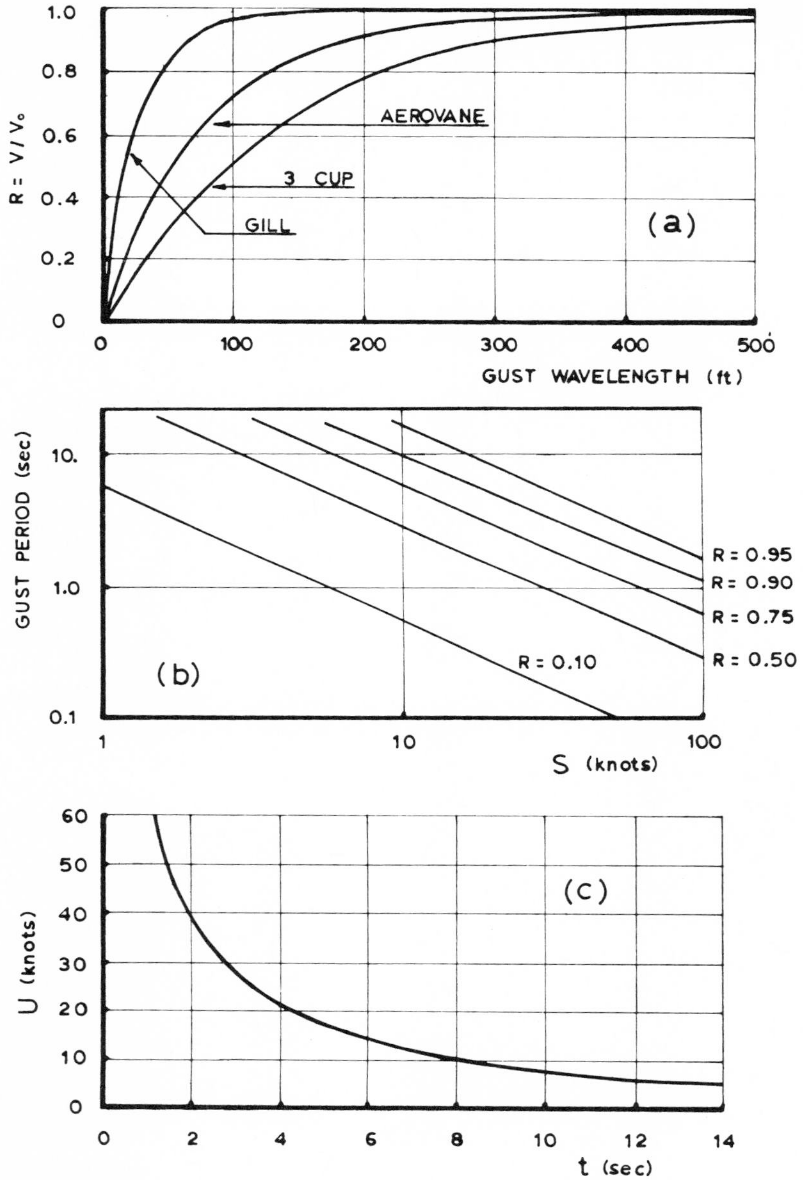

Fig. 12.2. Frequency response of several anemometers and vanes (taken from Carter[7]). (a) Speed sensor fractional responses to sinusoidal fluctuations of wind speed. (b) Fractional response for the airvane in (a) but including the mean wind speed as an additional variable. (c) Time required for the envelope of the vane response to decay to 10% of its initial value.

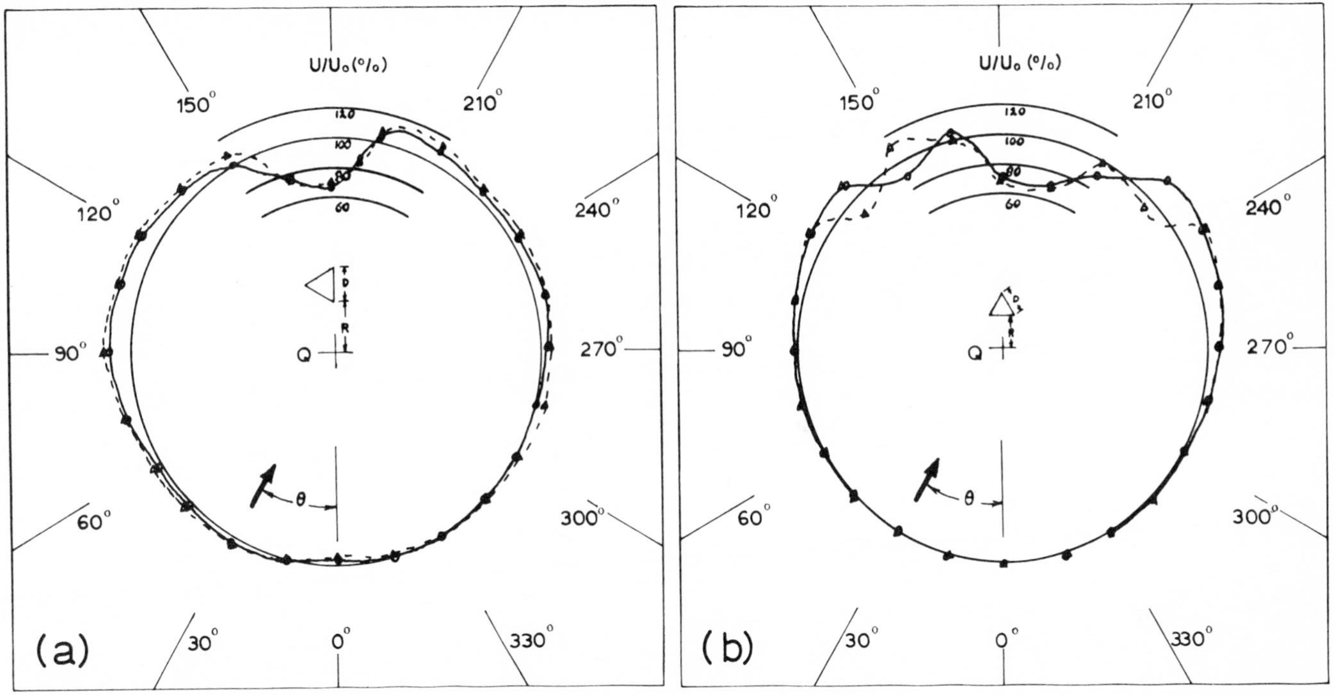

Fig. 12.3. Polar plots of percent true wind speed vs. wind direction for the anemometer, Q, located R in the direction shown from an open-structure triangular-cross-section tower of side width D. (a) Q on a line parallel to a side. $R/D = \frac{1}{2}$ (dashed); $R/D = 1$ (solid line). (b) Q on line of symmetry. Same R/D (adapted from Gill *et al.*[3]). Arrow shows wind direction.

velocity components before cross correlation (Kraus[12] and Dyer *et al.*[13]). This means removal of the lower frequencies in the horizontal velocity fluctuations and can reduce the error at least an order of magnitude below the upper value, given by Kraus,[11] of 100% per 1° tilt. A discussion of response of cup and propeller rotors and wind-direction vanes in turbulent wind fields is given by Acheson.[14] Wind measurements are discussed by Moses *et al.*[15]

12.2.3. Wave Propagation Methods

Wave propagation methods of measurement of turbulence are developing at a rapid rate. The most commonly known technique is the Doppler radar measurement of the radial velocity of backscatterers. An example of the unique capability is the measurement of two-dimensional velocity (with two radars) with spatial resolution of $(200\text{ m})^3$ in volumes as large as $(1.4\times10^4\text{m})^3$, rapidly and remotely. An unambiguous velocity range due to the online digital analysis system is often from −25 to 25 m/sec in the range of $30\times30\times3\text{ km}^3$. Some of the present limitations are great expense and upkeep, lack of mobility, restricted range and resolution for mesoscale, and larger and limited unambiguous-velocity range. The absence of backscattering tracer of airflow in all of clear air and parts of clouds is a severe restriction except for very-high-power radars. Release of chaff and of corner reflectors on balloons can provide data at a few selected lines in space and time for "clear" air.

Laser Doppler or tracker of sequential crossings of laser beam interference lines provides a high-spatial-resolution method of remote measurement of turbulence. Specialists expect rapid development of the method. At the moment the method is restricted to regions where aerosols or dust particles are suspended in the airflow in sufficient number. Most "Doppler" lasers work for a range up to about 1 m. Several in the country have a useful range of about 70 m.

Acoustic wave techniques are at the same time among the oldest and the newest for measurement of velocities. The following list indicates some of their major advantages:

a. Scattering centers (velocity and density fluctuations, droplets, e.g.) exist naturally in most of the atmospheres.
b. Equipment is low in cost.
c. Low-frequency (audio) techniques permit electronics to be simple compared to radar and laser techniques.

A simple two-dimensional wind instrument (Kelton and Bricout,[16]) is comprised of a slanting narrow sound beam from a 10-KHz whistle and two

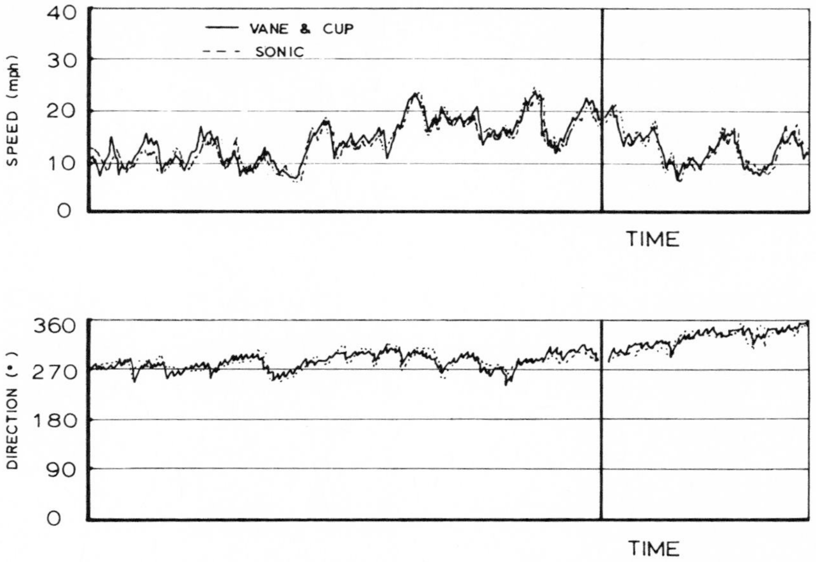

Fig. 12.4. A comparison of sonic measurement and conventional anemometer-vane measurement of wind speed and direction as a function of time. The sonic response to the speed of raindrops accounts for some of the differences noted. (Adapted from Kelton and Bricout.[16])

receivers, with frequency trackers set laterally so as to receive mutually orthogonal horizontal sound radiation from the beam at a point about 20 m away from the whistle. Average differences between 300 sonic measurements on winds from 1 to 10 m/sec and simultaneous measurements with a propeller-vane anemometer 20 m away were not larger than ±0.5 m/sec or ±0.5° in direction. A time-series comparison of this speed and direction measured by both techniques is shown in Figure 12.4.

Another sonic technique measures wind velocity with accuracy and especially with extremely rapid response. It is in fact a sonic-anemometer–thermometer–molecular-species meter, utilizing the speed of sound pulses between fixed pairs of send–receive–triggered piezoelectric crystals as follows (see Figure 12.5): A sound travels relative to still air with speed c and relative to the fixed emitter–receiver crystal A (or B) with speed $c' = c \pm U = (\gamma RT)^{1/2} \pm U$, where $\gamma = C_p/C_v$, R is the gas constant, and T is the temperature. The travel time of sound from A to B is

$$\Delta t_{AB} = L/c'_{AB} = L/(c - U) \qquad (12.1)$$

The travel time from B to A is

$$\Delta t_{BA} = L/(c + U) \qquad (12.2)$$

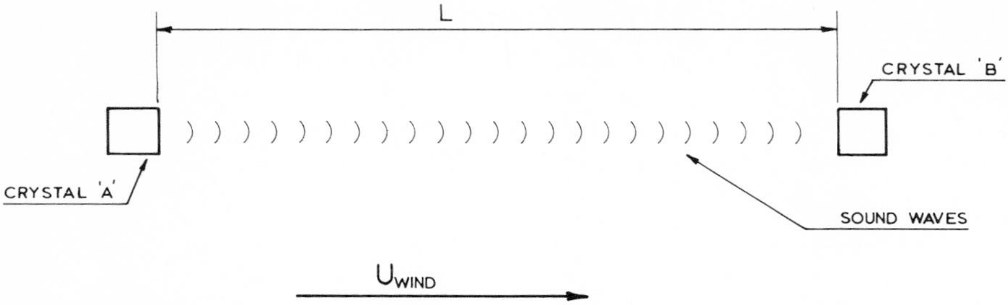

Fig. 12.5. Schematic of the travel of sound pulses between two piezoelectric sender–receiver crystals.

The electroacoustical circuitry is such that when the first pulse emitted by A reaches B, B pulses sound, which upon reaching A triggers a sound pulse in A, and repeated measurements of pairs of times of flight based upon c' and L are compared. Since L is a known constant we may write

$$c - U = L/\Delta t_{AB} \tag{12.3}$$

and

$$c + U = L/\Delta t_{BA} \tag{12.4}$$

Subtracting Equation (12.3) from Equation (12.4) we obtain

$$U = \frac{L}{2}\left(\frac{1}{\Delta t_{BA}} - \frac{1}{\Delta t_{AB}}\right) = \frac{L}{2}\left(\frac{\Delta t_{AB} - \Delta t_{BA}}{\Delta t_{AB}\Delta t_{BA}}\right) \tag{12.5}$$

For small L Equation (12.5) becomes

$$U \approx \frac{L}{2}\frac{\delta\Delta t}{\overline{\Delta t}^2} \tag{12.6}$$

where

$$\delta\Delta t \equiv \Delta t_{AB} - \Delta t_{BA}$$

The thermal and molecular components are removed from the measurement by this method. The air velocity is removed by addition of Equations (12.3) and (12.4), giving

$$c = \frac{L}{2}\left(\frac{1}{t_{BA}} + \frac{1}{t_{AB}}\right) \approx \frac{L}{\overline{\Delta t}} \tag{12.7}$$

The pulse rate of A or B is $f = \frac{1}{2}(1/\overline{\Delta t})$ and $\overline{\Delta t}$ is the fractional delay in the two-direction time of flight. Thus, both temperature and a component of wind speed are measured.

Recently remote sensing by pulsed "acoustic radar" has been developed to the point that three-dimensional wind velocity may be measured. Beam bending by wind is significant when the speed of air motion is large relative to the speed of sound. Temperature fluctuations degrade the measurements. Typical atmospheric boundary-layer conditions produce effects estimated by Beran *et al.*[17] to give errors in vertical speed of less than 0.1 m/sec. The vertical motion, w, is approximately calculable from

$$w = \frac{c}{2}\left(\frac{f_0}{f_s} - 1\right)$$

where c is the speed of sound relative to air, f_0 is the emitter frequency, and f_s is the frequency of sound waves returned by the scatterer. A Doppler spectrum is generated from each pulse. Figure 12.6 shows two aspects of vertical-beam acoustic sounder sensing of a drifting thermal plume (from Beran *et al.*[17]). The left-hand portion shows the distance-corrected intensity of backscatter as a function of vertical and time dimensions. The right-hand portion of the figure shows the corresponding vertical motion field derived from Doppler shift in the frequency of the backscattered sound.

Beran and Clifford[18] used an array of three acoustic sounders aimed at a common volume of the planetary boundary layer as indicated in Figure

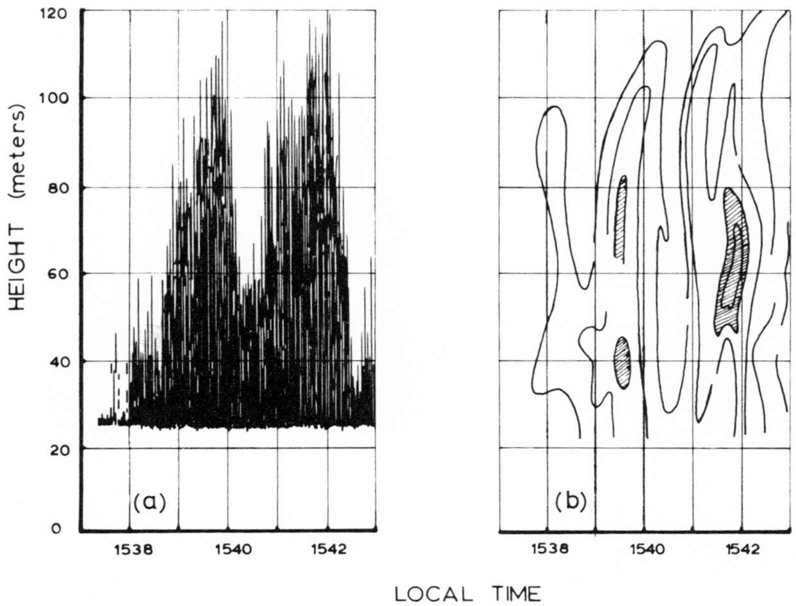

Fig. 12.6. (a) A facsimile record of acoustic echo intensity of atmospheric plumes. (b) Vertical velocity distribution for the same plumes as derived by Doppler-shift analysis of the echoes. (After Beran *et al.*[17])

12.7, adapted from their corresponding figure. The velocity-component resolution equation is added to the figure. Assuming negligible beam bending, the total velocity of the scattering volume may be resolved by simple geometrical and Doppler-shift considerations. Figure 12.8 shows as a solid line an example of the horizontal wind speed so measured. An independent measurement during the same time period at a somewhat displaced location using a cup anemometer suspended under a tethered balloon is shown as a dashed line. The data are lagged to correspond better to the Doppler data upwind and other differences in measurement are reduced by time-smoothing the data with a 1-min-time-constant low-pass filter. Figure 12.9a shows a z–t section of horizontal wind measured by Doppler acoustic radar adapted from Beran and Clifford.[18]

The nocturnal wind maximum is just above the temperature inversion, which extended to about 250 m above the ground. Figure 12.9b is derived from Figure 12.9a and shows "instantaneous" vertical profiles of the wind at the times marked t_1, t_2, and t_3 in Figure 12.9a.

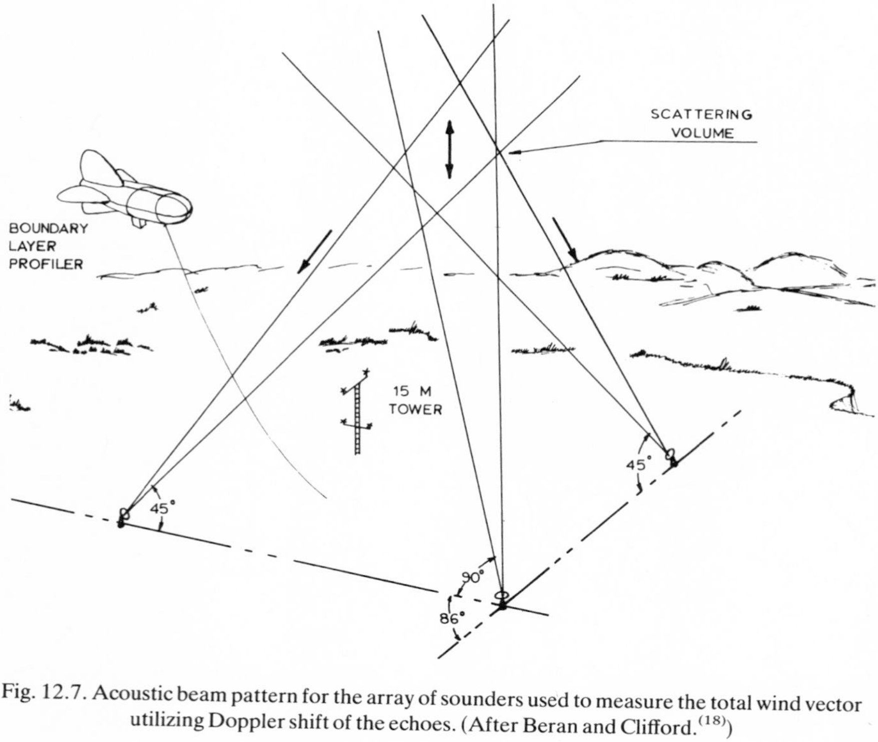

Fig. 12.7. Acoustic beam pattern for the array of sounders used to measure the total wind vector utilizing Doppler shift of the echoes. (After Beran and Clifford.[18])

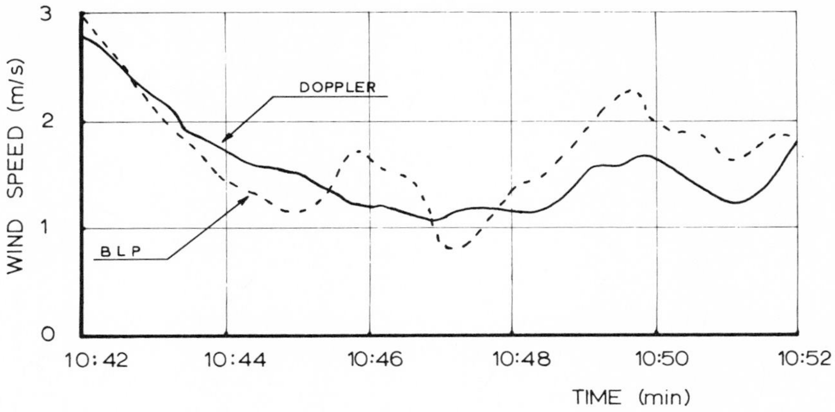

Fig. 12.8. A comparison of wind speed measured using the array of Doppler acoustic sounders and using a balloon-supported conventional anemometer. The balloon data are lagged and low-pass filtered with 1 min time constant to be more representative of the volume seen by the ACDARs. (After Beran and Clifford.[18])

Remote active wave propagation methods of measurement of turbulence open some new approaches to the analyses of turbulence. Some severe restrictions on the method are more than compensated by the new opportunities. As an example, laser Doppler has been used to measure winds at distances greater than a kilometer but only in the dark of night (Bennedetti-Michelangeli *et al.*[19]).

12.2.4. Other Measurement Techniques

12.2.4.1. Momentum Fluctuation. Before discussing the studies of turbulent flow over obstacles note should be taken of several other techniques for measuring turbulence and turbulent processes. Rapid-response local sensing is often required. Further, transport of properties such as heat, momentum, water vapor, aerosols, and particulates by turbulent fluctuations often is to be measured. Hot wires and hot films have proved quite useful for both purposes (Weiler and Burling[20]), although breakage and degradation from calibrated characteristics is often a problem. A suitable replacement may be found in the pressure-sphere or thrust anemometers, which utilize a spherical head coupled to strain gauges to detect the forces generated by wind (see, e.g., Pond *et al.*,[21] Wesely *et al.*,[22] and Thurtell *et al.*[23]).

12.2.4.2. Water Vapor Fluctuation. For measurement of eddy transport or water vapor with suitable resolution in time, the limiting factor in

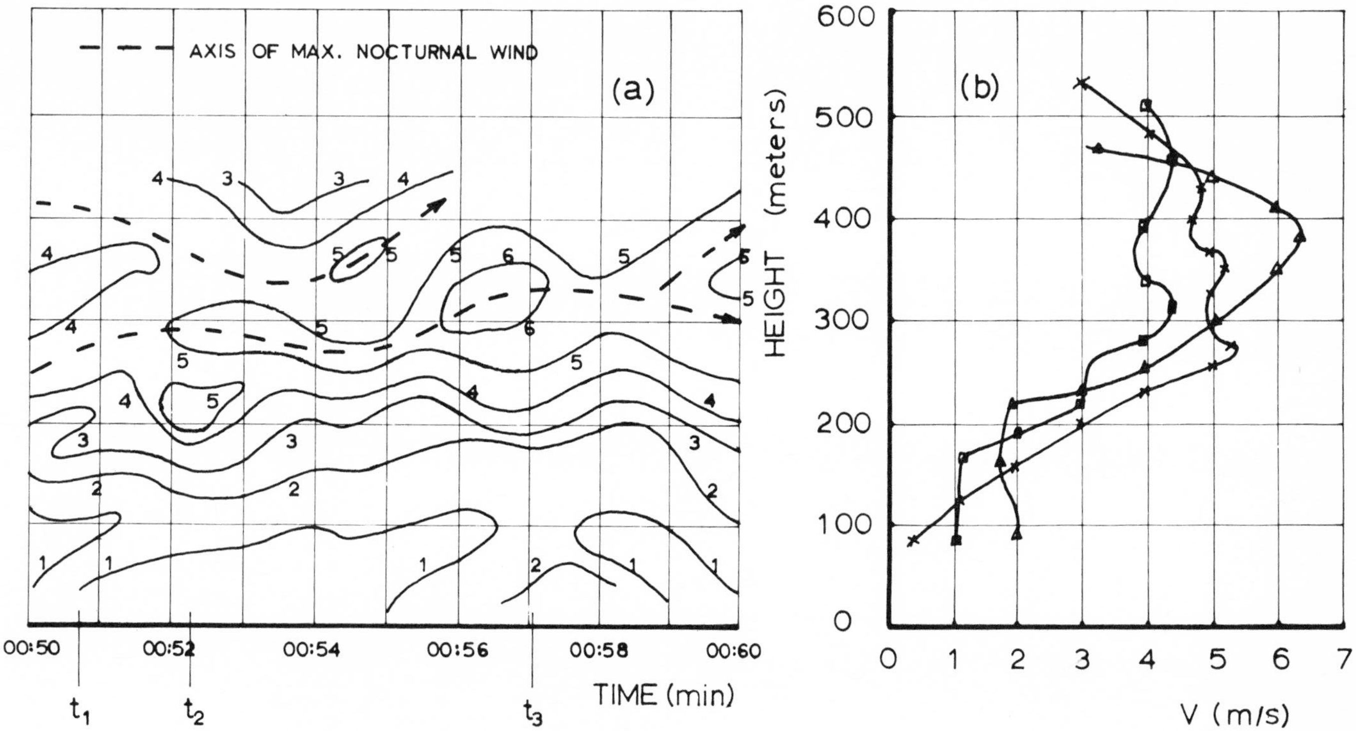

Fig. 12.9. A height–time section of wind speeds in the planetary boundary layer measured by the Doppler acoustic sounding technique. (a) The isotach (m/sec) distribution. (b) Three vertical profiles of wind speed at times $t_1 = 50$ m 45 sec; □, $t_2 = 52$ m 20 sec; ×, $t_3 = 57$ m 0 sec; △, derived from (a). (Adapted from Beran and Clifford.[18])

$\overline{w'Q'}$ has been the measurement of water vapor. The Lyman-alpha hygrometer has a time constant of about 10 msec. A somewhat slower response but better known absolute accuracy is offered by the dew-point, frost-point hygrometer. A comparison of data from these instruments is shown as Figure 12.10 (Buck[44]). Certainly the instrument is useful for fluctuation measurements. A microwave refractometer is another very-fast-response

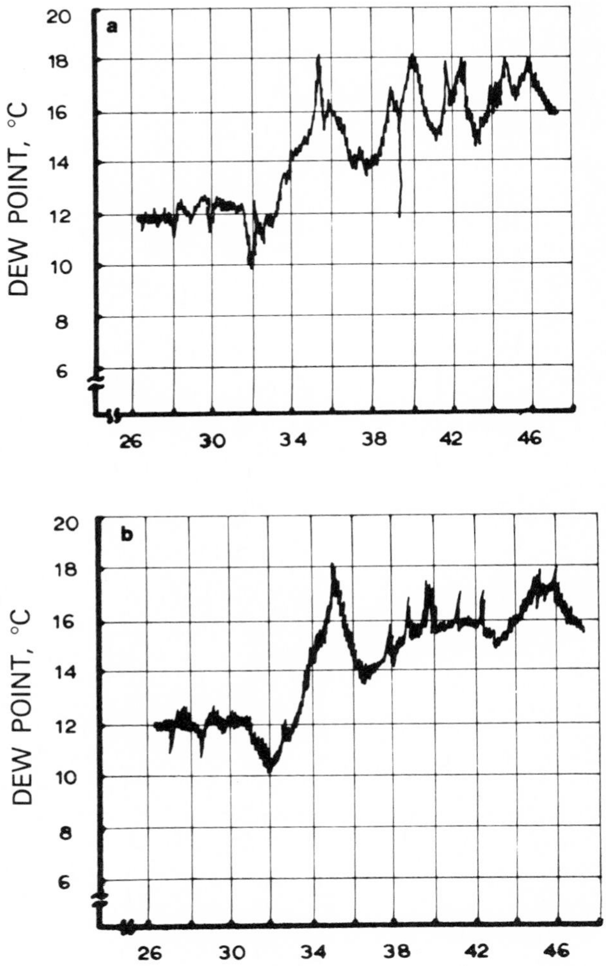

Fig. 12.10. A comparison of measurements by two instruments for atmospheric humidity. (a) A very-fast-response Lyman-alpha hygrometer. (b) A somewhat slower-response frost-point hygrometer, which has a better known absolute calibration (after Buck[44]).

instrument, which is sensitive to water vapor and also to other constituents such as dust.

12.2.4.3. Temperature Fluctuation. Measurement of rapid temperature fluctuations is achieved with very fine resistance wires (up to 60 Hz), with a sonic thermometer (up to 1000 Hz), or with several hot wires operated at different overheat ratios (up to 10 kHz). An intercomparison of measurement of the turbulent heat flux near the ground using three methods is made by Businger *et al.*[24]

12.3. Measurements from Aircraft

12.3.1. Introduction

Aircraft are essential in atmospheric measurements for two reasons. First, much data must be taken by local sensing because of the required sensitivity, response time, and spatial resolution, and sometimes because of the property to be measured. Secondly, the phenomena to be observed are geometrically large and/or occur somewhat unpredictably and with unsteady means (considerable trend) so that a rapidly moving platform with three-dimensional mobility and precision of placement is required. The piloted multiengine aircraft have proved to be the most generally suitable platforms for instruments for local and, sometimes, remote sensing.

12.3.2. Simple Techniques of Lower Accuracy

Several simple, crude methods of measuring turbulence have found continued use. The vertical motion of air with wavelengths between 200 m and 1 km and velocities greater than 2 m/sec may be measured by calibrating the aircraft itself as a sensor. The parameters generally considered in the calibration are (1) weight and balance of the aircraft, (2) engine manifold pressure, (3) indicated airspeed, (4) altitude, and (5) the rate of climb as indicated by a "leaky" aneroid barometer (dp/dt meter) (see "Variometers" in Middleton and Spilhaus[25]). The aircraft is calibrated in air having negligible vertical motion by letting engine manifold pressure be adjusted to give selected rates of climb. In measurement the manifold pressure and airspeed are maintained and the rate of climb is used as a measure of the vertical speed of airflow.

The component of velocity fluctuation along the axis of the Pitot tube of the aircraft may be readily extracted and analyzed once the filter function of the airspeed-measuring system is known. MacCready[26,45] has filtered and algebraically analog-operated on the basic $\bar{u}'^2$ signal within the inertial subrange to obtain an output proportional to the dissipation rate to the one-third power using Kolmogorov theory. This output is described as a universal indicator of turbulence intensity.

12.3.3. Higher-Accuracy Methods

If more precision is required in turbulence measurement by aircraft or if orthogonal components are to be sampled adequately over a range of scales, from, say, 4 m to 10 km, an inertial reference is required for determining angular orientation relative to the earth. The motion of the sensors relative to the earth may be computed if for every coordinate direction of inertial reference attained in the inertial platform an accelerometer is oriented to measure the corresponding acceleration. [A single two-axis gyroscope and a vertical accelerometer with or without Doppler navigation can provide a somewhat degraded version of this system (Dutton and Lenschow(27) and Warner and Telford(28)).] Integration of $\ddot{x}$, $\ddot{y}$, $\ddot{z}$ (accelerations) results in very accurate $\dot{x}$, $\dot{y}$, and correctable x, y. The vertical velocity $\dot{z}$ is generally small and its drift results in a need to correct z with pressure altimeter data except for short time intervals. This use of the inertial platform is discussed by Kelly,(29) Lenschow,(30) and Lilly and Lenschow.(1)

The appropriate equations for extracting the winds measured relative to the Pitot tube or hot-wire anemometer and vanes on a nose boom are given by Axford,(31) Dutton,(32) and Lenschow.(30) Errors existing after 6 hr flights from initial alignment of the platform are about 1–3 nautical miles in horizontal location and 1–1.5 m $\sec^{-1}$ in wind speed. Independent systems of measurement not having as significant a drift problem but having greater short-term error are used to correct these drifts.

Temperature measurement from an aircraft is difficult because a fast response is required, sometimes in a droplet or rain environment. It is necessary to prevent evaporation effects and sensor damage by impact from hydrometeors; consequently, shielding is required. Among the most common techniques employed are reverse-flow and vortex-thermometer shields. Although other inaccuracies in temperature at the sensor are induced, the main concern is the increase in time constant caused by the housings. Rodi and Spyers-Duran,(33) McCarthy,(34) and Acheson(35) analyze some of these errors. Time constants for reverse flow thermometers have magnitudes of several seconds, whereas a standard exposed resistance wire thermometer has a time constant of less than 0.1 sec.

An increasing number of studies of turbulence by aircraft methods may be found in the atmospheric science literature. Most often these results are achieved using well-instrumented large aircraft operated by federal and quasifederal agencies. However, several universities and private companies have produced excellent results for highly specialized problems with much less equipment. Some examples are marked with asterisks in the list of references.

12.3.4. Data Processing and Analysis of Errors

An analysis of some errors in a three-dimensional turbulence velocity measuring system is given by Brown *et al.*[37] and a list of instrumentation on one aircraft is given by Friedman *et al.*[38] Data processing procedures for research aircraft are described by Friedman *et al.*[39] Velocities may be resolved down to 20 cm/sec. Boom oscillation places the minimum resolvable wavelength at about 4 m. Long-term (30 min to 6 hr) means have errors that are partly removable by offline methods.

A broader scope of airborn measurements methods is provided in the paper "Airborn Remote-Sensing Symposium" in an NCAR Facilities Bulletin.[36]

12.4. Aircraft Measurement of Turbulent Airflow Downwind of a Mountain Range

In 1972 the author used the well-instrumented "Buffalo" research aircraft (NCAR[40]) from the National Center for Atmospheric Research to measure airflow and turbulence in the Saratoga Valley downwind of the Sierra Madre Range in Wyoming. The purpose was to assess the possibility that upwind turbulence and hydraulic jumps in the valley airflow due to flow over the Sierra Madre Range could mix cloud-seeding material generated at the valley floor deeply into the clouds snowing on the downwind Medicine Bow Range. Figure 12.11 is a map of the two ranges and the valley. The cloud base track of the aircraft is shown by the light dashed line, and a long dashed line A′A′ indicates the location of a vertical section of five levels flown by the aircraft from near the ground to cloud base. About 1000 feet above the valley floor the mean winds were measured to be flowing down the valley and around the north end of the Medicine Bow Range. Figure 12.12 shows the streamlines and isotachs in knots for this 8500 ft MSL height. Maximum speed centers occur in the lee of Blackhall (32 knots) and Elk Mountains (52 knots) and the Medicine Bow Range (52 knots).

At cloud base the air flowed across the mountain ranges and valley in gentle waves and with low level of turbulence. Wind-speed maxima reflected the influence of mountain peaks and lee waves (Figure 12.13).

The fluctuations of vertical velocity for scales less than 400 m long were calculated from the time series such as shown in Figure 12.14. The half-peak-to-peak values in m sec^{-1} for the valley flow are shown in Figure

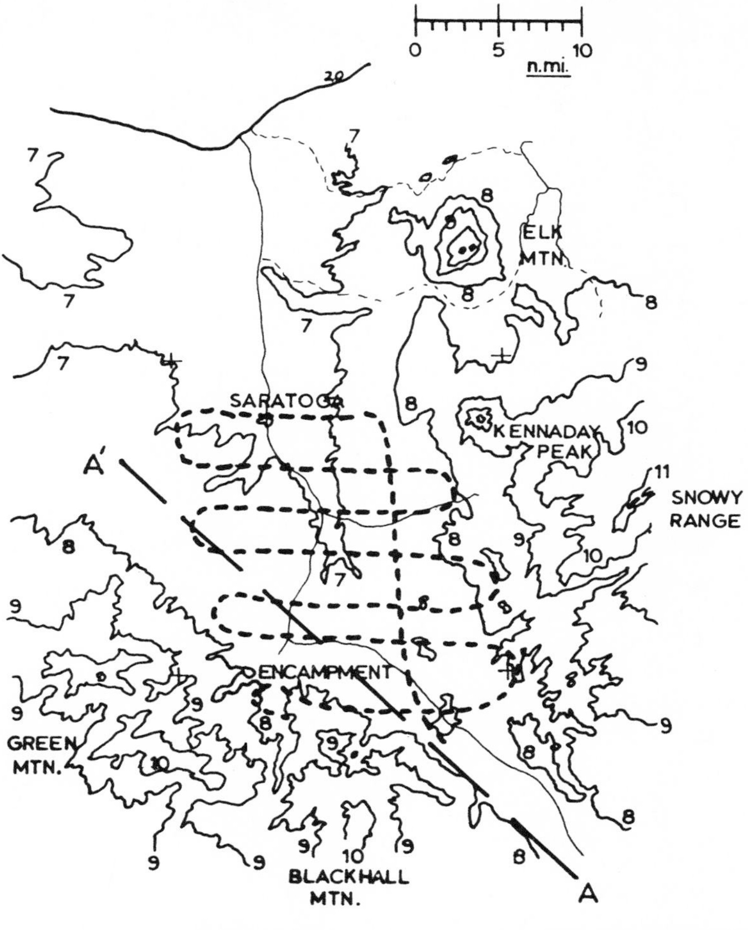

Fig. 12.11. A map of the Saratoga Valley, the westward Sierra Madre Range, and the eastward Medicine Bow Range. The dashed line indicates the track of a research aircraft at cloud-base altitude. The long-dash line A–A′ indicates where a vertical section composed of five levels of aircraft path was flown. Contours are in thousands of feet.

12.15a and for the cloud base flow in Figure 12.15b. The larger magnitudes again reflect the influence of near upwind mountain peaks.

The turbulent intensity of greatest magnitude was about $u'/U = 0.09$ at Elk Mountain and 0.06 at the north end of the Sierra Madre. The intensity was about 0.03 at the south and 0.01 in the middle of the valley. At cloud base turbulence intensity was generally from 0.01 to 0.02 with one spot having a value of 0.04.

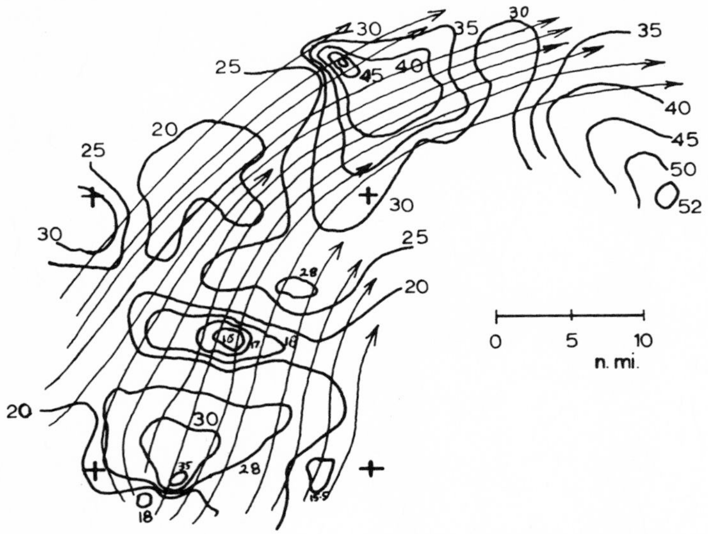

Fig. 12.12. The streamlines and isotachs [knots(kt)] of the windfield at 8500 feet MSL (about 1000 feet above the valley floor). Encampment—Elk Mountain, 12 February 1973. n.mi. stands for nautical miles.

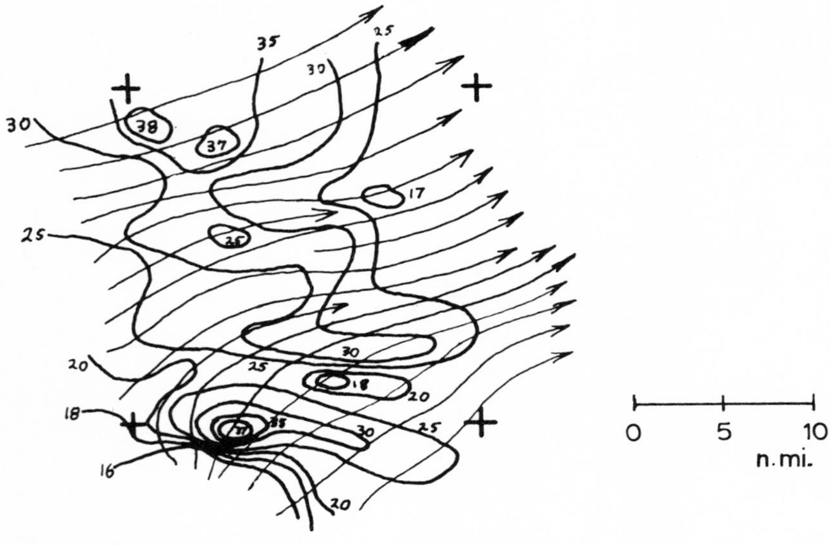

Fig. 12.13. The streamlines and isotachs (knots) of the windfield at cloud base (about 2500 feet above the valley floor).

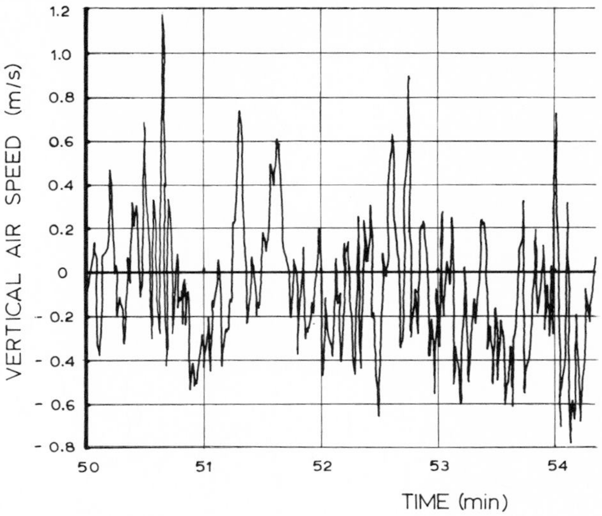

Fig. 12.14. A sample time series of the vertical wind measured by the research aircraft.

There was not much continuity of flow feature from one map level to the other except in the far southwest. This had earlier been identified as a possible prime site for location of a ground-based cloud "seeding" generator. A vertical section of horizontal aircraft tracks was executed through this region to better evaluate the vertical continuity. The mean vertical motion on scale lengths greater than 1 km in this section is shown as Figure 12.16a. One region of updraft was continuous up to cloud base at least with a w of $+1$ m/sec near the ground and $+4$ m/sec near cloud base. The "choice" seeding site is marked at the ground by an X in this region. Figure 12.16b shows the isotach of horizontal winds for the same vertical section. The regions of contiguous updraft is also the "fast" wind region.

A trajectory and spread for the aerosols released from a point source at position X was estimated using the fields of three-dimensional mean flow velocity and turbulence intensity. The result, shown in Figure 12.17, is that the "seeding" plume did not enter the cloud in appreciable concentrations over the Medicine Bow. Instead it rose only slightly as it moved toward the north end of the Medicine Bow near Kennedy Peak, an 11,000-ft mountain approximately 1000 m above the valley floor. Even the more extreme turbulence in the lee of Elk Mountain would not have been of appreciable value in the Saratoga Valley since air would quickly leave the highly turbulent region.

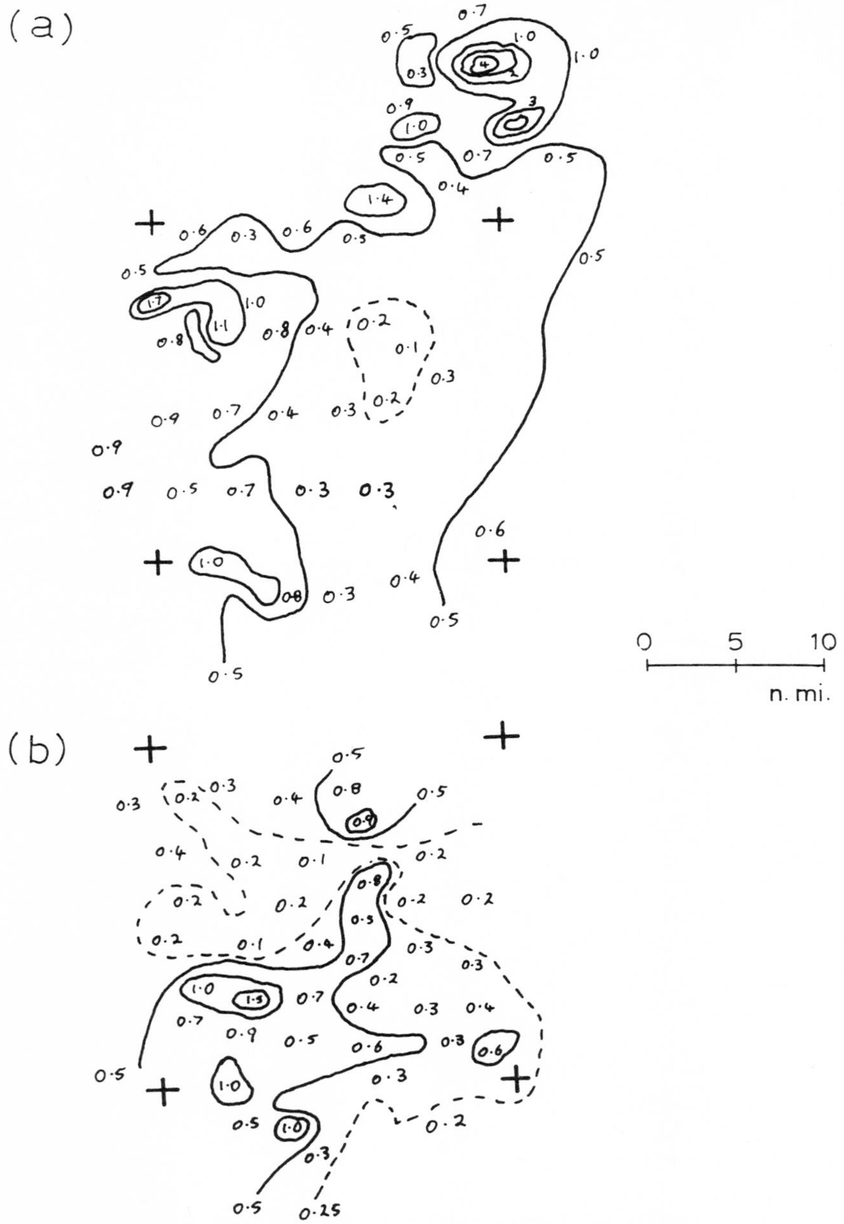

Fig. 12.15. Maps of horizontal distribution of vertical velocity fluctuation calculated as short-time mean half-peak-to-peak values (for scales less than 400 m long) in meters per second. (a) The valley turbulence (at 8500 feet MSL). (b) The cloud-base turbulence.

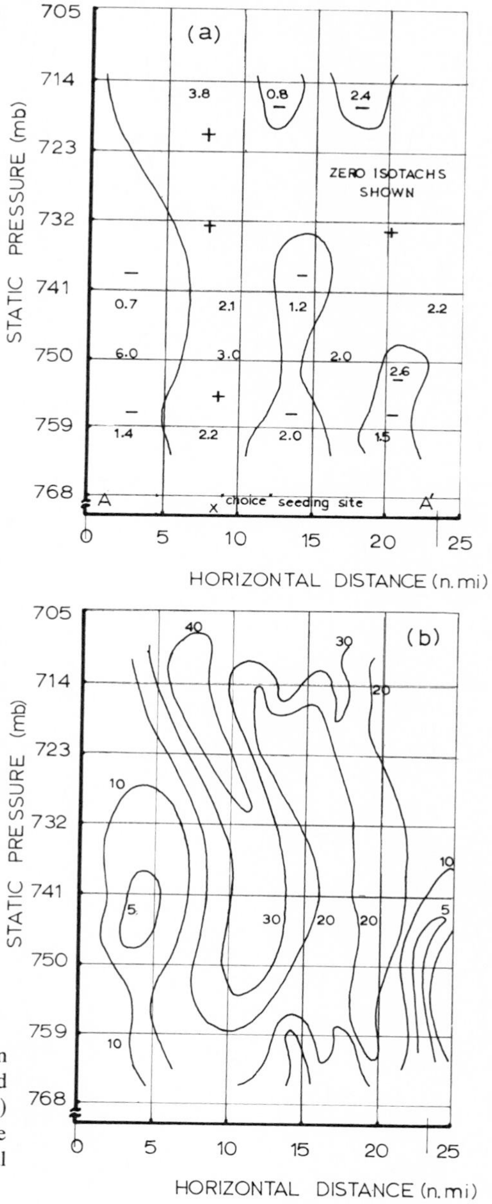

Fig. 12.16. (a) A vertical section of mean vertical velocity in meters per second upward (for scales greater than 1000 m) above the line marked A–A′ in Figure 12.11. (b) Same section but horizontal wind speed.

	Δt	LIFT	Δ WIDTH
A	1.7 min	+ 150 m	NOT CALC
B	15 min	+ 500 m	1.5 Km
C	15 min	0 → - 300m	0 → - 0.5 Km
D	15 min	- 100 m	NEGLIGIBLE

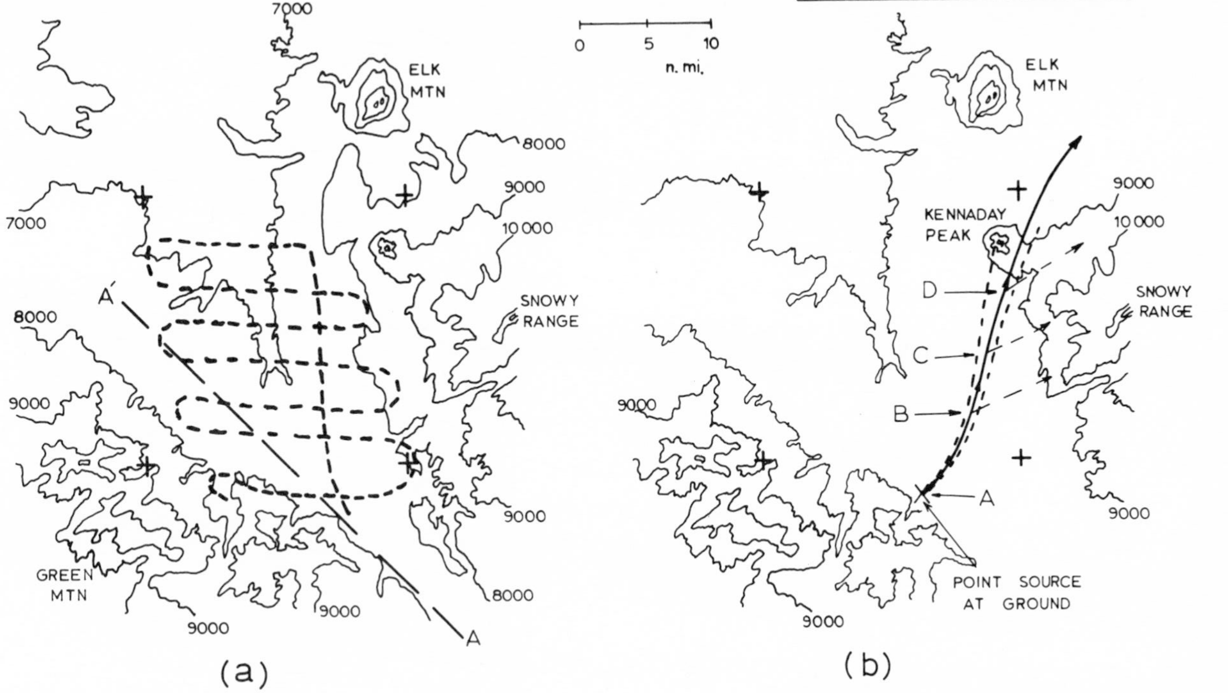

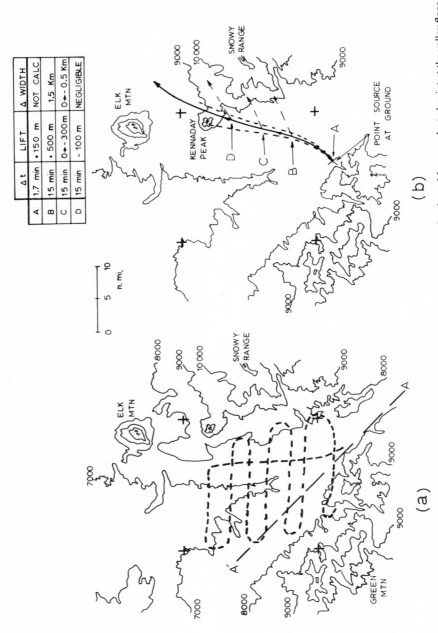

Fig. 12.17. A map of the Saratoga Valley showing the calculated geometry of a plume of aerosols released from a selected point at the valley floor based upon simple theory and observed mean winds and turbulence. (a) The base map as in Figure 12.11. (b) The plume: (1) estimated lateral diffusion about main trajectory is indicated by short dashes; (2) "sweep-away" at the top of the plume at cloud base is indicated by long dash arrows.

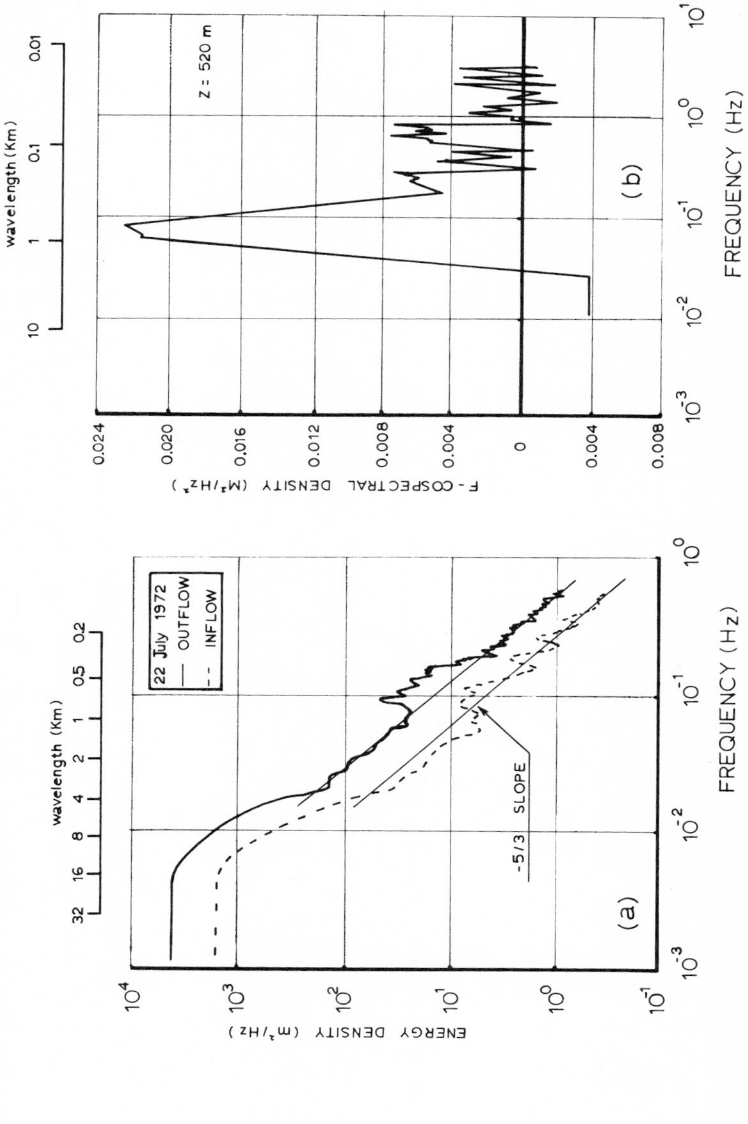

Fig. 12.18. Sample turbulence spectra computed from wind measurements taken with the NCAR N3 26D aircraft. (a) A horizontal velocity spectrum from flights near a thunderstorm (Foote and Faukhauser[41]). (b) A Reynolds stress spectrum for a trade wind flight (Pennell and LeMone[42]).

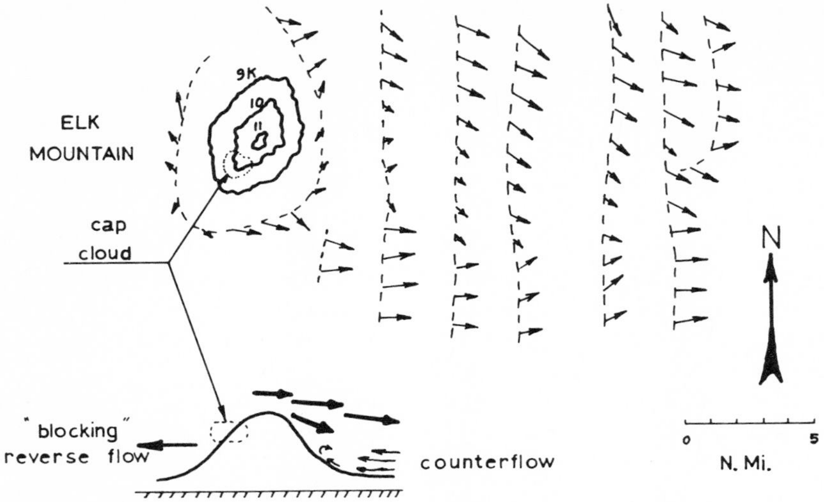

Fig. 12.19. A map of horizontal winds around and downwind of an isolated mountain measured with a research aircraft. Note the wake and the upwind counterflow. A schematic side view is inset.

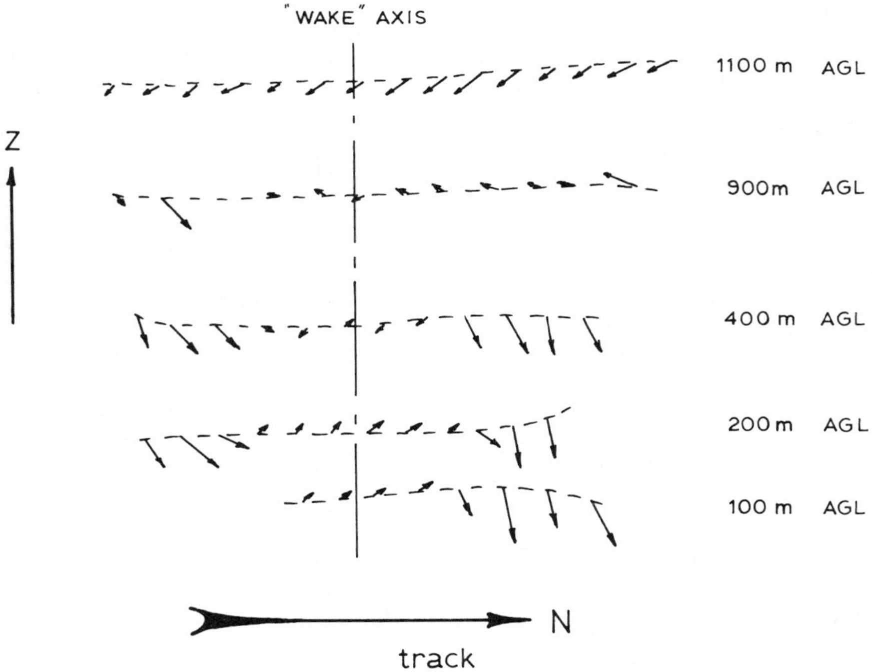

Fig. 12.20. A vertical section across the winds in the very near lee of an isolated mountain as measured by a research aircraft. *Note:* Wind from the north is from the top of the figure.

12.5. Elk Mountain PBL Profiles

Flights to investigate the planetary boundary-layer profiles of the airflow at Elk Mountain were undertaken. Spectral analyses of the turbulence data are yet to be completed. A velocity spectrum for the circumnavigated thunderstorm (Foote and Faukhauser[(41)]) and a Reynolds stress spectrum for a trade wind flight (Pennell and LeMone[(42)]) by the Buffalo aircraft are shown as Figures 12.18a and 12.18b to indicate the range of frequencies measured. Preliminary results of wind patterns for one flight at Elk Mountain show a wake and counterflow leeward and a blocking reverse flow upwind. See Figures 12.19 and 12.20.

Figure 12.21 shows an example of the mean winds found on one flight in the near lee of the Medicine Bow at 200 ft above the local ground. The analysis of turbulence is incomplete, but rough estimates of the intensity of turbulence are indicated. Considerable complexity is apparent. Long-time series are not achieved by the method and stationarity of the time series is not a proved condition. This restricts the methods of analysis somewhat, but more certainly it limits the extrapolation of the observations to other situations. It provides detail not otherwise available over such a large region.

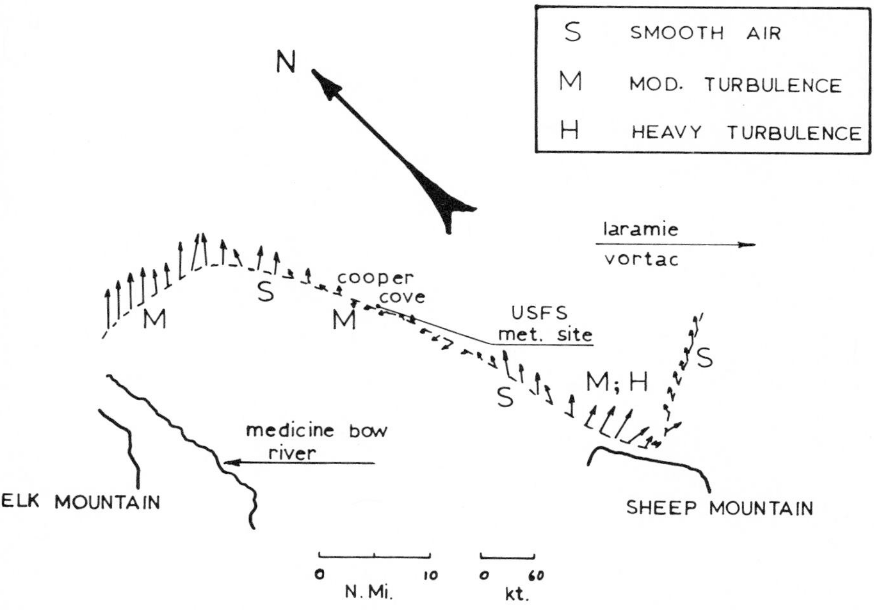

Fig. 12.21. Mean horizontal winds at 200 ft above the ground along the near lee of a mountain range as measured by aircraft. The turbulence levels, indicated in the categories smooth, light, moderate, heavy, are subjective. The data have not yet been analyzed.

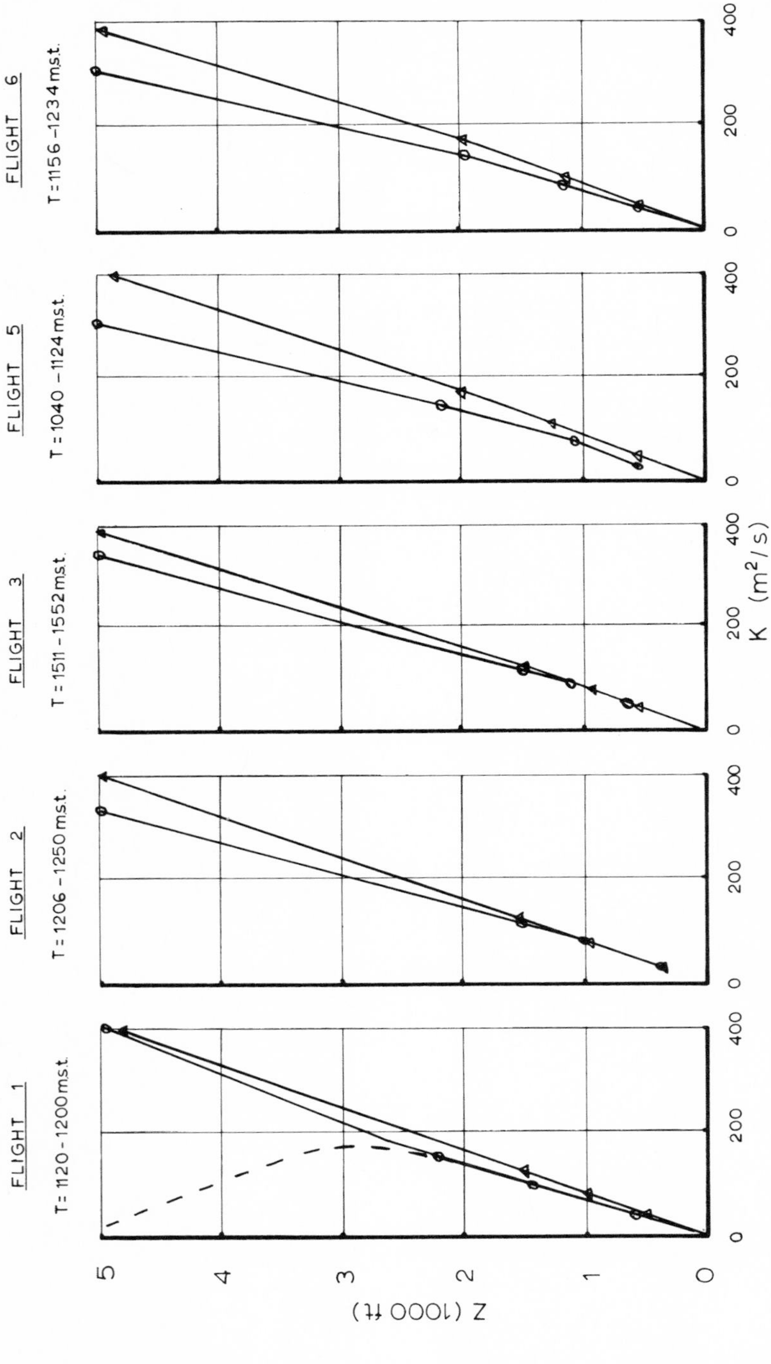

Fig. 12.22. Profiles of turbulent mixing coefficient K, in m^2/sec upwind and on the upwind slope of an isolated mountain. See the text for the method of computation from aircraft-measured wind fluctuations. The dashed line suggests how the curves would turn if the inversion height were incorporated into the computation. △—upwind; ○—downwind.

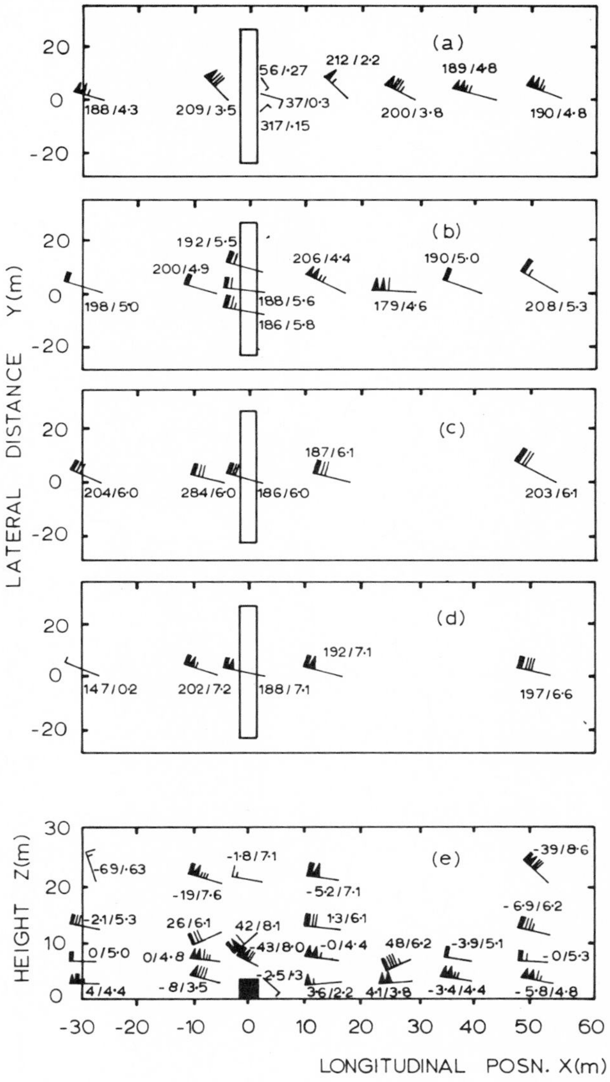

Fig. 12.23. Mean winds measured from meteorological towers for a transversely long building of height h. (a), (b), (c), and (d) are maps for heights above ground of $0.2h$, $1.8h$, $3.6h$, and $6.5h$, respectively. The numbers indicate (wind direction in degrees)/(wind speed in meters per second). The arrows point in the direction the wind blows. (e) is a vertical section down the tower array that shows the total wind (three-dimensional) speed and the slope of the wind.

12.6. Suppression of Mixing Coefficient by Forced Boundary-Layer Upward Curvature

A final intriguing bit of data from upwind PBL flights at Elk Mountain shows that the mixing coefficient is reduced by about 20% as airflow bends up onto the rising slope of the mountain (the curvature of the flow is concave). Figure 12.22 shows some estimated profiles of K in $m^2\ sec^{-1}$ using a longitudinal turbulence measurement from an aircraft Pitot–static tube and an analysis using the method of MacCready.[26]

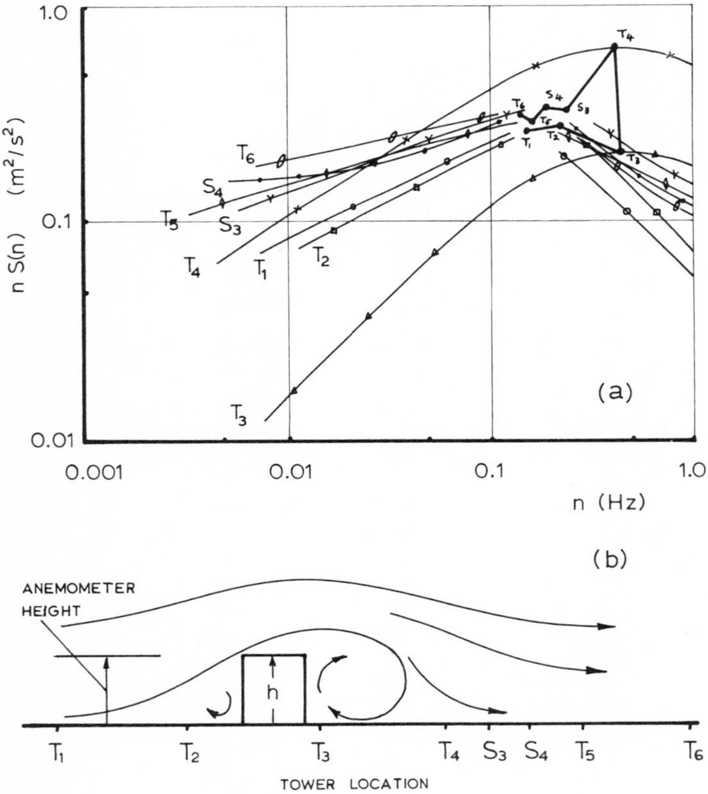

Fig. 12.24. (a) The frequency times power spectral density of the along-mean-wind velocity fluctuations at the 3-m height ($0.94h$) on all meteorological towers. The perturbation and recovery of the frequency and magnitude of the spectral peak is shown by the heavy solid line connecting tower-numbered spectral peaks in the sequence from far upwind to far downwind. (b) Schematic side view of tower locations and building wake.

12.7. Turbulent Airflow across a Building

An important area of turbulence study is being undertaken at UTSI in cooperation with NASA–Huntsville and funded by NSF. It is the flow of a turbulent boundary layer over a two-dimensional building. The data are comprised of three-dimensional velocities measured by cup anemometers and vanes and vertical-axis propeller anemometers. A first report of those data will be published soon. Figure 12.23 shows a set of wind fields for one case and Figure 12.24 shows the axial distribution of the longitudinal velocity spectra at the 3 m (0.9 building height) level. About 8.7 building heights downwind of the lee edge of the building, the wake gives way to conditions nearly identical to those upwind.

12.8. Concluding Remarks

The diversity of measurement requirements and techniques has been indicated. No effort at comprehensiveness has been made but a reasonable representative of several types of measurement of atmospheric turbulence has been given. Progress in research in atmospheric turbulence appears to be most rapid in aircraft-based measurements and in wave propagation methods. A review of progress is given by Lilly[43] covering the surface and planetary boundary layer, clear-air turbulence and gravity waves, cloud convection, and turbulence structure. An excellent list of references is included with Lilly's paper.

Notation

c	Sound speed	T	Temperature
f	Frequency	V	Velocity
k	Mixing coefficient	ω	Vertical velocity
L	Length	x, y, z	Rectangular coordinates
R	Gas constant	γ	Ratio of specific heats
t	Time	$\dot{}$	Time derivative
s	Wind	$\ddot{}$	Second time derivative

References

1. Lilly, D. K., and Lenschow, D. H., Aircraft measurements of the atmospheric mesoscales using an inertial reference system, *Facilities for Atmospheric Research*, **19**, National Center for Atmospheric Research, Boulder, Colorado (1971), pp. 2–8.
2. Starr, V., *Physics of Negative Viscosity Phenomena*, McGraw-Hill Book Company, New York (1968).

3. Gill, G. C., and Hexter, P. L., Some instrumentation definition for use by meteorologists and engineers, *Bull. Am. Meteorol. Soc.* **53**, 846–851 (1972).
4. Mazzarella, D. A., An inventory of specifications for wind measuring instruments, *Bull. Am. Meteorol. Soc.* **53**, 860–871 (1972).
5. MacCready, P. B., Dynamic response characteristics of meteorological sensors, *Bull. Am. Meteorol. Soc.* **46**, 533–538 (1965).
6. Gill, G. C., Olsson, L. E., Sela, J., and Suda, M., Accuracy of wind measurements on towers or stacks, *Bull. Am. Meteorol. Soc.* **48**, 665–674 (1967).
7. Carter, J. K., The meteorologically instrumented WKY-TV tower facility, report No. NOAA TM ERLTM-NSSL-50, National Severe Storms Laboratory, Norman, Oklahoma (1970).
8. Cermak, J. E., and Horn, J. D., Tower shadow effect, *J. Geophys. Res.* **73**, 1869–1876 (1968).
9. Deacon, E. L., The levelling error in Reynolds stress measurement, *Bull. Am. Meteorol. Soc.* **49**, 836 (1968).
10. Pond, S., Some effects of buoy motion on measurement of wind speed and stress, *J. Geophys. Res.* **73**, 507–512 (1968).
11. Kraus, E. B., What do we know about the sea-surface stress? *Bull. Am. Meteorol. Soc.* **49**, 247–253 (1968).
12. Kraus, E. B., Reply to Deacon, *Bull. Am. Meteorol. Soc.* **49**, 836 (1968). (See Reference 9.)
13. Dyer, A. J., Hicks, B. B., and Sitaraman, V., Minimizing the leveling error in Reynolds stress measurement by filtering, *J. Appl. Meteorol.* **9**, 532–534 (1970).
14. Acheson, D. T., Response of cup and propeller rotors and wind direction vanes to turbulent wind fields, *Meteorol. Monogr.* **11**, 252–261 (1970).
15. Moses, H., *et al.*, Meteorological instruments for use in the atomic energy industry, *Meteorology and Atomic Energy*, U.S. Atomic Energy Commission (1968), Chap. 6, pp. 257–298.
16. Kelton, G., and Bricout, P., Wind velocity measurements using sonic techniques, *Bull. Am. Meteorol. Soc.* **45**, 571–580 (1964).
17. Beran, D. W., Little, C. G., and Willmarth, B. C., Acoustic Doppler measurements of vertical velocities in the atmosphere, *Nature* **230**, 160–162 (1971).
18. Beran, D. W., and Clifford, S. F., Acoustic Doppler measurement of the total wind vector, *Second Symposium on Meteorological Observations and Instruments*, American Meteorological Society (1972).
19. Benedetti-Michelangeli, G., Congeduti, F., and Fiocco, G., Measurement of aerosol motion and wind velocity in the lower troposphere by Doppler optical radar, *J. Atmos. Sci.* **29**, 906–910 (1972).
20. Weiler, H. S., and Burling, R. W., Direct measurements of stress and spectra of turbulence in the boundary layer over the sea, *J. Atmos. Sci.* **24**, 653–664 (1967).
21. Pond, S., *et al.*, Spectra of velocity and temperature fluctuations in the atmospheric boundary layer over the sea, *J. Atmos. Sci.* **23**, 376–386 (1966).
22. Wesely, M. L., *et al.*, Three-dimensional pressure-sphere anemometer system, *J. Appl. Meteorol.* **9**, 379–385 (1970).
23. Thurtell, G. W., *et al.*, Eddy correlation measurements of sensible heat flux near the earth's surface, *J. Appl. Meteorol.* **9**, 379–385 (1970).
24. Businger, J. A., Miyake, M., Dyer, A. J., and Bradley, E. F., On the direct determination of the turbulent heat flux near the ground, *J. Appl. Meteorol.* **6**, 1025–1032 (1967).
25. Middleton, W. E. K., and Spilhaus, A. F., *Meteorological Instruments*, University of Toronto Press (1953), pp. 53–56.
26. MacCready, P. B., Standardization of gustiness values from aircraft, *J. Appl. Meteorol.* **3**, 439–449 (1964).

27. Dutton, J. A., and Lenschow, D. H., An airborne measuring system for micrometeorological studies, Annual Report, Contract No. DA-36-039-SC-80282, Department of Meteorology, University of Wisconsin, Madison, Wisconsin (1963).
28*. Warner, J., and Telford, J. W., On the measurement from an aircraft of buoyancy and vertical air velocity in cloud, *J. Atmos. Sci.* **19**, 415–420 (1962).
29. Kelly, N. D., Meteorological uses of inertial navigation, *Atmospheric Technology*, **1**, National Center for Atmospheric Research, 37–39 (1973).
30. Lenschow, D. H., The measurement of air velocity and temperature using the NCAR Buffalo aircraft measuring system, NCAR Technical Note No. NCAR-TN/EDD-4, National Center for Atmospheric Research (1972), p. 39.
31. Axford, D. N., On the accuracy of wind measurements using an inertial platform in an aircraft, and an example of a measurement of the vertical mesostructure of the atmosphere, *J. Appl. Meteorol.* **7**, 645–666 (1968).
32. Dutton, J. A., Clear-air turbulence, aviation and atmospheric science, *Rev. Geophys. Space Phys.* **9**, 613–657 (1971).
33. Rodi, A. R., and Spyers-Duran, P. A., Analysis of time response of airborne temperature sensors, *J. Appl. Meteorol.* **11**, 554–556 (1973).
34. McCarthy, J., A method for correcting airborne temperature data for sensor response time, *J. Appl. Meteorol.* **12**, 211–214 (1973).
35. Acheson, D. T., Comments on "A method for correcting airborne temperature data for sensor response time," *J. Appl. Meteorol.* **12**, 1089–1090 (1973).
36. NCAR, Instrumenting NCAR's "Buffalo" aircraft, *Facilities for Atmospheric Research*, No. 8, National Center for Atmospheric Research (1969), pp. 7–10.
37. Brown, W. J., McFadden, J. O., Hason, H. J., and Travis, C. W., Analysis of the Research Flight Facility gust probe system, *J. Appl. Meteorol.* **13**, 156–167 (1974).
38. Friedman, H. A., *et al.*, ESSA Research Flight Facility aircraft participation on the Barbados oceanographic and meteorological experiment, *Bull. Am. Meteorol. Soc.* **51**, 822–834 (1970).
39. Friedman, H. H., *et al.*, The ESSA Research Flight Facility: Data processing procedures, ESSA Technical Report No. ERL 132-RFF 2, Miami, Florida (1969), p. 64.
40. NCAR, Instrumenting NCAR's "Buffalo" aircraft, *Facilities for Atmospheric Research*, No. 8, National Center for Atmospheric Research (1969), pp. 9–12.
41*. Foote, G. B., and Fankhauser, J. C., Airflow and Moisture budget beneath a northeast Colorado hail storm, *J. Appl. Meteorol.* **12**, 1330–1353 (1973).
42. Pennell, W. T., and LeMone, M. A., An experimental study of turbulence structure in the fair-weather trade wind boundary layer, *J. Atmos. Sci.*, **31**, 1308–1323 (1974).
43. Lilly, D. K., Progress in research on atmospheric turbulence, International Union of Geodesy and Geophysics (1971), pp. 332–341.
44. Buck, A. L., Development of an improved Lyman-alpha hygrometer, *Atmospheric Technology*, No. 2, National Center for Atmospheric Research, Boulder (1973), pp. 43–46.
45. MacCready, P. B., Jr., An applications memorandum for the MRI universal indicated turbulence system, report No. MRI 70 M-917, Meteorological Research Inc., Altadena, California, 1970, p. 15.

Additional Annotated References on Turbulence Measurement

The references marked with asterisks discuss measurements with aircraft that have less than three axis inertial reference platforms.

* Lilly, D. K., and Lester, P. F., Waves and turbulence in the stratosphere, *J. Atmos. Sci.* **31**, 800–812 (1974). (Use of Doppler navigation and simple platform to reference vertical and aircraft-axis winds.)

* Waco, D. E., A statistical analysis of wind and temperature variables associated with high altitude clear air turbulence (HICAT), *J. Appl. Meteorol.* **9**, 300–309 (1970). (U-2 aircraft measurements of gust velocities and temperature fluctuations.)

* Prophet, D. T., Vertical extent of turbulence in clear air above the tops of thunderstorms, *J. Appl. Meteorol.* **9**, 320–321 (1970). (U-2 aircraft measurements of vertical gusts.)

* Robinson, F. L., and Konrad, T. G., A comparsion of the turbulent fluctuations in clear air convection measured simultaneously by aircraft and Doppler radar, *J. Appl. Meteorol.* **13**, 481–487 (1974). (Doppler radar from fine-scale refractivity fluctuations. Aircraft carried a hot-wire anemometer. Comparison of spectra of horizontal velocity fluctuations was good for wavelengths of between 50 and 1000 m. Only 2-min averaging with aircraft data.)

* Lilly, D. K., Waco, D. E., and Adelfang, S. I., Stratospheric mixing estimated from high-altitude turbulence measurements, *J. Appl. Meteorol.* **13**, 488–493 (1974). (Calculated dissipation, from U-2 measurements of longitudinal velocity fluctuation. Calculated diffusion coefficient for heat K_H, using ε and the Brunt–Vaisala frequency. Stratospheric values of K_H over mountains are of the order of 1 m^2/sec in winter. This is 20–100 times smaller than K_m—momentum diffusion coefficient—calculated similarly from measurements within 200 m of a mountain by Connell in 1972.)

Chiu, W.-C., and Crutcher, H. L., The spectrums of angular momentum transfer in the atmosphere, *J. Geophys. Res.* **71**, 1017–1032 (1966). (Synoptic scale to planetary scale horizontal transfer using data from rawinsonde balloon ascents for North and Central American stations.)

* Sheih, C. M., Tennekes, H., and Lumley, J. L., Airborn hot-wire measurements of the small-scale structure of atmospheric turbulence, *Phys. Fluids* **14**, 201–215 (1971). (Hot-wire on moving aircraft eliminated directional ambiguity and permitted linearized analysis of the hot-wire turbulence signal. No stable platform used. A pair of cross wires permitted estimation of two components of turbulence. Looked at dissipation end of spectrum. Found u and w computed dissipation rates to vary from 5 to 150 cm^2/sec^3.)

Horst, T. W., Corrections for response errors in a three-component propeller anemometer, *J. Appl. Meteorol.* **12**, 716–725 (1973). (Compared power spectral density computation, for each of the three wind components, as measured by propellers and by sonic anemometers taken as the reference. Considerable disagreement but especially at frequencies between 0.3 and 1 Hz (cutoff). Correction for noncosine response handles the uniform error and correction for inertial lag decreases the error remaining in the high-frequency range by 50%.)

CHAPTER 13

Optical and Acoustical Measuring Techniques

WILLIAM C. CLIFF

13.1. Introduction

In recent years considerable emphasis has been placed on developing techniques to remotely measure the velocity of confined and free fluid flows (such as atmospheric, oceanic, wind tunnel, blood, pipe, and channel flows, etc.). The laser Doppler and acoustic Doppler are two techniques that will be examined in this chapter. Both techniques are based on the Doppler effect. The Doppler effect is the fact that there is a change in frequency with which energy reaches a receiver when the receiver and the energy source are in motion relative to one another. In the cases of the laser and acoustic Doppler system, energy is transmitted to a moving scatterer (tracer), which then becomes a source, and the energy is transmitted to a receiver. The Doppler systems measure the velocity of these scattering sources and, therefore, the accuracy with which the Doppler systems measure fluid velocity is dependent upon the accuracy with which the imbedded scattering source follows the true velocity of the medium. Major efforts in the development of the laser and acoustic Doppler systems have been underway since the 1960's.[1-3] It is the intent of this chapter to skip the developmental years of these systems and present the techniques that are presently employed and the results that they are obtaining.

WILLIAM C. CLIFF • National Aeronautics and Space Administration, George C. Marshall Space Flight Center, Huntsville, Alabama 35812

13.2. Background and Basic Principles

The technique of using Doppler anemomometry is based on the fact that radiation, acoustic or electromagnetic (laser), passing through a fluid is scattered by tracers in the fluid. (In the case of laser Doppler systems, the scattering occurs because of particles suspended in the fluid. In the case of acoustic Doppler, the scattering may occur because of temperature or velocity gradients.) The scattered radiation contains information on the velocity of the tracers from which the radiation was scattered.

The information on the velocity of the tracer manifests itself by frequency-shifting the radiation striking the tracer. The amount that the source frequency is shifted, Δf, upon striking the tracer and returning to a receiver is called the Doppler shift or Doppler frequency and is expressed mathematically as

$$\Delta f = (\bar{V}n/\lambda)(\bar{e}_s - \bar{e}_i) \tag{13.1}$$

where $\bar{V}$ is the velocity vector of the tracer, λ is the wavelength of the source radiation, $\bar{e}_s$ is a unit vector along the scattered radiation (a unit vector from the scattering source to the receiver), and $\bar{e}_i$ is a unit vector along the incident radiation (a unit vector from the source to the tracer). In all cases the index of refraction of the medium, n, is assumed equal to unity and the term n will not be shown in the following equations.

From Equation (13.1) it is noted that a single Doppler system gives a one-dimensional velocity measurement and the velocity component measured lies along a vector bisecting the angle between the incident and scattered radiation. Figure 13.1 presents a three-dimensional view of a single-radiation Doppler system. In Figure 13.1 the axes are oriented so that the velocity component sensed by the system lies along the z axis. Thus, any motion in the x–y plane would not be detected by the Doppler system. The use of additional Doppler systems would allow two- or three-dimensional velocity measurements, however.

Figure 13.2 presents a plane view of a single Doppler system. Here again it is noted that the velocity sensed is parallel to the bisector of the incident and scattered radiation. In terms of the angles given in Figure 13.2, Equation (13.1) may be written

$$\Delta f = \frac{(\bar{V} \cos \beta)}{\lambda}\left(2 \sin \frac{\theta}{2}\right) \tag{13.2}$$

where β is the angle between the total velocity vector and the bisector of the incident and scattered radiation, and θ is the angle between the incident and scattered radiation.

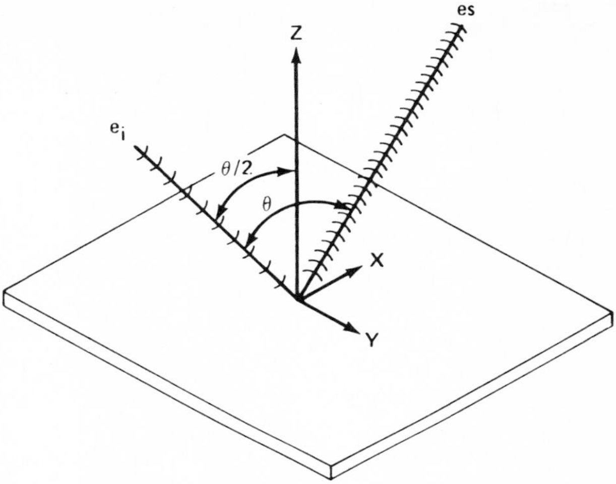

Fig. 13.1. A three-dimensional view of a single-radiation Doppler system. Note: The plane formed by the lines of the incident radiation and scattered radiation is normal to the x–y plane.

In terms of the velocity measured, V_m, Equation (13.2) may be written

$$V_m = \bar{V} \cos \beta = \frac{\lambda \Delta f}{2 \sin \frac{1}{2}\theta} \tag{13.3}$$

If one measures pure backscatter (i.e., the scattered radiation is measured along the same path as the radiation was sent out, $\theta = 180°$), Equation (13.3) becomes

$$V_m = \tfrac{1}{2}\lambda \Delta f \qquad \text{for } \theta = 180° \tag{13.4}$$

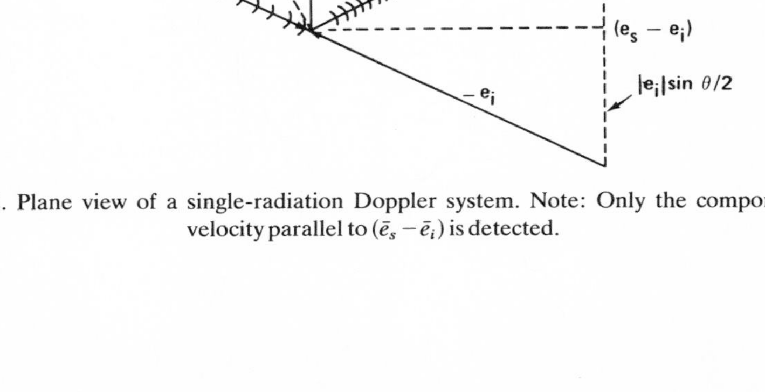

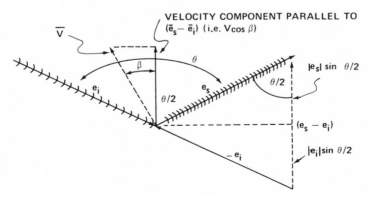

Fig. 13.2. Plane view of a single-radiation Doppler system. Note: Only the component of velocity parallel to $(\bar{e}_s - \bar{e}_i)$ is detected.

Equations (13.1)–(13.4) are the basic equations for a single Doppler system. As will be shown later, the use of three separate Doppler signals can give the investigator a means to directly measure the three-dimensional flow field.

13.3. Laser Doppler

The laser velocimeter, generally referred to as a laser Doppler velocimeter (LDV), measures the Doppler frequency shift in laser light caused by moving particles scattering the light. This frequency shift is related to the wavelength of the light, the geometrical direction of the scattered light, and the velocity of the particles producing the shift. Equations (13.1)–(13.4) give the Doppler frequency as a function of wavelength, direction, and velocity of the tracer. For a particular test configuration, the laser wavelength and geometrical scattering direction are given. Knowing the laser wavelength and test geometry, the measuring of the Doppler frequency shift allows the calculation of the scattering source's velocity. In the LDV case the scattering sources are generally particles imbedded within the flow. The particle velocity is then related to the fluid flow velocity in which the particle is suspended. If the particle is very small it is generally considered to move directly with the fluid surrounding the particle.

The general practice to get a measurable Doppler frequency is either to (1) mix the scattered laser light with some of the local oscillator (original laser) light on a photodetector or (2) split the original laser beam and make the dual beams cross, setting up a fringe pattern in space and sensing particle-scattering that comes from the fringe pattern on a photodetector. The system that mixes (or beats) the scattered light with some of the local oscillator light on a photodetector is called a reference-beam or local oscillator system. Figure 13.3a shows a forward-scatter reference-beam system.[4] Note that both the local oscillator (laser beam) and receiver optics are focused so that only scattering from the focal volume is sensed on the photodetector. In the case of Figure 13.3a, the photodetector is shown as a photomultiplier (PMT).

In the reference-beam system, the mixing (or beating) of the scattered light and local oscillator occurs on the photodetector. The mixing of two waves of different frequencies produces a function of the sum and difference of the two frequencies. Only the frequency difference is sufficiently low to be detected by the photodetector. This frequency difference, known as the Doppler frequency, is detected by the photodetector as the result of an apparent fringe pattern moving across the detector; that is, the detector area is impacted by radiation whose amplitude is driven by the Doppler frequency.

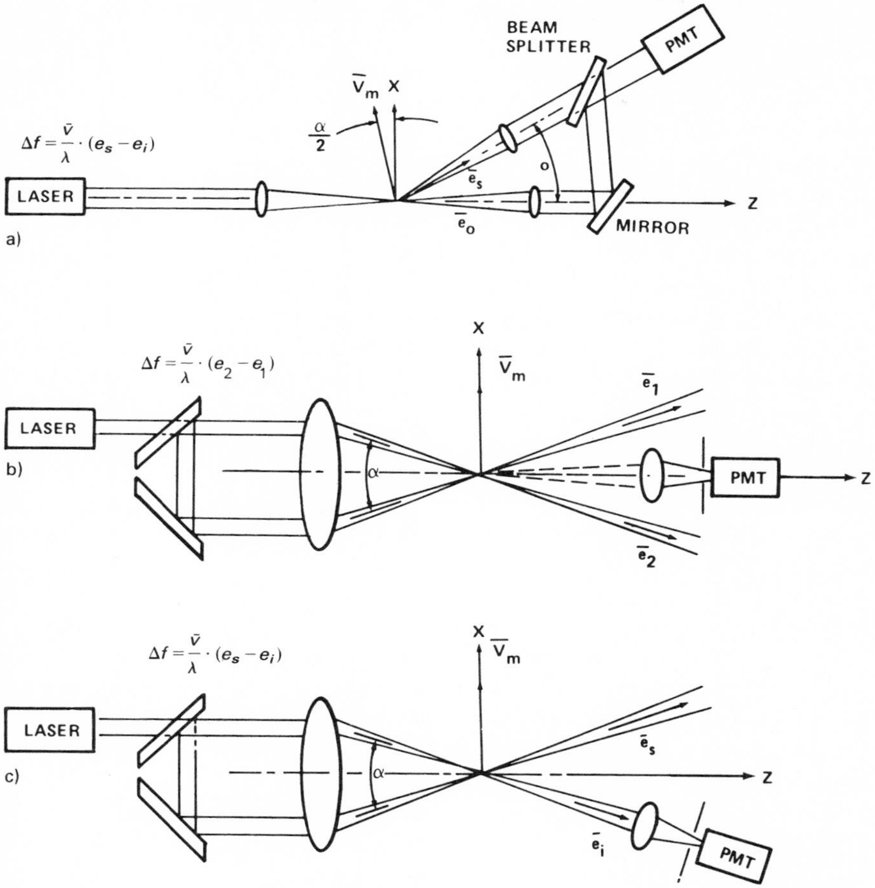

Fig. 13.3. Schematic of different LDV system arrangements. (a) Reference forward-scatter system. (b) Dual-beam forward-scatter system. (c) Dual-beam reference scatter system.

In the dual-beam system, the original laser beam is split, separated, and then both beams are focused so as to cross at some point in space. An apparent fringe pattern, which a particle encounters as it passes through the focal volume, is established where the beams cross.

As the particle passes through the fringe pattern, the intensity of the scattering is dictated by that portion of the fringe pattern the particle is in. Figure 13.3b presents a schematic of a forward-scatter fringe system. The receiver optics in the dual-beam system are focused on the volume where the two beams cross. Figure 13.3c presents a schematic of a dual-beam reference system which is a combination of the dual-beam and reference-beam systems.

In these first examples, we have beat two signals together to produce a difference frequency that was related to particle velocity. It should be pointed out that if the signals are beat together, as described to this point, an ambiguity in direction of 180° is present. That is, the difference in frequency of the scattered radiation and original laser radiation would be the same whether the particle was moving in a given direction or moving in a direction that is exactly opposite (180°). This direction ambiguity can be corrected, however, by frequency-shifting the local oscillator (original laser source light). Frequency-shifting (or translating) the local oscillator causes a particle with no velocity to produce a difference (Doppler) frequency equal to the amount that the local oscillator was shifted. Figure 13.4 presents the effect of

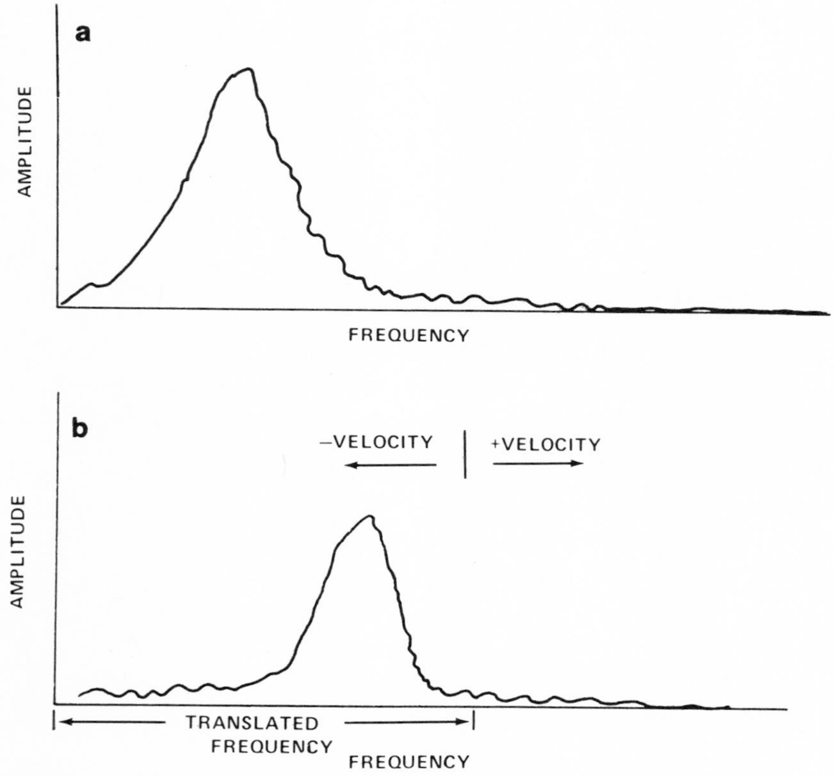

Fig. 13.4. Schematic of Doppler return (a) without frequency translator, and (b) with frequency translator.

frequency translation of the original laser light. The frequency translation may be accomplished by several methods such as scattering the local oscillator (original laser light) off a target moving at a constant rate or shifting the local oscillator by means of an acousto-optical translator. The latter appears to be the superior method. In the acousto-optical method, the local oscillator is passed through a crystal or liquid that is acoustically excited. Part of the local oscillator is shifted by some angle, δ, and translated in frequency. The most common of these systems is called a Bragg cell and is incorporated into many of the small commercial laser Doppler systems. The commercial systems generally operate in the visible spectrum and employ a helium–neon laser with a 0.6328-μ wavelength. Presently Marshall Space Flight Center (MSFC) has developed translators for their infrared laser Doppler systems (10.6-μ wavelength). The Doppler frequency is measured rather than actually measuring the light frequencies because the light frequencies are extremely large. For example, the frequency one would measure from an argon laser, a common laser used in laser Doppler research whose strongest line is at 0.5145μ (0.5145×10^{-6} m), would be equal to the speed of light divided by the wavelength, that is, (299,860,000 m/sec)/(0.5145×10^{-6} m) = 5.828×10^{14} Hz, which would be difficult to measure with the accuracy needed to define the flow measurement. The measured Doppler shift for a backscatter measurement, Equation (13.4), for a measured velocity of 1 m/sec would be

$$\Delta f = \frac{2V_m}{\lambda} = \frac{(2)\,1\ \text{m/sec}}{0.5145 \times 10^{6}\ \text{m}} = 3.89 \times 10^{6}\ \text{Hz}$$

Therefore, the measurement accuracy would have to be $\Delta f/f = 6.67 \times 10^{-9}$.

Another difficulty could arise in that the laser wavelength could change slightly with time. Because of these facts, the beating technique is considered to be the most practical at the present time. Another difficulty, common to most laser systems, is that the distance traveled between the local oscillator beam and the scattered light must be within the coherence limits of the laser. That is, light that is emitted and then mixed with light that was emitted at a time such that the difference in the path lengths is greater than the coherence length of the laser may not beat properly. The coherence length for a helium–neon laser may be from 20 cm to several kilometers[(5)]; an argon laser has about a 10-cm coherence length, which may be increased to about 10 m by employing an etalon; and a CO_2 laser's coherence length may be several kilometers. An etalon is an optical device that is generally placed in the laser cavity to better select the polarization and wavelength that the cavity amplifies and emits.

The detector in the local oscillator case has the scattered radiation and the local oscillator radiation impacting (beating) on its surface simultaneously. If the radiation from the local oscillator is given by $\sin 2\pi f_0 t$, the scattered radiation could be given by $\sin[2\pi f_0 t + 2\pi \Delta f_0 t + \phi]$. The current output from the photodetector, i_D, is proportional to the square of the incident radiation; that is,

$$i_D \propto [\sin 2\pi f_0 t + \sin(2\pi f_0 t + 2\pi \Delta f_0 t + \phi)]^2$$

$$= 4 \sin^2\left(\frac{4\pi f_0 t + 2\pi \Delta f_0 t + \phi}{2}\right) \cos^2\left(\frac{2\pi \Delta f_0 t + \phi}{2}\right) \tag{13.5}$$

$$\sin^2 \alpha = \frac{1 - \cos 2\alpha}{2}$$

However, $2\pi f_0$ is a light frequency too high to be followed by the detector; therefore, only the mean value of the first term on the right is seen by the detector, and Equation (13.5) is reduced to

$$i_D \propto 1 + \cos(2\pi \Delta f_0 t + \phi) \tag{13.6}$$

Thus the only frequency sensed by the detector is Δf_0, which is the difference or Doppler frequency.

13.3.1. General Types of Laser Doppler Systems

Brief descriptions of two configurations, local oscillator (reference beam) and dual beam, were presented earlier. These descriptions and uses will be expanded upon individually together with the scanning schemes presently employed.

13.3.1.1. Local-Oscillator-Focused Forward-Scatter (Continuous Wave) LDV Systems. Figure 13.3a gives a schematic of the local-oscillator (reference-beam) focused forward-scatter LDV system. Figure 13.5 presents another view of a forward-scatter reference system. From Equation (13.1) it can be seen that the LDV systems are linear functions of frequency. From this it may be shown that, for a statistically stationary flow, a single LDV system may be used to obtain two- or three-dimensional information.[4]

The method employed to get two- and three-dimensional information from a single LDV system is that statistical averages are made from several independent receiver locations at different times. Reference 4 shows that three independent measurements need to be taken to obtain three-dimensional mean information, six would be necessary to obtain all mean instantaneous cross products (such as $\overline{U'V'}$ and $\overline{U'^2}$, where U' and V' are the velocity fluctuations in the x and y directions, respectively), and nine

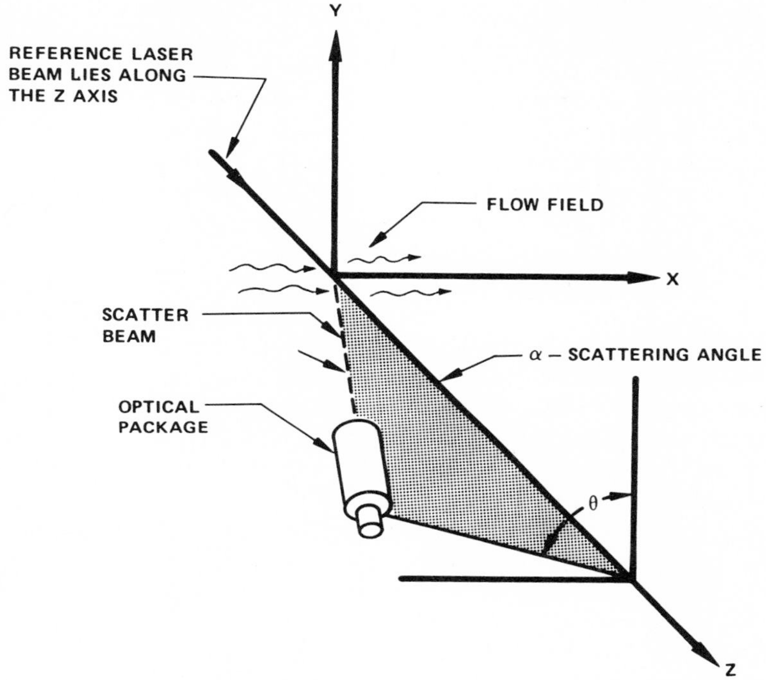

Fig. 13.5. Schematic of typical angular arrangement of a single-reference LDV system.

independent correlation measurements would be necessary for correlation calculations. The method used is simply that the Doppler frequency may be written as a linear function of the three velocity components, i.e.,

$$\lambda\,\Delta f = a_1U(t)+b_1V(t)+c_1W(t) \tag{13.7}$$

where λ is the wavelength of the laser; Δf is the Doppler frequency; a_1, b_1, and c_1 are constants determined by the geometrical configuration; and $U(t)$, $V(t)$, and $W(t)$ are instantaneous velocity components in the x, y, and z directions, respectively. Thus, in a statistically stationary flow one could make three separate mean measurements with the receiver at three independent locations, which would give the experimenter three equations with three unknowns. The three unknowns in this case are $\bar{U}$, $\bar{V}$, and $\bar{W}$, the mean velocity components in the x, y, and z directions, respectively. Higher-order velocity moments would require a greater number of independent observations as previously noted. The experimenter must be careful, however, in that the larger the number of equations needed, the greater is the required precision of the measurements. Figure 13.6 presents a schematic showing a cause of inaccuracy in LDV systems. It is shown that the angle sensed is never a line but some small angle, $d\alpha$, which means that the sensed direction is not really a line but some small cone with a central angle of $d\alpha/2$. This angular error is related to a "Doppler ambiguity" caused by finite transit time.(6) George and Lumley(6) also point out that other errors that arise in measuring with Doppler systems are mean velocity gradients and turbulent velocity gradients across the scattering volume together with normal electronic noise. It should be pointed out that any system with a finite sensing volume is subject to errors resulting from gradients across the sensing volume, from fluctuations smaller than the sensing volume, and from electronic noise. These types of errors can become significant if one attempts to carry the accuracy of the measurement too far; this could be the case in trying to define all second-moment correlations of a three-dimensional flow field with a single LDV. As previously noted this would require nine independent measurements.

One of the best scanning techniques for a single LDV system is to simply set the optics for a particular range and then move the entire LDV system, which, in turn, moves the sensing location.

It is suggested that two or three receivers be employed simultaneously to measure two- or three-dimensional flows. Figure 13.7 presents a schematic of a three-dimensional LDV system that was used at Marshall Space Flight Center.(7) (Note that three noncoplanar receivers are employed simultaneously.) The Doppler output from each detector may be written as a linear function of $U(t)$, $V(t)$, and $W(t)$, thus directly giving three equations with three unknowns, which may be solved on line giving $U(t)$, $V(t)$, and $W(t)$ directly.(8) If $\Delta f_1(t)$, $\Delta f_2(t)$, and $\Delta f_3(t)$ are the Doppler frequencies sensed at detectors 1, 2, and 3 at time t, the $U(t)$, $V(t)$, and $W(t)$ components

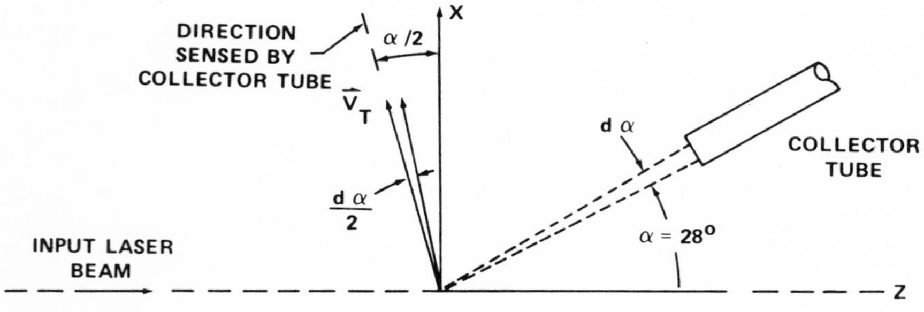

Fig. 13.6. Angular accuracy limitations in an LDV system.

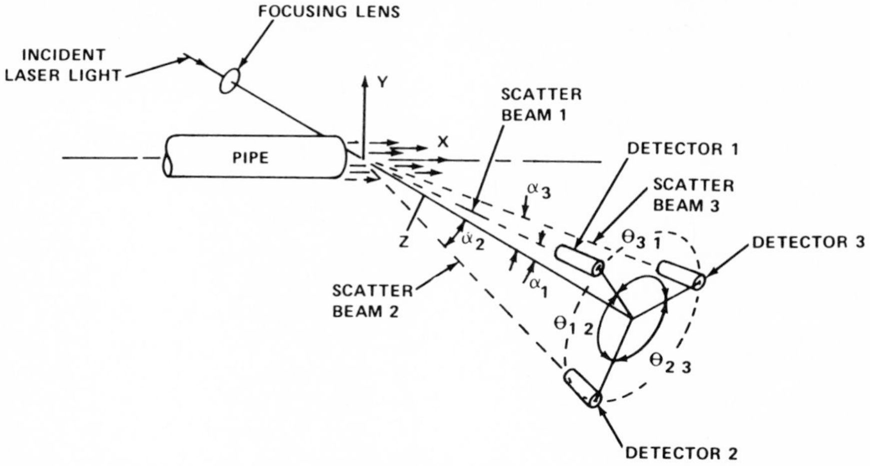

Fig. 13.7. Schematic of a three-dimensional LDV and its alignment relative to a pipe flow. Notes: the incident laser beam is in the x–z plane and at right angles to the pipe centerline. The velocity components U, V, and W are in the directions x, y, and z, respectively. Detector 3 is in the x–z plane.

can be written as

$$\begin{aligned} U(t) &= A_1\Delta f_1(t) + A_2\Delta f_2(t) + A_3\Delta f_3(t) \\ V(t) &= B_1\Delta f_1(t) + B_2\Delta f_2(t) + B_3\Delta f_3(t) \\ W(t) &= C_1\Delta f_1(t) + C_2\Delta f_2(t) + C_3\Delta f_3(t) \end{aligned} \tag{13.8}$$

where A_i, B_i, and C_i relate the velocity components to the respective Doppler shifts. Figure 13.8 presents a schematic of the electronic network for reducing the Doppler signals to the U, V, and W components together with comparing the U component of the LDV system with the U component measured with a hot-wire anemometer. Figure 13.9 presents an oscilloscope trace comparing the U component measured by a hot wire and the U component measured with the three-dimensional LDV system previously described. The laser used in the previously mentioned tests was an argon laser emitting at the 0.5145-μ wavelength.

The flow measured was at the exit of a fully developed pipe. The scanning technique employed was to crank the entire LDV system, thus moving the focal (sensing) volume to a desired location. The results of the test are presented in Reference 7. The results indicate excellent agreement when compared to conventional systems. The three-dimensional LDV system has the advantage over conventional systems that the experimenter

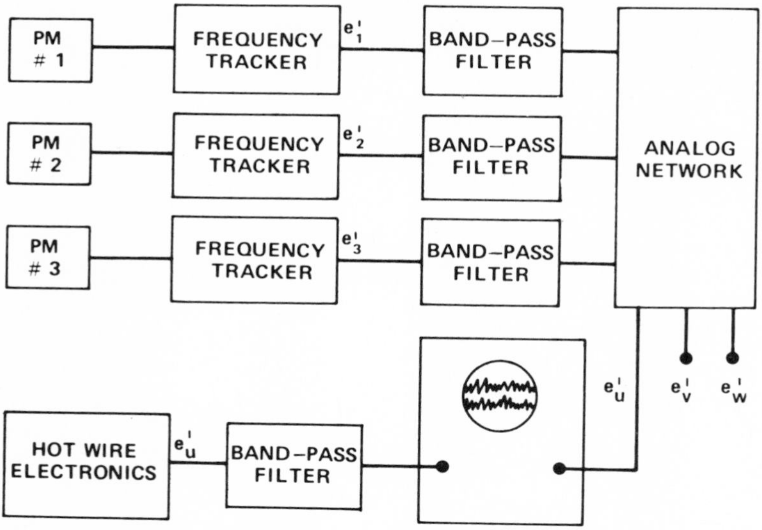

Fig. 13.8. Schematic diagram of electronic network for comparing the LDV and hot-wire signals. Note: e_i is a voltage representing the ith quantity.

obtains on line the three-dimensional velocity field at the focal volume. The focal volume in this experiment was found to be approximately 0.08 mm in diameter and approximately 0.27 mm long.(7) These dimensions are near those found in conventional hot-wire anemometry. Figure 13.10 is an on-line three-dimensional velocity display simultaneously with a hot-wire output. The hot wire was placed approximately 1 mm downstream from the sensing volume of the laser system.

13.3.1.2. Local Oscillator for Focused Backscatter (Continuous Wave) LDV Systems. The formula for the Doppler frequency shift for the backscatter mode is given by Equations (13.1)–(13.4). For the case where

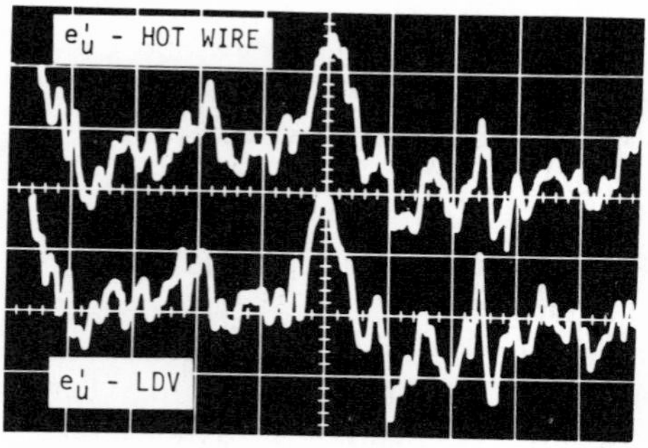

Fig. 13.9. Comparison of U component measured by an LDV system and a hot wire. Note: e'_U is the voltage representing the U component of velocity.

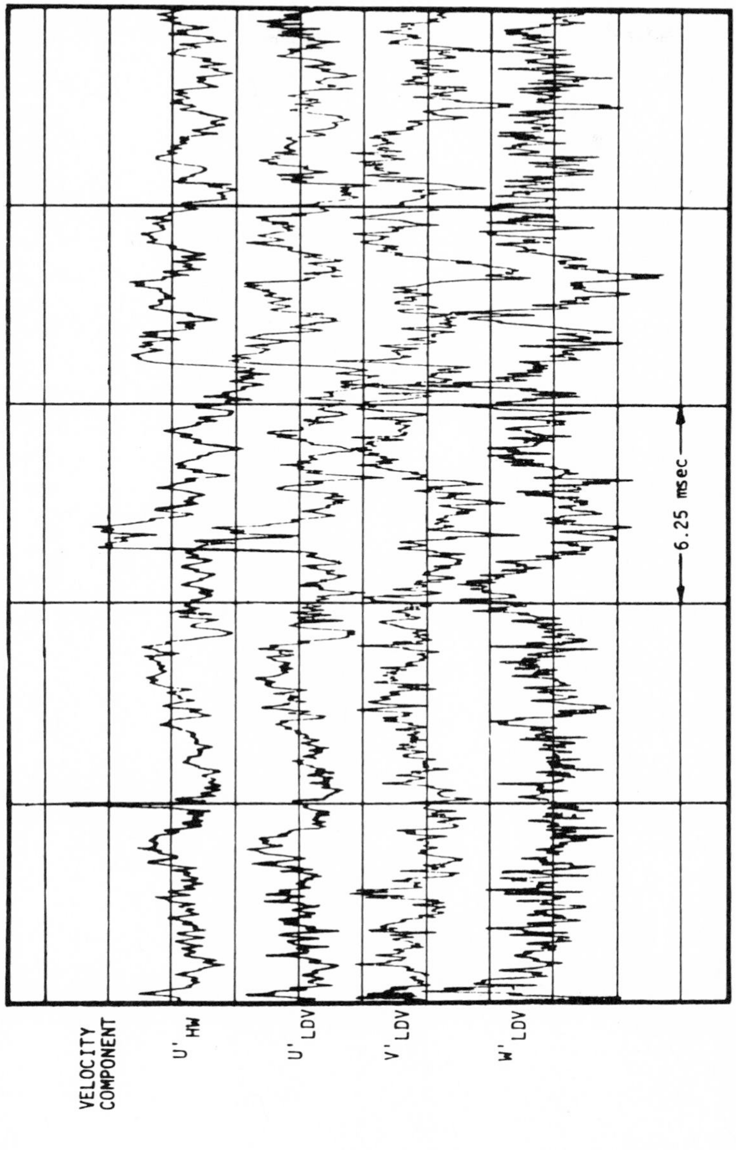

Fig. 13.10. Simultaneous comparison of hot-wire and LDV turbulent fluctuations at the center of a turbulent pipe flow. Pipe Reynolds number 150,000; filter bandpass 4 Hz to 10 kHz.

the receiver is separated from the source (bistatic), exactly the same approaches may be used as were employed with the forward-scatter local oscillator. It should be noted, however, that in many cases the size of particle tracers scattering the light is of the same order of magnitude as the wavelength of light or larger. When this condition prevails, the scattering is termed Mie scattering, and the scattered intensity from the particle is very much dependent on angle and is the most intense in the forward direction. The ratio of the intensity of forward scatter to backscatter (with the same angle from the line of incident radiation) may be two orders of magnitude. Since the techniques employed for measuring the Doppler signal are the same for the forward bistatic and backscatter bistatic, this section will address the systems employing coaxial (monostatic) backscatter (i.e., the receiver and source optics are the same). In the coaxial backscatter mode, the Doppler shift is defined by Equations (13.4). There are several advantages to using a coaxial (monostatic) focused backscatter system. Some of these are the following: (1) The same optics are used for focusing the transmitted and received radiation; (2) since the same optics are used for transmitting and receiving there is no need to align two beams to cross in space (dual-beam system) nor to align the receiver focal volume with the local oscillator focal volume (bistatic local oscillator); and (3) the Doppler

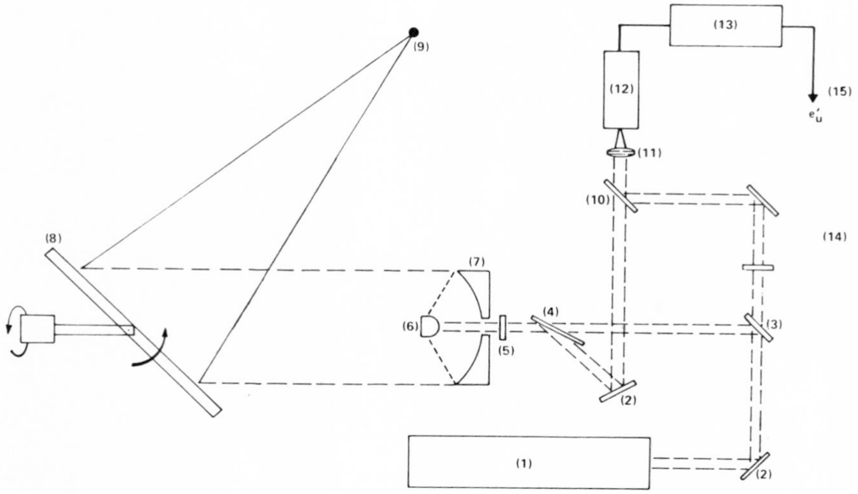

Fig. 13.11. Coaxial laser Doppler system. (1) CO_2 laser; (2) first surface mirror; (3) Ge beamsplitter; (4) Ge Brewster window; (5) quarter wave plate; (6) secondary mirror; (7) primary mirror; (8) beam-directing mirror; (9) focal volume; (10) recombining Ge beamsplitter; (11) collecting lens; (12) detector; (13) tracker; (14) Ge half-wave plate; (15) voltage out representing axial velocity.

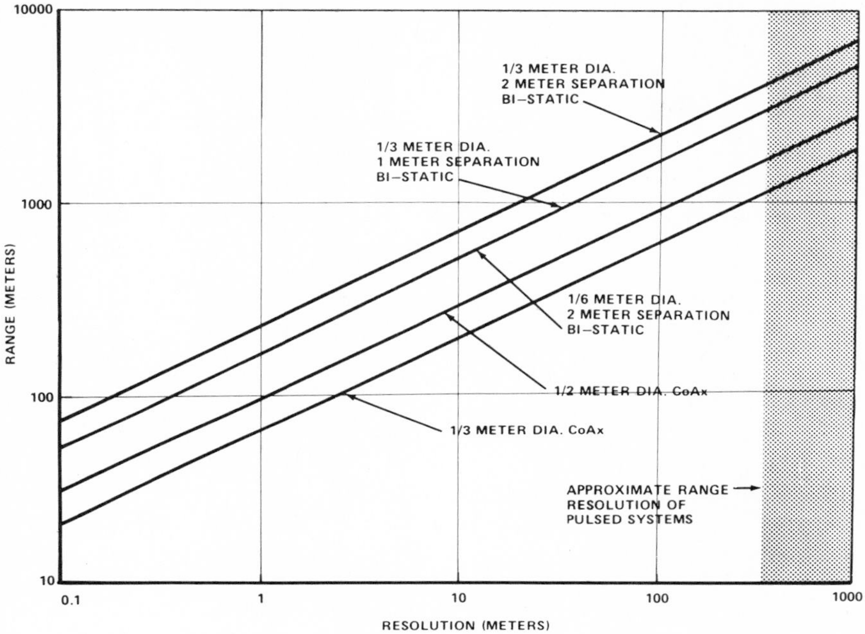

Fig. 13.12. Range resolution of CO_2 LDV systems.

shift comes from the component of velocity along the axis of the beam, and, as a result, there is no angular dependence of the Doppler shift.

Some disadvantages of the coaxial LDV system are as follows: (1) Spatial resolution is poorer than in the bistatic case and (2) more sophistication in optics is generally required. Figure 13.11 presents a schematic of a coaxial focused backscatter LDV system with scanning capabilities which will be explored later.

Figure 13.12 presents the comparison of the resolution of coaxial, bistatic, and pulsed CO_2 LDV systems.

Because of the long coherence length of the CO_2 laser, it has been used in the development of coaxial LDV systems for measuring atmospheric flows.[9,10]

Figures 13.13 and 13.14 present on-line comparisons of atmospheric velocities by a CO_2 coaxial LDV system and propeller anemometer at ranges of 60 and 200 m. It should be noted that the volume sensed by the cup anemometer and the CO_2 coaxial LDV system are, in general, different. The components of wind velocity measured with the anemometers and LDV system were the same.

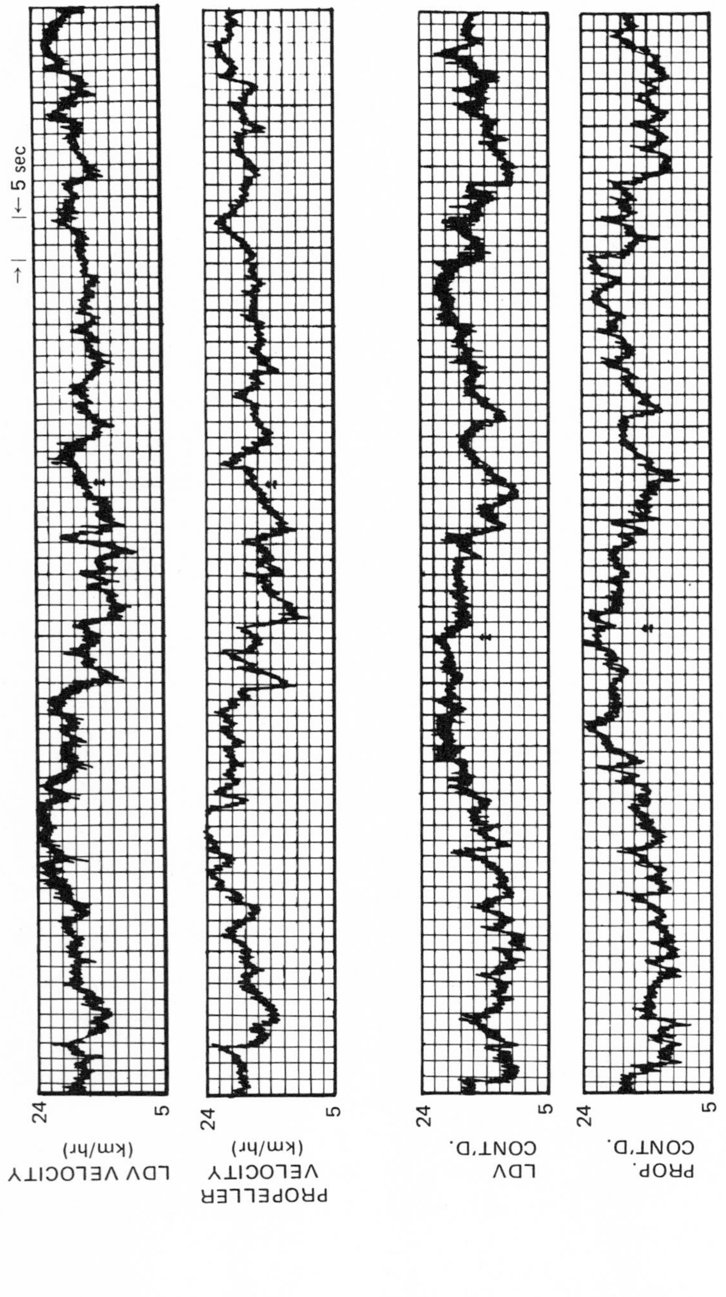

Fig. 13.13. Comparison of LDV radial velocity with a propeller anemometer at a range of 60 m.

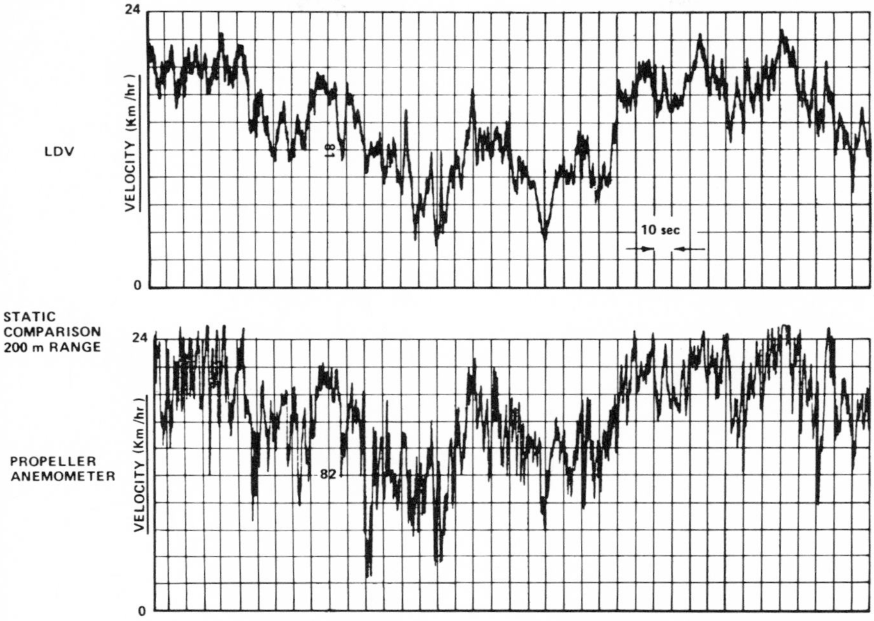

Fig. 13.14. Comparison of LDV radial velocity with a propeller anemometer at a range of 200 m.

Since the coaxial system gives only the axial component of velocity, three such systems could be slaved together to produce outputs that could give the three-dimensional velocity field at a given focal volume in space. This would be very similar to what was done for the forward-scatter three-receiver case previously discussed.

Another configuration of the coaxial CO_2 LDV system that has been employed at Marshall Space Flight Center is a conical scanning scheme.[9,10] A single electromagnetic system with a conical scanning scheme has been used previously with radar.[11] Figure 13.15 presents a schematic of a conical scan and the output from a conical scan.

Figure 13.16 presents a comparison of the horizontal wind measured with a conical scanning CO_2 coaxial LDV and a cup anemometer. The conical scan configuration appears to have great potential for measuring atmospheric velocities near the ground. The useful range of such a system is probably up to 500 m.

Ranging the focal volume in the coaxial system is accomplished by moving the secondary mirror. The ranging of the focal volume from 20 to 300 m can easily be performed in less than 1 sec. Presently MSFC is using coaxial CO_2 LDV systems and scanning in range and elevation while

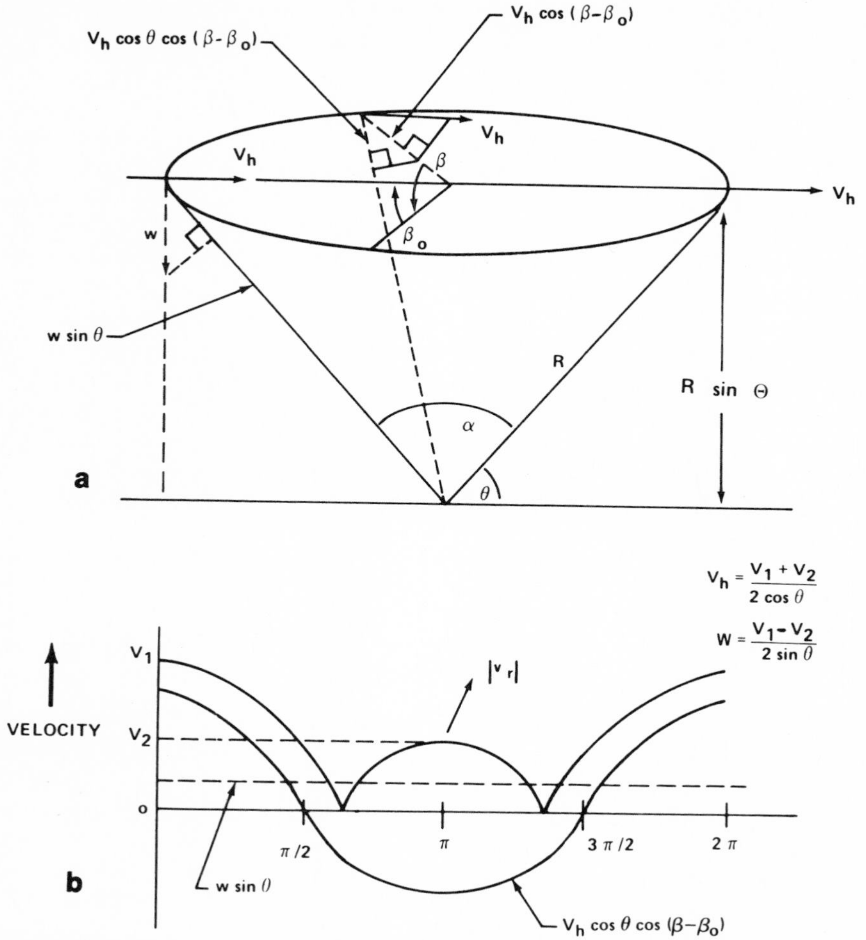

Fig. 13.15. Schematic of a conical scan and output from a conical scan (VAD ≡ Velocity Azimuth Display.) (a) VAD scan configuration. (b) Azimuth angle dependence of measured velocity component.

measuring atmospheric phenomena (such as airplane vortex location, wind fields, and dust devils).

13.3.1.3. Pulsed Coaxial LDV Systems. The pulsed coaxial LDV systems are similar to the coaxial systems previously mentioned except that range location is performed by gating the Doppler return. Figure 13.12 presents the approximate location where range resolution for the pulsed system becomes better than the coaxial or bistatic continuous wave (CW) configurations. MSFC has been developing a pulsed coaxial CO_2 LDV system to detect clear air turbulence. For the clear air detection system, the pulse length is between 4 and 10 μsec and is pulsed at a rate of 140–160 times per second.

13.3.1.4. Dual-Beam (CW) LDV Systems. The dual-beam system employs the use of two beams crossing in space establishing an apparent fringe pattern at the crossing point of the beams. When a particle passes through the fringe pattern, light is scattered in all directions, the frequency of which is determined by Equation (13.1). A general dual-beam configuration is shown in Figure 13.3. The dual-beam systems have the advantage that when the beams are crossed the Doppler frequency may be measured from any location in space. That is, the Doppler shift is a function of the angle of intersection of the beams. However, it should be noted that scattering intensity and, thus, intensity received at a photodetector may be very much dependent on spatial location of the detector. In general, the forward scatter is much stronger than the backscatter, as is indicated from Mie scattering theory. Another advantage of the dual-beam system is that it measures directly the transverse velocity component, whereas the coaxial system

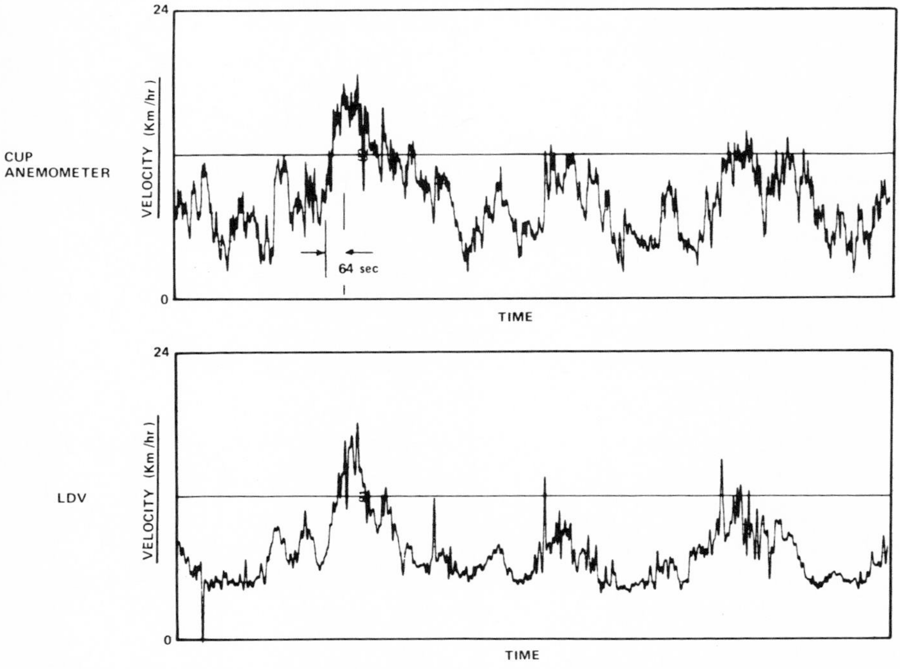

Fig. 13.16. Comparison of horizontal wind measured with a CO_2 coaxial LDV conical scan system and a cup anemometer at an elevation of 21 m (cone angle = 90°).

measures the axial component. These systems are generally employed in the laboratory where the range from the laser to the crossing point (fringe area) is small (generally less than 2 m). One of the disadvantages of the system is the fact that as you increase range it becomes increasingly difficult to align the system so that there is an actual crossing of the two beams. Also since the optical paths of the two systems are different, the local index of refraction along each beam path could change and cause the beams to wander slightly in time. Similar problems would occur if an attempt was made to focus a receiver and a laser at the same point in space. Because of these difficulties, it is easier to employ a coaxial system for atmospheric use at the present time.

13.3.1.5. Single-Particle Dual-Beam LDV Systems. Single-particle dual-beam Doppler systems are presently being developed at Ames Research Center, Langley Research Center, and Arnold Engineering Developmental Center as well as commercially. The concept is to examine the statistics of each particle that passes through the fringe pattern. Selective sampling electronics are used to eliminate signals produced by particles that do not pass directly through the fringe pattern or signals produced by multiple particles passing through the fringe pattern simultaneously. Since no continuous signal is achieved by this method, each particle velocity statistic is used to develop a probability density distribution. The statistics of the flow field are then calculated from the probability density distribution. That is, $\overline{U^n} = \int_{-\infty}^{\infty} U^n P(U)\, dU$, where U is the velocity of the particle, $P(U)$ is the probability distribution of the velocities, and the overbar indicates the mean. In the continuous output cases, time averages would be used. The advantage of a single-particle LDV system for wind-tunnel work is that no seeding of particles should be necessary. That is, the naturally occurring aerosols should be sufficient to operate the system. In wind-tunnel flows, seeding of the flow may be necessary to obtain a continuous signal from the sensing volume for continuous types of LDV systems. In water or in the atmosphere, there are generally sufficient particles to produce a continuous signal from a typical LDV focal volume.

13.3.2. Typical Wavelengths and Common Uses of Lasers Presently in Use in LDV Systems

(1) Helium–Neon:
  (a) Wavelength: 0.6328 μ (visible).
  (b) Uses: Dual-beam systems for short-range systems.
      Dual-beam single-particle short-range systems.

(2) Argon:
  (a) Wavelengths: 0.5145 and 0.4880 μ (visible). (Note: There are approximately eight usable lines in an argon laser. These two are the strongest.)

(b) Uses: Dual-beam systems for short-range systems. Dual-beam single-particle short-range systems. Local oscillator bistatic systems for short range.

(3) CO_2:

(a) Wavelength: 10.6 μ (infrared).

(b) Uses: Local oscillator bistatic and coaxial long-range (atmospheric to 500 m). Dual-beam medium range (10 m). Pulsed coaxial long range (atmospheric to 10 km).

13.3.3. Conclusions and Recommendations Concerning Laser Doppler Systems

(1) The laser Doppler system measures the velocity of particle tracers embedded within the flow. The particle velocity is then related to the fluid velocity. It is anticipated in many cases that the particles move with the same velocity as the fluid surrounding the particle, such as was shown in Figure 13.10.

(2) The basic laser Doppler systems presently in use are the following:

(a) Local oscillator, continuous wave.

(b) Local oscillator, pulsed.

(c) Dual-beam, continuous-wave–continuous-signal.

(d) Dual-beam, continuous-wave–single-particle-realization.

(3) Typical scanning methods are the following:

(a) Focusing the Doppler system at a point and then moving the entire laser Doppler system to move the focal point.

(b) With the coaxial-type system, range scanning is performed by manually or electronically moving the secondary mirror. A finger-type scan can be performed with this configuration. By continuously varying the pointing angle along with the range, a plane may be scanned.

(c) Using the coaxial system focused at a given range, the following scan patterns have been used:

(i) Conical scan—A rotating mirror moves the focal volume in a circular path in space. The three-dimensional velocity field at a given elevation is measured with this method. By ranging the focal volume as the beam is rotating, it is also possible to get a spiral conical scan that could give an altitude velocity profile.

(ii) Two-point scan—A mirror flips back and forth sending the focal volume back and forth between preselected points. The sum and difference of the velocities measured may be used to compute a longitudinal and lateral component of velocity. (This scheme has not proved very successful in the past.)

(iii) Ranging focal location with moving mirror—A mirror is set on an air track such that the mirror is moving at a constant velocity while the laser beam strikes the mirror (before the focal volume). The moving-mirror scheme thus moves the focal volume and gives the same effect as though the system were moving relative to the particles. MSFC used this scheme to investigate artificial fogs in a fog chamber belonging to Ames.

(4) The dual-beam systems are used frequently for wind-tunnels flows where a transverse one-dimensional velocity is needed.

(5) A three-receiver forward-scatter local oscillator LDV system has been used to measure (on line) the turbulent three-dimensional velocity field of a fully developed gaseous pipe flow.

(6) Direct comparison of velocities measured with a hot-wire and laser Doppler system has been performed and shown to be in good agreement. [This does not mean, however, that this is always the case. The experimenter has to be confident that the particle (tracer) velocity is the same as the velocity of the fluid around the particle.]

(7) A laser Doppler system without a frequency translator has a direction ambiguity of 180°. The Doppler frequency is the absolute difference between the frequency of the laser and scattered radiation. The use of a frequency translater allows the sense of the direction to be determined.

(8) In many wind-tunnel-type gas flows, seeding of particles may be necessary to retrieve a continuous signal from the LDV focal volume.

(9) The use of low-pass filters is recommended to eliminate high-frequency noise.

(10) Laser Doppler systems have shown excellent promise as a velocity-measuring tool. The laser systems are superior to acoustic systems for defining small-resolution volumes.

13.4. Acoustic Doppler

The acoustic Doppler, like the laser Doppler, works on the basic principle that radiation incident on a moving tracer is shifted in frequency in accordance with Equations (13.1)–(13.4). The tracers in the naturally occurring atmosphere that scatter the acoustic radiation are velocity and temperature gradients. That is, the acoustic Doppler systems measure the velocity at which velocity and temperature gradients are translated or convected through the atmosphere. The equation for the scatter of sound in dry air is given by Beran[12] as

$$dG = \frac{2\pi K^4 V \cos^2\beta \, d\Omega}{\{[E(K)/C^2] \cos^2 \frac{1}{2}\beta + \phi(K)/4T^2\}^{-1}} \tag{13.9}$$

where dG is the fraction of the incident acoustic power that is scattered by irregularities in volume, V (the scattering volume), through an angle β into a cone of solid angle $d\Omega$. The spectral intensities of the velocity and temperature fluctuations are given by $E(K)$ and $\phi(K)$, respectively, and the speed of sound and mean temperature are given by C and T, respectively. Equation (13.9) shows that the scattered acoustic radiation attributed to the velocity inhomogeneities is dependent on the scattering angle, $\beta(180-\theta)$, and the scattering angle divided by 2, $\beta/2$. The latter case shows that the velocity inhomogeneities produce no scatter in a complete backscatter mode, $\theta = 0$. Using the fact that pure backscatter contains no information from velocity inhomogeneities, Beran points out that an acoustic Doppler in pure backscatter mode has the disadvantage that some areas of the atmosphere may not have sufficient temperature gradients to produce a good Doppler return.

13.4.1. Types

The acoustic Doppler may be broken into two simple classes: pulsed and continuous wave.

13.4.1.1. Continuous Acoustic Doppler. The continuous acoustic systems, as the name implies, are operated using a continuous source. To use a continuous source, it is desirable to isolate a particular sensing volume by focusing just as was done with the laser Doppler systems. The focusing could conceivably be performed by employing focusing for the transmitter and receiver or possibly just focusing the receiver on some portion of the transmitted beam. A coaxial (pure backscatter) continuous acoustic Doppler system appears difficult at present as the returning and emitted acoustic radiation may not easily be separated as was the case with the laser Doppler.

Presently the most feasible configuration for a CW acoustic Doppler system is the bistatic configuration where the source and receiver are at spacially separated locations. Like the laser Doppler, more than one receiver could be employed to obtain two- and three-dimensional information. A schematic of a three-dimensional CW acoustic Doppler is shown in Figure 13.17.

The acoustic Doppler systems work in frequency ranges easily handled electronically (generally 400–4000 Hz); therefore, it is not necessary to use

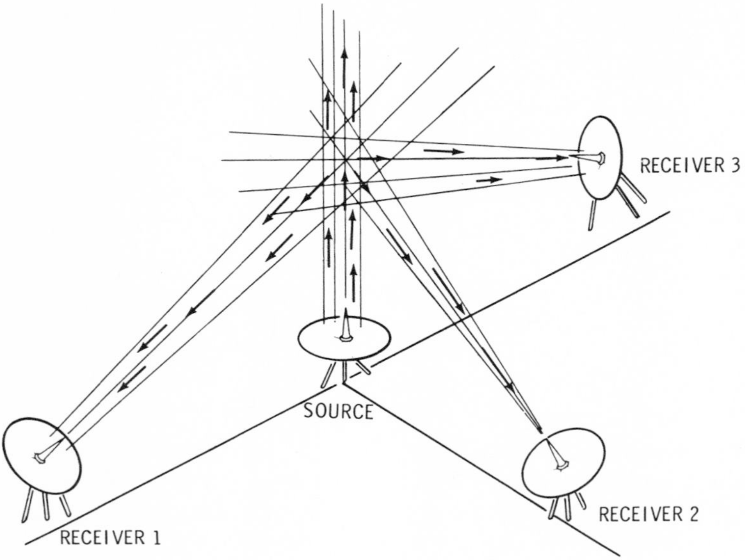

Fig. 13.17. Schematic of a three-dimensional CW acoustic Doppler.

the heterodyning process as was necessary with the laser Doppler systems. Instead it is desirable to measure the return frequency directly, which allows direct determination of the sense of the flow. In the laser Doppler case, it was pointed out that beating the scattered radiation with the source radiation created a beat frequency with an ambiguity of 180° in flow direction (it is noted that a frequency translator employed in the source radiation corrects the ambiguity problem). To date it has been difficult to sufficiently isolate the receiver of the CW acoustic Doppler system to eliminate source noise from swamping out the scattered radiation.

13.4.1.2. Pulsed Acoustic Doppler. The pulsed acoustic Doppler system is able to avoid the contamination of source noise by simply not sampling until after the source noise has passed the receiver. That is, since the path length taken by the scattered acoustic radiation is longer than the path length from the source to the receiver, the transit time for the scattered radiation to reach the receiver is longer than for the source noise. The elimination of source noise is shown in Figure 13.18.

As with the CW acoustic Doppler, the pulsed acoustic Doppler benefits from using a bistatic configuration instead of a coaxial system. This is due to the fact that in the coaxial mode only radiation scattered by temperature gradients is measured while in the bistatic mode scattering from velocity and temperature gradients contributes to the received radiation.[13–15] A pulsed

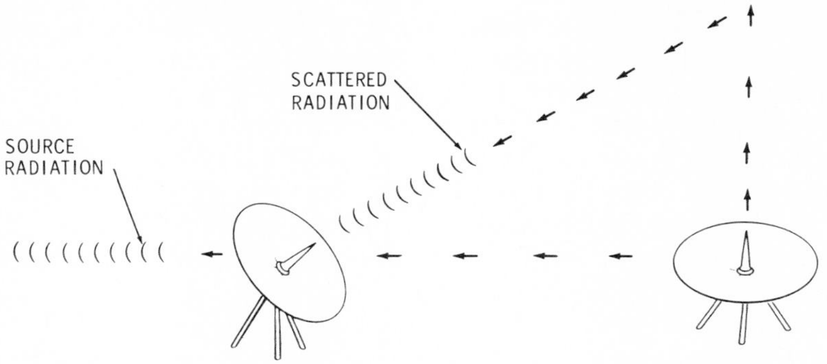

Fig. 13.18. Schematic of source noise elimination by pulsed acoustic Doppler system.

system with a configuration similar to that shown in Figure 13.17 could be used for three-dimensional velocity measurements. If one considers the average vertical velocity to be small, the horizontal two-dimensional velocity could be measured by using only two of the noncoplanar receivers shown in Figure 13.17. A system, as was just described, has been used in atmospheric research by Beran and Clifford of NOAA.[16] Figure 13.19 (taken from Reference 16, courtesy of Dr. D. W. Beran of NOAA, Boulder, Colorado) presents a comparison of the horizontal atmospheric wind sensed at an elevation of 150 m using a two-receiver pulsed acoustic Doppler system and a tethered kytoon (small blimp called Boundary Layer Profiler, BLP) which has sensors for measuring horizontal wind. The acoustic radiation was pulsed from a vertically pointing source. Two receivers forming a

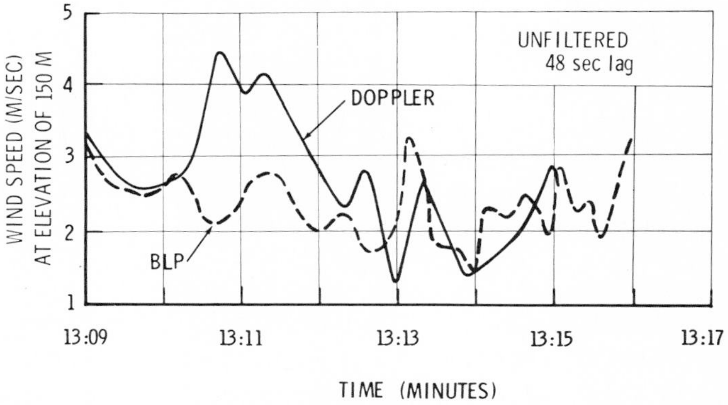

Fig. 13.19. Comparison of wind velocity measured with a pulsed acoustic Doppler and with a boundary-layer profiler (courtesy of Dr. D. W. Beran, NOAA, Boulder, Colorado).

right angle with the source were employed. The elevation of the sensing volume was determined by selectively gating the return such that only scattering from a particular elevation would be arriving at the receiver during the sensing time. Gating the return in this way permits the receiver to be nonfocused, merely accepting acoustic radiation from a general direction.

The NOAA systems as described by Beran[17] are presently employing 3-m parabolic dishes for receivers and sources. The parabolic configuration is used to develop a near parallel acoustical emission. The emission is 50–60 W. Present plans are to go to a 4×4 m array of 144 speakers with 500–600 W. The present resolution volume is approximately 30×30 m using a 0.5-sec pulse every 8 sec. The pulse duration may be shortened if there is sufficient power. The source frequency is generally 400–4000 Hz.

The use of acoustic Doppler systems is also being investigated for measuring blood flow.[18–20] The scattering source of interest in this case is the red blood cell, $\sim 10\,\mu$ in diameter. The ultrasonic Doppler is used to nondestructively penetrate the skin, tissue, and blood vessels, which also scatter sound at the interfaces but, since no motion is occurring, do not shift the frequency of the scattered acoustic radiation. Together with the red blood cells, the white blood cells and platelets will also be scattering sources but, owing to their smaller size and smaller percentage of mass, the white blood cells and platelets do not contribute significantly to the received Doppler return.[20] To analyze the depth, a coaxial pulsed system is generally used. Figure 13.20 presents a schematic of such a system. Comparison of

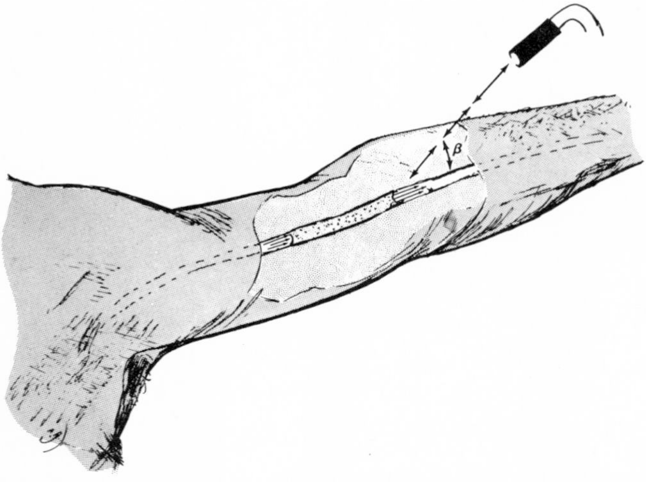

Fig. 13.20. Cut-away showing ultrasonic Doppler scheme for investigating blood flow.

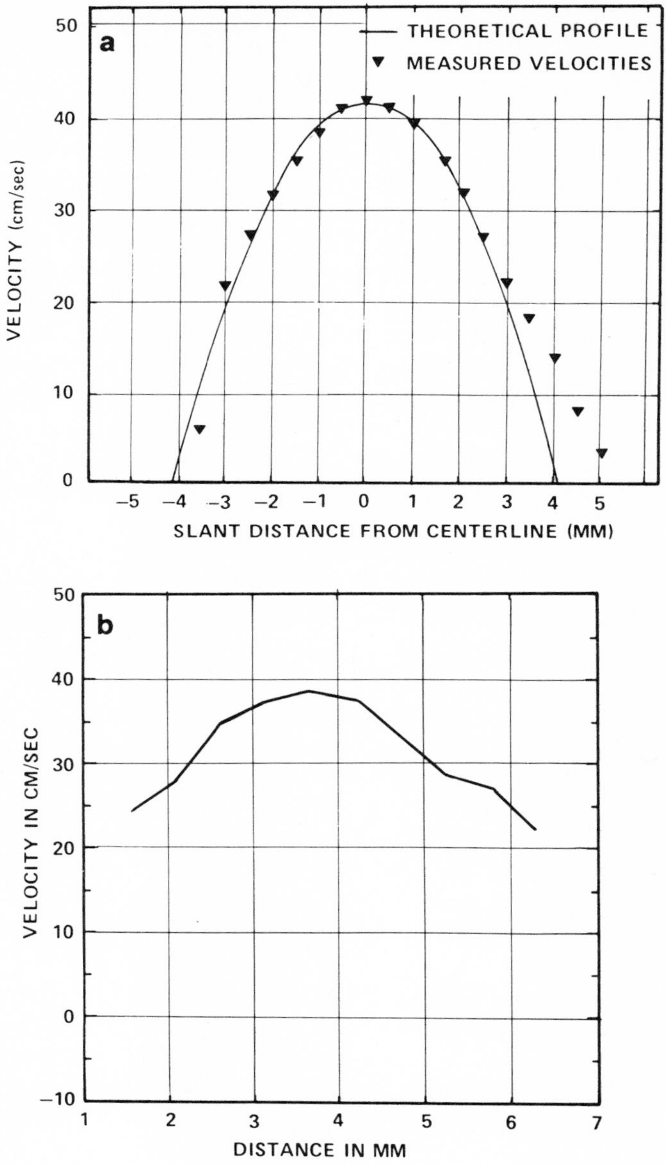

Fig. 13.21. Velocity profiles measured with a pulsed ultrasonic Doppler system (courtesy of Dr. M. Wells, Colorado State University). (a) Velocity profile in a 7.2-mm tube measured with an ultrasonic Doppler system. (b) Velocity profile in a horse's artery measured with a pulsed ultrasonic Doppler system.

the expected flow pattern versus the ultrasonic Doppler measured flow pattern for poiselle flow in a 7.2-mm-diam tube is presented in Figure 13.21a (courtesy of Dr. M. Wells, Colorado State University). Figure 13.21b presents a velocity profile in a horse's artery measured with an ultrasonic Doppler system (courtesy of Dr. M. Wells, Colorado State University).

To get adequate spatial resolution to interrogate flow in veins and arteries, ultrasonic frequencies are employed. Frequencies used range from 8 to 20 MHz, and may go as high as 40 MHz. Some problem areas associated with the use of a coaxial pulsed ultrasonic Doppler system for investigating blood flow are:

(1) the evaluation of the angle β (Figure 13.20); and

(2) getting a good measure of the cross-sectional area of the vessel to use with the velocity to estimate mass flow rate. The resolution of the system may not be small compared to the channel diameter.

13.4.2. Conclusions and Recommendations Concerning Acoustic Doppler Systems

(1) Presently the performance of the pulsed acoustic Doppler systems appears superior to the CW acoustic Doppler.

(2) Acoustic Doppler techniques appear to be viable candidates for the remote sensing of the motion of the temperature and velocity gradients in fluid motions.

(3) The ability of the temperature and velocity inhomogeneities to follow the mean motion is the criterion employed to interpret the results of acoustic Doppler system for use as a mean wind detector.

(4) Pulsed ultrasonic Doppler techniques are presently being used to investigate blood flow in animals. This appears to be an excellent area for future research in pulsed and CW acoustic Doppler because of the nondestructive penetrability of animal tissue by acoustics. The scattering source producing a Doppler shift in this case is the cellular flow within the arteries and veins.

(5) As with the laser Dopper systems the acoustic Doppler measures the velocity of a scattering source. The velocity of the scattering source is then studied by itself or is interpreted as a tracer of the surrounding medium and the motion and statistics of the tracers velocity are used to infer the velocity and statistics of the medium in which the tracer is embedded.

Notation

a, b, c	Constants	$E(k)$	Velocity spectral intensity
C	Speed of sound	f	Frequency
$\bar{e}_i$	Unit vector in the ith direction	G	Acoustic power

i	Photodetector current	V_m	Measured velocity
n	Index of refraction	$\bar{V}$	Velocity vector
t	Time	x, y, z	Cartesian coordinates
T	Temperature	λ	Wavelength
U, V, W	Instantaneous velocity components	β, θ	Angles
		$\phi(k)$	Thermal spectral intensity

References

1. Yeh, H., and Cummins, H. Z., Localized fluid flow measurements with an He–Ne laser spectrometer, *Appl. Phys. Lett.* **4**, 176–178 (1964).
2. Huffaker, R. M., Fuller, C. E., and Lawrence, T. R., Application of laser Doppler velocity instrumentation to the measurement of jet turbulence, International Automotive Engineering Congress, Society of Automotive Engineers, Detroit, Michigan, paper No. 690266 (1969).
3. Kelton, G., and Bricout, P., Wind velocity measurements using sonic techniques, *Bull. Am. Meteorol. Soc.* **45**, 571–580 (1964).
4. Cliff, W. C., and Fuller, Charles, E., III, Measurement capabilities of a one-dimensional LDV system, NASA report No. TM X-64774 (1973).
5. Lawrence, T. R., Lockheed, Huntsville, Alabama, private communication (1975).
6. George, W. K., and Lumley, J. L., The laser-Doppler velocimeter and its application to the measurement of turbulence, *J. Fluid Mech.* **60**, 321–362 (1973).
7. Fuller, C. E., III, Cliff, W. C., and Huffaker, R. M., Three-dimensional laser Doppler velocimeter turbulence measurements in a pipe flow, NASA report No. CR-129017 (1973).
8. Cliff, W. C., Fuller, C. E., III, and Sandborn, V. A., Simultaneous comparison of turbulent gas fluctuations by laser Doppler and hot wire, *AIAA J.* **11**, 748–749 (1973).
9. Cliff, W. C., and Huffaker, R. M., Application of a single laser Doppler system to the measurement of atmospheric winds, NASA report No. TM-X-64891 (1974).
10. Lawrence, T. R., Krause, M. C., Craven, C. E., Morrison, L. K., Thomson, J. A. L., Cliff, W. C., and Huffaker, R. M., A laser Doppler system for the remote sensing of boundary layer winds in clear air conditions, Preprints of the 16th Radar Meteorology Conference (1975).
11. Lhermitte, R. M., and Atlas, D., Precipitation motion by pulse Doppler radar, Proceedings of the 9th Weather Radar Conference, Boston, Massachusetts (1961).
12. Beran, D. W., Remote sensing wind and wind shear system, report No. FAA-RD-74-3, Department of Transportation, Federal Aviation Administration, Washington, D.C., 20591 (1974).
13. Beran, D. W., Little, C. G., and Willmarth, B. C., Acoustic Doppler measurements of vertical velocities in the atmosphere, *Nature* **230**, 160–162 (1971).
14. Browning, K. A., Beran, D. W., Quigley, M. J. S., and Little, C. G., Capabilities of radar, sonar, and lidar for measuring the structure and motion of the stably stratified atmosphere, *Boundary Layer Meteorol.* **5**, 195–200 (1973).
15. Little, C. G., Derr, V. E., Kleen, R. H., Lawrence, R. S., Lhermitte, R. M., Owens, J. C., and Thayer, G. D., Remote sensing of wind profiles in the boundary layer, ESSA report No. ERL 168-WPL 12.
16. Beran, D. W., and Clifford, S. F., Acoustic Doppler measurements of the total wind vector, preprints, Second Symposium on Meteorological Observations and Instruments, San Diego, California, AMS (1972), pp. 100–110.

17. Beran, D. W., NOAA, Boulder, Colorado, private communication (1975).
18. Baker, D. W., Pulsed ultrasonic Doppler blood flow sensing, *IEEE Trans. Sonics Ultrason.* **SU-17**, 170–185 (1970).
19. Histand, M. B., Miller, C. W., and McLeod, F. D., Transcutaneous measurement of blood velocity profiles and flow, *Cardiovas. Res.* **7**, 703–712 (1973).
20. Wells, M. K., Colorado State University, private communication (1975).

CHAPTER 14

Monte Carlo Turbulence Simulation

G. H. FICHTL, MORRIS PERLMUTTER, and WALTER FROST

14.1. Introduction

The subject of this chapter is the simulation of turbulence with emphasis on atmospheric turbulence simulation. What does the word simulation mean? In the context of this chapter it is interpreted to be the generation of a random process that possesses prescribed statistical attributes of turbulence. The attributes consist of a restricted set of turbulence statistics determined either experimentally or theoretically (usually the former). If the simulation scheme yields random functions that upon statistical data reduction yield the prescribed attributes of atmospheric turbulence, we accept the simulation scheme as being a "good" one and it can thus be used to predict the behavior of turbulence and its interaction with other media, i.,e., airplanes, bridges, wind-generated water waves, etc., within the limitations imposed by the finite number of selected attributes and the underlying assumption involved. We shall refer to the simulation scheme that yields random functions that satisfy prescribed statistical attributes of turbulence as being a model. We use the adjective "Monte Carlo" with the word "simulation" in the sense of

G. H. FICHTL • Marshall Space Flight Center, Huntsville, Alabama
MORRIS PERLMUTTER and WALTER FROST • The University of Tennessee Space Institute, Tullahoma, Tennessee 37388

Kendall and Buckland,[1] namely, "to construct an artificial stochastic model of the mathematical (or rather physical*) process and the *to perform sampling experiments upon it.*"

The approach to turbulence simulation described herein is based on the principles of control-system theory. The idea is to develop, synthesize, construct, or "conjure up" a system characterized by a system function such that upon "exciting" the system with prescribed noise processes the output of the system is a realization of a random process that possesses the desired attributes of the physical process that was to be simulated.

Monte Carlo simulation has been growing rapidly. It is being used to solve problems in a wide variety of areas. These include problems in scheduling, economic systems, rarefied gases, control systems, turbulent flows, radiation, etc. We intend here to develop a unifying basis for simulation using a control-system approach with special emphasis in the field of meteorological turbulence.

Initially, we will describe various techniques of generating random signals with desired statistical characteristics. This will include both Gaussian and non-Gaussian signals. The results will then be extended from one-dimensional random signals to two-dimensional random signals using the same concepts. Descriptive examples will be given.

14.2. Control-System Simulation

In control-system simulation a random Gaussian white noise signal $I(t)$ is input to a control-system that has an impulse response function $h(t)$ designed so that the signal output $y(t)$ has the desired statistical behavior. This relationship can be represented using the convolution integral as

$$y(t)=\int_{-\infty}^{+\infty} h(\tau)I(t-\tau)\,d\tau \equiv h(t)*I(t) \tag{14.1}$$

This equation can be Fourier-transformed using the transform pair

$$\begin{aligned}\hat{y}(\omega)&=\int_{-\infty}^{+\infty} y(t)\,e^{-i\omega t}\,dt = FT[y(t)]\\ y(t)&=\int_{-\infty}^{+\infty} \hat{y}(\omega)\,e^{i\omega t}\,d\omega = FT^{-1}[\hat{y}(\omega)]\end{aligned} \tag{14.2}$$

to give

$$\hat{y}(\omega)=H(\omega)\hat{I}(\omega) \tag{14.3}$$

* The authors' words.

where the system function is given by

$$H(\omega) = FT[h(t)] \tag{14.4}$$

we can then write

$$\langle \hat{y}(\omega)\hat{y}^*(\omega)\rangle = H(\omega)H^*(\omega)\langle \hat{I}(\omega)\hat{I}^*(\omega)\rangle \tag{14.5}$$

where * refers to the complex conjugate and ⟨ ⟩ denotes the ensemble average. Equation (14.5) is equivalent to the statement

$$\phi_y(\omega) = H(\omega)H^*(\omega)\phi_I(\omega) \tag{14.6}$$

where ϕ refers to the power spectrum. Since a white Gaussian noise spectrum is constant we can, without loss of generality, set it equal to unity and obtain the output signal spectrum as

$$\phi_y(\omega) = H(\omega)H^*(\omega) \tag{14.7}$$

We know $\phi_y(\omega)$, since it is the desired power spectrum for our signal, and we therefore wish to find the system function H that will satisfy Equation (14.7).

14.3. Use of Standard System Function Elements

One method of obtaining a system function that will give an output signal with a desired spectrum is to fit standard control-system elements together as in the Bode chart synthesis. This procedure was followed in Reference 2, where the problem was to simulate the turbulent wind as a function of altitude. Detailed Jimsphere measurements of the wind velocity v as a function of altitude z were taken. It was found that the random wind velocity could be transformed into a homogeneous function by normalizing the variables as follows:

$$y(\zeta) = \frac{v(\zeta)}{\sigma_v(\zeta)}, \qquad \zeta = \frac{z}{L_v(z)} \tag{14.8}$$

where L_v is the length scale and σ_v the standard deviation.

The resulting autocorrelation of the Jimsphere data, based upon an integral average rather than upon the ensemble average, namely,

$$R_y(l) = \overline{y(\zeta)y(\zeta + l)} = \frac{1}{\zeta}\int_0^{\zeta} y(\zeta')y(\zeta' + l)\, d\zeta' \tag{14.9}$$

all fell within the darkened area of Figure 14.1 and so were considered homogeneous since there was no direct dependence on altitude, only on altitude difference.

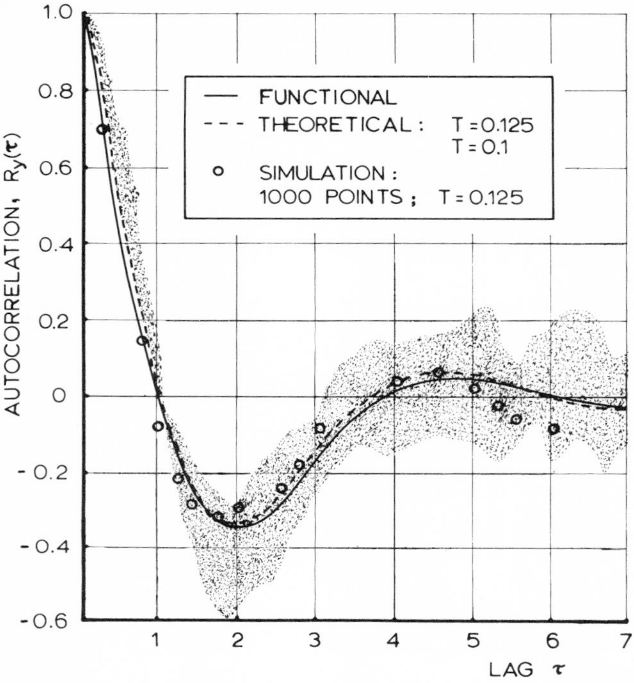

Fig. 14.1. Sample autocorrelation function simulation. Shaded area shows data from Jimsphere.

14.3.1. Fitting the Empirical Autocorrelation

The experimental autocorrelation was approximated by a second-order system function of the form

$$R_y(\tau) = \langle y(t)y(t+\tau)\rangle$$
$$= [\exp(-D|\tau|)][\cos(B\tau) - (D/B)\sin(B|\tau|)] \qquad (14.10)$$

where B and D are empirically determined constants having the values

$$B = 1.122, \qquad D = 0.539 \qquad (14.11)$$

and τ is the autocorrelation lag. A comparison of the experimental (shaded area) and functional form [Equation (14.10)] of the autocorrelation in Figure 14.1 shows good agreement.

By taking the Fourier transform of the autocorrelation, the power spectrum ϕ_y is obtained and, since R_y is an even function, the power spectrum can be written as

$$\phi_y(\omega)=2\int_0^\infty R_y(\tau)\cos\omega\tau\,d\tau=\frac{4D\omega^2}{[D^2+(B-\omega)^2][D^2+(B+\omega)^2]} \tag{14.12}$$

14.3.2. The System Function

As in Equation (14.7), the output power spectrum of the system can now be written as

$$\phi_y(\omega)=H(\omega)H^*(\omega) \tag{14.13}$$

where H must be given by

$$H(s)=\frac{(2D^{1/2})s}{(s+D-iB)(s+D+iB)} \tag{14.14}$$

where $s=i\omega$. The system equation can now be written as

$$y(s)=H(s)\,I(s) \tag{14.15}$$

where $y(s)$ is the dimensionless wind velocity function having the autocorrelation given by Equation (14.10) and $I(s)$ is the Gaussian white noise input.

14.3.3. The State Space System

It can be shown that Equation (14.15) is the Laplace transformation of a second-order, ordinary linear differential equation, which in turn can be transformed into a set of first-order differential equations. Thus we can define the system[2] in terms of a state variable X_i given by the following equation, where the Einstein summation convention is implied by repeated indices:

$$\frac{dX_i}{dt}=a_{ij}X_j+d_iI \tag{14.16}$$

where for our system there are two state variables (i.e., $i, j=1, 2$).

The system output is given by

$$y=e_iX_i \tag{14.17}$$

Following Reference 2, the flow graph state model is produced as shown in Figure 14.2, where

$$b_1=2D^{1/2},\qquad a_1=2D,\qquad \text{and}\quad a_0=D^2+B^2 \tag{14.18}$$

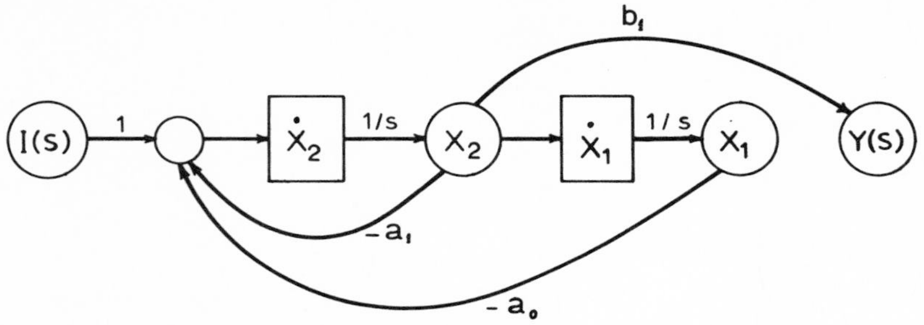

Fig. 14.2. Flow graph state model.

Then the coefficients of Equations (14.16) and (14.17) can be exhibited in the matrix form

$$[a_{ij}] = \begin{pmatrix} 0 & 1 \\ -a_0 & -a_1 \end{pmatrix} \tag{14.19}$$

$$[d_i] = \begin{pmatrix} 0 \\ 1 \end{pmatrix} \tag{14.20}$$

$$[e_i] = \begin{pmatrix} 0 \\ b_1 \end{pmatrix} \tag{14.21}$$

14.3.4. The Discrete State Space System

For use on digital computers the state equations must be converted to a discrete time system. One procedure for achieving this is given in Reference 2. In this procedure the input signal $I(t)$ is passed through a zero-order holding device that samples the signal at unit intervals of time and holds the signal value constant between samples (Figure 14.3). The procedure for converting the sample to a discrete time system is discussed next.

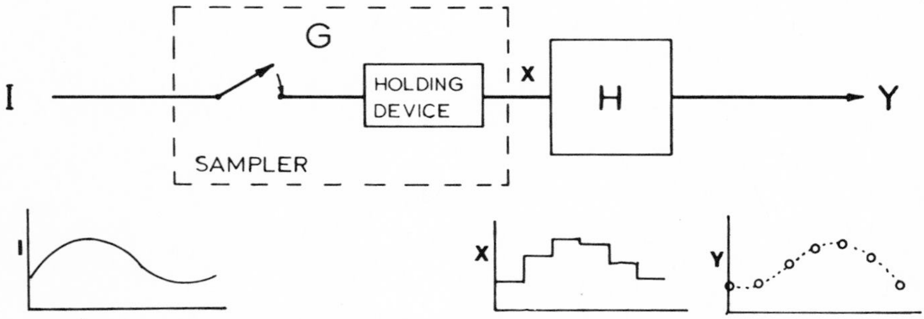

Fig. 14.3. Control system with sampler and holding device.

Equation (14.16) can be integrated[2] to give

$$X_i(t)=\chi_{ij}(t-t_0)X_j(t_0)+\int_{t_0}^{t}\chi_{ij}(t-\tau)d_jI(\tau)\,d\tau \qquad (14.22)$$

where $\chi_{ij}(t)=e^{a_{ij}t}$ is known as the fundamental matrix. Since $I(k)$ is now a discrete-valued function constant over the interval T, Equation (14.22) can be evaluated at time $t=(k+1)T$ and $t_0=kT$ to obtain

$$X_i(k+1)=\chi_{ij}(T)X_j(k)+\Lambda_i(T)I(k) \qquad (14.23)$$

where

$$\Lambda_i(T)=\int_0^T\chi_{ij}(\tau')\,d_j\,d\tau' \qquad (14.24)$$

and the indicial notation designates discrete values. Following the usual procedures, the fundamental matrix can be evaluated using Laplace transforms. If L represents the Laplace transform operation, then,

$$L[\chi_{ij}(T)]=[s\delta_{ij}-a_{ij}]^{-1} \qquad (14.25)$$

where

$$\delta_{ij}=\begin{cases}0 & \text{if } i\neq j\\ 1 & \text{if } i=j\end{cases}$$

This equation can be solved to give

$$[\chi_{ij}(T)]=\begin{pmatrix} e^{-DT}\left(\cos BT+\dfrac{D}{B}\sin BT\right) & \dfrac{1}{B}e^{-DT}\sin BT\\ -\dfrac{a_0}{B}e^{-DT}\sin BT & e^{-DT}\left(\cos BT-\dfrac{D}{B}\sin BT\right)\end{pmatrix} \qquad (14.26)$$

Taking the limit for small T there results

$$[\chi_{ij}(T)]=\begin{pmatrix}1 & T\\ -a_0T & 1-2DT\end{pmatrix}=[A_{ij}] \qquad 0<T\ll 1 \qquad (14.27)$$

Then, from Equation (14.24), we obtain

$$[\Lambda_i(T)]=\int_0^T\begin{pmatrix}\tau\\ 1-2D\tau\end{pmatrix}d\tau=\begin{pmatrix}0\\ T\end{pmatrix}\equiv[D_i], \qquad 0<t\ll 1 \qquad (14.28)$$

This gives the result for the discrete case with

$$X_i(k+1)=A_{ij}X_j(k)+D_iI(k) \qquad (14.29)$$

This same relationship can be obtained by using a forward finite-difference technique. In this method Equation (14.16) is written as

$$\frac{X_i(k+1)-X_i(k)}{T}=a_{ij}X_j(k)+d_iI(k) \tag{14.30}$$

This expression can then be rewritten to give the same results as Equation (14.29):

$$\begin{aligned} X_i(k+1)&=(Ta_{ij}+\delta_{ij})X_j(k)+Td_iI(k) \\ &=A_{ij}X_j(k)+D_iI(k) \end{aligned} \tag{14.31}$$

14.3.5. Effect of Digitizing on the Autocorrelation

The effect of digitizing the input noise signal $I(k)$ requires that a correction factor be applied to the output signal to give the true autocorrelation. This is developed as follows: In Figure 14.3 it is shown that a zero-order holding device has been added to the system. Therefore, the input into the continuous system will have a discrete form consisting of a stair function with the steps of Gaussian height and of width T instead of a continuous Gaussian signal (see Figure 14.3). The correction effect can be obtained by finding the spectrum of the discrete input. The autocorrelation of the discrete input is given by

$$\begin{aligned} R_{I'}(\tau)&=\langle I'(t)I'(t+\tau)\rangle \\ &=\sigma^2\,\Pr[A] \end{aligned} \tag{14.32}$$

where $\sigma^2=1.0$ is the variance of the Gaussian noise input, and $\Pr[A]$ is the probability that points t and $t+\tau$ of the input in the discrete form both occur between the times kT and $(k+1)T$. Primes and indicial notation have been used interchangeably to denote discrete values. The probability of A occurring (Reference 2) can be seen to be

$$\Pr[A]=\begin{cases} \dfrac{T-|\tau|}{T}, & |\tau|\le T \\ 0, & |\tau|>T \end{cases} \tag{14.33}$$

Then the input autocorrelation is given by

$$R_{I'}(\tau)=\begin{cases} 1-\dfrac{|\tau|}{T}, & |\tau|\le T \\ 0, & |\tau|>T \end{cases} \tag{14.34}$$

This can be Fourier-transformed to give the input power spectrum

$$\phi_{I'}=H_sH_s^*\phi_I=T\left[\frac{\sin(\omega T/2)}{\omega T/2}\right]^2\approx T, \qquad 0<T\ll 1 \tag{14.35}$$

where H_s refers to the system function of the zero-order holding device. Thus, the output spectrum of the discrete input, I', is now given by

$$\phi_{y'} = H_s H_s^* H H^* \phi_I \approx T H H^* \approx T\phi_y \tag{14.36a}$$

and the digitized autocorrelation is therefore given as

$$R_{y'}(\tau) = T R_y(\tau) \tag{14.36b}$$

Thus to obtain the correct continuous spectrum from the sampled signal, it must be normalized as follows:

$$y = y'/T^{1/2} \tag{14.37}$$

14.3.6. Discrete Autocorrelations

The theoretical autocorrelation can be calculated for the discrete equation following the procedures in Reference 2 by writing Equation (14.29) as

$$\begin{aligned} X_i(k+1) &= A_{ij} X_j(k) + D_i I(k) \\ X_i(k+2) &= A_{ij} X_j(k+1) + D_i I(k+1) \\ &= A_{ik}^{(2)} X_k(k) + A_{ik} D_k I(k) + D_i I(k+1) \end{aligned} \tag{14.38}$$

where

$$A_{ik}^{(2)} = A_{ij} A_{jk} \tag{14.39}$$

Similarly

$$X_i(k+3) = A_{ik}^{(3)} X_k(k) + A_{ik}^{(2)} D_k I(k) + A_{ik} D_k I(k+1) + D_i I(k+2) \tag{14.40}$$

and, in general,

$$X_i(k+n) = A_{im}^{(n)} X_m(k) + \sum_{r=k}^{k+n-1} A_{ij}^{(k+n-1-r)} D_j I(r) \tag{14.41}$$

where

$$A_{ij}^{(0)} D_j = D_i \tag{14.42}$$

Letting $k + n = \nu$, and, as $k \to -\infty$, $X_m(-\infty) = 0$, then

$$X_i(\nu) = \sum_{r=-\infty}^{\nu-1} A_{ij}^{(\nu-1-r)} D_j I(r) \tag{14.43}$$

This gives a nonrecursive form for obtaining $X_i(\nu)$. If $m = \nu - 1 - r$ then the equation becomes

$$X_i(\nu) = \sum_{m=0}^{m=\infty} A_{ij}^{(m)} D_j I(\nu - m - 1) \tag{14.44}$$

since $\langle I(j)\rangle = 0$ then $\langle X_i \rangle = 0$. The autocorrelation can be found using Equation (14.44) as follows. Beginning with the cross correlation

$$\langle I(\nu)X_i(\nu+l)\rangle = R_{IX_i}(l) = \sum_{m=0}^{m=\infty} A_{ij}^{(m)} D_j \langle I(\nu)I(\nu+l-m-1)\rangle \qquad (14.45)$$

and since $I(j)$ is discrete Gaussian white noise, it follows that

$$\langle I(\nu)I(\nu+l-m-1)\rangle = \delta(l-m-1)$$

where

$$\delta = \begin{cases} 0 & \text{when } l-m-1 \neq 0 \\ 1 & \text{when } l-m-1 = 0 \end{cases} \qquad (14.46)$$

So that Equation (14.45) becomes

$$R_{IX_i}(l) = \begin{cases} A_{ij}^{(l-1)} D_j & \text{for } l \neq 0 \\ 0 & \text{for } l = 0 \end{cases} \qquad (14.47)$$

Similarly,

$$\begin{aligned} R_{X_iX_k}(l) &= \langle X_i(\nu)X_k(\nu+l)\rangle \\ &= \sum_{m=0}^{\infty} A_{ij}^{(m)} D_j \langle I(\nu-m-1)X_k(\nu+l)\rangle \\ &= \sum_{m=0}^{\infty} A_{ij}^{(m)} D_j R_{IX_k}(l+m+1) \end{aligned} \qquad (14.48)$$

or

$$R_{X_iX_k}(l) = \sum_{m=0}^{\infty} A_{ij}^{(m)} D_j A_{k\nu}^{(l+m)} D_\nu = \sum_{m'=0}^{\infty} f(m)f(m+l) \qquad (14.49)$$

where $f(m) = A_{ij}^{(m)} D_j$. The equation for the autocorrelation of the system output $y(\nu)$ becomes

$$R_y(l) = \frac{\langle y(\nu)y(\nu+l)\rangle}{\sigma_{y(\nu)}^2} = \frac{b_1^2 \langle X_2(\nu)X_2(\nu+l)\rangle}{\sigma_{y(\nu)}^2} = \frac{b_1^2 R_{X_2X_2}(l)}{\sigma_{y(\nu)}^2} \qquad (14.50)$$

Then, by combining Equations (14.49) and (14.47), we obtain

$$R_y(l) = \frac{1}{\sigma_{y(\nu)}^2} b_1^2 T^2 \sum_{m=0}^{\infty} A_{22}^{(m)} A_{22}^{(m+l)} = \sum_{m=0}^{\infty} A_{22}^{(m)} A_{22}^{(m+l)} \Big/ \sum_{m=0}^{\infty} (A_{22}^{(m)})^2 \qquad (14.51)$$

where the normalizing factor is given by

$$\sigma_{y(\nu)}^2 = \langle y^2(\nu)\rangle = b_1^2 R_{X_2X_2}(0) = b_1^2 T^2 \sum_{m=0}^{\infty} (A_{22}^{(m)})^2 \qquad (14.52)$$

The analytical result for the discrete autocorrelation $R_y(l)$ is shown in Figure 14.1 for $T=0.125$ and for $T=0.1$ and is in good agreement with the experimentally derived autocorrelation. It should be noted that the solution is insensitive to the time increment and the curves for the different values of T lie upon one another. The values found for $\sigma^2_{y(\nu)}$ were 0.145 for the value $T=0.125$, and 0.111 for $T=0.1$. This indicates that the system is stable since the variance is finite. The correction factor found earlier in Equation (14.36b) was $\sigma^2_{y(\nu)}=T$, which in the present case gives close agreement with the more detailed discrete results, Equation (14.52).

14.3.7. Computer Signal Output

Discrete Gaussian white noise can be generated on the computer using readily available programs. Inputting this into the recursive equation for X_i [Equation (14.29)] results in a set of discrete values of $y(k)$, $k=1, 2, 3, \ldots$. This result can be normalized by

$$S_y=\left(\frac{1}{N}\sum_{k=1}^{N} y^2(k)\right)^{1/2} \tag{14.53}$$

The resulting autocorrelation was calculated from

$$R_y(l)=\left[\frac{1}{N}\sum_{k=1}^{N} y(k)y(k+l)\right]\Big/S_y^2 \tag{14.54}$$

For a time increment of $T=0.125$ and for 1,000 samples, the result shown with the circles in Figure 14.1 was obtained. This result is in good agreement with the desired autocorrelation. The value obtained for S_y^2 when $T=0.125$ was 0.142, which is in good agreement with the theoretical result of 0.145 (Section 14.3.6).

14.4. Digital Filter Simulation

Another method of simulation is to solve the convolution integral by filter techniques. This method was used in Reference 3 to simulate ocean waves. The chief advantage over the previous technique is that fitting the spectrum with some combination of simple system functions is not required. Standard methods for the design of low-pass, high-pass, and bandpass digital filters are readily available and can be used to shape a white noise signal to give a desired spectrum.

As shown in Equation (14.7), we can write the power spectrum as

$$\phi(\omega)=H(\omega)H^*(\omega) \tag{14.55}$$

The system function can be written in terms of an amplitude $A(\omega)$ and a phase angle $\theta(\omega)$:

$$H(\omega)=A(\omega)\,e^{i\theta(\omega)} \tag{14.56}$$

Letting $\theta(\omega)=0$, we obtain

$$H(\omega)=A(\omega)=[\phi(\omega)]^{1/2} \tag{14.57}$$

Then the output signal can be expressed as

$$\hat{y}(\omega)=[\phi(\omega)]^{1/2}\hat{I}(\omega) \tag{14.58}$$

The impulse response is given by

$$h(t)=FT^{-1}[\phi(\omega)^{1/2}] \tag{14.59}$$

and the signal output is

$$y(t)=h(t)*I(t) \tag{14.60}$$

Thus by generating white Gaussian noise and convolving it with the appropriate impulse response a random signal with the appropriate spectrum $\phi(\omega)$ can be obtained.

14.4.1. Discretizing the Convolution Integral

For use with digital computers the convolution integral must be discretized. The procedure for achieving this is to have the input signal $I(t)$ pass through a zero-order holding device as described in Section 14.3.5. The discrete sampled output signal is denoted by y', and it must be corrected as in Equation (14.37).

The convolution integral with the sampler can be written at time t_0 as

$$y'(t_0)=\int_{-\infty}^{+\infty} h(\tau)I'(t_0-\tau)\,d\tau \tag{14.61}$$

where I' is the sampled white noise. Then

$$\begin{aligned}y'(t_0)=I'(t_0)&\int_{-T/2}^{+T/2} h(\tau)\,d\tau+I'(t_0-T)\int_{T/2}^{3T/2} h(\tau)\,d\tau\\&+I'(t_0-2T)\int_{3T/2}^{5T/2} h(\tau)\,d\tau\\&+I'(t_0-3T)\int_{5T/2}^{7T/2} h(\tau)\,d\tau+\cdots+I'(t_0+T)\int_{-3T/2}^{-T/2} h(\tau)\,d\tau\\&+I'(t_0+2T)\int_{-5T/2}^{-3T/2} h(\tau)\,d\tau+\cdots\end{aligned} \tag{14.62}$$

For small values of T the above equation becomes

$$\begin{aligned}y'(t_0)=I'(t_0)h(0)T+I'(t_0-T)h(T)T+I'(t_0-2T)h(2T)T\\+I'(t_0-3T)h(3T)T\\+\cdots+I'(t_0+T)h(-T)T+I'(t_0+2T)h(-2T)T+\cdots\end{aligned} \tag{14.63}$$

which can be rewritten as

$$y'(t_0)=T\sum_{l=-N}^{N}I'(t_0-lT)h(lT) \tag{14.64}$$

Suppressing the symbol T for convenience [i.e., writing $h(lT)=h(l)$] and letting t_0 equal mT, we obtain

$$y(m)=T\sum_{l=-N}^{+N}I(m-l)h(l) \tag{14.65}$$

where the bracketed indices, again denoting discrete values, are interchangeable with the primes. Applying the correction due to the digitizing process gives

$$y(m)=T^{1/2}\sum_{l=-N}^{N}I(m-l)h(l) \tag{14.66}$$

This gives the digital filter $h(l)$, which is the impulse response function evaluated at discrete values. It is then multiplied by the Gaussian white noise signal and summed over l to give the mth value of the output signal. The collection of $y(m)$'s is the random output having the prescribed spectrum of the process to be simulated.

14.4.2. Theoretical Correlation for the Control-System Simulation

It is of interest, as before, to find the statistical moments given by the digital filter approach. It can be seen that

$$\langle y(m)\rangle=T^{1/2}\sum_{l=-N}^{N}h(l)\langle I(m-l)\rangle=0 \tag{14.67}$$

that is, the mean value is zero. Then

$$\begin{aligned}R_{I,y}(l)&=\langle I(m)y(m+l)\rangle\\&=T^{1/2}\sum_{r=-N}^{+N}h(r)\langle I(m)I(m+l-r)\rangle\\&=\begin{cases}0 & \text{when } r\neq l\\ T^{1/2}h(l) & \text{when } r=l\end{cases}\end{aligned} \tag{14.68}$$

Similarly

$$R_{yy}(\lambda)=\langle y(l)y(l+\lambda)\rangle=T^{1/2}\sum_{m=-N}^{+N}h(m)\langle I(l-m)y(l+\lambda)\rangle$$

$$=T\sum_{m=-N}^{+N}h(m)h(m+\lambda) \qquad (14.69)$$

In the limits of small T and large N this equation becomes equal to the integral

$$R_{yy}(\tau)=\int_{-\infty}^{+\infty}h(t)h(t+\tau)\,dt=h(t)*h(t+\tau)$$

$$=FT^{-1}[H(\omega)H^*(\omega)]$$

$$=FT^{-1}[\phi_y(\omega)] \qquad (14.70)$$

This result shows that the autocorrelation of the output from the digital filter indeed satisfies the desired power spectrum, providing a check on the simulation scheme.

14.5. Discrete Fourier Series

The random signals can be written in terms of Fourier series with random coefficients. The random coefficients can be generated so that the statistical moments of the signal will have the desired values. Also recently developed fast Fourier transforms allow very rapid manipulation of signals from the time domain into the frequency domain and back in a very efficient manner. These processes allow the input random noise to be generated, and then, using the fast Fourier transform, the Fourier spectra can be obtained. The Fourier spectra can then be multiplied by the appropriate system function and the result back-transformed to give the desired output signal as a function of time. The results obtained in this way will be described in the following section.

14.5.1. Discrete Fourier Transform

When a signal $w(t)$ is to be analyzed on a digital computer it is the discrete Fourier spectrum rather than the continuous spectrum that must be considered. If the Fourier spectrum $\hat{w}(f)$ is band-limited so that $\hat{w}(f)\approx 0$ for $f>f_M$ we can expand $\hat{w}(f)$ in a Fourier series as

$$\hat{w}(f)=\sum_{n=-\infty}^{+\infty}C_n e^{-i2\pi nf/f_p}, \qquad f_p>2f_M, \frac{-f_p}{2}<f<\frac{f_p}{2} \qquad (14.71)$$

where

$$C_n = \frac{1}{f_p}\int_{-f_p/2}^{+f_p/2} \hat{w}(f)\, e^{i2\pi nf/f_p}\, df \tag{14.72}$$

Since $\hat{w}(f) = 0$ for $f > f_M$ we obtain the signal at discrete values of $t = n/f_p$, i.e.,

$$C_n f_p = FT^{-1}[\hat{w}] = w_n(t = n/f_p) \tag{14.73}$$

Thus knowing w at discrete values of t allows the discrete Fourier transform $\hat{w}$ to be evaluated from Equations (14.71) and (14.73) as a continuous function of f. Then by taking the inverse continuous Fourier transform of $\hat{w}$, we obtain w at all times. Since we assumed $t = n/f_p = n\,\Delta t$, we can see that

$$\Delta t = \frac{1}{f_p} \le \frac{1}{2f_M} \tag{14.74}$$

which is sometimes called the Nyquist interval.

If we take $N+1$ samples of w_n and write $f = k\,\Delta f$ and $f_p = N\Delta f$ we arrive at Equation (14.71) as

$$\hat{w}_k(k\,\Delta f) = \frac{1}{f_p}\sum_{n=-N/2}^{+N/2} w_n\, e^{-i2\pi nk/N} \tag{14.75}$$

Since $f_p = 1/\Delta t = N/T_p$ we can write the above as

$$\frac{\hat{w}_k}{T_p} = \frac{1}{N}\sum_{n=-N/2}^{+N/2} w_n\, e^{-i2\pi nk/N} \tag{14.76}$$

As shown in Reference 4, this can be considered as a discrete Fourier transform. The inverse discrete Fourier transform can then be used as the Fourier spectrum for the Fourier series expansion of w_n, i.e.,

$$w_n = \sum_{k=-N/2}^{N/2} \left(\frac{\hat{w}_k}{T_p}\right) e^{i2\pi kn/N} \tag{14.77}$$

so that Equations (14.76) and (14.77) form a discrete Fourier transform pair that are approximations to the continuous Fourier transform pair.

Thus, one method of generating the random signal would be to first generate a discrete random Gaussian white noise $I_1, I_2, \ldots, I_n$. Then upon taking the fast Fourier transform, we would obtain the spectrum of the noise

$$\frac{\hat{I}(f = k\,\Delta f)}{T_p} = \frac{1}{N}\sum_{n=-N/2}^{N/2} I(t = n\,\Delta t)\, e^{-i2\pi nk/N} \tag{14.78}$$

but since

$$\hat{y}(f) = H(f)\hat{I}(f) \tag{14.79}$$

it follows that

$$\hat{y}(f = k\,\Delta f) = H(f = k\,\Delta f)\hat{I}(f = k\,\Delta f) \tag{14.80}$$

Thus we have generated the Fourier spectra for the desired signal $\hat{y}$. Then by taking the inverse fast Fourier transform of $\hat{y}$ we can obtain the desired signal $y(n\,\Delta t)$.

14.5.2. Discrete Fourier Series Using Randomly Chosen Coefficients

Another approach commonly used in random signal simulation is to directly generate random coefficients of the discrete Fourier series. This section describes the procedure for first relating the Fourier spectrum to the power spectrum and then showing that the Fourier coefficients have a Gaussian distribution with zero mean and a variance related to the power spectrum at a given frequency. Finally a procedure for sampling the Fourier coefficients is given.

14.5.3. Relationship of the Fourier Spectrum to the Power Spectrum

To relate the Fourier spectrum to the power spectrum, utilizing the method set down in Reference 4, the following equation can be written:

$$\Omega_{\eta\eta}(\tau)=\lim_{T\to\infty}\frac{1}{T}\int_{-T/2}^{+T/2}\eta(t)\eta(t+\tau)\,dt \tag{14.81}$$

The Fourier transform of this expression is

$$FT[\Omega_{\eta\eta}(\tau)]=\lim_{T\to\infty}\frac{1}{T}\int_{-T/2}^{T/2}\eta(t)\left[\int_{-\infty}^{\infty}\eta(t+\tau)\,e^{i\omega\tau}\,d\tau\right]dt \tag{14.82}$$

Hence

$$FT[\Omega_{\eta\eta}(\tau)]=\lim_{T\to\infty}\frac{1}{T}\int_{-T/2}^{T/2}\eta(t)\hat{\eta}(\omega)\,e^{-i\omega t}\,dt$$

$$=\lim_{T\to\infty}\frac{1}{T}\hat{\eta}_T^*(\omega)\hat{\eta}(\omega) \tag{14.83}$$

where the subscript T denotes a finite time interval and $\hat{\eta}$ is the Fourier spectrum.

The ensemble average of this equation shows that the power spectrum $\phi(\omega)$ is given by

$$\phi(\omega)=FT[R_{\eta\eta}]=FT[\langle\Omega_{\eta\eta}\rangle]=\lim_{T\to\infty}\frac{1}{T}\langle\hat{\eta}_T^*(\omega)\hat{\eta}(\omega)\rangle \tag{14.84}$$

where $R_{\eta\eta}$ is the autocorrelation,

$$R_{\eta\eta}(\tau)=\langle\eta(t)\eta(t+\tau)\rangle \tag{14.85}$$

Thus the power spectrum is given in terms of the Fourier spectra $\hat{\eta}(\omega)$ and the required relationship is established.

14.5.4. Discrete Fourier Series Simulation

As was shown in Section 14.51, Equation (14.77), we can write a signal as

$$y(\tau)=\sum_{n=-N/2}^{+N/2} \alpha_n e^{-i\omega_n\tau} \tag{14.86}$$

where

$$\alpha_n = \hat{y}_n/T_p$$

This expression can be expanded to give

$$y(\tau)=\sum_{-N/2}^{+N/2} (\alpha_{R,n}+i\alpha_{I,n})[\cos(\omega_n\tau)-i\sin(\omega_n\tau)] \tag{14.87}$$

Since y is a real quantity, the imaginary term in this equation must be zero. This results because $\alpha_{R,n}$ is an even function and $\alpha_{I,n}$ is an odd function about the condition $\omega_n=0$. Then an expansion of Equation (14.87) gives

$$y(\tau)=\sum_{-N/2}^{+N/2} [\alpha_{R,n}\cos(\omega_n\tau)+\alpha_{I,n}\sin(\omega_n\tau)] \tag{14.88}$$

Since both $\alpha_{R,n}$ and $\cos(\omega_n\tau)$ are even functions, their product is also even. Similarly, since $\alpha_{I,n}$ and $\sin(\omega_n\tau)$ are both odd functions, their product is an even function. Thus, the above equation can be written as

$$y(\tau)=\alpha_{R,0}+2\sum_{n=1}^{N/2} [\alpha_{R,n}\cos(\omega_n\tau)+\alpha_{I,n}\sin(\omega_n\tau)] \tag{14.89}$$

It has been shown in Reference 4 that $\alpha_{R,n}$ and $\alpha_{I,n}$ are random variables having the Gaussian probability

$$f(\alpha_{R,n},\alpha_{I,n})\,d\alpha_{R,n}\,d\alpha_{I,n}=\frac{1}{2\pi\langle\alpha_n^2\rangle}\exp\left[\frac{-(\alpha_{R,n}^2+\alpha_{I,n}^2)}{2\langle\alpha_n^2\rangle}\right] \tag{14.90}$$

and

$$\langle\alpha_{R,n}\rangle=\langle\alpha_{I,n}\rangle=0 \tag{14.91}$$

Furthermore, as shown in Chapter 4, to assure stationarity

$$\langle\alpha_{R,n}^2\rangle=\langle\alpha_{I,n}^2\rangle=\langle\alpha_n^2\rangle \tag{14.92}$$

From Section 14.5.1, the following result is obtained:

$$\langle\alpha_n\alpha_n^*\rangle=\langle\alpha_{R,n}^2\rangle+\langle\alpha_{I,n}^2\rangle=2\langle\alpha_n^2\rangle=(1/\tau_p^2)\langle\hat{y}(\omega_n)\hat{y}^*(\omega_n)\rangle \tag{14.93}$$

Then, from Section 14.5.3

$$2\langle\alpha_n^2\rangle=\lim_{\tau_p\to\infty}\frac{\phi(\omega_n)}{\tau_p}=\lim_{\tau_p\to\infty}\phi(\omega_n)\frac{\Delta\omega}{2\pi} \qquad \text{when } n\neq 0 \tag{14.94}$$

where $\phi(\omega_n)$ is the power spectrum evaluated at ω_n and $1/\tau_p = \Delta\omega/2\pi$. However, when $\omega_n = 0$, $\langle\alpha_{I,n}^2\rangle = 0$, the above expression becomes

$$\langle\alpha_0^2\rangle = \lim_{\tau_p\to\infty} \frac{\phi(0)}{\tau_p} = \lim_{\tau_p\to\infty} \phi(0)\frac{\Delta\omega}{2\pi} \tag{14.95}$$

If the transformations

$$\begin{aligned} \alpha_{R,n} &= \alpha_{\rho,n} \cos \varepsilon_n \\ \alpha_{I,n} &= \alpha_{\rho,n} \sin \varepsilon_n \end{aligned} \tag{14.96}$$

are used, then it follows that

$$y(\tau) = \alpha_{\rho,0} \cos \varepsilon_0 + 2 \sum_{n=1}^{N/2} \alpha_{\rho,n} \cos(\omega_n \tau - \varepsilon_n) \tag{14.97}$$

where the joint distribution of $\alpha_{n\rho}$ and ε_n is obtained by substituting Equation (14.96) into Equation (14.90) to obtain

$$f(\alpha_{\rho,n}, \varepsilon_n)\, d\alpha_{\rho,n}\, d\varepsilon_n = f(\alpha_{\rho,n})\, d\alpha_{\rho,n} f(\varepsilon_n)\, d\varepsilon_n \tag{14.98}$$

where

$$f(\alpha_{\rho,n})\, d\alpha_{\rho,n} = \frac{\alpha_{\rho,n}}{\langle\alpha_n^2\rangle} e^{-\alpha_{\rho,n}^2/2\langle\alpha_n^2\rangle}\, d\alpha_{\rho,n} \tag{14.99}$$

and

$$f(\varepsilon_n)\, d\varepsilon_n = (1/2\pi)\, d\varepsilon_n \tag{14.100}$$

Note $d\alpha_{R,n}\, d\alpha_{I,n} = \alpha_{\rho,n}\, d\alpha_{\rho,n}\, d\varepsilon_n$. Equations for randomly sampled values taken from Equations (14.99) and Equation (14.100) can be derived[4] by using the uniformly distributed random number R with a range of values between 0 and 1 in the following equations:

$$\varepsilon_n = 2\pi R_{\varepsilon_n} \tag{14.101}$$

$$\alpha_{\rho,n} = \left[2\langle\alpha_n^2\rangle \ln\left(\frac{1}{R_{\alpha,n}}\right)\right]^{1/2} \tag{14.102}$$

where $\langle\alpha_n^2\rangle$ is given by Equations (14.94) and (14.95).

14.5.5. Theoretical Statistical Moments for Discrete Fourier Series Simulation

It is of interest to calculate the statistical moments given by the discrete Fourier series simulation. The mean value of y is obtained by taking the ensemble average of Equation (14.97) and using the fact that α and ε are statistically independent:

$$\langle y\rangle = \langle\alpha_{\rho,0}\rangle\langle\cos \varepsilon_0\rangle + 2 \sum_{n=1}^{N/2} \langle\alpha_{\rho,n}\rangle\langle\cos(\omega_n \tau - \varepsilon_n)\rangle \tag{14.103}$$

Using Equation (14.100), we obtain

$$\langle \cos \varepsilon_0 \rangle = \int_0^{2\pi} \cos \varepsilon_0 \frac{d\varepsilon_0}{2\pi} = 0 \tag{14.104}$$

$$\langle \cos(\omega_n \tau - \varepsilon_n) \rangle = \int_0^{2\pi} \cos(\omega_n \tau - \varepsilon_n) \frac{d\varepsilon_n}{2\pi} = 0 \tag{14.105}$$

so that $\langle y \rangle = 0$.

The second-order statistics can be similarly obtained. Using Equation (14.97) again at point τ, and at point $\tau + \Delta\tau$, and taking ensemble averages, the following result is obtained:

$$\begin{aligned} \langle y(\tau) y(\tau + \Delta\tau) \rangle = \Bigg\langle &\left[\alpha_{\rho,0} \cos \varepsilon_0 + 2 \sum_{n=1}^{N/2} \alpha_{\rho,n} \cos(\omega_n \tau - \varepsilon_n) \right] \\ &\times \left\{ \alpha_{\rho,0} \cos \varepsilon_0 + 2 \sum_{l=1}^{N/2} \alpha_{\rho,l} \cos[\omega_l(\tau + \Delta\tau) - \varepsilon_l] \right\} \Bigg\rangle \end{aligned} \tag{14.106}$$

This reduces to

$$R_{\rho\rho}(\Delta\tau) = \langle \alpha_{\rho,0}^2 \rangle \langle \cos^2 \varepsilon_0 \rangle + 4 \sum_{n=1}^{N/2} \langle \alpha_{\rho,n}^2 \rangle \langle \cos[\omega_n \tau - \varepsilon_n] \cos[\omega_n(\tau + \Delta\tau) - \varepsilon_n] \rangle \tag{14.107}$$

since the cross-product terms are zero owing to the fact that α and ε are statistically independent. Using Equation (14.99)

$$\langle \alpha_{\rho,n}^2 \rangle = \int_0^{\infty} \alpha_{\rho,n}^2 \frac{\alpha_{\rho,n}}{\langle \alpha_n^2 \rangle} (e^{-\alpha_{\rho,n}^2/2 \langle \alpha_n^2 \rangle}) \, d\alpha_{\rho,n} = 2 \langle \alpha_n^2 \rangle \tag{14.108}$$

$$\langle \cos^2 \varepsilon_0 \rangle = \int_0^{2\pi} (\cos^2 \varepsilon_0) \frac{d\varepsilon_0}{2\pi} = \frac{1}{2} \tag{14.109}$$

and

$$\begin{aligned} &\langle \cos[\omega_n \tau - \varepsilon_n] \cos[\omega_n(\tau + \Delta\tau) - \varepsilon_n] \rangle \\ &\quad = \tfrac{1}{2} \langle \cos[\omega_n 2(\tau + \Delta\tau) - 2\varepsilon_n] \rangle + \tfrac{1}{2} \langle \cos[\omega_n \Delta\tau] \rangle \\ &\quad = \tfrac{1}{2} \cos[\omega_n \Delta\tau] \end{aligned} \tag{14.110}$$

To confirm that the statistics give the correct simulation the autocorrelation can be computed as follows:

$$R_{yy}(\Delta\tau) = \langle \alpha_0^2 \rangle + 4 \sum_{m=1}^{N/2} \langle \alpha_n^2 \rangle \cos[\omega_n \Delta\tau] \tag{14.111}$$

Referring to Equations (14.94) and (14.95), Equation (14.111) becomes

$$R_{yy}(\Delta\tau)=\frac{1}{2\pi}\phi(0)\Delta\omega+\frac{1}{\pi}\sum_{m=1}^{N/2}\phi(\omega_n)\cos[\omega_n\Delta\tau]\Delta\omega$$

$$=\lim_{\substack{\Delta\omega\to 0\\ N\to\infty}}\frac{1}{2\pi}\int_{-\infty}^{+\infty}\phi(\omega)\cos(\omega\,\Delta\tau)\,d\omega$$

$$=\frac{1}{2\pi}\int_{-\infty}^{+\infty}\phi(\omega)\,e^{-i\omega\Delta\tau}\,d\omega \tag{14.112}$$

This shows that the autocorrelation of the output for this case will satisfy the desired spectrum.

14.6. Non-Gaussian Simulation

In some cases, we are interested in generating a random signal with a non-Gaussian probability distribution. To do so we must first generate a Gaussian signal, and then transform it into the desired non-Gaussian signal. A Gaussian signal $w(t)$ with a specified spectrum is subjected to a suitable nonlinear, no-memory transformation such that signal γ is obtained having the desired probability density function p_γ. This transformation can be written as $\gamma = g(w)$. We can relate the distribution functions of γ and w by the well-known relation

$$\int_{g(-\infty)}^{g(w)} p_\gamma\,d\gamma=\int_{-\infty}^{w} p_w\,dw \tag{14.113}$$

The power spectrum of the generated signal $\phi_{ww}(\omega)$ must be specified so that the nonlinear transformation, $\gamma = g(w)$, will have the desired power spectrum $\phi_{\gamma\gamma}(\omega)$.

The autocorrelation of the generated w signal is obtained in terms of the desired γ signal autocorrelation as follows.

Generated Signal Autocorrelation. We can write the γ signal autocorrelation as follows[5]:

$$R_{\gamma\gamma}=\langle\gamma_1(t_1)\gamma_2(t_2)\rangle=\int_{-\infty}^{+\infty} g(w_1)g(w_2)p_{w_1w_2}\,dw_1\,dw_2 \tag{14.114}$$

where $p_{w_1w_2}$ is the normal bivariate distribution given by

$$p_{w_1w_2}=\frac{\exp[-(w_1^2-2\rho_{ww}w_1w_2+w_2^2)/2\sigma_w^2(1-\rho_{ww}^2)]}{2\pi\sigma_w^2(1-\rho_{ww}^2)^{1/2}} \tag{14.115}$$

ρ_{ww} is the normalized autocorrelation R_{ww}/σ_w^2 and σ_w^2 is the variance.

It has been shown[5] that the function $R_{\gamma\gamma}$ can be written as an algebraic function of ρ_{ww}. Then for a known Gaussian signal probability, p_w, and known transformation, $\gamma = g(w)$, the autocorrelation ρ_{ww} of the input Gaussian signal, after a nonlinear transformation, will give the desired autocorrelation $R_{\gamma\gamma}$.

A simpler method of calculating ρ_{ww} is given[6] for the case where the nonlinear transformation between γ and w is given in the form

$$\gamma = g(w) = \frac{1}{K\sigma_w\alpha(2\pi)^{1/2}} \int_0^w e^{-u^2/2\sigma_w^2\alpha^2}\, du$$
$$= \frac{1}{K(\pi)^{1/2}} \int_0^{w/(2)^{1/2}\sigma_w\alpha} e^{-\eta^2}\, d\eta \qquad (14.116)$$

The K and α are adjustable parameters that allow a wide variety of relationships between γ and w to be approximated.

The plots of γ versus w given by Equation (14.116) are shown in Figure 14.4. Although all values of w are possible, the values of γ are restricted between $\pm 0.5/K$. The value of K thus determines the cutoff amplitude of γ. For $\alpha = 0$, all values of w transform to $\pm 0.5/K$, giving a rectangular wave for γ. As α becomes large, Equation (14.116) reduces to $K\gamma \to w/[(2\pi)^{1/2}\alpha\sigma_w]$ so that $\gamma \propto w$.

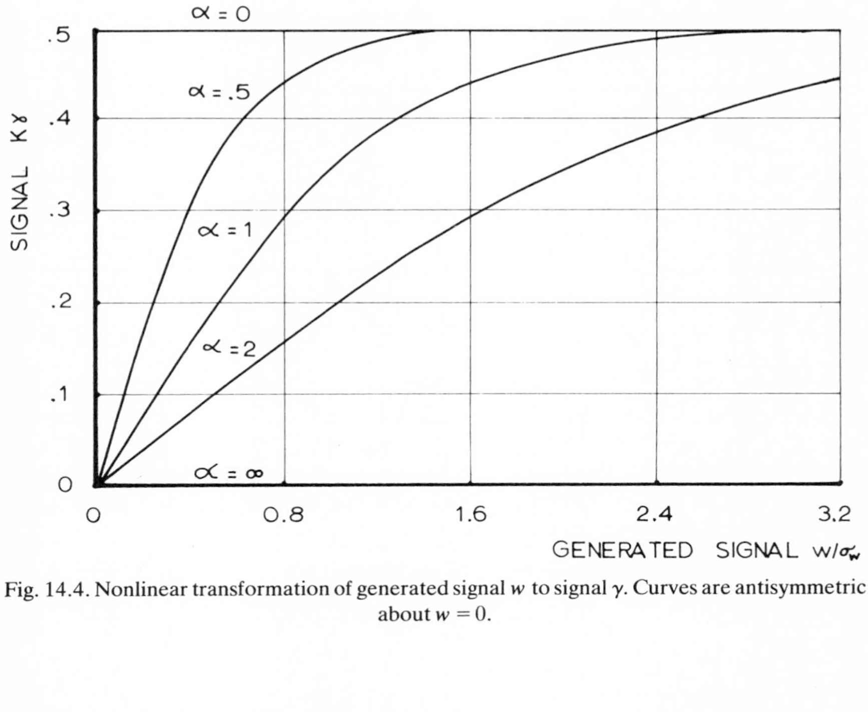

Fig. 14.4. Nonlinear transformation of generated signal w to signal γ. Curves are antisymmetric about $w = 0$.

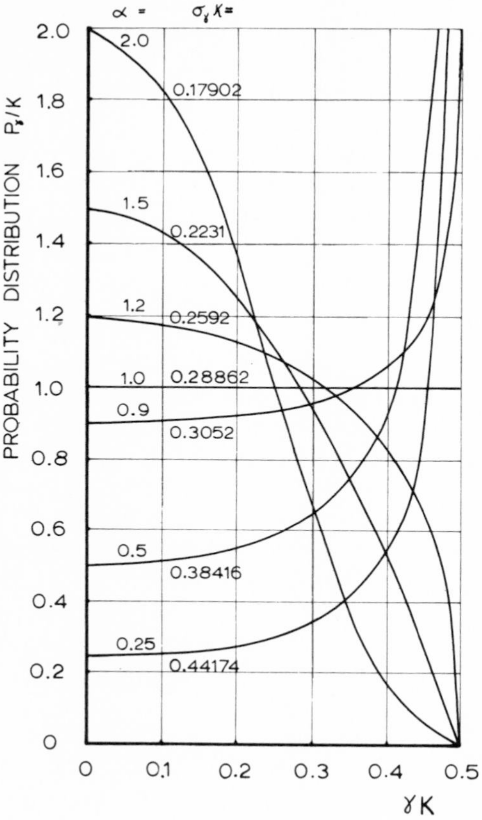

Fig. 14.5. Probability distribution of γ signal. Shows influence of parameter α and standard deviation $\sigma_\gamma \kappa$.

From Equations (14.113) and (14.116), we can transform the distribution of γ as

$$p_\gamma/K = \alpha\, e^{-w(1-\alpha^{-2})2\sigma_w^2} \tag{14.117}$$

Equation (14.117) can be evaluated by finding w for a specified value of γ using Equation (14.116). Then p_γ/K can be evaluated from Equation (14.117). The plot of p_γ/K versus γK is shown in Figure 14.5. Significantly different probability distributions p_γ can be obtained for different values of α and K. If $\gamma = g(w)$, given by Equation (14.116), is substituted into Equation (14.114), then it can be shown[6] that $R_{\gamma\gamma}$ is given by Equation (14.116):

$$R_{\gamma\gamma} = \frac{1}{2\pi K^2} \sin^{-1}\left(\frac{\rho_{ww}}{1+\alpha^2}\right) \tag{14.118}$$

This can be evaluated at $\tau = 0$ to give

$$R_{\gamma\gamma}(\tau = 0) = \frac{\zeta}{2\pi K^2}$$

where

$$\zeta = \sin^{-1}\left(\frac{1}{1+\alpha^2}\right) \tag{14.119}$$

We can now write

$$\rho_{\gamma\gamma} = \frac{1}{\zeta}\sin^{-1}\left(\frac{\rho_{ww}}{1+\alpha^2}\right) \tag{14.120}$$

which can be inverted to

$$\rho_{ww} = (1+\alpha^2)\sin(\zeta\rho_{\gamma\gamma}) \tag{14.121}$$

This gives the needed Gaussian autocorrelation ρ_{ww} so that the nonlinear transformation given by Equation (14.116) will give the desired autocorrelation coefficient $\rho_{\gamma\gamma}$. As an example we may wish to generate a non-Gaussian signal with an exponential form for the autocorrelation

$$R_{\gamma\gamma} = \sigma_\gamma^2 e^{-\gamma|\tau|} \tag{14.122}$$

Then the Gaussian signal input into the nonlinear system function from Equation (14.121) is given by

$$\rho_{ww} = (1+\alpha)^2 \sin(\zeta e^{-\lambda|\tau|}) \tag{14.123}$$

The power spectrum of ρ_{ww} is then given by

$$\begin{aligned}\phi_{ww} &= (1+\alpha)^2 \int_{-\infty}^{+\infty} \sin(\zeta e^{-\lambda|\tau|})\, e^{-iw\tau}\, d\tau \\ &= (1+\alpha)^2 \sum_{n=1,3,5,\ldots} \frac{\zeta^n}{n!} \frac{(-1)^{(n+3)/2}(2\lambda n)}{[(n\lambda)^2 + w^2]}\end{aligned} \tag{14.124}$$

Thus to generate a non-Gaussian signal with a probability distribution in one of the forms shown in Figure 14.5 with a exponential autocorrelation as given in Equation (14.122), we first generate a Gaussian signal with a power spectrum given by Equation (14.124), then transform it using Equation (14.116). Greater detail of this procedure is given in Reference 5.

14.7. Multidimensional Simulation

Much of the previous material can be generalized to multidimensional problems. As before, we can define a two-dimensional Fourier transform

$$\eta(\chi, \psi) = \frac{1}{(2\pi)^2} \int\int_{-\infty}^{+\infty} \hat{\eta}(\kappa_\chi, \kappa_\psi)\, e^{i(\kappa_\chi \chi + \kappa_\psi \psi)}\, d\kappa_\chi\, d\kappa_\psi \tag{14.125}$$

and its inverse

$$\hat{\eta}(\kappa_\chi, \kappa_\psi) = \iint_{-\infty}^{\infty} \hat{\eta}(\chi, \psi)\, e^{-i(\kappa_\chi \chi + \kappa_\psi \psi)}\, d\chi\, d\psi \tag{14.126}$$

As before [see Equation (14.3)] we can write

$$\hat{\eta}(\kappa_\chi, \kappa_\psi) = H(\kappa_\chi, \kappa_\psi)\hat{I}(\kappa_\chi, \kappa_\psi) \tag{14.127}$$

where

$$\eta(\chi, \psi) = FT^{-2}[\hat{\eta}(\kappa_\chi, \kappa_\psi)] \tag{14.128}$$

and I represents a two-dimensional Gaussian white noise signal and FT^{-2} represents a two-dimensional inverse Fourier transform. The H in this case is the two-dimensional system function. The power spectrum can be represented by

$$\langle \hat{\eta}(\kappa_\chi, \kappa_\psi)\hat{\eta}^*(\kappa_\chi, \kappa_\psi)\rangle = H(\kappa_\chi, \kappa_\psi)H^*(\kappa_\chi, \kappa_\psi)\langle \hat{I}(\kappa_\chi, \kappa_\psi)\hat{I}^*(\kappa_\chi, \kappa_\psi)\rangle \tag{14.129}$$

which then gives

$$\phi(\kappa_\chi, \kappa_\psi) = H(\kappa_\chi, \kappa_\psi)H^*(\kappa_\chi, \kappa_\psi)\phi_I(\kappa_\chi, \kappa_\psi) \tag{14.130}$$

where ϕ represents the two-dimensional power spectrum. Since the input signal is a two-dimensional Gaussian white noise with unit spectrum, the above becomes

$$\phi(\kappa_\chi, \kappa_\psi) = H(\kappa_\chi, \kappa_\psi)H^*(\kappa_\chi, \kappa_\psi) \tag{14.131}$$

As in the previous procedure [see Equation (14.56)] let

$$H(\kappa_\chi, \kappa_\psi) = A(\kappa_\chi, \kappa_\psi)\, e^{i\theta(\kappa_\chi, \kappa_\psi)} \tag{14.132}$$

Letting the phase angle be zero gives

$$H(\kappa_\chi, \kappa_\psi) = A(\kappa_\chi, \kappa_\psi) = \phi^{1/2}(\kappa_\chi, \kappa_\psi) \tag{14.133}$$

since H and H^* are the same for zero phase angle. Then the output signal can be represented by

$$\hat{y}(\kappa_\chi, \kappa_\psi) = [\phi(\kappa_\chi, \kappa_\psi)]^{1/2}\hat{I}(\kappa_\chi, \kappa_\psi) \tag{14.134}$$

The impulse response function h can be written as

$$h(\chi, \psi) = FT^{-2}[\phi^{1/2}(\kappa_\chi, \kappa_\psi)] \tag{14.135}$$

Then the output signal can be expressed as

$$y(\chi, \psi) = FT^{-2}\hat{y}(\kappa_\chi, \kappa_\psi) = h(\chi, \psi) * I(\chi, \psi) \tag{14.136}$$

where $*$ refers to the convolution integral

$$y(\chi, \psi) = \int_{-\infty}^{+\infty} h(\chi', \psi') I(\chi - \chi', \psi - \psi')\, d\chi'\, d\psi' \qquad (14.137)$$

This result is similar to that obtained for the one-dimensional case [see Equation (14.60)], and other relationships and solutions similar to those found in that case can be used here. Thus a two-dimensional filter can be obtained to transform the two-dimensional white noise Gaussian input into a two-dimensional signal y having the desired two-dimensional spectrum.

14.8. Nonhomogeneous Atmospheric Boundary-Layer Simulation

The case of nonhomogeneous atmospheric boundary-layer turbulence simulation will be used as an example of how the concepts previously discussed can be utilized. Many techniques have been used to simulate atmospheric turbulence.[7] The majority of these techniques are based on the Dryden hypothesis of the Eulerian spectral density function of the turbulence[8] and Taylor's frozen-eddy hypothesis.[9] A major criticism of the existing engineering turbulence simulation schemes for the atmospheric boundary-layer is that they fail to account for the statistically nonhomogeneous character of the atmospheric boundary layer turbulence along the vertical direction. It is this nonhomogeneous feature that results in the turbulence appearing as a nonstationary process relative to an aircraft during takeoff and landing. A general technique is given here for simulating atmospheric turbulencelike random processes that are statistically homogeneous along the horizontal direction and nonhomogeneous along the vertical direction. This technique is general in the sense that it can be used for a broad class of similar problems. Like the other presently available schemes the technique presented herein is based on the Dryden hypothesis (note that the method is not limited to the Dryden spectrum and any suitable autospectrum can be used) and Taylor's frozen-eddy hypothesis; however, it goes a step further by utilizing certain self-similarity properties of the Dryden spectral density function which permit the development of height-invariant filters. These filters are, in turn, used to generate vertically homogeneous (statistically) random processes from which turbulence at any specified level in the boundary layer can be simulated, thus facilitating the simulation of a nonstationary turbulence process along any path without the necessity of generating the entire turbulence field. The entire field can be generated, however, if the application requires.

14.8.1. Definition of the Problem

A one-dimensional Fourier decomposition of the longitudinal component of turbulence $u(x, y, z, t)$ along the x axis (directed along the mean wind vector) in a coordinate system moving with the mean wind is given by

$$u(x, y, z, t) = \int_{-\infty}^{\infty} \hat{u}(\kappa, y, z, t)\, e^{i\kappa x}\, d\kappa \tag{14.138}$$

where $\hat{u}(\kappa, y, z, t)$ is the Fourier transform of $u(x, y, z, t)$ at wave number κ (rad m^{-1}). The y and z axes are orthogonal to the x axis with the z axis directed along the upward vertical. Of course the turbulence energy spectrum has been assumed sufficiently continuous to avoid the use of the proper Fourier–Stieltjes integral form of Equation (14.138). In the discussion that follows, we will be concerned with the simulation of turbulence fields in the (x, z) plane at a given instant, so we shall not carry y and t along in the analysis. Although the discussion is confined to the longitudinal component of turbulence, the analysis is sufficiently general so as to be applicable to the simulation of the lateral and vertical components of turbulence.

Let us now construct the longitudinal correlation function by evaluating Equation (14.138) at (x, z) and its complex conjugate at $(x + r, z - \Delta)$, multiplying the resulting relationship, and then averaging over the ensemble of products to obtain

$$\begin{aligned} R(r, \Delta, x, z) &= \langle u(x, z) u(x + r, z - \Delta)\rangle \\ &= \int_{-\infty}^{\infty} \int_{-\infty}^{\infty} \langle \hat{u}(\kappa, z) \hat{u}^*(\kappa', z - \Delta)\rangle\, e^{i(\kappa - \kappa')x - i\kappa' r}\, d\kappa'\, d\kappa \end{aligned} \tag{14.139}$$

The analysis that follows will be restricted to the case of a horizontally homogeneous turbulence process, so that correlation of the Fourier amplitudes must be of the form (see Section 4.2.4)

$$\langle \hat{u}(\kappa, z) \hat{u}^*(\kappa', z - \Delta)\rangle = \phi(\kappa', z - \Delta, z)\, \delta(\kappa' - \kappa) \tag{14.140}$$

where $\delta(\kappa' - \kappa)$ is the Dirac delta function and $\phi(\kappa', z - \Delta, z)$ is the interlevel spectral density function. Combination of Equations (14.139) and (14.140) yields the horizontally homogeneous correlation function

$$R(r, z - \Delta, z) = \int_{-\infty}^{\infty} \phi(\kappa, z - \Delta, z)\, e^{-i\kappa r}\, d\kappa \tag{14.141}$$

We now express the interlevel cross-spectral density function $\phi(\kappa, z - \Delta, z)$ in terms of the nondimensionalized magnitude called the spectral coherence and phase angle, namely,

$$\operatorname{coh}(\kappa, z-\Delta, z)=\frac{\phi(\kappa, z-\Delta, z)\phi^*(\kappa, z-\Delta, z)}{\phi(k, z, z)\phi(k, z-\Delta, z-\Delta)} \tag{14.142}$$

$$\theta(\kappa, z-\Delta, z)=\tan^{-1}\left\{-\frac{\operatorname{Im}[\phi(\kappa, z-\Delta, z)]}{\operatorname{Re}[\phi(\kappa, z-\Delta, z)]}\right\} \tag{14.143}$$

so that

$$\phi(\kappa, z-\Delta, z)=[\phi(\kappa, z, z)\phi(\kappa, z-\Delta, z-\Delta)\operatorname{coh}(\kappa, z-\Delta, z)]^{1/2} e^{-i\theta(\kappa,z-\Delta,z)} \tag{14.144}$$

where $\phi(\kappa, z, z)$ is the autospectral density function. Thus specification of the functions $\phi(\kappa, z, z)$, $\operatorname{coh}(\kappa, z-\Delta, z)$, and $\theta(\kappa, z-\Delta, z)$ serves to completely specify the function $\phi(\kappa, z-\Delta, z)$ and hence the second-moment statistics of $u(x, z)$ in view of Equation (14.141). In the development that follows we seek a filter such that an input of white noise produces an output $v(x, z)$ that possesses the second-moment statistics that resemble those of the actual process $u(x, z)$. The synthesis of the filter in the following is based on the functions $\phi(\kappa, z, z)$, $\operatorname{coh}(\kappa, z-\Delta, z)$, and $\theta(\kappa, z-\Delta, z)$.

14.8.2. Filter Synthesis

The one-dimensional Fourier decomposition of the random process field $v(x, z)$ is given by

$$v(x, z)=\int_{-\infty}^{\infty} \hat{v}(\kappa, z)\, e^{i\kappa x}\, d\kappa \tag{14.145}$$

We begin the simulation with n independent horizontally homogeneous Gaussian white noise processes $I_k(x)$ $(k=1, 2, \ldots, n)$ with the Fourier transform of the kth noise process being $\hat{I}_k(\kappa)$. We pass the kth noise process through a filter with response function $H_k(\kappa, z)$ and generate a composite process $v_c(x, z)$—see Figure 14.6. The composite process is then passed

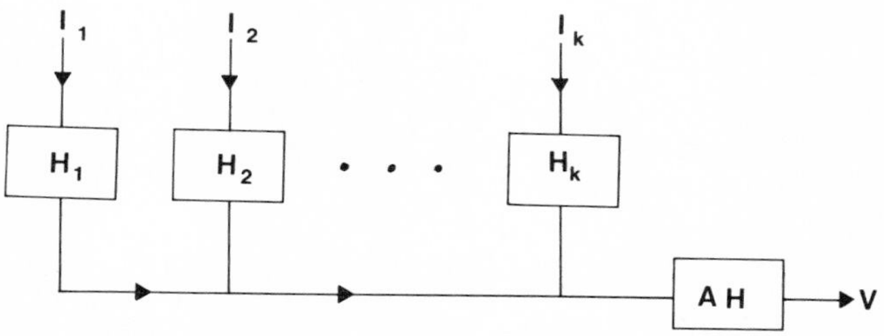

Fig. 14.6. Simulation control system.

through a linear filter with response function $AH(\kappa, z)$ to obtain the desired process $v(x, z)$ with Fourier transform given by

$$\hat{v}(\kappa, z) = AH(\kappa, z)\hat{v}_c(\kappa, z)$$
$$= AH(\kappa, z) \sum_{k=1}^{n} H_k(\kappa, z)\hat{I}_k(\kappa) \tag{14.146}$$

where A is a multiplier which will be specified later. Multiplying Equation (14.146) by its complex conjugate evaluated at κ' and $z-\Delta$, and ensemble-averaging the resulting relationship yields

$$\langle \hat{v}(\kappa, z)\hat{v}(\kappa', z-\Delta)\rangle = |A|^2 H(\kappa, z)H^*(\kappa', z-\Delta) \times \sum_{k=1}^{n} \sum_{m=1}^{n} \langle \hat{I}_k(\kappa)\hat{I}_m(\kappa')\rangle H_k(\kappa, z)H_m^*(\kappa', z-\Delta) \tag{14.147}$$

The noise processes are statistically independent, and their Fourier transforms are orthogonal with unit spectral density, so that

$$\langle \hat{I}_k(\kappa)\hat{I}_m(\kappa')\rangle = \delta(\kappa'-\kappa)\delta_{km} \tag{14.148}$$

where δ_{km} is the Kronecker delta. In view of the fact that we require the second-moment statistics of $v(x, z)$ to be identically equal to those of $u(x, z)$ we may write

$$\langle \hat{v}(\kappa, z)\hat{v}(\kappa', z-\Delta)\rangle = \phi(\kappa', z-\Delta, z)\delta(\kappa'-\kappa) \tag{14.149}$$

so that combination of Equations (14.147)–(14.149) yields the interlevel spectral density function of $u(x, z)$ in terms of the response functions of the components of the filter, that is

$$\phi(\kappa, z-\Delta, z) = |A|^2 H(\kappa, z)H^*(\kappa, z-\Delta) \sum_{k=1}^{n} H_k(\kappa, z)H_k^*(\kappa, z-\Delta) \tag{14.150}$$

or

$$\phi(\kappa, z-\Delta, z) = |A|^2 \phi_H(\kappa, z-\Delta, z)\phi_S(\kappa, z-\Delta, z) \tag{14.151}$$

where

$$\phi_H(\kappa, z-\Delta, z) = H(\kappa, z)H^*(\kappa, z-\Delta) \tag{14.152}$$

$$\phi_S(\kappa, z-\Delta, z) = \sum_{k=1}^{n} H_k(\kappa, z)H_k^*(\kappa, z-\Delta) \tag{14.153}$$

14.8.3. Coherence Matching

In this subsection we seek the functions $H_k(\kappa, z)$, $k = 1, 2, \ldots, n$. Substitution of Equation (14.150) into Equation (14.142) yields the spectral coherence

$$\operatorname{coh}(\kappa, z-\Delta, z) = \frac{\phi_S(\kappa, z-\Delta, z)\phi_S^*(\kappa, z-\Delta, z)}{\phi_S(\kappa, z, z)\phi_S^*(\kappa, z-\Delta, z-\Delta)} \tag{14.154}$$

or

$$\operatorname{coh}(\kappa, z-\Delta, z) = \frac{\sum_{k=1}^n H_k(\kappa, z)H_k^*(\kappa, z-\Delta)\sum_{m=1}^n H_m^*(\kappa, z)H_m(\kappa, z-\Delta)}{\sum_{k=1}^n H_k(\kappa, z)H_k^*(\kappa, z)\sum_{m=1}^n H_m^*(\kappa, z-\Delta)H_m(\kappa, z-\Delta)} \tag{14.155}$$

Note that the coherence is independent of $\phi_H(\kappa, z-\Delta, z)$, as one should expect, so that the function $H(\kappa, z)$ can be reserved for the spectra in the homogeneous plane and phase angle matching. Thus, the functions $H_k(\kappa, z)$ will be used to match the filter coherence of $v(x, z)$ with the coherence of $u(x, z)$.

We shall now specialize to a class of processes with coherence function independent of z and dependent only on differences in altitude Δ. The motivation for this specialization results from the fact that the turbulence coherence near the ground ($z \leq 150$ m) appears to behave in this manner. We shall return to this point later.

To formulate a method of satisfying an arbitrary coherence function $\operatorname{coh}(\kappa, \Delta)$ we can proceed by assuming that $H_k(\kappa, z)$ is a pure lag, that is

$$H_k(\kappa, z) = C_k e^{-i\kappa z d_k}, \qquad k = 1, 2, \ldots, n \tag{14.156}$$

Thus, from Equation (14.153) we obtain

$$\phi_S(\kappa, \Delta) = \sum_{k=1}^{n} C_k^2 e^{-i\kappa\Delta d_k} \tag{14.157}$$

or

$$\operatorname{coh}(\kappa, \Delta) = \frac{\left\{\sum_{k=1}^{n} b_k^2 + 2\sum_{m=2}^{n}\sum_{k=1}^{m-1} b_k b_m \cos[(d_m - d_k)\kappa\Delta]\right\}}{\left(\sum_{k=1}^{n} b_k\right)^2} \tag{14.158}$$

where

$$b_k = C_k^2 \tag{14.159}$$

Examination of the coherence function given by Equation (14.158), which is the sum of a finite series of cosine functions, suggests that we might determine the b's and d's by expanding the coherence in an even Fourier series on the interval $-\varepsilon_0 \leq \kappa\Delta \leq \varepsilon_0$, where ε_0 is the largest value of $\kappa\Delta$ of interest in the simulation. Identification of the coefficients and arguments of the trigonometric functions of the said expansion with the corresponding ones in Equation (14.158) yields the values of b_k and d_k. The Fourier expansion of coherence is given by

$$\mathrm{coh}(\xi) = \sum_{j=0}^{\infty} A_j \cos(j\pi\xi) \tag{14.160}$$

where

$$\xi = \kappa\Delta/\varepsilon_0 \tag{14.161}$$

$$A_j = \int_{-1}^{1} \mathrm{coh}(\xi)\cos(j\pi\xi)\, d\xi \tag{14.162}$$

We now set

$$d_k = \frac{\pi(k-1)}{\varepsilon_0} + \beta \tag{14.163}$$

where β is a constant to be determined through consideration of the phase angle θ. Substitution of Equation (14.163) into Equation (14.158) and a term-by-term comparison of Equation (14.158) with the first n terms of Equation (14.160) yields

$$A_0 = \left[1 + \sum_{k=2}^{n} \left(\frac{b_k}{b_1}\right)\right]^{-2} \left[1 + \sum_{k=2}^{n} \left(\frac{b_k}{b_1}\right)^2\right] \tag{14.164}$$

$$A_j = 2\left[1 + \sum_{k=2}^{n} \left(\frac{b_k}{b_1}\right)\right]^{-2} \left[\frac{b_{1+j}}{b_1} + \sum_{k=2}^{n-j} \frac{b_k b_{k+j}}{b_1^2}\right], \quad j = 1, 2, \ldots, n-1 \tag{14.165}$$

We thus have n equations in the $n-1$ unknowns, b_k/b_1, $k = 2, \ldots, n$. The quantity b_1 is arbitrary, and can be set equal to unity with no loss of generality. A number of methods are available for determining values for the remaining b's. One way is to directly solve the first $n-1$ equations for b_k, $k = 2, \ldots, n$. The difficulty with this approach is that for values of $n \geq 4$ the algebraic manipulations involved with determining the solution of the equations becomes unwieldy. A second approach is to determine a set of values for b_k, $k = 2, \ldots, n$, that satisfies the n-equations given by Equations (14.164) and (14.165) subject to the constraint that the positive definite function

$$J = \sum_{j=0}^{n-1} (A_j - A_j')^2 \tag{14.166}$$

takes on a minimum value, where A_j denotes the coefficients of the Fourier expansion, Equation (14.160), and A_j' denotes the right-hand side of

Equations (14.164) and (14.165). The minimization procedure was to guess initial values of the b's, then calculate J. An arbitrary value of b_i was then incremented and a new value of J was calculated. The incrementing of b_i continued until a minimum in J occurred. The process was repeated with the next value of b until the minimum was found.

14.8.4. Autospectral Density Matching

In this subsection we seek to determine that part of $H(\kappa, z)$ that is responsible for yielding a turbulence simulation $v(x, z)$ that possesses an assigned functional form for the autospectral density function $\phi(\kappa, z, z)$. Evaluation of Equation (14.151) at $\Delta = 0$ yields

$$\phi(\kappa, z, z) = |A|^2 \phi_H(\kappa, z, z)\phi_S(\kappa, z, z) = |H(\kappa, z)|^2 \qquad (14.167)$$

where we set the multiplier A such that

$$|A|^2 = \phi_S^{-1}(\kappa, z, z) = \left[\sum_{k=1}^{n} C_k^2\right]^{-1} \qquad (14.168)$$

Factorization of the left-hand side of Equation (14.167) yields $H(\kappa, z)$ to within an arbitrary phase angle. Thus, for example, if $\phi(\kappa, z, z)$ is a meromorphic function, then we may write

$$\phi(\kappa, z, z) = B^2 \left[\frac{(1+i\alpha_0\kappa)(1-i\alpha_0\kappa)\cdots(1+i\alpha_\lambda\kappa)(1-i\alpha_\lambda\kappa)}{(1+i\beta_0\kappa)(1-i\beta_0\kappa)\cdots(1+i\beta_\nu\kappa)(1-i\beta_\nu\kappa)}\right] \qquad (14.169)$$

where λ is not necessarily equal to ν and B is a real positive constant. Factorization of Equation (14.169) yields

$$H(\kappa, z) = H_p(\kappa, z)H_\lambda(\kappa, z) \qquad (14.170)$$

where

$$H_p(\kappa, z) = \frac{(1+i\alpha_0\kappa)\cdots(1+i\alpha_\lambda\kappa)}{(1+i\beta_0\kappa)\cdots(1+i\beta_\nu\kappa)} \qquad (14.171)$$

and

$$H_\gamma(\kappa, z) = e^{-i\theta_\gamma(\kappa, z)} \qquad (14.172)$$

where θ_γ is a function of z and κ that will be determined by requirements on the cross-spectral phase angle θ. Proper selection of the constant α's and β's permits the process $v(x, z)$ to satisfy the autospectral density function of the process $u(x, z)$. Later an example of the selection of the constants is given.

14.8.5. Phase Angle Matching

In this subsection we develop the appropriate matching of the phase angles between the filter and the turbulence data [see Equation (14.143)]. If we express H_γ, H_p, and ϕ_s in polar form ($e^{-i\theta}$) and employ Equations (14.144), (14.151), and (14.170) we arrive at

$$[\phi(\kappa, z, z)\phi(\kappa, z-\Delta, z-\Delta)\,\mathrm{coh}(\kappa, z-\Delta, z)]^{1/2}\, e^{-i\theta(\kappa, z-\Delta, z)}$$
$$= |H_p(\kappa, z)|e^{-i\theta_p(\kappa,z)}|H_\gamma(\kappa, z)|e^{-i\theta_\gamma(\kappa,z)}|H_p^*(\kappa, z-\Delta)|e^{i\theta_p(\kappa,z-\Delta)}$$
$$\times |H_\gamma^*(\kappa, z-\Delta)|e^{i\theta_\gamma(\kappa,z-\Delta)}|\phi_s(\kappa, \Delta)|e^{-i\theta_s(\kappa,\Delta)}$$

Equating the exponents gives

$$\theta(\kappa, \Delta, z) = \theta_\gamma(\kappa, z) - \theta_\gamma(\kappa, z-\Delta) + \theta_p(\kappa, z) - \theta_p(\kappa, z-\Delta) + \theta_s(k, \Delta) \tag{14.173}$$

The quantity θ_s is given by

$$\theta_s(\kappa, \Delta) = \tan^{-1}\left\{\frac{\left[\sum_{k=1}^{n} b_k \sin\left(\frac{\kappa\Delta}{\varepsilon_0}\pi(k-1)+\Sigma\right)\right]}{\left[\sum_{k=1}^{n} b_k \cos\left(\frac{\kappa\Delta}{\varepsilon_0}\pi(k-1)+\Sigma\right)\right]}\right\} \tag{14.174}$$

The b's have been determined to match the coherence, and the quantity θ_p is a known function of κ and z [Equation (14.171)]. The $\theta(\kappa, z-\Delta, z)$ is known from the physical data, and the quantity Σ and the function $\theta_\gamma(\kappa, z)$ can be chosen so as to satisfy the phase angle matching requirements, Equation (14.173). Having now developed the theory, a sample application is given. This first requires, however, the appropriate turbulence input, which is described in the following.

14.8.6. Longitudinal Gust Statistics

The spectral data required to implement the simulation scheme outlined in the previous sections consist of longitudinal autospectra and the coherence and phase angle associated with the interlevel longitudinal cross spectra. In the following sections we shall discuss these statistics.

14.8.7. Longitudinal Autospectra

The simplest longitudinal spectral density that has been widely used for simulation of atmospheric turbulence is the Dryden spectrum[10]:

$$\phi(x, z, z) = \frac{\sigma^2 L}{\pi}\frac{1}{1+(\kappa L)^2} \tag{14.175}$$

where σ^2 is the variance, $\langle u^2 \rangle$, and L is the integral scale length. This is chosen for purposes of illustration; however, any other desired spectrum can be employed.

14.8.8. Standard Deviation and Integral Scale of Turbulence

In the neutral surface layer, as shown in References 11 and 12,

$$\sigma = 2.5 u_{*0} \tag{14.176}$$

$$L = K_1 z \tag{14.177}$$

where $K_1 = 4.08$ and u_{*0} is the surface friction velocity.

14.8.9. Coherence and Phase

The interlevel longitudinal spectra, or equivalently the interlevel coherence and phase angle, Equations (14.142) and (14.143), have only been studied very recently so that available data are sparse in relation to the amount of data on longitudinal autospectra at a given level. Reference 13 and other available data on longitudinal coherence and phase angles appear to show that in the neutrally stable surface layer the following laws are valid:

$$\text{coh}(\kappa, \Delta, z) = e^{-(a/2\pi)\kappa\Delta} \tag{14.178}$$

$$\frac{\theta(\kappa, \Delta, z)}{\kappa\Delta} = 1.0 \tag{14.179}$$

where $a \approx 19.0$.

14.8.10. Longitudinal Gust Simulation and Application

In the following sections we shall combine theoretical development and empirical longitudinal gust statistics presented in the previous sections to demonstrate the practical aspects of the simulation of turbulence. In our discussion we shall use the Dryden approximation to the longitudinal spectral density function, Equation (14.175) with σ and L given by Equations (14.176) and (14.177) and the interlevel spectral coherence and phase functions given by Equations (14.178) and (14.179). The latter equations show that the interlevel spectral coherence and phase angle of the random process $v(x, z)$ depends on $\kappa\Delta$ only; therefore in the neutral stability case the dependence of the coherence and phase angle on z vanishes. We shall restrict the analysis in this section to the neutral case.

14.8.11. Coherence Determination

In view of the fact that κ and Δ are present only as a product in the neutral case, it is clear that Equation (14.178) satisfies the scaling law given by Equation (14.160). Transformation of $\kappa\Delta$ in Equation (14.178) to the dependent variable ξ [see Equation (14.161)] yields

$$\mathrm{coh}(\xi) = e^{-\xi/\xi_0} \tag{14.180}$$

where

$$\xi_0 = 2\pi/a\varepsilon_0 \tag{14.181}$$

Substitution of Equation (14.180) into Equation (14.162) yields

$$A_j = \frac{2\xi_0[e^{-1/\xi_0}(-1)^{j+1}+1]}{1+(j\pi\xi_0)^2} \tag{14.182}$$

Figure 14.7 is an example of the coherence of the simulation model in which $\xi_0 = 0.1$. In this particular case we truncated the Fourier expansion after the tenth term. In Table 14.1 we have indicated the values of b_k ($k = 1, 2, \ldots, 10$) associated with the values of the A's derived from Equation (14.182). We have also indicated in the figure typical 95% confidence limits of coherence data from many sites. It is apparent from the figure that a ten-term expansion results in an adequate representation for engineering applications.

Figure 14.8 is an example of the coherence for $\xi_0 = 0.01$, where we have again used a ten-term expansion. In this particular case the simulation model fails to have coherence fidelity. To remedy this situation, an increase in the number of terms (noise sources) in the Fourier expansion of the coherence by a factor of 5–10 would be required. However, since the coherence in Figure 14.8 is close to zero over most values of ξ we could assume that the velocity at different levels was effectively uncorrelated and that the velocity signals at each level could be generated independently.

14.8.12. Autospectra Factorization

In view of Equations (14.171) and (14.175) the function of $H_p(\kappa, z)$ in this particular example is given by

$$H_p(\kappa, z) = \sigma\left(\frac{L}{\pi}\right)^{1/2} \frac{1}{1+i\kappa L} \tag{14.183}$$

where $L = K_1 z$. This result will be used in the following sections to determine phase angles and for system self-similar considerations.

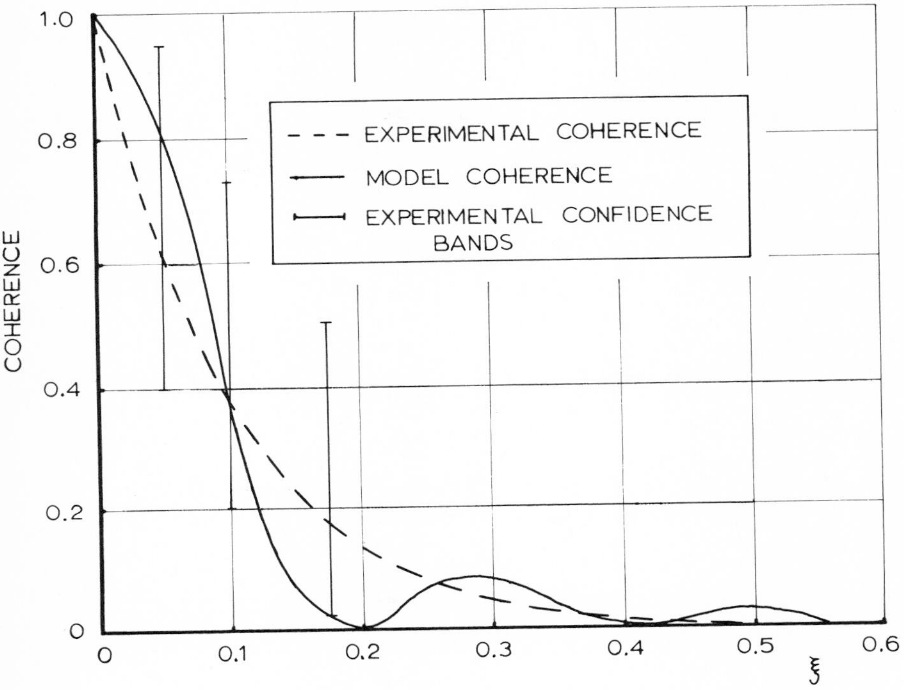

Fig. 14.7. Experimental and model ($\xi_0 = 0.1$) coherence (expectation) as a function of ξ.

Table 14.1. Values for the Fourier Coefficients b_i

$\xi_0 = 0.1$ (Figure 14.7)	$\xi_0 = 0.01$ (Figure 14.8)
$b_1 = 1.00$	$b_1 = 1.000$
$b_2 = 0.85$	$b_2 = 0.830$
$b_3 = 1.00$	$b_3 = 0.645$
$b_4 = 1.00$	$b_4 = 0.496$
$b_5 = 1.00$	$b_5 = 0.440$
$b_6 = 0.85$	$b_6 = 0.475$
$b_7 = 0.90$	$b_7 = 0.490$
$b_8 = 1.30$	$b_8 = 0.450$
$b_9 = 1.30$	$b_9 = 0.315$
$b_{10} = 1.70$	$b_{10} = 1.960$

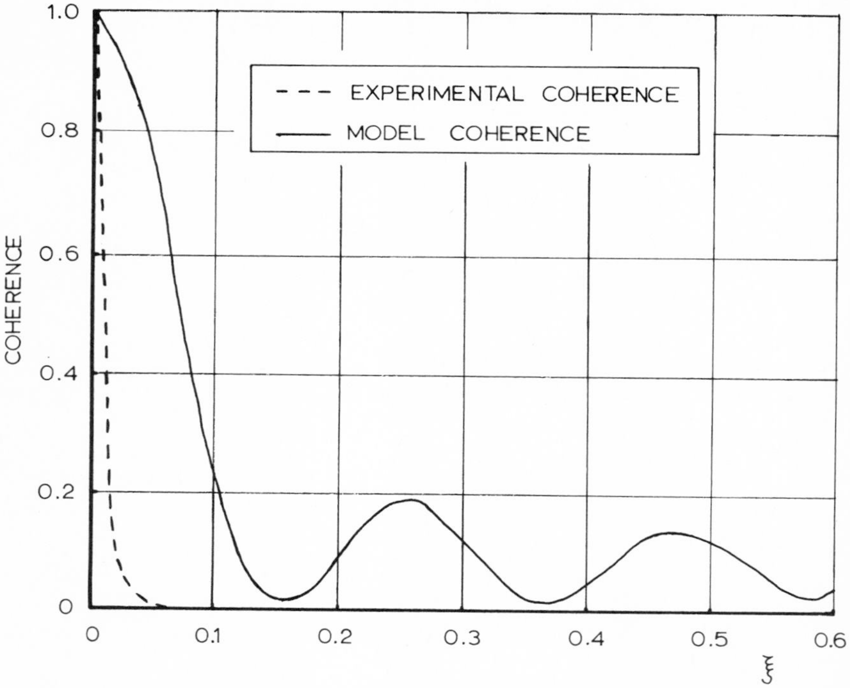

Fig. 14.8. Experimental and model ($\xi_0 = 0.01$) coherence (expectation) as functions of ξ.

14.8.13. Phase Angle Determination

According to Equations (14.173) and (14.179), for the neutral case we can write

$$\theta(\kappa, z) = \kappa\Delta = \theta_p(\kappa, z) - \theta_p(\kappa, z-\Delta) + \theta_\gamma(\kappa, z) - \theta_\gamma(k, z-\Delta) + \theta_s(\kappa\Delta) \tag{14.184}$$

where $\theta_s(\kappa\Delta)$ is given by Equation (14.174) and is a known function of $\kappa\Delta$. The functions $\theta_s(\kappa\Delta)$ for $\xi_0 = 0.1$ and 0.01 are given in Figure 14.9. It is readily seen that for engineering applications $\theta_s(\kappa\Delta)$ can be approximated as a linear function of $\kappa\Delta$, i.e.,

$$\theta_s(\kappa\Delta) = C(n, \varepsilon_0)\kappa\Delta \tag{14.185}$$

where $C(n, \varepsilon_0)$ is a function of the noise source number, n, and the nondimensional cutoff wave number, ε_0. We found this linear approximation to be valid for $3 \le n \le 10$ and $0.1 \le \varepsilon_0 \le 0.01$.

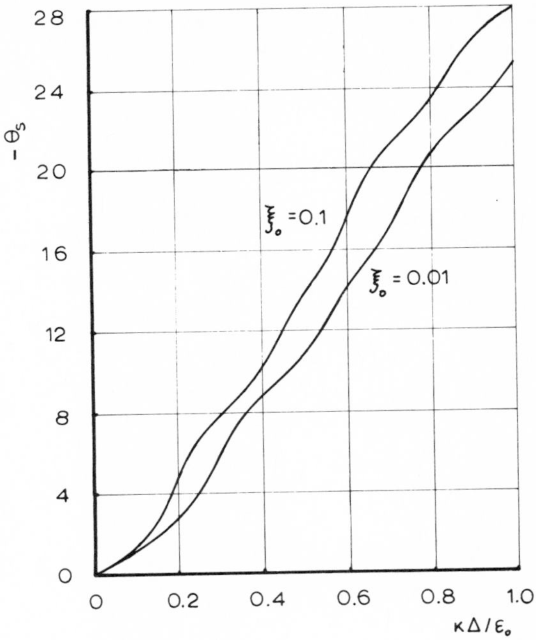

Fig. 14.9. The function $-\theta_s$ vs. $\kappa\Delta/\varepsilon_0$ for $\xi_0 = 0.1$ and 0.01.

The quantity $\theta_p(\kappa, z)$ by definition is

$$\theta_p(\kappa, z) = \tan^{-1}(\kappa L) = \tan^{-1}(\kappa, \kappa, z) \tag{14.186}$$

which is derivable from Equation (14.183).

Substitution of Equations (14.185) and (14.186) into Equation (14.184) yields

$$\theta_\gamma(\kappa, z) = \kappa z[1 - C(n, \varepsilon_0)] - \tan^{-1}(\kappa K_1 z) + G_0(\kappa, z_r) \tag{14.187}$$

where $G_0(\kappa, z_r)$ is an arbitrary function of κ and a reference height z_r. For definitiveness we shall set this function equal to zero.

14.9. Self-Similar Simulation

In the previous sections we derived the appropriate response function for the simulation of the longitudinal component of turbulence in the context of a first-order Dryden autospectrum. The resulting mathematical system permits the simulation of the random function field $v(x, z)$. This function field is useful in many applications; however, in aeronautical and aerospace vehicle applications we desire the function

$$v(t)=v[X(t), Z(t)] \tag{14.188}$$

where $X(t)$ and $Z(t)$ are the horizontal and vertical displacements of the vehicle at time t along its trajectory. To generate the function $v(t)$ we thus evaluate $v(x, z)$ along the trajectory of the vehicle. This could be impractical in the sense that the entire random function field is needed to determine the function $v(t)$. However, in the present analysis we can circumvent these problems by appealing to the self-similar properties of the model under discussion.

14.9.1. Inverse Fourier Transformation

Combination of Equations (14.146), (14.156), (14.170), (14.183), and (14.187) yields the result

$$\frac{\hat{v}(\kappa z)}{\sigma(z)L^{1/2}(z)}=\left(\pi \sum_{k=1}^{n} b_k\right)^{-1/2}[1+(\kappa L)^2]^{-1/2}\sum_{k=1}^{n} b_k^{1/2} \times e^{-ikz(1+d_k-c)}\hat{I}_k(\kappa) \tag{14.189}$$

where $L(z)=K_1 z$. Taking the inverse Fourier transform of $\hat{v}(\kappa, z)$ yields

$$\frac{v(x, z)}{\sigma(z)L^{1/2}(z)}=\left(\pi \sum_{k=1}^{n} b_k\right)^{-1/2}\sum_{k=1}^{n} b_k P_k[x-z(1+b_k-c)] \tag{14.190}$$

where

$$P_k(\xi)=\frac{1}{2\pi}\int_{-\infty}^{\infty}\frac{\hat{I}_k(\kappa)}{[1+(K_1 z\kappa)^2]^{1/2}}e^{i\kappa\xi}\,d\kappa \tag{14.191}$$

or

$$P_k(\xi)=\frac{1}{K_1\pi z}\int_{-\infty}^{\infty} K_0\left(\frac{|\xi'|}{zK_1}\right)I_k(\xi-\xi')\,d\xi' \tag{14.192}$$

where K_0 is the zero-order modified Bessel function of the second kind. Let us now map the noise process $I_k(x)$ from the x domain to the x/z domain and denote this mapping by $M_k(x/z)$, which is still a Gaussian white noise process with unit spectral density. The transformation permits us to express Equation (14.190) in the form

$$\frac{v(x, z)}{\sigma(z)L^{1/2}(z)}=\left(\pi \sum_{k=1}^{n} b_k\right)^{-1/2}\sum_{k=1}^{n} b_k P'_k\left[\frac{x}{z}-(1+b_k-c)\right] \tag{14.193}$$

where P'_k is a "universal" function of x/z, given by

$$P'_k\left(\frac{x}{z}\right)=\frac{1}{K_1\pi}\int_{-\infty}^{\infty} K_0\left(\frac{|\xi'|}{K_1}\right)M_k\left(\frac{x}{z}-\zeta'\right)d\zeta' \qquad (14.194)$$

and thus generates a white noise process $M_k(\zeta)$. Convolving this according to Equation (14.194), then lagging it in x/z space by a distance $(1+b_k-c)$, and finally summing over the n noise processes according to Equation (14.193) yields a "universal" random function $F(x/z)$ for the random variable $v(x,z)/\sigma(z)L^{1/2}(z)$. In short we have transformed the random function field $v(x,z)$ of two independent variables to a random function process $F(x/z)$ with one independent variable, and in this sense we speak of $F(x/z)$ being a self-similar simulation of the wind process $u(x,z)/\sigma(z)L^{1/2}(z)$.

14.9.2. Transformation to Vehicle Time Domain

As noted earlier, to determine the process $v(t)$, Equation (14.188), along the trajectory of an aeronautical or aerospace vehicle we merely evaluate Equation (14.193) at the location of the vehicle $[X(t), Z(t)]$. This yields

$$\frac{v(t)}{\sigma[Z(t)]L^{1/2}[Z(t)]}=\left(\pi\sum_{k=1}^{n} b_k\right)^{-1/2}\sum_{k=1}^{n} b_k P'_k\left[\frac{X(t)}{Z(t)}-(1+b_k-c)\right] \qquad (14.195)$$

Thus, it is possible to perform a simulation of atmospheric turbulence that retains the nonstationary character of turbulence relative to a vehicle ascending or descending in the atmospheric boundary layer for the trajectory only and thus avoid explicit evaluation of the entire two-dimensional velocity field.

14.10. Conclusions

The concepts of nonstationary turbulence simulation, especially the self-similar aspects of the subject, represent what we believe to be a new viable approach to the solution of a long-outstanding problem. Although the development was restricted to atmospheric turbulent random process fields, we believe that the technique is sufficiently general so as to be applicable to other two-dimensional random function fields. In these other applications, the self-similar properties may not be available; however, the mathematical machinery prior to the self-similar developments here should be applicable.

The technique developed here is a practical one, and the implementation of the model should be no more difficult than currently available procedures, which are capable of simulating a process that only satisfies the autospectra properties of atmospheric turbulence.

The analysis serves to indicate areas of deficiencies relative to the observational aspects of atmospheric turbulence. Improvements in the fidelity of the model can only be accomplished through better empirical definition of the autospectra, coherence, and phase angle θ. Of the three ingredients to the model, the latter two require further study.

We have attempted to give a unifying and elementary description of simulation techniques in an attempt to illustrate the ultimate capabilities of these methods. Further research in this field in respect to turbulence problems especially in the field of meteorology is needed if the capability of the method is to be fully developed and utilized.

Notation

a_{ij}, A_{ij}	Coefficients
B, D	Constants
C_n	Coefficients
coh( )	Coherence function
d_i, e_i, D_i	Coefficients
$f(\ )$	Probability density function
f	Frequency
FT	Fourier transform
$h(t)$	Impulse response function
$H(\omega)$	System response function
$I(t)$	Random Gaussian white noise
k, l, m, n, r	Integers
L_v	Length scale
M_k	Mapping of I_k
N	Number of samples
P_r	Probability
r	Separation distance in x direction
R	Autocorrelation function
s	$i\omega$
S_y	Root-mean-square function [Equation (14.53)]
T	Time interval
t	Time
u	Longitudinal velocity component
$v(t)$	Random function field
V	Velocity
$\hat{w}(t)$	Gaussian signal
w	Fourier spectrum
x, y, z	Rectangular coordinates
X_i	State variable
$y(t)$	Output signal
z	Altitude
$(\ ) * (\ )$	Convolution integral
$\langle\ \rangle$	Ensemble average
$\hat{\ }$	Fourier-transformed variable
$(\ )^*$	Conjugate function
α_i	Coefficient
β	Constant
γ	Non-Gaussian signal $= g(w)$
δ	Dirac delta function
Δ	Separation distance in the z direction
ε	Angle defined by Equation (14.96)
Λ	Function defined in Equation (14.24)
θ	Phase angle
ρ	Random signal
$\eta(t)$	Random variable
λ	Constant
τ	Time increment
ϕ	Power spectrum
σ	Standard deviation
ω	Angular frequency
ξ	Function defined in Equation (14.161)
ζ	Normalized altitude
$'$	Designates discrete function

References

1. Kendall, M. G., and Buckland, W. R. (eds.), *Dictionary of Statistical Terms*, third edition, Hafner Press, New York (1971).
2. Perlmutter, M., Simulation of random wind fluctuations, NASA report No. CR-120561 (1974).
3. Perlmutter, M., Stochastic simulation of ocean waves for SRB simulation, Northrop Services Inc., report No. TR 230-1446 (1975).
4. Perlmutter, M., Randomly fluctuating flow in a channel due to randomly fluctuating pressure gradients, NASA report No. TN D 6213 (1971).
5. Perlmutter, M., Effect of randomly fluctuating pressure gradients, with arbitrary specified power spectrum and probability density on flow in channels, The Symposium on Turbulence in Liquids, University of Missouri at Rolla, Missouri (1971).
6. Gujar, U. G., and Kavanagh, R. J., Generation of random signals with specified probability density functions and power density spectra, *IEEE Trans. Autom. Control* **AC-13** (1968).
7. Reeves, P. M., A Non-Gaussian turbulent simulation, report No. AF FDL-TR-69-67, Wright-Patterson Air Force Base, Ohio (1969).
8. Daniels, G. E. (ed.), Terrestrial environment (climatic) criteria guidelines for use in space shuttle vehicle development, NASA report No. TM-X-64589 (1971).
9. Lumley, J. L., and Panofsky, H. A., *The Structure of Atmospheric Turbulence*, Interscience Publishers, New York (1964).
10. Chalk, C. R., Neal, T. R., Harris, T. M., Pritchard, F. E., and Woodcock, R. J., Background information and user guide for MIL-F-8785B (ASG), "Military Specification—Flying qualities of piloted airplanes" report No. AF FDL-TR-69-72, Wright-Patterson Air Force Base, Ohio (1969).
11. Panofsky, H. A., Pielke, R. A., and Mares, E. V., Properties of wind and temperature at Round-Hill South Dartmouth, Mass., report No. ECOM-0335-F, U.S. Army Electronics Command, Fort Huachuca, Arizona (1967).
12. Fichtl, G. H., and McVehil, G. E., Longitudinal and lateral spectra of turbulence in the atmospheric boundary layer at Kennedy Space Center, *J. Appl. Meterorol.* **9**, 51–63 (1970).
13. Dutton, J. A., Panofsky, H. A., Deavan, D. C., Kerman, B. R., and Washington, V. M., Statistical properties of turbulence at the Kennedy Space Center for Aerospace Vehicle Design, NASA report No. CR-1889 (1971).
14. Fichtl, G. H., and Perlmutter, M., Nonstationary atmospheric boundary layer turbulence simulation, AIAA Seventh Fluid and Plasma Dynamics Conference, paper No. 74-587 (1974).

CHAPTER 15

Wind, Turbulence, and Buildings

R. C. ELSTNER

I am honored to make a contribution to the work of this distinguished group of scientists. However, because all of you know more about the subject than I, I feel like a cat that has just won the blue ribbon at a dog show. With this thought in mind, let me establish a few ground rules for our discussion. I am a structural engineer, not a scientist, or a mathematician. My field is buildings, and my interest in wind and turbulence is related only to the horizontal load that they impart onto my buildings.

It may surprise some of you to learn that structural engineering is not a pure science. At best, it can be described as an art based upon scientific principles. When I was an undergraduate, one of my professors described civil engineering in the following manner: "Anyone can build a bridge or an office building, but only a trained engineer can build one economically with just enough factor of safety." Safety and economy oppose each other, and engineering is the art of attaining the fine point of balance between them. We generally think of factor of safety in terms of protection from overload. However, it protects against other inherent deficiencies as well. These include errors in construction, deficiencies in materials, and limitations in engineering knowledge. As quality control of materials has improved, as construction practices have improved, as engineer design has become more

R. C. ELSTNER • Wiss, Janney, Elstner, and Associates, Northbrook, Illinois refined, especially with the advent of the computer, we have materially reduced the required factor of safety in structures.

Progress in the last seventy years has been tremendous. To illustrate this point, I like to compare two Chicago buildings, the Monadnock and the Sears Tower. The Monadnock Building is considered the world's first high-rise, dating back to the turn of the century. In the lower levels, the thickness of the exterior walls can be measured in feet, and windows were just large enough to permit necessary light. The structure is so heavy that overturning due to wind forces cannot be a bothersome problem. Walls are so thick and windows so small that the turbulence problem is confined to the problem of a word that can easily be misspelled.

In seventy years our art has improved, particularly since the end of World War II. We have higher-strength concrete, higher-strength steel, monstrous construction equipment, and computer analysis, all of which permits reduction in weight and justifies reduction in the factor of safety.

The old, heavy Monadnock Building has given way to the modern, tall, slim, graceful high-rise, magnificently represented by the Sears Tower.

There is no question that the modern high-rise is structurally safe, but we now have other problems, which had been under control in the older buildings without our having had to think about them. These problems include undesirable deflection of floors, human discomfort from building sway, and leaks and broken glass in the curtainwall or the skin of the structure. The latter two are new problems in my profession, and they are, of course, related to wind and turbulence.

As I have indicated, wind is a relatively new subject for structural engineers, though the present-day literature adequately discusses wind profiles, pressure coefficients, shape factors, etc. Building codes are rapidly being updated to present a logical approach to the prediction of wind forces on buildings. On the other hand, turbulence is a brand new subject with just enough recent reference in the literature to throw the fear of God into the average structural engineer.

Furthermore, glass has become a major structural component in high-rise structures. Although glass was the first material manufactured by human beings, we know very little about its structural properties. My textbook on engineering materials devotes 145 pages to steel, 221 pages to concrete, 67 pages to wood, but says not one word about glass.

You only have to look at a few modern buildings to realize that glass has become a favorite material with architects. We find large windows, insulated glass, tinted glass, reflective glass, more than a score of different glazing materials, and several score of different glazing details. There is no question that the use of glass in high-rise structures has outdistanced design knowledge.

I suspect by now you will agree that structural engineering is not an exact science, and that the old simple rules of thumb regarding forces on buildings due to wind are no longer applicable.

About fifteen years ago, we well realized the necessity of wind-tunnel testing. Herein lies a very interesting story: at that point, the aeronautical discipline had outgrown the wind tunnel and was abandoning them. Our firm uses the wind tunnel at Purdue University. I personally met with the President of Purdue about ten years ago to beg him not to tear down their wind tunnel. Since that time, over fifty structural models have been tested in the Purdue wind tunnel.

Since wind tunnels were under the control of aeronautical engineers, we had difficulty getting them out of the uniform laminar flow at 10,000 ft down to the profiled, turbulent wind from ground level up to 1,500 ft.

A few months ago, a movie featuring the life of the Wright Brothers was shown on television. I vividly remember the pictures of their crude wind tunnel and the problems they were having in obtaining laminar flow. For the next fifty years, aerodynamists kept improving the laminar flow characteristics; now for the last ten years, with emphasis on buildings, we are trying to replace laminar flow with turbulence.

I am sure some of you are aware that in the last few years a few wind tunnels have been designed and constructed specifically for the testing of models of structures. I mention Davenport's tunnels at Western Ontario University and Cermak's tunnel at Colorado State, as well as a small tunnel at Illinois Institute of Technology, and there may be others of which I am not aware. Whether these tunnels create turbulence-simulating prototype conditions is yet to be proved. However, we are attempting to do the best we can with the tools at hand.

Our firm attempted its first wind tunnel model about ten years ago involving the design of the First National Bank Building in the Chicago Loop. It was quite an experience, not unlike the first pioneers crossing the Donner Pass. Our measurements were confined to static pressures. Gusts were handled by multiplying static pressures by a factor of 1.3. Wind flow was determined by means of tufts. We recognized the problem of Reynolds number and solved it, rightly or wrongly, by roughening the surface of the model. We attempted to profile the approaching wind with screens and the addition of buildings surrounding the test model.

By today's standards these techniques were crude. However, I am beginning to wonder whether our present finesse in wind-tunnel studies is really justified. I have a gut feeling that the answers coming from wind-tunnel studies are, because of certain interpretations, ultraconservative.

This leads to the primary problem now facing our profession, which is the problem of correlating wind-tunnel data with prototype data. Such

Fig. 15.1. Wind model of Southland Life Building, Dallas, Texas.

Fig. 15.2. Peachtree Center, Atlanta, Georgia (photo courtesy of Professor George Palmer, Purdue University).

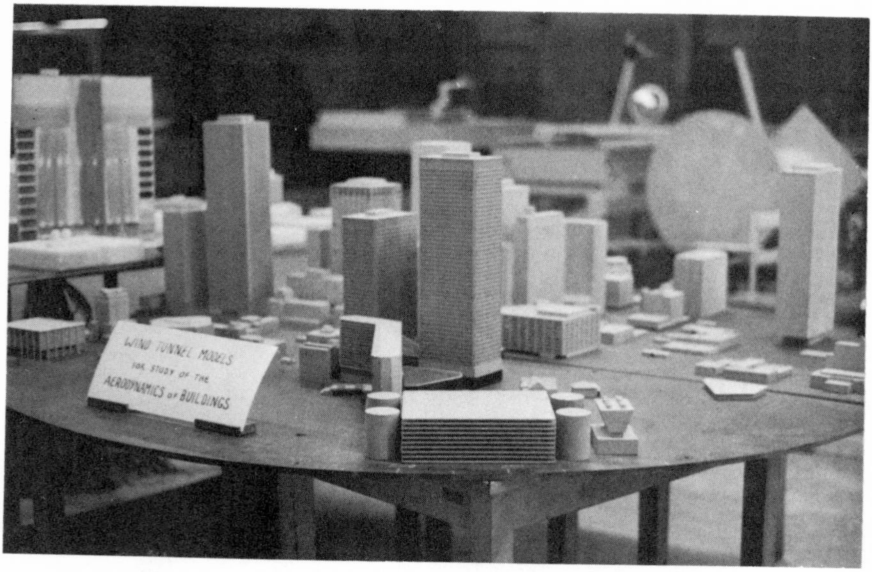

Fig. 15.3. Wind model of a portion of the city of Houston, Texas (photo courtesy of Professor George Palmer, Purdue University).

Fig. 15.4. Wind model of area surrounding RCA Building, New York, New York.

Fig. 15. 5. Wind model of Bank Complex and portions of adjacent Albuquerque, New Mexico.

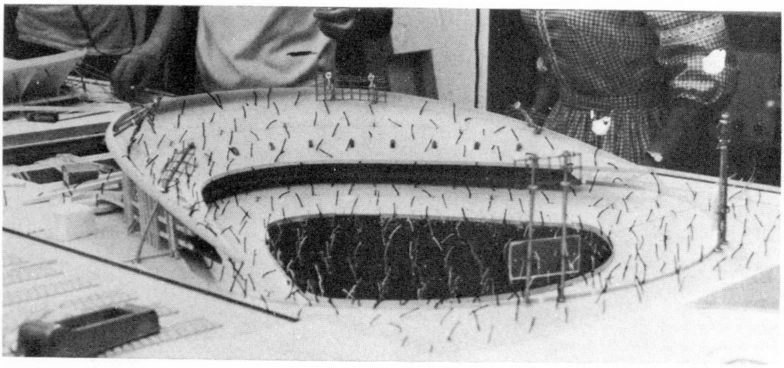

Fig. 15.6. Wind model of Kansas City Sports Complex, Kansas City, Missouri.

correlation can only be accomplished by comprehensive measurements on actual structures and must include measurements of pressures and velocities as well as the profile of the approaching wind. Data must be gathered over a long period of time and treated statistically. Our firm, in conjunction with Northwestern University, has submitted a proposal to the National Science Foundation to instrument six buildings in Chicago. The price tag is $1,500,000, and, in my opinion, it represents only a good start in answering the problem.

Although the structural engineering and fluid mechanics disciplines seem miles apart on the surface (no pun intended), we seem to be asking the same general questions about wind, gusts, turbulence, statistical analysis, etc. It is gratifying to find that structural engineers are not behind the times, but then again terrifying to find that we are on the firing line.

Figures 15.1–15.6 are photographs of various wind tunnel models of structures. Please note some of the detail we build into the small model as well as surrounding structures. Some of the studies were devoted to measurements of pressure; some, like the shopping center in Albuquerque and the baseball stadium in Kansas City, were devoted to wind flow as related to human comfort.

Author Index

Numbers in parentheses denote pages where the indicated name occurs as a reference. Boldface page numbers indicate a chapter in this book.

Subject Index

Many section headings included in the Contents are not repeated here, so the Contents should be used in conjunction with this index to locate specific material.

ROCKET PROPULSION ESTABLISHMENT LIBRARY

8/9

Please return this publication, or request a renewal, by the date stamped below.

SHORT LOANS

CANCELLED

Name	Date
A. S. Wilson	23. 10. 78
D. E. Jensen	16. 11. 78.
M. G. Drube.	13. 11. 78
J. Birchley	15. 11. 78
J. M. Cousins	13. 12. 78
	LONG LOANS
~~A. S. Wilson~~	14. 1. 79
~~J. Cousins~~	~~7. 3. 79~~
C. Mace	11. 4. 79